Sensors in Agriculture

Sensors in Agriculture

Volume 1

Special Issue Editor

Dimitrios Moshou

MDPI • Basel • Beijing • Wuhan • Barcelona • Belgrade

MDPI

Special Issue Editor
Dimitrios Moshou
Aristotle University of Thessaloniki
Greece

Editorial Office
MDPI
St. Alban-Anlage 66
4052 Basel, Switzerland

This is a reprint of articles from the Special Issue published online in the open access journal *Sensors* (ISSN 1424-8220) from 2017 to 2018 (available at: https://www.mdpi.com/journal/sensors/special_issues/agriculture)

For citation purposes, cite each article independently as indicated on the article page online and as indicated below:

LastName, A.A.; LastName, B.B.; LastName, C.C. Article Title. *Journal Name* **Year**, *Article Number*, Page Range.

Volume 1
ISBN 978-3-03897-412-3 (Pbk)
ISBN 978-3-03897-413-0 (PDF)

Volume 1-2
ISBN 978-3-03897-858-9 (Pbk)
ISBN 978-3-03897-859-6 (PDF)

Contents

About the Special Issue Editor

Dimitrios Moshou has been a full Professor at AUTH since 2018 and Lab Head of the Laboratory of Agricultural Engineering since 2012. He has a PhD from the Departments of Electrical Engineering and Biosystems, Faculty of Engineering, K.U. Leuven, Belgium; an MSc in Control Systems from the University of Manchester, UK; and an MSc in Electrical Engineering from Democritus University of Thrace, Greece. His research interests include the theory and applications of bio-inspired information processing, neuroscience, self-organization, and computational intelligence. He is interested in applications of these techniques in intelligent control, pattern recognition, data fusion, food safety, traceability, and cognitive robotics. He is the author of one research monograph on self-organizing networks and learning (Dimitrios Moshou, 'Artificial Neural Maps', ISBN: 978-3639150568, 2009) and more than 200 papers (with a total of 2600+ citations). He has contributed in research and management tasks of 40 local and EU research projects.

Preface to "Sensors in Agriculture"

Agriculture requires technical solutions for increasing production while lessening environmental impact by reducing the application of agro-chemicals and increasing the use of environmentally friendly management practices. A benefit of this is the reduction of production costs. Sensor technologies produce tools to achieve the abovementioned goals. The explosive technological advances and developments in recent years have enormously facilitated the attainment of these objectives, removing many barriers for their implementation, including the reservations expressed by farmers. Precision agriculture and 'smart farming' are emerging areas where sensor-based technologies play an important role. Farmers, researchers, and technical manufacturers are joining their efforts to find efficient solutions, improvements in production, and reductions in costs. This book brings together recent research and developments concerning novel sensors and their applications in agriculture. Sensors in agriculture are based on the requirements of farmers, according to the farming operations that need to be addressed.

Dimitrios Moshou
Special Issue Editor

![sensors logo] *sensors*

MDPI

Article

A Wireless Sensor Network for Growth Environment Measurement and Multi-Band Optical Sensing to Diagnose Tree Vigor

Shinichi Kameoka [1,†], Shuhei Isoda [1], Atsushi Hashimoto [1], Ryoei Ito [1], Satoru Miyamoto [2], Genki Wada [3], Naoki Watanabe [3], Takashi Yamakami [4], Ken Suzuki [4] and Takaharu Kameoka [1,*,†]

[1]   Graduate School of Bioresources, Mie University, 1577 Kurimamachiya-cho, Tsu City 514-8507, Mie, Japan; 515d201@m.mie-u.ac.jp (S.K.); 107isoda@gmail.com (S.I.); hasimoto@bio.mie-u.ac.jp (A.H); itou-r@bio.mie-u.ac.jp (R.I.)
[2]   Sumitomo Precision Products Co., Ltd., 1-10 Fuso-cho, Amagasaki City 660-0891, Hyogo, Japan; miyamo-s@spp.co.jp
[3]   Tomi no Oka Winery, Suntory Wine International Limited, 2786, Ohnuta Kai-shi 400-0103, Yamanashi, Japan; genki_wada@suntory.co.jp (G.W.); naoki_watanabe@suntory.co.jp (N.W.)
[4]   Department of Agriculture, Kisyu Agricultural Extension Center, Mie Prefectural Government, 371 Idomachi, Kumano City 519-4300, Mie, Japan; yamakt04@pref.mie.jp (T.Y.); suzukk07@pref.mie.jp (K.S.)
*   Correspondence: kameoka@mie-u.ac.jp; Tel./Fax: +81-59-231-9248
†   These authors contributed equally to this work.

Academic Editor: Dimitrios Moshou
Received: 8 March 2017; Accepted: 23 April 2017; Published: 27 April 2017

Abstract: We have tried to develop the guidance system for farmers to cultivate using various phenological indices. As the sensing part of this system, we deployed a new Wireless Sensor Network (WSN). This system uses the 920 MHz radio wave based on the Wireless Smart Utility Network that enables long-range wireless communication. In addition, the data acquired by the WSN were standardized for the advanced web service interoperability. By using these standardized data, we can create a web service that offers various kinds of phenological indices as secondary information to the farmers in the field. We have also established the field management system using thermal image, fluorescent and X-ray fluorescent methods, which enable the nondestructive, chemical-free, simple, and rapid measurement of fruits or trees. We can get the information about the transpiration of plants through a thermal image. The fluorescence sensor gives us information, such as nitrate balance index (NBI), that shows the nitrate balance inside the leaf, chlorophyll content, flavonol content and anthocyanin content. These methods allow one to quickly check the health of trees and find ways to improve the tree vigor of weak ones. Furthermore, the fluorescent x-ray sensor has the possibility to quantify the loss of minerals necessary for fruit growth.

Keywords: wireless sensor network (WSN); Wi-SUN; vine; mandarin orange; thermal image; fluorescent measurement; X-ray fluorescence spectroscopy

1. Introduction

Agricultural plants are extremely sensitive to climate change. Higher temperatures eventually reduce yields of desirable crops, while encouraging weed and pest proliferation. Changes in precipitation patterns increase the likelihood of short-run crop failures and long-run production declines. Today, the necessity of support of cultivation has been increasing with the escalation of issues such as the decrease in the number of people engaged in agriculture and the aging of this population.

The development of science-based agriculture is desired in order to adapt to the changing climate and to promote environmentally friendly smart agriculture with energy saving strategies. To that end, two kinds of measurements are indispensable; the first one is establishing a periodic acquisition system of meteorological and soil information at the field and creating cultivation indices by using this information; the second one is establishing the method for examining the tree vigor or balance of nutrition contents in the plant.

In regard to the first kind of measurement, a Wireless Sensor Network (WSN) is a methodology for acquiring growing environmental information. The WSN is a wireless network of small, low-cost sensors used for monitoring the physical environment at remote locations [1]. Therefore, since 2009, we have been using WSN in a mandarin orange orchard and a vineyard to promote smart cultivation management practices [2–4]. Based on these two kinds of fruit-growing examples, we have obtained some knowledge related to the issues of a field sensor network as well as the installation of weather stations and soil moisture sensors. There were also some problems; it takes too much time to restore the WSN system because the sensors and weather station were not homemade; the communication range was limited due to the frequency of radiowaves and the tipping bucket rain gauge needs regular maintenance.

In terms of the second objective, quality evaluation and control of agricultural products are very important to provide consistently high quality for the cultivation and postharvest management and marketing. Valuable information on plant nutrition needs to be addressed not only in the contents, but also in their balance in plant organs over the entire period of plant growth and the postharvest process. Understanding the change in the balance of elements at the level of field cultivation is another important factor in fruit cultivation, and acquiring the correct information on the balance of elements enables control of the amount of fertilizer and the plant's environment. Focusing on the relation between recent unstable climate and crops, it is also very important at the level of field cultivation because external environmental factors, such as abnormal climate and air pollution, produce a large change in the balance of the nutritional state in a plant, leading to a decrease in its yield.

So far, we have been developing integrated investigations on the multiband optical sensing of metabolites, biological systems and foodstuffs by using color imaging, and on the applications of such sensing techniques to the measurements of plants and agricultural materials at the field using infrared (IR) spectroscopy, thermal imaging and X-ray fluorescence (XRF) spectroscopy because optical sensing enables the simple, non-destructive, simultaneous, chemical-free, and rapid measurement of plants [4–8]. Especially, element measurement in the leaf (such as K, Ca, P and S) by using XRF spectroscopy and nitrogen measurement in the leaf by using mid-infrared (MIR) spectroscopy show high possibility for quantitative measurement [4,7,8]. In addition, measurement of the leaf temperature by using thermography camera is implemented [5,6]. In recent years, handy types of XRF sensors and fluorescence sensors for pigment analysis have been developed and we could apply these sensors in the field. Therefore, by using these portable sensors, we could achieve the non-disruptive and real-time measurement of the elements and pigments present in the leaves.

By the way, in Japan, the research for changing from agricultural ICT to agricultural IoT are undertaken as a project supported by the Ministry of Agriculture, Forestry and Fisheries. Agricultural IoT system refers to the whole system that includes the WSN in the field, data cloud containing growth environmental information and a web application service for farmers [9]. By integrating the growth environmental information and tree vigor information in the data cloud, it is possible to provide a more useful and effective service for farmers.

In this study, we developed a revised WSN system at the Tomi-no-oka vineyard toward the next generation of WSN for cultivation management based on several basic concepts. The first one is a locally produced scalable WSN with an affinity similar to the previously used eKo (Crossbow Technology Inc., Milpitas, CA, USA) system. The second one is a locally produced sensor with good support as a general rule. For the third monitoring items in farms are determined from the view point of phenology, plant physiology and synecology. Lastly, the WSN is correctly placed based on the

Interoperable Agricultural Information Platform structure. In this sense, some cultivation indices for wine grape are established and used in Europe; we used a vineyard for wine that can verify the validity of these indices. This paper also focuses on the vigor measurement of mandarin orange leaves by using thermal images, fluorescence and fluorescence X-ray methods in the field. Since there is a lot of available information about nutrient measurement of mandarin orange tree in Japan, a mandarin orange field was used in this experiment.

2. Object Fields and Methods

2.1. Object Fields

We have now been conducting demonstration experiments of agricultural WSN applications for more than three years using two farms. One is a mandarin orange grove (north latitude: 33.8634418°, east longitude: 136.05652822°), located at the Kanayama pilot farm in Kumano City, Mie Prefecture, Japan and the other is a vineyard (north latitude: 35.7103912885258°, east longitude: 138.5118197255086°), the Suntory Tomi-no-oka Winery in Kai city, Yamanashi Prefecture, Japan.

2.2. Growing Environmental Measurement by Using WSN

An eKo wireless sensor network has been used at the vineyard in the Suntory Tomi-no-oka Winery for climate and soil moisture measurement. In order to update the WSN for the next generation WSN, there are four points of modification from our previous study [4].

First, because of its maintenance, the eKo WSN should be replaced with another system that is domestically produced. Although the former eKo WSN system had been working effectively for more than five years, the plastic and battery inside the eKo deteriorate due to photoreactions and we should address this issue. In addition, once the system was broken, it took too much time since the eKo could not be repaired in Japan. In order to maintain the durability of the WSN system for a long time, a domestic system is favored in the viewpoint of quick and smooth restoration. Thus, the eKo WSN was replaced with a new system (Sumitomo Precision Products Co., Ltd., Amagasaki, Japan).

Secondly, the sensors used in the new WSN should be domestically made or maintenance free from the point of view of operation and maintenance. The soil water potential sensors were homemade and the soil moisture sensors were replaced with locally made products (ARP Co., Ltd., Hadano, Japan). This WD-3-WET-5Y TDR type soil water potential sensor can simultaneously measure volumetric water content (VWM), electric conductivity (EC) and temperature of the soil. Furthermore, this sensor is rated IP68; this rating means protection from contact with harmful dust and immersion in water with a depth of more than 1 m. Currently for the WSN, the soil water potential seems to be the most suitable measurement item whose aim is irrigation control [10]. For measuring the water potential in soil, it is necessary to measure it at different points in a field, because soil moisture varies even within the same field. Therefore, we also designed and developed a low-cost water potential sensor and connected it to the wireless sensor network. This sensor is produced experimentally to translate soil water content into voltage variation and is composed of a gypsum block and a simple electric circuit. Regarding the ground environment, the weather station and solar radiation sensor were replaced with a German-made weather station. The modified WSN in this study consists of a weather station, three soil moisture sensors and a soil water potential sensor.

Third, the data acquired from the WSN should be standardized in order to provide information services, especially when we integrate various data sources. Agricultural IoT is necessary for the next generation WSN and science-based agriculture so that the data obtained from the WSN are standardized and the standardized data are modified or combined to create indices that are related to the plant growth phases and useful for farmers.

Last, the service provided in this study should be useful for farmers to cultivate high quality fruits. Along with this purpose, we developed two kinds of indices for cultivation; the first one is the "primary index" and the second one is the "secondary index". The primary index is the monitored raw data that is made into a graph, or made into an at-a-glance list. The secondary index is the one that is made by modifying or combining the monitored raw data based on plant physiological theory. Therefore, the secondary index should be useful for farmers to improve their daily work.

So far, some secondary indices are recognized, especially about the grapevine cultivation. In our web service, there are six indices; the Accumulated Growing Degree Days (AGDD) is the index for predicting the growing stage of fruit; the Growing Season Temperature (GST) is the index for deciding the species suitable in the temperature of the place; the Coolnight Index (CI) is the index that indicates how much secondary metabolite is contained in the wine grape; the Heliothermal Index (HI) is the index which uses the daily temperature and solar irradiance to evaluate the mass of photosynthetic products; the Biologically Effective Degree Days (BEDD) is the index which uses the daily temperature difference between highest and lowest in addition to the daily temperature as parameter to predict maturity of the wine grape; and the Dryness Index (DI) is the index which uses the temperature, relative humidity, precipitation, wind speed and solar duration as parameters to recognize the soil dryness state [11]. In addition to these indices, our service provides the solar duration data using the algorithmic program invented by Slob and Monna [12].

2.3. Diagnosis of Fruit Tree Vigor Using Optical Sensing

In the test field in the mandarin orange grove, a mulch and drip irrigation system was deployed four years ago. The Mulch sheet (Shibataya kakohshi Co., Ltd., Niigata, Japan) is a waterproof, moisture-vapor permeable sheet. Because of this Mulch sheet, most of the rainfall does not go into the soil so that meaningful irrigation control is possible. In addition, the growing condition was monitored by the WSN [3].

2.3.1. Thermal Image Acquisition of Mandarin Orange Leaves

A series of laboratory scale experiments was carried out to determine the emissivity of mandarin orange leaves [5]. The thermal image of each mandarin orange leaf was taken by a Thermoshot thermography camera (Nippon Avionics Co., Ltd., Tokyo, Japan) from two trees with greatly different tree vigor under the same cultivation environment selected in the orange grove (Figure 1a,b). This thermography camera can measure from −20 to 100 °C and its resolution is 0.1 °C. The spectral range of this camera is from 8 to 13 μm, and its thermal image pixels are 160 × 120 pixels in size. The control area was marked on the thermal image using aluminum tape of which the emissivity greatly differed in order to facilitate the comparison between the visible and thermal images (Figure 1c).

Figure 1. Aluminum as the control point in the field experiment: (**a**) Tree whose vigor strength is good; (**b**) Tree whose vigor strength is bad; (**c**) The usage of aluminum tape.

The cultivation environment is shown in Figure 2. The emissivity determined in the laboratory (0.95) was applied in this experiment. A leaf template is needed to make a precise comparison of the temperature distribution between two leaves. Ten leaves with different shapes shown in Figure 3 were sampled in the orange grove and the average shape of the Satsuma Mandarin leaf was determined using the shape analytical method based on r-θ-ϕ coordinate system, which we have already developed [13]. This shape analysis was applied to the Satsuma Mandarin leaf.

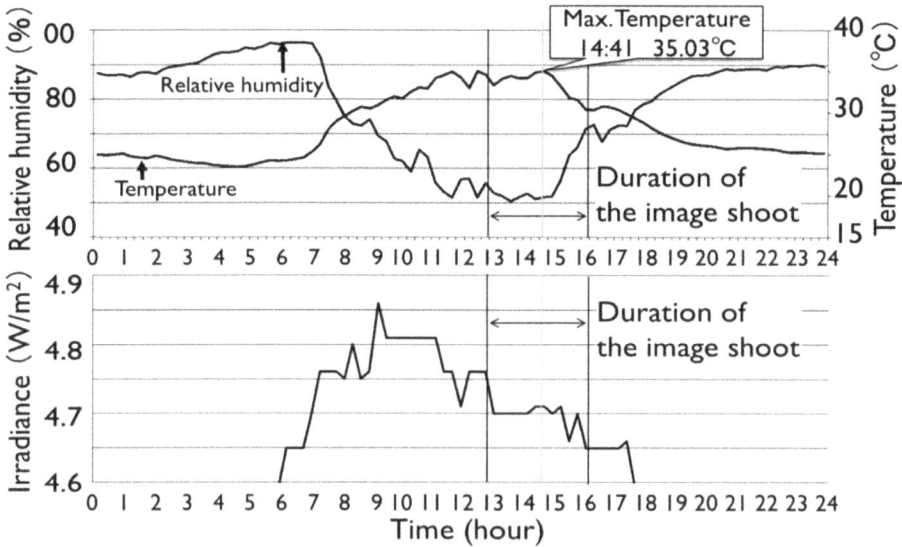

Figure 2. Cultivation condition of orange trees.

Figure 3. Sampled leaves used in this study.

Shape analysis was performed by the projection of a binary image of a leaf on tangent coordinate system. The origin was set on the center of gravity. A vector from the origin to a point on the orbital was auxiliary drawn as an arrow. Let r, θ and ϕ be the vector, angle of the vector and angle between the vector and tangent line at the point along the orbital, respectively. The orbital can then be projected into the tangent coordinate system with parameters θ and ϕ. Each obtained thermal image was then mapped to the template leaf and a comparison of temperature distribution was made between the strong leaf and weak one.

2.3.2. Vigor Measurement of Mandarin Orange Leaves by Fluorescence and Fluorescence X-ray Methods

Two types of leaves of which the vigor greatly differed in the same cultivation environment were selected from the orange grove. Ten leaves were taken from three weak vigor trees and another 10 leaves were taken from three trees with strong vigor (Figure 4). These leaves were taken from shoots without fruits. The evaluation criteria of tree vigor are based on the tree size, the number of leaves per tree, and the leaf color. The collected leaves were sent cool to the laboratory at Mie University using a courier and measurements by fluorescence and fluorescence X-ray methods were performed in the laboratory. As for the fluorescent measurement, a Dualex Scientific+ instrument (Force-A, Orsay, France) was used. It provided indices of the flavonols (FLAV), anthocyanins (ANTH), and chlorophyll (CHL) [14]. The leaf chlorophyll content was assessed by measuring the light transmission at 710 nm, absorbed by chlorophyll, and in the near-infrared (NIR) at 850 nm to take into account the effects of the leaf structure.

(a)   (b)

Figure 4. Sampled leaves: (**a**) Leaves of strong tree; (**b**) Leaves of weak tree.

The chlorophyll Dualex index is given by the formula:

$$CHL = [(I_{850}/I_{0,850})/(I_{710}/I_{0,710})] - 1 \tag{1}$$

where I and I_0 are the signals measured with and without the leaf sample in the leaf clip, respectively. The Dualex (Dx) measures the leaf epidermal flavonols or anthocyanins at 375 and 520 nm, respectively, using the chlorophyll fluorescence (ChlF) screening method and equalizing the ChlF signals under these excitation wavelengths and that under a red excitation at 650 nm as a reference.

Compounds present in the epidermis of the leaves attenuate the incident radiation before this can reach the first chlorophyll layer present in the mesophyll, depending on their absorption spectrum. Flavonols are the main flavonoids in dicotyledons absorbing UV radiation at 375 nm; therefore, the intensity of the ChlF induced by this radiation (ChlF_UV) will be inversely proportional to the epidermal flavonol concentration. Using a red light excitation, not attenuated by flavonols, a ChlF signal (ChlF_R) independent of the flavonol concentration is obtained. This signal is used as a reference. By comparing the ChlF signals from the two different excitations, the index of the flavonols can be calculated (in accordance with the Beer-Lambert law) as the logarithm of the ratio between the ChlF under red light and that under UV radiation:

$$FLAV = log(ChlF\_R/ChlF\_UV) \tag{2}$$

The same concept applies to the determination of anthocyanins using a green light, absorbed by anthocyanins, instead of UV radiation. In addition to the above indices, the Dx sensor calculates the nitrogen balance index (NBI).

$$NBI = CHL/FLAV \tag{3}$$

as the ratio between the chlorophyll and flavonol indices that can be used as a proxy of the crop leaf nitrogen level. Fluorometric analysis with the measurements of the chlorophyll, flavonol and anthocyanin Dualex indices (CHL, FLAV, ANTH) was performed using 10 leaves of good color and another 10 leaves of bad color. The measurements were made at three areas of each leaf shown in Figure 5 in order to find the pigment distribution inside the leaf.

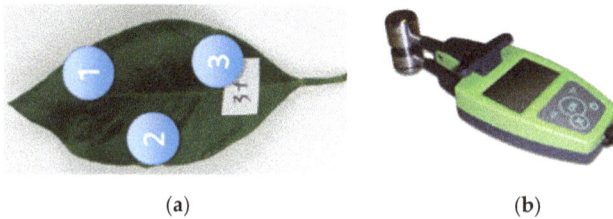

(a)   (b)

Figure 5. Fluorescent sensor used and measuring area of the leaf: (**a**) Fluorescent sensor used in this study; (**b**) Measurement areas in each leaf.

As for the fluorescence X-ray measurement, an Innov-X DELTA Premium Handheld XRF (hXRF) Analyzer (Olympus Corp., Tokyo, Japan) was used [15]. Although this hXRF sensor is portable, the specification of this sensor is as good as that of the Rayny EDX-700 (Shimadzu Corporation, Kyoto, Japan) for the laboratory use that was used in previous study [16]. Weighing roughly 2 kg, the instrument is equipped with a built-in camera/collimator mounted in the vicinity of the probe and a rechargeable Li-ion battery for easy field operation. We used the low beam energies of hXRF for the light elements, which are specifically those lighter than Mg since our target is the plant. In this study, although the hXRF analyzer was originally developed for use in the field, we mostly used the benchtop configuration of this analyzer. We used the same leaves shown in Figure 4 and analyzed them in the same way as the fluorescence method shown in Figure 5.

We monitored the internal instrument stability by measuring the Fe K-a counts on a 316 stainless steel coin every day using the Delta Docking Station (DDS). In addition, we made a special measurement station for the leaf with modifications that replace the iron coin to a titanium one because titanium is not present in the plant but is included in both the high and low beam energies (Figure 6a). Then, the leaf is set on the station and measured by the XRF sensor (Figure 6b,c)

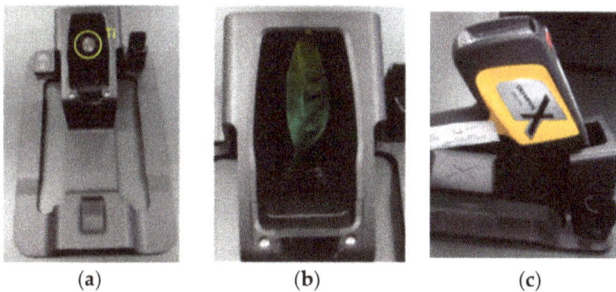

(a)   (b)   (c)

Figure 6. Special measurement station for leaf: (**a**) Titanium coin used instead of iron; (**b**) special measurement station for the leaves; (**c**) hXRF analyzer used in this study.

XRF analysis was performed to 10 leaves of good color and another 10 leaves of bad color. The measurements were made at three areas of each leaf shown in the figure according to the built-in camera information in order to find the element distribution inside each leaf.

3. Results and Discussions

3.1. Growing Environment Measurement by the WSN

3.1.1. Growing Environment Data Acquired by the WSN

The new WSN system with the selected weather station (WS700-UMB, Lufft Inc., Berlin, Germany) and TDR type soil water potential sensors (WD-3-WET-SE, ARP Co., Ltd.) were designed and deployed in the vineyard. The weather station can acquire air temperature, relative humidity, air pressure, wind speed, wind direction, 1 min precipitation and solar irradiance data, while the soil water potential sensor can acquire the soil volumetric water content, soil temperature and soil electrical conductivity. These data are collected by gateway via 920 MHz radio and sent to the data link server that is connected to the gateway (Figure 7).

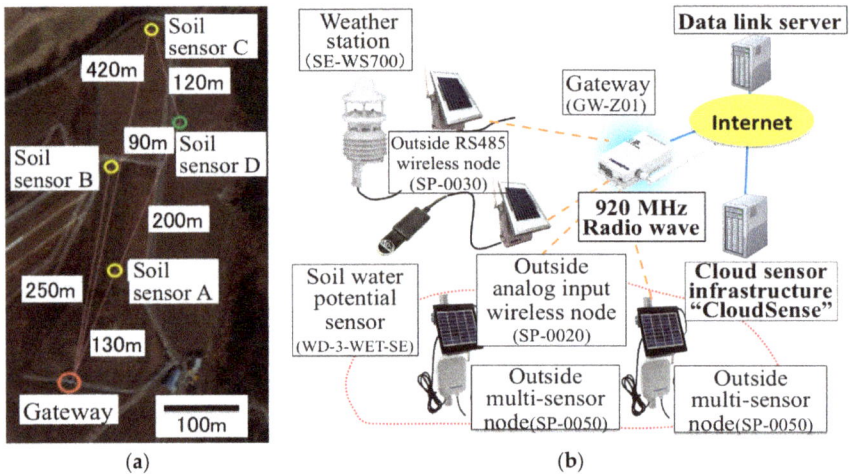

Figure 7. Structure of the WSN in Tomi-no-oka vineyard: (**a**) Installation points of sensors; (**b**) System configuration of WSN.

To send these acquired data to the cloud sensor infrastructure, the data link server uses HTTP-GET API defined by the cloud sensor infrastructure. In the revised WSN system, we use the SPPNet protocol produced by Sumitomo Precision Products Co., Ltd. The weather station is connected to the sensor node (SP-0030) by a two-wire RS485 network. Connecting the node to the RS485 serial interface devices can make a 1 to 1 or 1 to N see Duplex wireless communication system.

The power source is the AC 100 V commercial power supply existing in the vineyard which is converted to DC 24 V. On the other hand, each soil moisture sensor is connected to the analog sensor node (SP-0020) with specifications similar to those of the weather station. Its electricity source is DC from a solar panel deployed above the soil moisture sensor. These sensor nodes were developed in this study in order to connect the sensor to this WSN system (Table 1). The data link server of this WSN system has a customized web application based on SPP's commercial product 'EcoWizard' which has HTTP Server capability, DB and Flash Application.

Table 1. WSN Devices.

Product Name	Description	Sensor Interface	RF	Power	Battery Life w/o Sunlight	Waterproof
SP-0020	Used for sensing analog sensors; Prototyping Model for PoC	Arduino UNO, which has analog (0–5 V) inputs		Solar Panel (2.15 W) Rechargeable batteries (9600 mAh)	<3 days	IP65
SP-0030	Used for sensing RS485 sensors; Prototyping Model for PoC	SPP's original RS485 interface board	SPP's original 920 MHz RF module	Solar Panel (2.15 W) Rechargeable batteries (9600 mAh)	<3 days	IP65
SP-0050	Used for sensing analog, digital sensors; SPP's Commercial Model, developed based on SP-0020	SPP's original board which has multiple interfaces such as analog (0–5 V, 4–20 mA), digital inputs		Solar Panel (1.4 W) Rechargeable batteries (3200 mAh)	10 days *Low power mode	IP66
GW-Z01	Used for connecting WSN to the internet; SPP's Commercial Model	N/A		AC Adaptor 5 V/1.6 A	N/A	N/A Deployed in Waterproof Box

Regarding the data communication, the Modbus protocol is applied and the frequency of the radio used in this WSN system is a 920 MHz radio wave based on the Wireless Smart Utility Network which enables long-range wireless communication. Thanks to the 920 MHz, wave attenuation caused by the plant is not significant compared to the 2.4 GHz used in the previous WSN. In addition, this SPPNet protocol can provide the wireless transmission system of sensors compatible with Modbus protocol easier than in the eKo sensor network. All environmental data collected by the sensors in the WSN are sent to the Modbus master (Gateway GW-Z01), then stored in the data link server. The Over-The-Air throughput of this network is more than 100 kbps in the ideal state, while the actual throughput is more than 24 kbps. The MAC of this network is CSMA/CA system and comply with the transmission time restrictions defined in Association of Radio Industries and Business in Japan (ARIB) STD-T108. The routing protocol of this system ad-hoc mesh networking protocol, self-organizing and selt-healing. In addition, Transmission latency of this system is from 2 to 912 ms/hop. Other information about this system is shown in Figure 8.

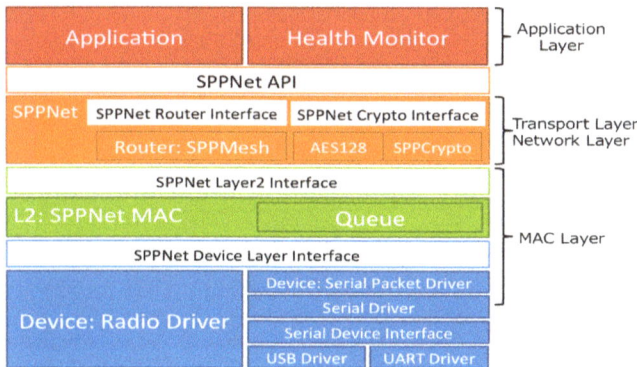

Figure 8. The information about SPPNet protocol used in this study.

This WSN consists of a weather station and three soil moisture sensors and a soil water potential sensor (Table 2, Figure 9). The soil water potential sensor was manufactured for this study to translate the soil water content into a voltage variation, and composed of a gypsum block and simple electric circuit to reduce material costs. As a result, it costs only about 5000 yen to produce this sensor and the process is so simple that farmers can do it by themselves. As the sensor for measuring the other soil information, the WD3-WET-SE soil moisture sensor was selected. It can measure the soil volumetric moisture content, soil electric conductivity and soil temperature at the same time. In addition, this sensor is worthy of the IP68 code that means it is completely resistant to both water and dust. Thus, we will be able to use this sensor in the ground for a long period. As the weather station, a SE-WS700 system was selected. This weather station can acquire the data of air temperature, relative humidity, air pressure, wind speed, wind direction, 1 min_precipitation and solar irradiance at the same time, which enables a cheaper WSN system than in the previous work [3]. In addition, this weather station can measure each environmental data point every one minute. The important point is the accuracy of each measurement acquired by this weather station, especially the wind speed measured by the ultrasonic anemometer and precipitation measured by the Doppler radar rain gauge. Furthermore, its water sampler requires minimal maintenance [17]. Thus, this weather station has the potential to solve the problem raised in the previous work.

Table 2. Sensor information used in this study.

Sensor Name	Model Number	Note
Weather station	WS700-UMB	• Adoption of ultrasonic anemometer. • Adoption of doppler radar rain gauge. • Free from maintenance.
Soil moisture sensor	WD-3-WET-SE	• Being able to measure EC (0–1 V), VWC (0–1 V) and temperature (0–1.2 V) of soil. • Worthy of IP68 code.

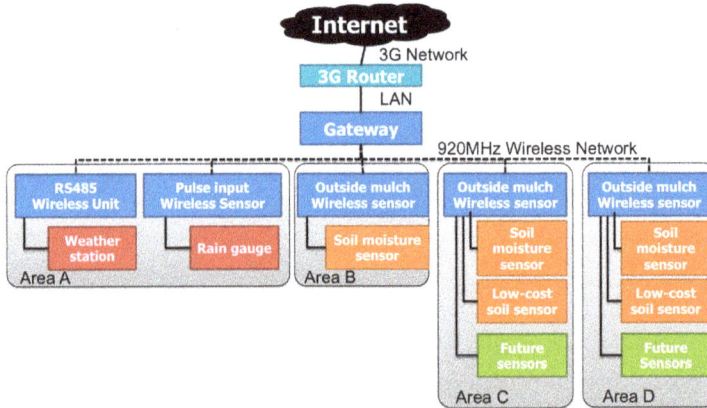

Figure 9. The structure of the WSN used in this study.

In order to provide information services, especially when we integrate various data sources, interoperability of the data and system is important. Therefore, we used the Sensor Observation Service (SOS) to standardize the terminology and Units of Measure (UOM) (Table 3).

SOS is a web service to query real-time sensor data and sensor data time series (http://www.opengeospatial.org/standards/sos). This service applies XML encoding for observation & measurement (O & M) data originating from the sensors. It defines a web service interface that allows querying observations, sensor metadata, as well as representations of observed features.

Table 3. Sensor name, physical quantity and UOM.

Sensor Name	Physical Quantity	UOM
air_temperature	Temperature	Cel
air_pressure	Air pressure	hPa
wind_speed	Wind speed	m/s
10 min_maximum_wind_speed	Wind speed	m/s
10 min_minimum_wind_speed	Wind speed	m/s
10 min_average_wind_speed	Wind speed	m/s
wind_direction	Aind direction	deg
1 min_precipitation	Precipitation	mm
solar_irradiance	Global radiation	W/m^2
10 min_maximum_solar_irradiance	Global radiation	W/m^2
10 min_minimum_solar_irradiance	Global radiation	W/m^2
10 min_average_solar_irradiance	Global radiation	W/m^2
relative_humidity	humidity	%

Following the OGC standard, defines the means to register new sensors, to remove existing ones and to insert new sensor observations. The cloud sensor infrastructure called "cloudSense" has a sensor backend service based on the SOS, thus making it possible for us to add and modify sensors and automatically reflect the change in the sensor configuration without program modification [18,19].

In order to send these environmental data from the data link server to the cloudSense, the data link server has HTTP-GET API defined by SOS. The terminology, physical quantity and UOM of each observed data point are standardized and stored in the database in cloudSense (Figure 10).

Figure 10 shows the data flow of the IoT service in this study. The obtained data goes to the local server for WSN. The dataset then goes to the cloudSence for standardization. The standardized data are back to the Database of the local server. The primary information and the secondary information service were then delivered to the farmers at the field.

The standardized data stored in cloudSense are also acquired using HTTP-GET API defined by SOS. In the HTTP-GET API, there are several kinds of requesting methods. First, "getCapability" is the method that can confirm the information about each observed data. Secondly, "getObservation" is the method that can acquire the monitored value of the selected observed property. When you use the getObservation request, you should set "eventTime", that sets both the start time and end time, and "observedProperty" which decides what sensor to choose, then one can obtain the monitored data described by the O & M language based on XML. Using these methods, the standardized data are sent to the server located at the Mie University, then XML tags are parsed and the extracted data are stored in the database constructed in this study (Figure 11).

Verifying the existence and accuracy of the weather data obtained from the weather station is necessary. Therefore, we used the "getObservation" method of cloudSense to make sure whether this system can acquire information about the growing environment from the WSN and the connection between the data link server and cloudSense, then we could confirm that the data obtained from the WSN are stored in the cloudSense. Concerning the values from the soil moisture sensor, there are some deficit points. In our web application, each missing value is replaced with an average value of the value former day and the value of the following day. In addition, we also compare the data from the SE-WS700 weather station and the data from the reference weather station. The reference weather station is added to the WSN and deployed next to the SE-WS700. As a result, almost all the observed data were correct, except for the precipitation and air temperature data [20,21].

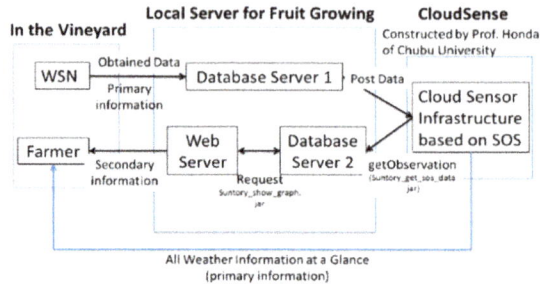

Figure 10. Design of agricultural IoT as a service for farmers

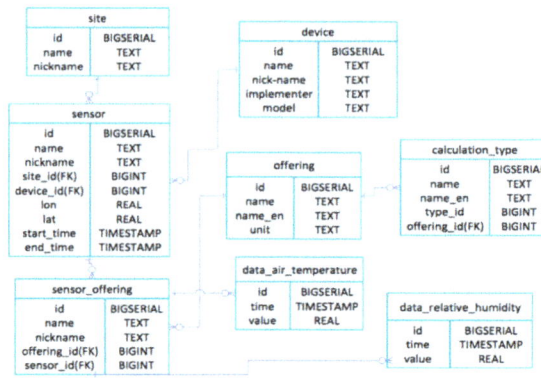

Figure 11. ER chart of database structure.

3.1.2. IoT service for Farmers

By getting the standardized data from the cloudSense and storing them in our server, we can create a web service that offers various kinds of cultivation indices to the farmers in the field. In this web site, growing environment information, as a primary index, is displayed as a list of numeric values (Figure 12). All the data acquired by the weather station are simultaneously displayed every one hour.

					Export to xlsx	Export to csv		Export to txt		
Date	Air_temp (°C)	Relative humidity (%)	Air pressure (hPa)	Wind speed (m/s)	max wind speed (m/s)	min wind speed (m/s)	avr wind speed (m/s)	Wind direction (°)	Precipi- tation (mm)	Solar irradiance (W/m²)
2016-12-05 00:00:00.0	7.6	96.8	950.2	0.6	1.8	0.0	0.8	249.4	3.33	0.0
2016-12-05 01:00:00.0	7.5	98.2	949.6	0.0	1.5	0.0	0.4	114.4	1.14	0.0
2016-12-05 02:00:00.0	7.5	98.5	949.4	0.7	1.9	0.0	0.7	244.2	0.21	0.0
2016-12-05 03:00:00.0	7.2	98.8	948.3	1.6	2.8	0.9	1.9	103.7	0.0	0.0
2016-12-05 04:00:00.0	7.2	98.7	948.0	0.8	1.5	0.0	0.7	132.6	0.0	0.0
2016-12-05 05:00:00.0	7.1	98.6	948.0	0.0	1.6	0.0	0.8	66.5	0.0	0.0
2016-12-05 06:00:00.0	6.6	98.7	948.2	2.1	3.9	0.3	1.8	6.3	0.0	0.0
2016-12-05 07:00:00.0	7.6	93.4	948.3	1.1	2.5	1.0	1.7	47.4	0.0	12.4
2016-12-05 08:00:00.0	7.1	95.4	949.0	1.0	2.7	0.5	1.5	178.4	0.0	94.4
2016-12-05 09:00:00.0	7.3	98.4	949.3	1.4	1.9	0.0	0.6	113.3	0.0	73.1

Figure 12. Primary index as a list displayed in web service.

In addition to the primary index, our web service also provides several secondary indices; AGDD, GST, CI, HI, BEDD, and DI. They are calculated using growing environment data standardized by cloudSense. The validation of the solar duration data that is necessary for the DI calculation validity was made to verify the certainty of this algorithmic program for calculating the solar duration by comparing it with the true value obtained from the solar duration sensor next to the weather station. Figure 13 shows the result of the solar duration data verification. When comparing in the winter, the difference between the true value and the value obtained from the algorithmic program is only 3 to 5%, while when comparing in the summer, the difference between these two values is about 10%. Based on this result, we should modify the solar duration data used in this study.

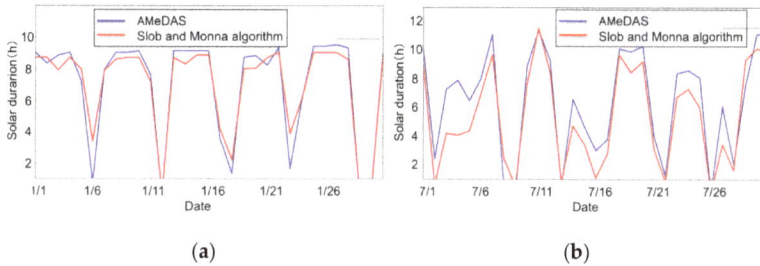

(a)   (b)

Figure 13. Data verification about solar duration: (a) The data acquired in January; (b) The data acquired in July.

In the service for secondary indices shown in Figures 14 and 15, users can select the observation place, type of sensor, observed property, type of index and time period that you want to observe (Figure 14). After they choose, the result of both the chosen secondary index and primary index is graphically shown (Figure 15). Figure 15 shows the result of the chosen primary and secondary indices in the web application developed in this study. In order to verify the accuracy of each secondary index; AGDD, GST, CI, HI, BEDD, DI, we compared each data point about the CI and HI obtained from this service and corresponding data obtained from previous research, then the validity of the secondary index calculated by our service is proved [22].

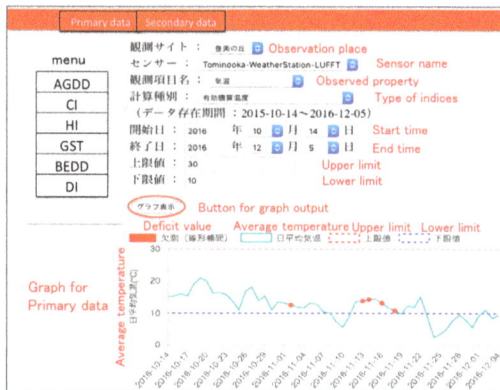

Figure 14. Secondary index displayed in web application. Since this application is written in Japanese, English annotation is mentioned with red letter.

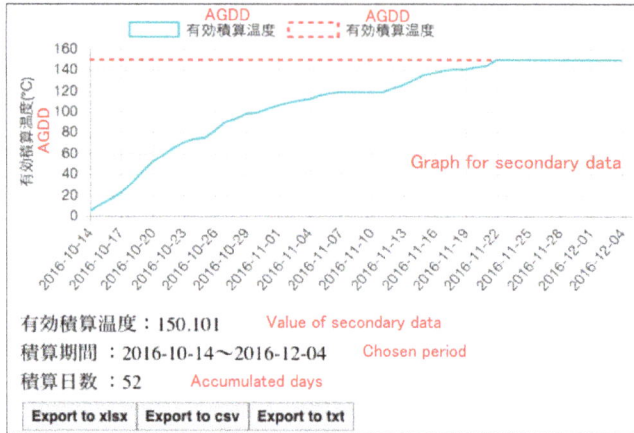

Figure 15. Polygonal line graph shown in this web application.

3.1.3. Vigor Sensing of Mandarin Orange Leaves Using Thermal Image

The results of the transformation are shown in Figure 16. The bundle of Figure 16a corresponds to the shapes of ten leaves and the line in the bundle indicates the average. The line shown in Figure 16b was adopted as a template shape because a symmetrical shape would be better for easy analysis. A reverse transformation was made for the line in Figure 16b and the template shape shown in Figure 16c was derived.

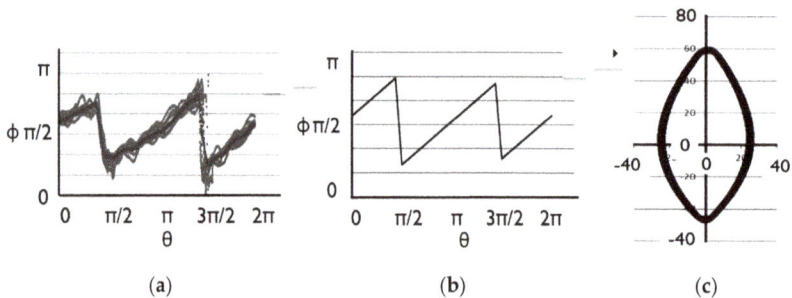

Figure 16. Transformation of leaves: (**a**) Transformed data; (**b**) Average of bundle; (**c**) Inverse transformation.

The coordinate system was defined for the original shape as well as the template shape, which was explained by using the pattern diagram shown in Figure 17. Figure 17a shows the determination method of the coordinates for the template while Figure 17b is for the original shape. For both, the center of gravity corresponds to the origin. The *X*-axis corresponds to the straight line via the leaf apex center of gravity and leaf base for the template image while the parabola is for the original image. The *Y*-axis is perpendicular to the *X*-axis at the center of gravity. Each pixel of the original image was transformed in manner similar to Figure 17b.

Figure 18 show the two sets of the original image and transformed one for the strong tree and the weak one. The thermal image in Figure 18a has the temperature distribution lower than that in Figure 18b, which coincides with the fact that the strong tree has a larger amount of transpiration than the weak tree. It can be found that the normalized shape can produce the distribution of temperature difference within the leaf surface as is shown in Figures 18c and 19.

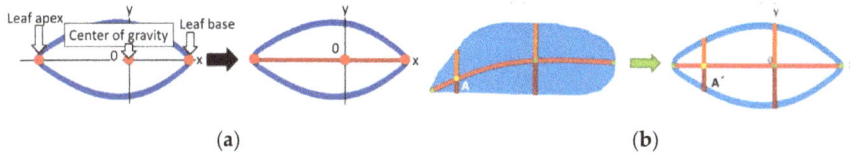

Figure 17. Mapping of the leaf to the template: (**a**) Sample of the mapping; (**b**) apply this mapping method to real leaf.

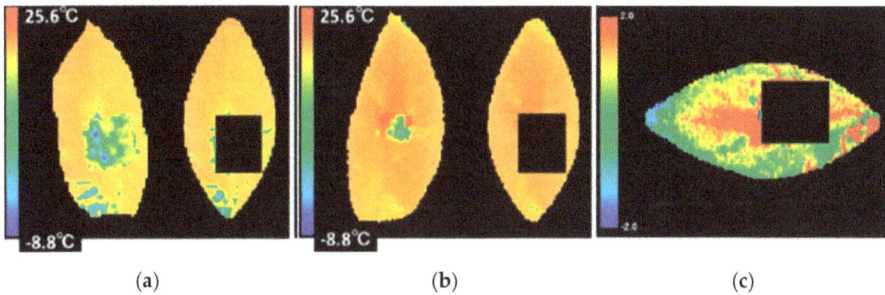

Figure 18. Thermal images of the leaves: (**a**) Strong leaf; (**b**) Weak leaf; (**c**) Temperature difference between strong leaf and weak one.

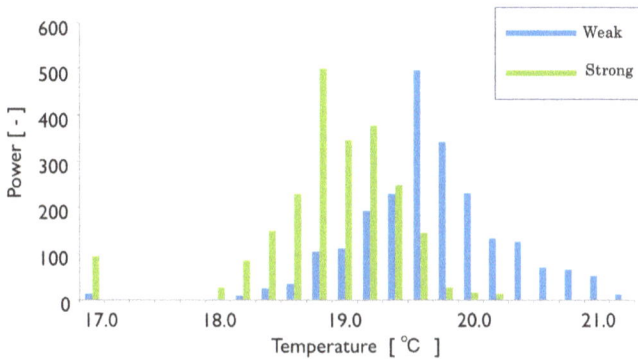

Figure 19. Comparison of temperature distribution between weak leaf and strong one.

3.1.4. Vigor Measurement of Mandarin Orange Leaves by Using Fluorescent Method

The measurements were made at three areas in each leaf shown in the Figure 5b in order to find the pigment distribution inside the leaf. First, the distribution of pigments inside each leaf was examined for 10 leaves taken from three trees of weak vigor (weak leaf) and for another 10 leaves taken from three trees of strong vigor (strong leaf). Each bar in the chart of Figure 20 shows the difference among the three areas for the good color leaf of strong vigor, bad color leaf of weak vigor and good color of weak vigor. Figure 20a corresponds to the result of chlorophyll (CHL), Figure 20b corresponds to that of flavonol (FLAV), Figure 20c corresponds to that of anthocyanin (ANTH), and Figure 20d corresponds to that of nitrogen balance index (NBI).

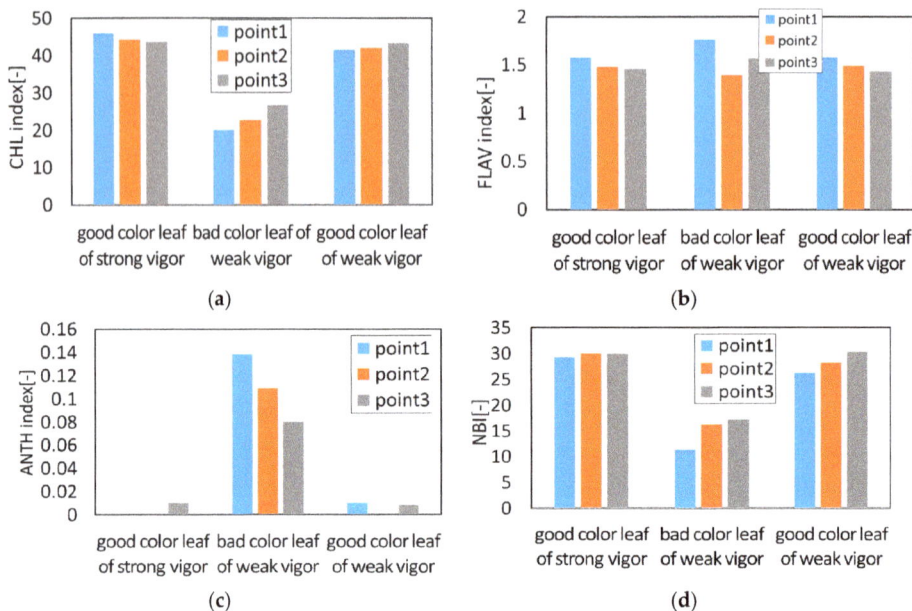

Figure 20. Comparison of each index of various leaves: (**a**) The rusult of chlorophyll; (**b**) The result of flavonol; (**c**) The result of anthocyanin; (**d**) The result of NBI.

As a result, the difference in each value among the three areas was hardly found in the case of the strong leaf. Meanwhile, as is shown in Figure 20, in the case of the weak leaves, it was found that a leaf with a good color has a similar tendency as the strong leaf, and the leaf with bad color has a difference among the three areas inside the leaf for CHL and NBI. Although there is a little problem in the weak leaf, the average value of the three areas was used as a representative value of each leaf in the later discussion. According to the results of the CHL, FLAV, ANTH and NBI for 10 weak leaves and for another 10 strong leaves, the CHL is distributed from 22 to 42 for the weak leaves, while it is distributed from 38 to 52 for the strong leaves. The ANTH is around 0.1 for the weak leaves and the existence of anthocyanin was confirmed while it was almost 0 in the strong leaves and no anthocyanin was found. The FLAV is distributed from 1.4 to 1.8 for the weak leaves while it is distributed from 1.3 to 1.6 for the strong ones. The tendency that the FLAV of the weak leaf is slightly higher than that of the strong leaf was observed. The NBI showed a similar tendency to the CHL. That is, the NBI is distributed from 13 to 30 for the weak leaves while it is distributed from 30 to 42 for the strong ones.

Figure 21 shows the relationship between ANTH and NBI for strong leaves and weak leaves. As for the strong leaves, a weak negative correlation was found between the ANTH and NBI (Figure 21a). Meanwhile, in the case of the weak leaf, a strong negative correlation was found and the correlation coefficient was 0.95 (Figure 21b).

Generally, photosynthesis is caused by a light reaction (electron transfer system) that generates biochemical energy such as ATP and NADPH and a dark reaction (Calvin cycle) that immobilizes the carbon dioxide using these biochemical energies. The energy of sunlight is entirely used to immobilize the carbon dioxide when the light reaction is balanced with the dark reaction. However, the surplus energy that is not used for immobilization of the carbon dioxide contributes to active oxygen generation in the plant body and tissue destruction occurs when the solar energy uptake by the light reaction exceeds the carbon dioxide fixation by the dark reaction. As one of the defense mechanisms to this phenomenon, it is known that resolving the chlorophyll that takes part in the light reaction and generating anthocyanins reduces the surplus sunlight, and the light reaction is made to balance with the dark reaction [23]. The negative correlation between the ANTH and CHL obtained in our experiments seems to be consistent with this fact.

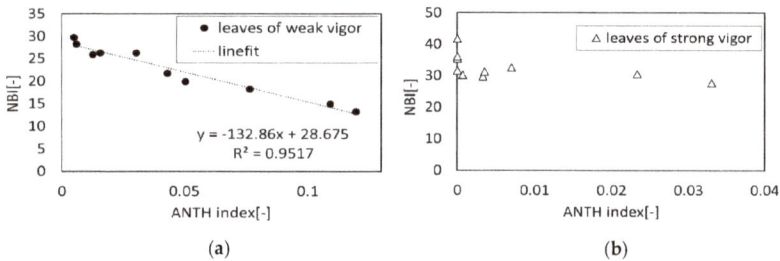

(a)   (b)

Figure 21. Correlation between ANTH and NBI: (**a**) Leaves of strong vigor; (**b**) Leaves of weak vigor.

3.1.5. Vigor Measurement of Mandarin Orange Leaves Using Fluorescent X-ray Method

It is necessary to determine the suitable beam exposure time for the fluorescence X-ray analysis of a leaf. Firstly, after changing the irradiation time to 30, 60, 90, 120 and 150 s we carried out FXR analysis of the leaf, using two kinds of beams, beam 1 and beam 2. As a result, the characteristic peak heights for the typical elements were almost the same in all the data. Therefore, the irradiation time for beam 1 was selected to be 30 s, and that for beam 2, as 60 s and these irradiation times were used for all the experiments. The measured values of the fluorescence X-rays were normalized using the peak intensity of RhLα (approximately 2.71 keV) generated by irradiation [8]. The measurements were then made at three areas of each leaf shown in the Figure 5b in order to find the element distribution inside the leaf. The distribution of elements inside each leaf was examined for 10 leaves taken from three trees of weak vigor (weak leaf) and for another 10 leaves taken from three trees of strong vigor (strong leaf) by fluorescence X-ray analysis the same as the fluorometric analysis.

As a result, both the weak and strong leaves have a slight difference in the element contents among the three areas. Figure 22 shows the FXR spectroscopic data of three areas for the strong leaf. Although there is a slight distribution inside the leaf, the average value of the three areas was used as a representative value of each leaf in the later discussion. Table 4 shows the average element values of 10 weak leaves and that of 10 strong values. The kind of the detected element was quite different between these two groups. Twelve kinds of elements were detected in the weak leaf while 24 kinds were found in the strong leaf.

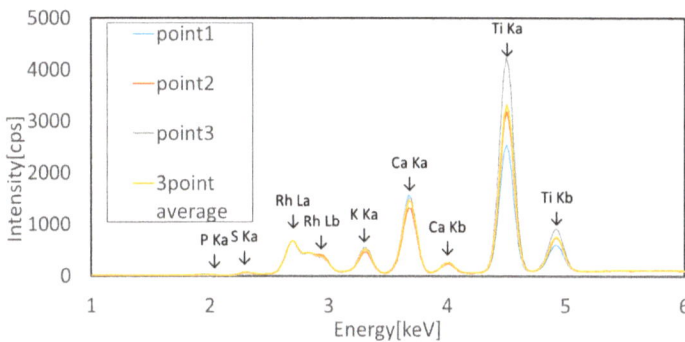

Figure 22. Normalised fluorescence X-ray spectroscopic data of strong leaf.

Table 4. Average element values of strong and weak values.

Metal	Average Content (%, ppm)	Sample Number	Evaluation	Metal	Average Content (%, ppm)	Sample Number	Evaluation
LE (%)	73.7	30		LE (%)	71.9	30	
Ti (%)	20.3	30		Ti (%)	22.4	30	
Ca (%)	3.30	29	Optimum	Ca (%)	3.19	30	Optimum
K (%)	1.33	29	Optimum	K (%)	1.01	30	Optimum
S (%)	0.230	29		S (%)	0.201	30	
P (%)	0.112	26	A little deficiency	P (%)	0.0952	30	Deficiency
Si (ppm)	855	29		Si (ppm)	764	30	
Mn (ppm)	380	30	Excess	Mn (ppm)	375	30	Excess
Fe (ppm)	138	30	Optimum	Fe (ppm)	155	30	A little excess
Cu (ppm)	87.2	30	A little excess	Cu (ppm)	72.0	19	A little excess
Zn (ppm)	49.5	14	Optimum	Zn (ppm)	37.4	8	Optimum
Mo (ppm)	22.3	22	Excess	Mo (ppm)	23.5	19	Excess
Ni (ppm)	20.2	6	A little excess	Ni (ppm)	42.7	4	Excess
Mg		0		Mg		0	
Co		0		Co		0	

The meaning of the amount of each detected element for the tree vigor was then examined using "The Shizuoka Prefecture Soil Fertilizer Handbook" as a reference (Table 5).

Table 5. Information on the appropriate kinds and contents of elements inside a leaf.

Crop Name	Content	Concentration in Dry Matter (%)				
		Nitrogen (N)	Phosphorus (P)	Potassium (K)	Calcium (Ca)	Magnesium (Mg)
mandarin orange	deficiency	under 2.3	under 0.10	under 0.7	under 2.0	under 0.10
	optimum	2.9~3.4	0.16~0.20	1.0~1.6	3.0~6.0	0.30~0.60
	excess	over 4.0	-	over 1.8	over 7.0	-

Crop Name	Content	Concentration in Dry Matter (ppm)							
		Boron (B)	Manganese (Mn)	Iron (Fe)	Zinc (Zn)	Copper (Cu)	Molybdenum (Mo)	Nickel (Ni)	Cobalt (Co)
mandarin orange	deficiency	under 30	under 30	under 35	under 10	under 4	under 0.05	-	-
	optimum	30~100	30~100	50~150	30~100	10~50	0.2~3.0	2.0~15	5~20
	excess	over 170	over 150	over 250	over 200	over 150	-	over 25	over 30

Table 5 shows the information on the appropriate kinds and contents of the elements inside the mandarin orange leaf. A comparison was made between the actual measurement results in Table 4 and a desirable amount of the element listed in Table 5. As a result, we had a diagnostic outcome that the content of Ca, K, Fe, Zn, and Cu are suitable, P is a somewhat small amount, and Ni and Mo are in excess for the strong leaf, while Ca and K are suitable, Fe is in a somewhat small amount, P is lacking, and Mn and Ni are in excess for the weak leaf (Table 4). In this diagnosis, however, the elements of N and B cannot be detected by the hXRF among the essential elements shown in Table 5, and the proper quantity of Mg for the plant, 0.3 to 0.6%, was not evaluated because the hXRF techniquer can only detect weight percent concentrations of Mg of no less than 1% in a sample.

3.2. Future Prospects of the Cultivation Indices and Tree Vigor Sensing

In this study, we developed the cultivation management system that provides cultivation indices for farmers at the vineyard. On the other hand, we also verified the effectiveness of mandarin orange tree vigor diagnosis using multi-band optical sensing.

The next objective of our study has four topics: firstly, to confirm the availability of the cultivation indices for grapevine in Japan, secondly, to replace the WSN in the orange grove and introduce a cultivation management system similar to the system in the vineyard and to develop appropriate cultivation indices for mandarin orange in Japan, thirdly, to verify the availability of the tree vigor diagnosis using multi-band optical sensing of vines and lastly, to store multi-band optical sensing data related to tree vigor in cloudSense and to standardize the terminology and UOM.

Since the main aim of this kind of service is to improve farmers' cultivation management practices, growing environmental information and tree vigor information must be combined from the plant physiological point of view in order to provide more useful cultivation indices. Recently, several studies about sap flow sensing and translocation in the phloem have been started. Especially, sap flow sensing research has been carried out in the Tomi-no-oka vineyard and we plan to connect a sap flow sensor to the SP-0050 node in the WSN [24]. On the other hand, research on the sensing of translocation in the phloem by using IR-spectroscopy is also ongoing [25,26]. This information could play an important role to improve cultivation indices.

4. Conclusions

We have developed a new WSN with a weather station and soil moisture sensors. The integrated system consisting of the WSN system and cloudSense will enable us to develop an effective web application for farmers. In future studies, there are three tasks, such as improvement of the durability of the WSN and accuracy of the sensors, a web application service that shows indexes of plant growth useful for farmers and a suitable cloud design for agricultural service, and design of a low-cost and energy-saving WSN.

The results in this paper could play a very important role in the health diagnosis of plants for agriculture based on optically sensed results. As for the optical and spectroscopic sensing, the combination of some sensing techniques could extract more precious quality information. It is necessary to totally understand the integrative quality of agricultural products, and the optically sensed information should be represented in the information to be easily understood, and the development of visualization technology for the complex sensing information of agricultural products is also desired.

In the near future, the advanced sensing technology such as sap flow sensing could be linked to WSN and the physiologically-based knowledge of agricultural plants. By the assimilation and optimization of such subjects, the comprehensive vigor of agricultural plants could be evaluated by the science-based system for the sustainable supply of high quality agricultural materials. Furthermore, multi-band optical sensing data relating with tree vigor should be stored in cloudSense in order to provide more useful cultivation indices for farmers at the field. We shall also apply WSN knowledge acquired in vineyard to the mandarin orange grove and also apply the multi-band optical sensing knowledge acquired in mandarin orange grove to the vineyard.

Acknowledgments: The authors sincerely thank the Project "Integration research for agriculture and interdisciplinary fields" by the Ministry of Agriculture, Fisheries and Forests, Japan, for the financial support of this research.

Author Contributions: This work was carried out in collaboration between all authors. Takaharu Kameoka defined the research theme. Shuhei Isoda and Shinichi Kameoka carried out the experiments about fluorescent and fluorescent X-ray. Atsushi Hashimoto and Ryoei Ito supported the experiments about multi-band optical sensing. Satoru Miyamoto constructed the structure of wirelesss sensor network in the point of view of low power consumption. Genki Wada and Naoki Watanabe helped us choose and create the secondary indices used in this study. Takashi Yamakami and Ken Suzuki provided the leaves of various conditions used in this study. Shinichi Kameoka wrote the paper. All authors have contributed to the reviewing and revision of the manuscript.

Conflicts of Interest: The authors declare no conflict of interest.

References

1. Rasyid, M.U.H.A.; Nadhori, I.U.; Sudarsono, A.; Alnovinda, Y.T. Pollution Monitoring System Using Gas Sensor Based on Wireless Sensor Network. *Int. J. Eng. Technol. Innov.* **2016**, *6*, 79–91.
2. Togami, T.; Yamamoto, K.; Hashimoto, A.; Watanabe, N.; Takata, K.; Nagai, H.; Kameoka, T. A wireless sensor network in a vineyard for smart viticultural management. In Proceedings of the 2011 SICE Annual Conference, Tokyo, Japan, 13–18 September 2011.
3. Togami, T.; Hashimoto, A.; Ito, R.; Kameoka, T. Agro-Environmental Monitoring Using A Wireless Sensor Network to Support Production of High Quality Mandarin Oranges. *Agric. Inf. Res.* **2011**, *20*, 110–121. [CrossRef]

4.   Kameoka, T.; Hashimoto, A. A Sensing Approach to Fruit-Growing. In *Wireless Sensor Networks and Ecological Monitoring*; Subhas, C., Mukhopadhyay, S.C., Jiang, J.-A., Eds.; Smart Sensors, Measurement and Instrumentation; Springer: Berlin/Heidelberg, Germany, 2013; Volume 3, pp. 217–246.

5.   Ohtani, Y.; Togami, T.; Kimura, Y.; Hashimoto, A.; Kameoka, T. Thermal image utilization on the vigor diagnosis of mandarin orange tree. In Proceedings of the 2010 SICE Annual Conference, Taipei, Taiwan, 18–21 August 2010.

6.   Vadivambal, R.; Jayas, D.S. Applications of Thermal Imaging in Agriculture and Food Industry—A Review. *Food Bioprocess Technol.* **2011**, *4*, 186–199. [CrossRef]

7.   Hashimoto, A.; Suehara, K.; Kameoka, T. Applications of Infrared Spectroscopic Techniques to Quality Evaluation in Agriculture and Food Process. In Proceedings of the 13th International Workshop on Advanced Infrared Technology & Applications, Pisa, Italy, 29 September–2 October 2015.

8.   Kameoka, T.; Hashimoto, A. Effective Application of ICT in Food and Agricultural Sector-Optical Sensing is Mainly Described—. *IEICE Trans. Commun.* **2015**, *98*, 1741–1748. [CrossRef]

9.   Jayaraman, P.; Yavari, A.; Georgakopoulos, D.; Morshed, A.; Zaslavsky, A. Internet of things platform for smart farming: Experiences and lessons learnt. *Sensors* **2016**, *16*, 1884. [CrossRef] [PubMed]

10.   Irmak, S.; Payero, J.O.; Eisenhauer, D.O.; Kranz, W.L.; Martin, D.L.; Zoubek, G.L.; Rees, J.M.; van de Walle, B.; Christiansen, A.P.; Leininger, D. *Watermark Granular Matrix Sensor to Measure Soil Matric Potential for Irrigation Management*; University of Nebraska–Lincoln: Lincoln, NE, USA, 2006; pp. 1–8.

11.   Gladstones, J.S. *Wine, Terroir and Climate Change*; Wakefield Press: Adelaide, Australia, 2015; pp. 5–26.

12.   Vivar, M.; Fuentes, M.; Norton, M.; Makrides, G.; De Bustamante, I. Estimation of sunshine duration from the global irradiance measured by a photovoltaic silicon solar cell. *Renew. Sustain. Energy Rev.* **2014**, *36*, 26–33. [CrossRef]

13.   Motonaga, Y.; Kondou, H.; Kameoka, T.; Hashimoto, A. Determination of the Standard Shape and Color of Agricultural Products. In Proceedings of the 4th International Conference on Quality Control by Artificial Vision (QCAV'98), Kagawa, Japan, 10–12 November 1998; pp. 29–34.

14.   Agati, G.; Tuccio, L.; Kusznierewicz, B.; Chmiel, T.; Bartoszek, A.; Kowalski, A.; Grzegorzewska, M.; Kosson, R.; Kaniszewski, S. Nondestructive Optical Sensing of Flavonols and Chlorophyll in White Head Cabbage (*Brassica oleracea* L. var. capitata subvar. alba) Grown under Different Nitrogen Regimens. *J. Agric. Food Chem.* **2016**, *64*, 85–94. [CrossRef] [PubMed]

15.   Cerovic, Z.; Ghozlen, N.B.; Milhade, C.; Obert, M.; Debuisson, S.; Le Moigne, M. Nondestructive Diagnostic Test for Nitrogen Nutrition of Grapevine (*Vitis vinifera* L.) Based on Dualex Leaf-Clip Measurements in the Field. *J. Agric. Food Chem.* **2015**, *63*, 3669–3680. [CrossRef] [PubMed]

16.   Hashimoto, A.; Niwa, T.; Yamamura, T.; Suehara, K.; Kanou, M.; Kameoka, T.; Kumon, T.; Hosoi, K. X-Ray Fluorescent and Mid-Infrared Spectroscopic Analysis of Tomato Leaves. In Proceedings of the 2006 SICE-ICASE International Joint Conference, Busan, Korea, 18–21 October 2006; pp. 3359–3562.

17.   De Haij, M.; Wauben, W.F. Investigations into the improvement of automated precipitation type observations at KNMI. In Proceedings of the WMO Technical Conference on Meteorological and Environmental Instruments and Methods of Observation 2010, Helsinki, Finland, 2–8 September 2010.

18.   Honda, K.; Chinnachodteeranun, R.; Witayangkurn, A. Sensor Observation Service for Connecting Heterogeneous Field Sensor Platforms to Applications. In Proceedings of the 2016 WCCA-AFITA, Suncheon, Korea, 21–24 June 2016.

19.   Chinnachodteeranun, R.; Honda, K. Sensor Observation Service API for Providing Gridded Climate Data to Agricultural Applications. *Future Int.* **2016**, *8*, 40–47. [CrossRef]

20.   Kameoka, S.; Hashimoto, A.; Ito, R.; Kameoka, T. Wireless Sensor Network System for Fruit-Growing Environment at the Field. In Proceedings of the 2016 WCCA-AFITA, Suncheon, Korea, 21–24 June 2016.

21.   Kameoka, S.; Isoda, S.; Hashimoto, A.; Ito, R.; Miyamoto, S.; Wada, G.; Watanabe, N.; Kameoka, T. Deployment of a Wireless Sensor Network for the Drowing Environment Information Acquisition at the Field. *Agric. Inf. Res.* **2017**, *26*, 11–25.

22.   Tonietto, J.; Carbonneau, A. A multicriteria climatic classification system for grape-growing regions worldwide. *Agric. For. Meteorol.* **2004**, *124*, 81–97. [CrossRef]

23.   Taiz, L.; Zeiger, E. *Plant Physiology*, 4th ed.; Sinauer Associates, Inc.: Sunderland, UK, 2010; p. 114.

24. Kameoka, T.; Nishioka, K.; Motonaga, Y.; Kimura, Y.; Hashimoto, A.; Watanabe, N. Smart sensing in a vineyard for advancedviticultural management. In Proceedings of the 2014 International Workshop on Web Intelligence and Smart Sensing, Saint Etienne, France, 1–2 September 2014; pp. 1–8.

25. Xiu, J.; Tanida, W.; Kimura, Y.; Hashimoto, A.; Kameoka, T. Quantitative analysis of squeezed liquid of sorghum by using FTIR spectroscopy. In Proceedings of the 2015 54th Annual Conference of the Society of Instrument and Control Engineers of Japan (SICE), Hangzhou, China, 28–30 July 2015; pp. 100–103.

26. Tanida, W.; Hashimoto, A.; Ito, R.; Ozaki, S.; Kameoka, S.; Kameoka, T. Measurement of Sugar in Phloem Sap by Using IR Spectroscopy. In Proceedings of the 2016 WCCA-AFITA, Suncheon, Korea, 21–24 June 2016.

Article

Spectroscopic Diagnosis of Arsenic Contamination in Agricultural Soils

Tiezhu Shi [1], Huizeng Liu [1,2], Yiyun Chen [3], Teng Fei [3,4], Junjie Wang [1] and Guofeng Wu [1,*]

[1]   Key Laboratory for Geo-Environmental Monitoring of Coastal Zone of National Administration of Surveying, Mapping and GeoInformation & Shenzhen Key Laboratory of Spatial Smart Sensing and Services & College of Life Sciences and Oceanography, Shenzhen University, 518060 Shenzhen, China; tiezhushi@whu.edu.cn (T.S.); zhongzheng0512@126.com (H.L.); wjjlight@whu.edu.cn (J.W.)
[2]   Department of Geography, Hong Kong Baptist University, Kowloon Tong, Kowloon, Hong Kong, China
[3]   School of Resource and Environmental Sciences, Wuhan University, 430079 Wuhan, China; chenyy@whu.edu.cn (Y.C.); feiteng@whu.edu.cn (T.F.)
[4]   Suzhou Institute of Wuhan University, 215000 Suzhou, China
*   Correspondence: guofeng.wu@szu.edu.cn

Academic Editor: Dimitrios Moshou
Received: 24 March 2017; Accepted: 2 May 2017; Published: 4 May 2017

Abstract: This study investigated the abilities of pre-processing, feature selection and machine-learning methods for the spectroscopic diagnosis of soil arsenic contamination. The spectral data were pre-processed by using Savitzky-Golay smoothing, first and second derivatives, multiplicative scatter correction, standard normal variate, and mean centering. Principle component analysis (PCA) and the RELIEF algorithm were used to extract spectral features. Machine-learning methods, including random forests (RF), artificial neural network (ANN), radial basis function- and linear function-based support vector machine (RBF- and LF-SVM) were employed for establishing diagnosis models. The model accuracies were evaluated and compared by using overall accuracies (OAs). The statistical significance of the difference between models was evaluated by using McNemar's test (Z value). The results showed that the OAs varied with the different combinations of pre-processing, feature selection, and classification methods. Feature selection methods could improve the modeling efficiencies and diagnosis accuracies, and RELIEF often outperformed PCA. The optimal models established by RF (OA = 86%), ANN (OA = 89%), RBF- (OA = 89%) and LF-SVM (OA = 87%) had no statistical difference in diagnosis accuracies (Z < 1.96, $p < 0.05$). These results indicated that it was feasible to diagnose soil arsenic contamination using reflectance spectroscopy. The appropriate combination of multivariate methods was important to improve diagnosis accuracies.

Keywords: visible and near-infrared reflectance spectroscopy; heavy metal contamination; spectral pre-processing; feature selection; machine-learning

1. Introduction

Soil heavy metal contamination demands effective methods for diagnosing suspected contaminated areas and controlling the rehabilitation process. There is increasing interest in using visible and near-infrared reflectance spectroscopy (VNIRS, 350–2500 nm) to measure soil heavy metal contents and to map its spatial distribution [1], since this technique provides a non-destructive, rapid, and cost-effective method for measuring several soil properties from a single scan, and requires minimal sample preparation and hazardous chemicals [2].

The spectroscopic measurement of heavy metals is usually feasible because of their indirect relationships with some spectral feature soil properties, such as organic matter, iron-oxides or clays [1]. Therefore, the spectral information for soil heavy metal estimations is weak, indirect, and non-specific. Moreover, the spectral features of soil properties in visible/near-infrared spectra are largely overlapping, while other factors, such as surface roughness, moisture content, and organic matter of soil, also weaken the spectroscopic measurement of soil properties [3]. Thus, the analysis of visible/near-infrared spectra requires the use of multivariate chemometric techniques to mathematically extract useful information for soil property estimations.

Pre-processing techniques are commonly used to reduce the random noise, baseline drift and multiple scattering effects in the spectra [1]. For instance, Savitzky-Golay (SG) smoothing is adopted to increase the spectral quality by eliminating random noise. Derivative transformation can remove background interferences, resolve overlapping spectra and minimize the baseline drift caused by the differences in grinding and optical setups [4]. Multiplicative scatter correction (MSC) and standard normal variate (SNV) seek to eliminate the multiplicative interferences of scattering and particle size [5]. Moreover, data enhancement algorithms, such as mean centering (MC) and normalization, are able to highlight the diversities of spectral data, reduce redundant information, and simplify calibration models [5]. For soil reflectance spectroscopy, the type and amount of pre-processing required are data-specific; no single or combination of pre-processing techniques will work well with all data sets [6].

Feature selection techniques, such as successive projection algorithm (SPA), uninformative variables elimination (UVE) and genetic algorithm (GA), are often applied to remove uninformative spectral bands and to select optimal spectral variable subsets for establishing regression models [7,8]. SPA is a forward feature selection technique, and it uses a simple projection operation in a vector space to minimize the collinearity problem [9]. UVE detects uninformative spectral variables based on a stability analysis of regression coefficients (b-coefficient) [10]. GA uses a probabilistic, non-local search process to randomly select an initial spectral data-set and to optimize this data set by considering many combinations of spectral variables and their interactions [10]. In soil spectroscopy, GA always results in better performances than SPA and UVE for soil property estimates [7,8].

These feature selection methods are designed to select features to improve the estimation of numerical variables, such as soil property contents, and they are inappropriate to reduce dimensionality and select features for classifying nominal variables, such as heavy metal contamination levels. Principal component analysis (PCA) and the RELIEF algorithm have been widely applied for feature selection in the classification applications, such as image classification and text categorization [11]. However, as far as we know, PCA and RELIEF have rarely been employed to select features for diagnosing soil heavy metal contamination from soil reflectance spectra.

From a large data-set using trained models, data mining techniques automatically or semi-automatically uncover patterns, which are used on a new data-set for prediction [12]. Various data mining techniques, such as principal component regression (PCR) [13], partial least squares regression (PLSR) [14–16], artificial neural network (ANN) [4], multivariate adaptive regression splines (MARS) [17] and support vector machine (SVM) [18–20] were employed to train models from spectral data for estimating soil properties, including heavy metals. The 'training model' process is synonymously described as 'machine-learning', which can be defined as the process of discovering the relationships between predictor and response variables using computer-based statistical methods [21]. In soil science, machine-learning techniques have been used to classify soil types, soil depth classes, and soil drainage classes [22]. However, few studies have adopted machine-learning techniques to diagnose soil heavy metal contamination from soil reflectance spectroscopy [23].

Several studies have adopted multivariate chemometric techniques to quantitatively predict heavy metal contents in agricultural soils by using reflectance spectroscopy. For example, Ren et al. [24] used PLSR to establish a quantitative relation between reflectance spectra and As, and Cu contents in agricultural soils; Wu et al. [13] predicted Hg concentration in suburban agricultural soils of the Nanjing region by using PCR and reflectance spectra within the visible-near-infrared region. By reviewing the literature on soil heavy metal predictions, it is found that the prediction accuracies of soil heavy metal contents usually cannot reach a good quantitative level (the recommended R^2 of 0.81 or above for soil analysis [25]) because of the indirect prediction mechanisms. For practical applications, such as soil heavy metal monitoring, contamination remediation, or digital soil mapping, the diagnosis of soil heavy metal contamination may be sufficient rather than accurate heavy metal content estimations. However, at present, soil reflectance spectroscopy is rarely employed to qualitatively diagnose soil heavy metal contamination. To the best of our knowledge, Bray et al. [23] were the first to employ an ordinal logistic regression technique to diagnose Cd, Cu, Pb and Zn contamination in urban soils from reflectance spectra. Therefore, it is interesting and necessary to extend the knowledge about the diagnosis of soil heavy metal contamination by using soil reflectance spectroscopy.

In China, arsenic content has continuously increased in agricultural soils during the past 30 years, because of some anthropogenic activities, such as chemical fertilizers, arsenic-bearing pesticides, animal manures, mining, smelting, and irrigation with arsenic-contaminated water [26]. Excessive arsenic accumulation in agricultural soils can hinder the crops' growth and decrease the yield and quality of agricultural products. Moreover, as a potent carcinogen, arsenic might pose a serious health threat to the human body, such as malignant arsenical skin lesions, respiratory disease, gastrointestinal disorder, liver malfunction, nervous system disorder and haematological diseases [27].

Given the importance of monitoring arsenic contamination in agricultural soils, this study aimed to compare the abilities of pre-processing techniques (derivative transformations, MSC, SNV, MC) and machine-learning techniques (random forests (RF), ANN, and SVM) in diagnosing soil arsenic contamination from soil reflectance spectroscopy, and to investigate whether the feature selection approaches (PCA and RELIEF) could improve the diagnosis accuracy by using different machine-learning methods. The result of this study is expected to establish a technical process for diagnosing soil heavy metal contamination by using soil reflectance spectroscopy.

2. Materials and Methods

2.1. Soil Samples

In total, 195 historical soil samples collected in Yixing and Zhongxiang regions were used for this work. Yixing (Figure 1b) is located in the south of Jiangsu Province, China, with an annual temperature of 15.7 °C and a mean annual precipitation of 1177 mm. Zhongxiang (Figure 1c) is situated in the middle of Hubei Province, China, and its mean annual temperature is 15.0 °C with a mean annual precipitation of 961 mm. Yixing's dominant soil types are dystric cambisols, lixisols, anthrosols, alisols, calcaric fluvisols, calcisols, cambisols and gleysols for different crop cultivation [20]. The soils collected from Zhongxiang mainly belong to anthrosols for rice planting [28]. At each sample site, surface soils (0–10 cm) were collected. The industrial wastewater, exhaust gas or waste residues produced by local chemical factories are the major causes of arsenics contamination in agricultural soils in the Zhongxiang region [28]; in Yixing, the contamination may mostly result from sewage irrigation, parent materials or vehicle exhausts [29].

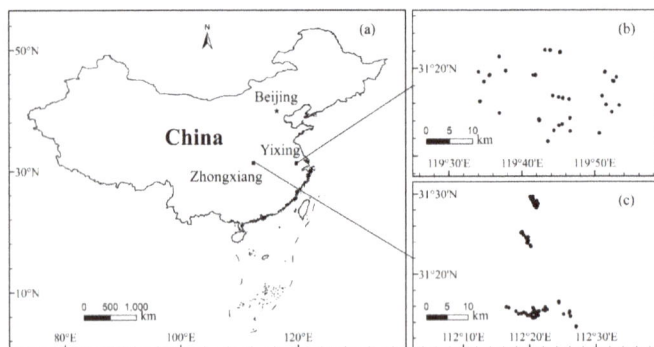

Figure 1. Study areas (**a**) and spatial distribution of soil samples in Yixing (**b**) and Zhongxiang (**c**).

2.2. Laboratory Spectrum and Soil Arsenic Content Measurement

Soil samples were air-dried and ground in a mechanical agate grinder to a particle size of ≤2 mm. The diffuse reflectance spectra were measured by using the FieldSpec3 portable spectroradiometer (ASD Inc., now PANalytical Company, Boulder, CO, USA) with a spectral range of 350 to 2500 nm. The spectral measurements were conducted in a dark room. The air-dried and ground soil sample was placed in a 10 cm diameter petri dish with a thickness of approximately 15 mm. A 50 W halogen lamp was used as the light source, which was positioned 30 cm away from soil sample, with a 15° zenith angle [20]. The optical probe was installed about 15 cm above the soil sample. A Spectralon panel (Labsphere, North Sutton, NH, USA) was used for white referencing once every six measurements.

After spectral measurement, soil samples were further ground, and passed through a 100-mesh sieve (0.15 mm). The finely ground soil samples were digested by HF-HClO4-HNO3. The arsenic contents of digested samples were then analyzed by using a hydride generation atomic fluorescence spectrometry (HG-AFS) method [30]. Certified soil reference materials (GBW 07401, GBW 07402, and GBW 07407, National Research Center for Certified Reference Materials of China) were used to verify the precision of HG-AFS method.

For the purpose of diagnosis, the measured soil arsenic contents were coded into binary 0 or 1, describing uncontaminated or contaminated samples, respectively. The index of geo-accumulation (I_{geo}) [31] was applied to assess the arsenic contamination in the soils:

$$I_{geo} = \log_2 \frac{M_{As}}{1.5B_{As}} \tag{1}$$

where M_{As} is the measured arsenic contents in the soils, B_{As} is the geochemical background value of arsenic (13 mg·kg^{-1}), the constant of 1.5 was used to eliminate fluctuations caused by regional differences and anthropogenic influences [31]. $I_{geo} \leq 0$ indicates practically uncontaminated, whereas $I_{geo} > 0$ means contaminated [31].

2.3. Pre-Processing Transformations

The whole measured soil arsenic content data and their corresponding spectral data were divided into training ($n = 98$) and test ($n = 97$) data sets using a Kennard-Stone algorithm [32], which is effective for selecting spectra-representative samples for model development. The reflectance spectra were first reduced to 400–2450 nm to remove the wavelengths with high noise effects at the spectral edges. The reflectance spectra were then SG smoothed with a moving window of 9 nm. The smoothed spectra were resampled to 10 nm intervals (e.g., 400, 410, and 420 nm, etc.) to eliminate the data redundancy by using a Gaussian model [4]. Moreover, first and second derivatives, MSC, SNV and MC of reflectance spectra were performed for soil spectra to enhance spectral features and to further establish robust diagnosis models. Reflectance spectra were transformed into log(1/Reflectance) before MSC and SNV were performed.

2.4. Feature Selection

PCA and the RELIEF algorithm were applied to extract features from spectral variables of the training data-set. PCA was an optimal linear scheme for extracting several principle components (PCs) from high dimensional variables, and the extracted components can hold the majority of the variables' information. The RELIEF algorithm, first described by Kira and Rendell [33], was used as a simple, fast and effective approach to weigh variables, and its output is the ranking weights between −1 and 1 for spectral variables, in which the more positive weights indicate more predictive spectral variables. In this study, PCA and the RELIEF algorithm were implemented in Weka (Waikato Environment for Knowledge Analysis). The number of PCs was determined by the diagnosis accuracy of the calibration. The threshold for the RELIEF weight value was set to 0, and the scattered spectral bands with local extreme weights were selected as spectral features to avoid the multicollinearity among RELIEF-selected features.

2.5. Multivariate Diagnosis Analysis

Machine-learning methods, such as RF, ANN and SVM, were employed for calibrating diagnosis models using the training data set. For brevity, the summaries of these techniques were provided, and some key references were cited. Interested readers may find more details about these techniques in these references. In this study, the machine-learning methods were implemented by using a R-based Rattle package developed by Williams [34].

2.5.1. Random Forests (RF)

RF, introduced by Breiman [35], is an ensemble learning method that constructs a multitude of decision trees. For the RF learner, each tree is independently trained from a randomized bootstrap sample of the entire training data set, and a subset of explanatory variables is randomly selected for the node-splitting rules in each tree [36]. In classification, trees are voted by majority [35]. The RF depends only on two user-defined parameters: the number of variables in each random subset (nv) and the number of trees in the forest (nt). In this study, the nv was optimized from 1 to the total number of variables with increments of 1, and nt from 0 to 1000 by increments of 10. The variable that is important for RF modeling can be determined by mean decrease GINI values.

2.5.2. Artificial Neural Network (ANN)

The concept of ANN learner may date back to 1940s when McCulloch and Pitts [37] initially planned to develop a virtual "central nervous system" for computer modeling. The design of ANN simulates the data processing in biological nervous systems. The structure of an ANN consists of a set of interconnected neurons. Some neurons are adopted for the reception of information, others for its forwarding and storage, and another group for the outward release of information [38]. Neurons are connected to each other through weighted synapses. In an ANN, the number of hidden layers and neurons in each hidden layer ought to be optimized [21]. In this study, the number of hidden layers was optimized by iterating this parameter from 1 to 20, and the number of neurons in each layer was set as the total number of variables.

2.5.3. Support Vector Machine (SVM)

SVM is a kernel-based machine learning method developed on the basis of statistical learning theory [39]. SVM applies a kernel function to map training data into a higher dimensional feature space, and computes separating hyperplanes that achieve maximum separation (margin) between the classes [40]. The maximum separation hyperplane is the training data on the margin, which are called support vectors. The quality of the SVM classifier is affected by the type of kernel function, kernel width (γ) and regularization parameter (C) [40]. In this study, radial basis function (RBF) and linear function (LF) were adopted as kernel functions, respectively.

2.6. Validation and Comparison of Diagnosis Models

The calibrated models were applied for diagnosing the contaminated and uncontaminated soil samples of the test data-set. The overall accuracy (OA, Equation (2)) [38] of the test data-set was calculated and employed for comparing the diagnosis abilities of multivariate methods. The same computer environment was kept for running different machine-learning algorithms.

$$OA = \frac{pp + nn}{pp + np + pn + nn} \tag{2}$$

where the meanings of *pp*, *np*, *pn* and *nn* are displayed in Table 1.

Table 1. Confusion matrix of observed and diagnosed soil samples for calculating overall accuracy [1].

		Observed	
Allocation		Contaminated (Positive, Value = 1)	Uncontaminated (Negative, Value = 0)
Predicted	Contaminated (positive, value = 1)	*pp*	*np*
	Uncontaminated (negative, value = 0)	*pn*	*nn*

[1] *pp*: number of correctly diagnosed contaminated soil samples; *np*: number of falsely diagnosed uncontaminated soil samples; *pn*: number of falsely diagnosed contaminated soil samples; *nn*: number of correctly diagnosed uncontaminated soil samples.

The statistical significance of the difference between diagnosis models was evaluated by using McNemar's test [41], which is based on a binary distinction between correct and incorrect class allocations (Table 2). McNemar's test is also based on the standardized normal test statistic expressed in Equation (3):

$$Z = \frac{f_{12} - f_{21}}{\sqrt{f_{12} + f_{21}}} \tag{3}$$

Therefore, the test is focused on the cases that are correctly diagnosed by one classifier but misdiagnosed by the other. Two diagnosis models may exhibit different accuracies at the 95% level of confidence if $Z > |1.96|$.

Table 2. Assessment of the statistical significance of the difference between two diagnosis models using McNemar's Test [1].

		Diagnosis Model 2	
Allocation		Correct	Incorrect
Diagnosis model 1	Correct	f_{11}	f_{12}
	Incorrect	f_{21}	f_{22}

[1] f_{12}: the test soil samples that are correctly diagnosed by diagnosis model 1 but misdiagnosed by diagnosis model 2; f_{21} the test soil samples that are correctly diagnosed by diagnosis model 2 but misdiagnosed by diagnosis model 1.

3. Results

3.1. Soil Arsenic and the Spectra

The percent mean standard error of the HG-AFS method for arsenic determination was 2.9%. The descriptive statistics of soil arsenic of the 195 soil samples are shown in Table 3. For the total data set, the soil arsenic contents varied from 1.91 to 133.36 mg·kg^{-1}, with a mean of 18.13 mg·kg^{-1} and a standard deviation of 18.67 mg·kg^{-1}. Considering I_{geo} values, 27%, 26% and 29% of samples were contaminated by arsenic in total, training and test data sets, respectively.

Table 3. Statistical descriptions for the arsenic contents (mg·kg^{-1}) and the percent value of contaminated samples (per %) [1].

	No.	Minimum	Maximum	Mean	Std.	Per %
Total data set	195	1.91	133.36	18.13	18.67	27
Training data set	98	1.91	106.10	12.70	16.81	26
Test data set	97	4.40	133.36	19.00	20.43	29

[1] No.: number of samples; Std.: standard deviation.

The mean value and standard deviation of original and pre-processed spectra for contaminated and uncontaminated soil samples are shown in Figure 2. Three prominent absorption peaks around 1400, 1900 and 2000 nm are visibly water absorption features [42] (Figure 2a); MC centered the reflectance spectra on 0 values (Figure 2b); SNV (Figure 2c) and MSC (Figure 2d) had similar spectral curves, and served the same purpose to remove the multiple scattering effects in the reflectance spectra; first (Figure 2e) and second (Figure 2f) derivatives minimized the baseline drift and highlight the minor absorption features of reflectance spectra. These demonstrated that the original reflectance and pre-processed spectra of uncontaminated and contaminated soil samples were overlapped, which indicates that there might exist a nonlinear relationship between spectra and soil arsenic contamination.

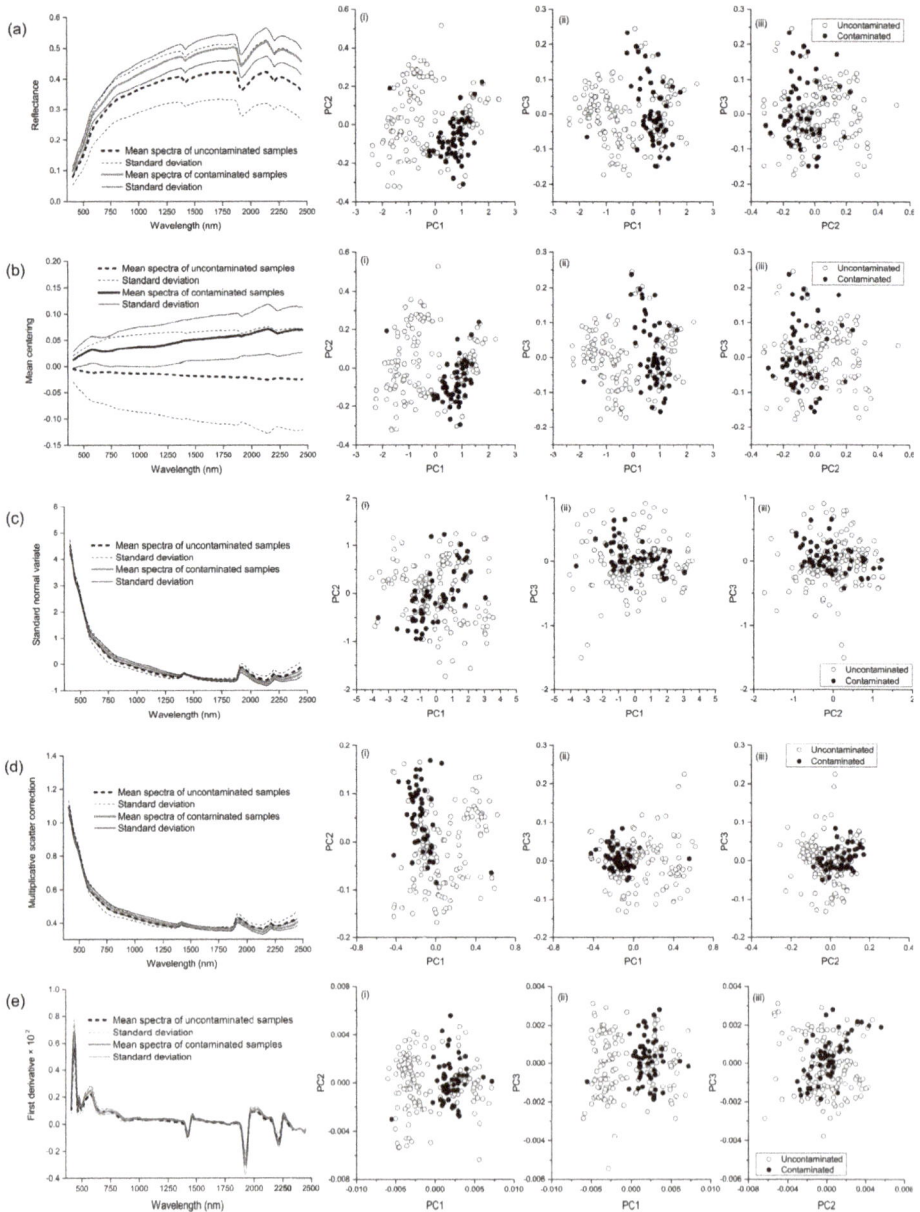

Figure 2. *Cont.*

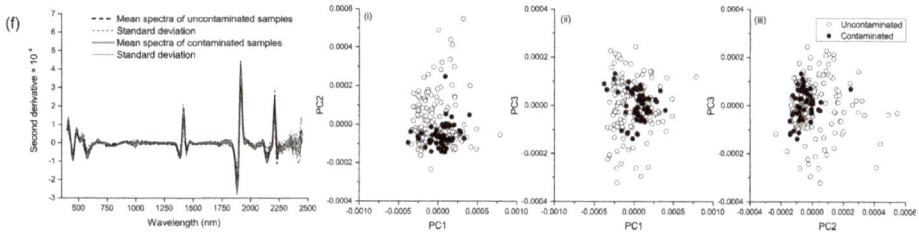

Figure 2. The reflectance spectra and the three first principal components (PC1, PC2 and PC3) for the contaminated and uncontaminated soil samples: (**a**) original reflectance spectra, (**b**) mean centering spectra, (**c**) standard normal variate spectra, (**d**) multiplicative scatter correction spectra, (**e**) first derivative spectra, and (**f**) second derivative spectra.

3.2. Principal Components and RELIEF Selected Features

The first three loadings of the PCA analysis for original reflectance and pre-processed spectra were displayed in Figure 2. The score plots showed that the spectral space of the contaminated samples fell into those of uncontaminated samples. This meant that the linear classifier might be unable to effectively diagnose contaminated or uncontaminated soil samples by using principal components.

The RELIEF weights and the selected spectral features are displayed in Figure 3. The RELIEF weights of the MC spectra (Figure 3b) had the same values as those of original reflectance spectra (Figure 3a), thus the same spectral variables at 400, 470, 930, 1090, 1840, 2140, 2350 and 2400 nm were selected as spectral features for original reflectance and MC spectra. The RELIEF weights of SNV (Figure 3c) and MSC (Figure 3d) processed spectra showed the same tendency, and the same spectral variables at 470, 1100, 1420, 1780, 1910 and 2120 nm were identified as spectral features. Spectral variables at 410, 490, 540, 640, 820, 1210, 1300, 1460, 1940 and 2210 nm (Figure 3e), and variables at 570, 670, 750, 810, 990, 1290, 1400, 1570, 1890, 1990, 2150 and 2220 nm (Figure 3f) were selected as spectral features for first and second derivatives, respectively. Compared with the original reflectance, MC, SNV and MSC spectra, first and second derivatives resulted in more spectral features with higher RELIEF weights.

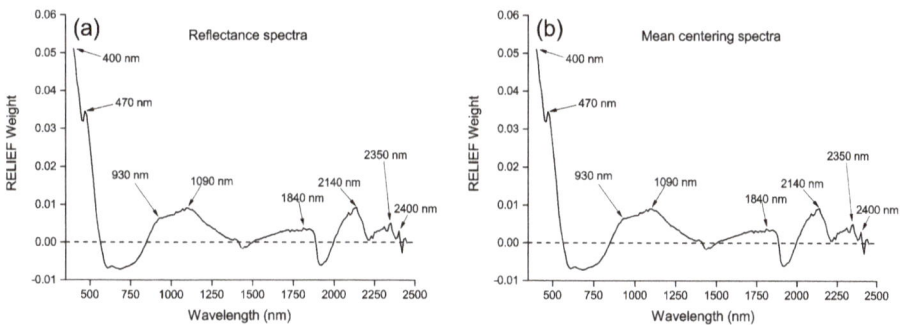

Figure 3. *Cont.*

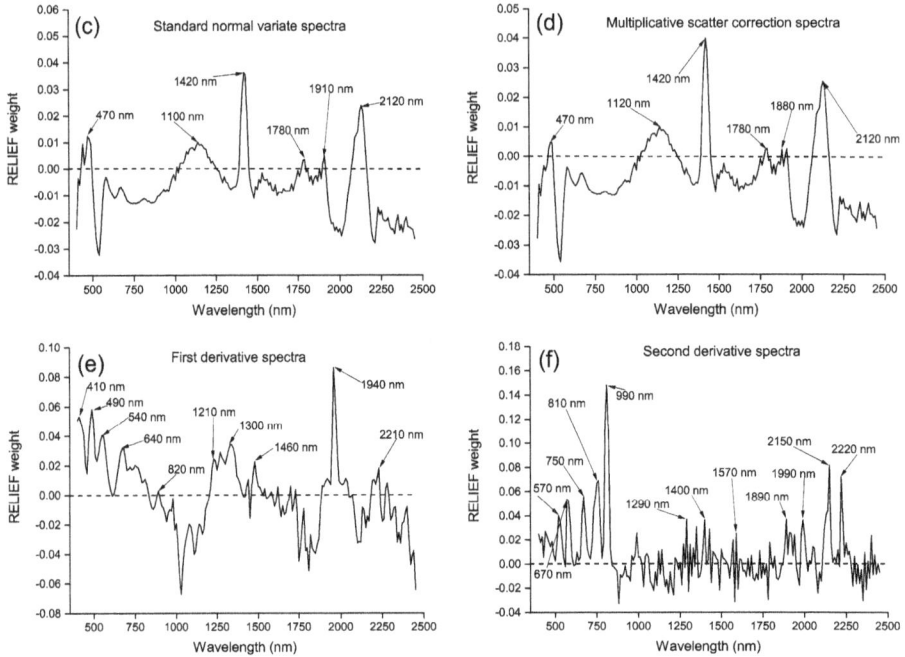

Figure 3. RELIEF weights and the selected spectral features for original reflectance spectra (**a**), mean centering spectra (**b**), standard normal variate spectra (**c**), multiplicative scatter correction spectra (**d**), first derivative spectra (**e**), and second derivative spectra (**f**). The threshold of RELIEF weight was set to 0 (horizontal dashed lines).

3.3. Comparison of the Abilities of Different Methods

The operation times, parameter setting, and validated OAs for diagnosis models by using different methods are illustrated in Table 4. The results showed that (1) the suitable combination of pre-processing and feature selection was vital to improve OAs of each machine-learning method; (2) feature selection methods, PCA and RELIEF, could improve modeling accuracies and decrease operation times of modeling, and RELIEF often outperformed PCA; (3) derivative transformation often resulted in the best diagnosis models. The optimal models for RF, ANN, LF and RBF-SVM were described as follows:

Table 4. The operation times, parameter setting, and overall accuracies for diagnosis models by using different pre-processing, feature selection and machine-learning methods [1].

Machine-Learning Methods	Pre-Processing Methods	No Feature Selection Parameters	time (s)	OA (%)	PCA Parameters	time (s)	OA (%)	RELIEF Parameters	time (s)	OA (%)
RF	none	nt = 70, nv = 5	0.32	80	nt = 60, nv = 7, nPC = 5	0.22	82	nt = 150, nv = 5, nfeature = 8	0.04	85
	MC	nt = 270, nv = 4	0.27	74	nt = 160, nv = 2, nPC = 7	0.17	83	nt = 130, nv = 3, nfeature = 8	0.03	71
	SNV	nt = 290, nv = 3	0.32	84	nt = 20, nv = 2, nPC = 7	0.05	70	nt = 60, nv = 3, nfeature = 6	0.03	82
	MSC	nt = 150, nv = 4	0.25	71	nt = 30, nv = 2, nPC = 6	0.03	71	nt = 30, nv = 2, nfeature = 6	0.03	71
	1st	nt = 50, nv = 2	0.25	77	nt = 80, nv = 4, nPC = 8	0.05	79	nt = 30, nv = 3, nfeature = 10	0.03	81
	2nd	nt = 200, nv = 2	0.28	85	nt = 50, nv = 4, nPC = 6	0.05	71	**nt = 50, nv = 2, nfeature = 12**	**0.05**	**86**
ANN	none	nlayer = 1	0.34	86	nlayer = 9, nPC = 6	0.05	71	nlayer = 3, nfeature = 8	0.02	84
	MC	nlayer = 2	0.48	76	nlayer = 2, nPC = 8	0.04	71	nlayer = 10, nfeature = 8	0.05	76
	SNV	nlayer = 1	0.27	81	nlayer = 2, nPC = 6	0.03	64	nlayer = 6, nfeature = 6	0.03	86
	MSC	nlayer = 1	0.28	29	nlayer = 2, nPC = 8	0.03	40	nlayer = 3, nfeature = 6	0.02	52
	1st	nlayer = 3	0.67	87	**nlayer = 3, nPC = 8**	**0.03**	**89**	nlayer = 1, nfeature = 10	0.03	81
	2nd	nlayer = 1	0.30	82	nlayer = 2, nPC = 5	0.03	62	nlayer = 1, nfeature = 12	0.03	75
RBF-SVM	none	C = 0.01, γ = 0.04, nSV = 32	0.11	80	C = 1, γ = 0.04, nSV = 32, nPC = 7	0.05	85	C = 1, γ = 0.17, nSV = 32, nfeature = 8	0.02	82
	MC	C = 0.01, γ = 0.08, nSV = 32	0.14	70	C = 1, γ = 0.08, nSV = 35, nPC = 7	0.05	87	C = 1, γ = 0.38, nSV = 31, nfeature = 8	0.03	76
	SNV	C = 0.01, γ = 0.04, nSV = 36	0.09	81	C = 1, γ = 0.28, nSV = 42, nPC = 9	0.04	66	C = 1, nSV = 36, nfeature = 6	0.03	80
	MSC	C = 0.01, γ = 0.23, nSV = 37	0.08	71	C = 1, γ = 0.31, nSV = 38, nPC = 5	0.03	71	C = 1, nSV = 37, nfeature = 6	0.02	71
	1st	C = 0.01, γ = 0.05, nSV = 46	0.06	79	C = 1, γ = 0.09, nSV = 43, nPC = 8	0.05	75	C = 1, nSV = 33, nfeature = 10	0.33	82
	2nd	C = 0.01, γ = 0.07, nSV = 53	0.08	81	C = 1, γ = 0.06, nSV = 41, nPC = 5	0.03	71	**C = 1, nSV = 42, nfeature = 12**	**0.05**	**89**
LF-SVM	none	C = 1, nSV = 36	0.16	84	C = 1, nSV = 35, nPC = 7	0.05	81	C = 1, nSV = 37, nfeature = 8	0.05	80
	MC	C = 1, nSV = 36	0.12	85	C = 1, nSV = 35, nPC = 7	0.05	85	C = 1, nSV = 35, nfeature = 8	0.03	79
	SNV	C = 1, nSV = 33	0.11	86	C = 1, nSV = 27, nPC = 5	0.06	56	C = 1, nSV = 39, nfeature = 6	0.06	72
	MSC	C = 1, nSV = 34	0.11	29	C = 1, nSV = 39, nPC = 5	0.06	29	C = 1, nSV = 39, nfeature = 6	0.04	73
	1st	C = 1, nSV = 26	0.09	80	C = 1, nSV = 27, nPC = 8	0.05	80	**C = 1, nSV = 29, nfeature = 10**	**0.05**	**87**
	2nd	C = 1, nSV = 36	0.10	76	C = 1, nSV = 30, nPC = 4	0.06	63	C = 1, nSV = 26, nfeature = 12	0.05	81

[1] RF: random forests; ANN: artificial neural network; SVM: support vector machine; RBF: radial basis function; LF: linear function; MC: mean centering; SNV: standard normal variate; MSC: multiplicative scatter correction; 1st: first derivative; 2nd: second derivative; PCA: principle component analysis; time: operation time for calibration; OA: validated overall accuracy; nPC: number of principle components; nfeature: number of RELIEF selected features; nt: number of trees; nv: number of variables; nlayer: number of layers; nSV: number of support vectors; C: regularization parameter; γ: kernel width. The results of selected models are emphasized in bold.

3.3.1. RF

The optimal pre-processing method for the RF model was second dervative. The best RF model was calibrated by using 12 RELIEF-selected spectral features, and the optimized *nv* and *nt* of the RF model were 3 and 50, respectively. The mean decrease GINI values (Figure 4) showed the importance of the spectral features for RF modeling in descending order as 2150, 810, 1400, 670, 1890, 2220, 1290, 570, 990, 750, 1570 and1990 nm. The validated OA for the RF model was 86%, which mean that the RF model correctly diagnosed 86% of soil samples in the test data-set (Figure 5a).

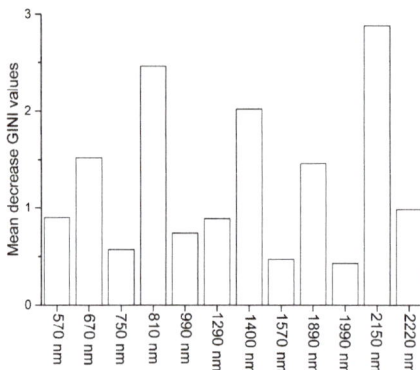

Figure 4. Mean decrease GINI values for RELIEF-selected spectral features.

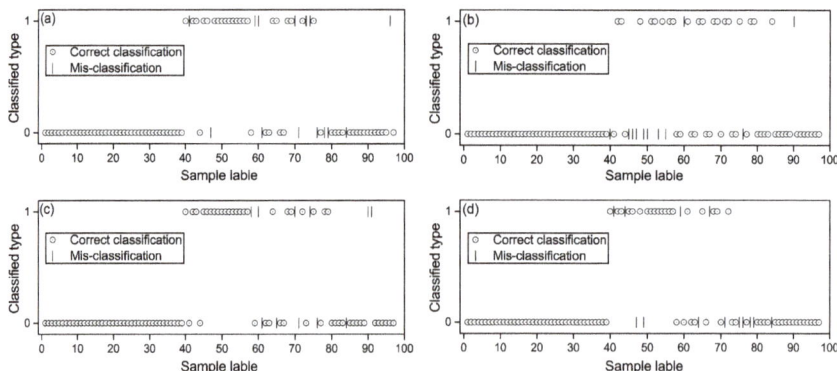

Figure 5. Values of samples predicted by using: (**a**) second derivative spectra (second), RELIEF and random forests; (**b**) first derivative spectra (first), principle component analysis and artificial neural network; (**c**) second, RELIEF and radial basis function-based support vector machine (SVM); and (**d**) first, RELIEF and linear function-based SVM. Value 1 indicates contaminated, and value 0 indicates uncontaminated. The correctly-diagnosed and misdiagnosed samples are displayed in the figures.

3.3.2. ANN

The optimal pre-processing method employed for ANN modeling was first derivative; PCA was selected as the feature selection method, and the number of hidden layers was three. The factor number for modeling was eight, and the first eight PCs explained approximately 99% of the variation of the spectral data. The ANN model correctly diagnosed 89% of soil samples in the test data-set (Figure 5b).

3.3.3. SVM

Second derivative was the optimal pre-processing method for RBF-SVM, and first derivative was the optimal pre-processing method for LF-SVM. The optimized C and γ for RBF-SVM were 1 and 0.06, respectively, while the optimized C for LF-SVM was 1. By adopting 12 RELIEF-selected spectral features, the RBF-SVM model correctly diagnosed 89% of soil samples in the test data-set (Figure 5c); and the LF-SVM model correctly diagnosed 87% of soil samples by using the RELIEF-selected spectral features (Figure 5d).

3.3.4. Model Comparison

Figure 5 displayed the predicted values of samples in the test dat-set by using three optimal diagnosis models. McNemar's test applied to these diagnosis models showed that the Z values were all less than 1.96 (Table 5), which indicated that there was no statistical difference in the diagnosis abilities of these optimal diagnosis models ($p < 0.05$).

Table 5. Z Values of McNemar's test between the optimal diagnosis models.[1]

	Second + RELIEF + RF	First + PCA + ANN	Second + RELIEF + RBF-SVM
First + PCA + ANN	0.24		
Second + RELIEF + RBF-SVM	0.90	0.00	
First + RELIEF + LF-SVM	0.30	0.26	0.41

[1] Second: second derivative spectra; First: first derivative spectra; PCA: principle component analysis; RF: random forests; ANN: artificial neural network; SVM: support vector machine; RBF: radial basis function; LF: linear-function.

4. Discussion

In this study, with the combination of pre-processing, feature selection and machine-learning methods, the OAs for soil arsenic contamination diagnosis achieved a satisfactory level (OA > 85%). This result demonstrated that VNIRS could be applied to diagnose soil arsenic contamination, although in the process of developing diagnosis models, VNIRS technology depended on conventional methods for providing the ground-truth of soil heavy metal contamination. Compared with conventional methods, this study confirmed that VNIRS might allow for faster and cheaper classification of soil heavy metal contaminants in an increased spatial coverage, which has been suggested by Bray, Viscarra Rossel and McBratney [23].

This study demonstrated that, to establish robust diagnosis models, the trial and error of various pre-processing methods was vital. Pre-processing methods, including SNV, MSC, first derivative, and second derivative, can be employed to eliminate the baseline drift caused by the difference in particle size and optical setups [6]. Derivative transformations also enhance the minor absorption features which may be useful to improve the diagnosis abilities of models. Nevertheless, derivative transformation will add noises into the spectral data, generating more noises with the increase of derivative orders [20]. Therefore, derivative transformations are often applied in conjunction with a smoothing algorithm to amplify noise [6]. Our research suggested that, compared with other pre-processing methods, derivative transformation was a more suitable pre-processing method for developing diagnosis models.

Feature selection methods could improve modeling accuracies by eliminating uninformative spectral variables and increase modeling efficiency by reducing the independent variables for modeling [10]. PCA extracted principle components from spectral variables without consideration of dependent variables (i.e., soil arsenic contamination in this study). However, RELIEF-selected spectral features based on their contributions to the classification of dependent variables [33]. Therefore, the results in this study indicated that RELIEF always outperformed PCA for diagnosing soil arsenic contamination from hyperspectral spectra. We considered that, based on these factors, the RELIEF algorithm was a more suitable method to select spectral features. Moreover, van Groenigen et al. [43] demonstrated that pre-processing methods could strongly influence the reflectance spectra, and they will therefore have an impact on the spectral features. Therefore, in this study, the results indicated that pre-processing methods affected the RELIEF-selected spectral features (Figure 3).

The establishment of robust diagnosis models by using different machine-learning methods (i.e., RF, ANN, LF-SVM, RBF-SVM) depends on the selection of appropriate pre-processing and feature selection methods. In addition, our study demonstrated that these optimal models for machine-learning methods had no statistical difference in diagnosis abilities; moreover, RF was superior to other machine-learning methods because of its ability to simplify parameter optimization and its better models explanatory. In this study, based on mean decrease GINI values, wavelengths at 2150, 810, 1400, and 670 nm can be identified as the first four important wavelengths for diagnosing arsenic contamination with the RF model. Wavelengths near 2150 and 810 nm relate to organic matter, and spectral features near 1400 and 670 nm coincide with wavelengths related to mostly iron oxides [42]. This might demonstrate that the diagnosis of arsenic contamination might depend on its surrogated correlations with organic matter and iron oxides.

Over-fitting is a common problem for modeling. It means that the best diagnosis model for the training data-set will not work well for the test data-set. RF is robust against over-fitting. Breiman [35] observed that the error associated with the error of RF converged to a limit with the increase in the number of trees in a forest. Nevertheless, in the case of ANN, over-fitting is a serious problem [40]. RF is easily accessible to non-specialists because of its simplicity in parameters optimization. However, for SVM, a number of hyper-parameters need to be optimized for each kernel function [40], while its parameters optimization also requires considerable knowledge of the frequently non-trivial underlying mathematics [40]. Moreover, complex machine-learning algorithms, such as SVM and ANN, were not easily interpretable to present relationships between independent and dependent variables [44]. However, RF, a method that performs a majority vote of tree-based classifiers, is explicit and comprehensible, revealing the important spectral variables for modeling [40]. Variable importance in RF can be evaluated by the increase in prediction error when the out-of-bag data are permuted for a certain variable, while keeping all other data constant. Considering these advantages, we regarded RF as a more efficient machine-learning method for modeling soil arsenic contamination levels.

This study investigated the abilities of laboratory reflectance spectroscopy to diagnose soil arsenic contamination. The field and air-/space-borne imaging spectroscopy have the potential to rapidly map heavy metal contamination over large areas [17,45,46]. Compared with laboratory spectroscopy, the application of field or imaging spectroscopy faces some constraints, such as soil surface, atmospheric and illumination conditions [47]. Therefore, the principles of this study should be further tested with field and imaging data.

5. Conclusions

The spectroscopic diagnosis of soil arsenic contamination is feasible, and the appropriate combination of pre-processing, feature selection and machine-learning methods is important for diagnosis accuracies. The RELIEF algorithm is a simple and efficient method to extract spectral features to improve modeling efficiency and diagnosis accuracy. Compared with ANN and SVM, RF is a more optimal machine-learning method for developing diagnosis models, because of its ability to simplify parameter optimization and its better models explanatory.

Acknowledgments: This study was supported by the China Postdoctoral Science Foundation (No. 2016M602521), by Science and Technology Bureau of Suzhou (No. SYN201309), the Scientific Research Foundation for Newly High-End Talents of Shenzhen University, the Basic Research Program of Shenzhen Science and Technology Innovation Committee (No. JCYJ20151117105543692), and Shenzhen Future Industry Development Funding Program (No. 201507211219247860).

Author Contributions: Tiezhu Shi and Guofeng Wu conceived and designed the experiments; Tiezhu Shi, Huizeng Liu and Yiyun Chen performed the experiments; Junjie Wang and Teng Fei analyzed the data; Tiezhu Shi wrote the paper. All authors conctributed, reviewed and improved this manuscript.

Conflicts of Interest: The authors declare no confict of interest.

References

1. Shi, T.Z.; Chen, Y.Y.; Liu, Y.L.; Wu, G.F. Visible and near-infrared reflectance spectroscopy—An alternative for monitoring soil contamination by heavy metal. *J. Hazard. Mater.* **2014**, *265*, 166–176. [CrossRef] [PubMed]
2. Viscarra Rossel, R.A.; Walvoort, D.J.J.; McBratney, A.B.; Janik, L.J.; Skjemstad, J.O. Visible, near infrared, mid infrared or combined diffuse reflectance spectroscopy for simultaneous assessment of various soil properties. *Geoderma* **2006**, *131*, 59–75. [CrossRef]
3. Ben-Dor, E.; Irons, J.R.; Epema, G.F. *Remote Sensing of the Earth Sciences: Manual of Remote Sensing*; Wiley: New York, NY, USA, 1999.
4. Kemper, T.; Sommer, S. Estimate of heavy metal contamination in soils after a mining accident using reflectance spectroscopy. *Environ. Sci. Technol.* **2002**, *36*, 2742–2747. [CrossRef] [PubMed]
5. Chu, X.L.; Yuan, H.F.; Lu, W.Z. Progress and application of spectral data pretreatment and wavelength selection methods in nir analytical technique. *Prog. Chem.* **2004**, *16*, 528–542. (In Chinese).
6. Stenberg, B.; Viscarra Rossel, R.A.; Mouazen, A.M.; Wetterlind, J. *Visible and Near Infrared Spectroscopy in Soil Science*; Agronomy, A.I., Ed.; Academic Press: Burlington, VT, USA, 2010; Volume 107, pp. 163–215.
7. Vohland, M.; Emmerling, C. Determination of total soil organic c and hot water-extractable c from vis-nir soil reflectance with partial least squares regression and spectral feature selection techniques. *Eur. J. Soil Sci.* **2011**, *62*, 598–606. [CrossRef]
8. Shi, T.Z.; Chen, Y.Y.; Liu, H.Z.; Wang, J.J.; Wu, G.F. Soil organic carbon content estimation with laboratory-based visible-near-infrared reflectance spectroscopy: Feature selection. *Appl. Spectrosc.* **2014**, *68*, 831–837. [CrossRef] [PubMed]
9. Araújo, M.C.U.; Saldanha, T.C.B.; Galvão, R.K.H.; Yoneyama, T.; Chame, H.C.; Visani, V. The successive projections algorithm for variable selection in spectroscopic multicomponent analysis. *Chemom. Intell. Lab. Syst.* **2001**, *57*, 65–73. [CrossRef]
10. Zou, X.B.; Zhao, J.W.; Malcolm, J.W.; Holmes, M.; Mao, H.P. Variables selection methods in near-infrared spectroscopy. *Anal. Chim. Acta* **2010**, *667*, 14–32.
11. Guyon, I.; Elisseeff, A. An introduction to variable and feature selection. *J. Mach. Learn. Res.* **2003**, *3*, 1157–1182.
12. Witten, I.H.; Frank, E. *Data Mining: Practical Machine Learning Tools and Techniques*; Elsevier: San Francisco, CA, USA, 2005.
13. Wu, Y.Z.; Chen, J.; Ji, J.F.; Tian, Q.J.; Wu, X.M. Feasibility of reflectance spectroscopy for the assessment of soil mercury contamination. *Environ. Sci. Technol.* **2005**, *39*, 873–878. [CrossRef] [PubMed]
14. Kooistra, L.; Wehrens, R.; Leuven, R.S.E.W.; Buydens, L.M.C. Possibilities of visible-near-infrared spectroscopy for the assessment of soil contamination in river floodplains. *Anal. Chim. Acta* **2001**, *446*, 97–105. [CrossRef]
15. Wu, Y.Z.; Chen, J.; Wu, X.M.; Tian, Q.J.; Ji, J.F.; Qin, Z.H. Possibilities of reflectance spectroscopy for the assessment of contaminant elements in suburban soils. *Appl. Geochem.* **2005**, *20*, 1051–1059. [CrossRef]
16. Chen, T.; Chang, Q.R.; Clevers, J.G.P.W.; Kooistra, L. Rapid identification of soil cadmium pollution risk at regional scale based on visible and near-infrared spectroscopy. *Environ. Pollut.* **2015**, *206*, 217–226. [CrossRef] [PubMed]
17. Wu, Y.Z.; Zhang, X.; Liao, Q.L.; Ji, J.F. Can contaminant elements in soils be assessed by remote sensing technology: A case study with simulated data. *Soil Sci.* **2011**, *176*, 196–205. [CrossRef]
18. Tan, K.; Ye, Y.Y.; Du, P.J.; Zhang, Q.Q. Estimation of heavy metal concentrations in reclaimed mining soils using reflectance spectroscopy. *Spectrosc. Spectr. Anal.* **2014**, *34*, 3317–3322.
19. Lucà, F.; Conforti, M.; Castrignanò, A.M.; Matteucci, G.; Buttafuoco, G. Effect of calibration set size on prediction at local scale of soil carbon by vis-nir spectroscopy. *Geoderma* **2017**, *288*, 175–183. [CrossRef]
20. Shi, T.Z.; Cui, L.J.; Wang, J.J.; Fei, T.; Chen, Y.Y.; Wu, G.F. Comparison of multivariate methods for estimating soil total nitrogen with visible/near-infrared spectroscopy. *Plant Soil* **2013**, *366*, 363–375. [CrossRef]
21. Heung, B.; Ho, H.C.; Zhang, J.; Knudby, A.; Bulmer, C.E.; Schmidt, M.G. An overview and comparison of machine-learning techniques for classification purposes in digital soil mapping. *Geoderma* **2016**, *265*, 62–77. [CrossRef]
22. Brungard, C.W.; Boettinger, J.L.; Duniway, M.C.; Wills, S.A.; Edwards, T.C. Machine learning for predicting soil classes in three semi-arid landscapes. *Geoderma* **2015**, *239–240*, 68–83. [CrossRef]
23. Bray, J.G.P.; Viscarra Rossel, R.A.; McBratney, A.B. Diagnostic screening of urban soil contaminations using diffuse reflectance spectroscopy. *Aust. J. Soil Sci.* **2009**, *47*, 433–442. [CrossRef]

24. Ren, H.Y.; Zhuang, D.F.; Singh, A.N.; Pan, J.J.; Qiu, D.S.; Shi, R.H. Estimation of as and cu contamination in agricultural soils around a mining area by reflectance spectroscopy: A case study. *Pedosphere* **2009**, *19*, 719–726. [CrossRef]

25. Vohland, M.; Besold, J.; Hill, J.; Frund, H.C. Comparing different multivariate calibration methods for the determination of soil organic carbon pools with visible to near infrared spectroscopy. *Geoderma* **2011**, *166*, 198–205. [CrossRef]

26. Huang, R.Q.; Gao, S.F.; Wang, W.L.; Staunton, S.; Wang, G. Soil arsenic availability and the transfer of soil arsenic to crops in suburban areas in fujian province, southeast China. *Sci. Total Environ.* **2006**, *368*, 531–541. [CrossRef] [PubMed]

27. Santra, S.C.; Samal, A.C.; Bhattacharya, P.; Banerjee, S.; Biswas, A.; Majumdar, J. Arsenic in foodchain and community health risk: A study in gangetic west bengal. *Procedia Environ. Sci.* **2013**, *18*, 2–13. [CrossRef]

28. Shi, T.Z.; Liu, H.Z.; Wang, J.J.; Chen, Y.Y.; Fei, T.; Wu, G.F. Monitoring arsenic contamination in agricultural soils with reflectance spectroscopy of rice plants. *Environ. Sci. Technol.* **2014**, *48*, 6264–6272. [CrossRef] [PubMed]

29. Wang, J.J.; Cui, L.J.; Gao, W.X.; Shi, T.Z.; Chen, Y.Y.; Gao, Y. Prediction of low heavy metal concentrations in agricultural soils using visible and near-infrared reflectance spectroscopy. *Geoderma* **2014**, *216*, 1–9. [CrossRef]

30. Guo, J.H.; Ma, H.; Wang, S.F. Determination of arsenic in national standard reference soil and stream sediment samples by atomic fluorescence spectrometry. *Rock Miner. Anal.* **2009**, *28*, 182–184. (In Chinese).

31. Loska, K.; Wiechula, D.; Korus, I. Metal contamination of farming soils affected by industry. *Environ. Int.* **2004**, *30*, 159–165. [CrossRef]

32. Kennard, R.W.; Stone, L.A. Computer aided design of experiments. *Technometrics* **1969**, *11*, 137–148. [CrossRef]

33. Kira, K.; Rendell, L. A practical appoach to feature selection. In *The Ninth International Workshop on Machine Learning*; Sleeman, D., Edwards, P., Eds.; Morgan Kaufmann: Aberdeen, UK, 1992.

34. Williams, G.J. Rattle: A data mining gui for R. *R J.* **2009**, *1*, 45–55.

35. Breiman, L. Random forests. *Mach. Learn.* **2001**, *45*, 5–32. [CrossRef]

36. Breiman, L. Bagging predictors. *Mach. Learn.* **1996**, *24*, 123–140. [CrossRef]

37. McCulloch, W.S.; Pitts, W. A logical calculus of the ideas immanent in nervous activity. *Bull. Math. Biophys.* **1943**, *54*, 115–133. [CrossRef]

38. Behrens, T.; Förster, H.; Scholten, T.; Steinrücken, U.; Spies, E.D.; Goldschmitt, M. Digital soil mapping using artificial neural networks. *J. Plant Nutri. Soil Sci.* **2005**, *168*, 21–33. [CrossRef]

39. Vapnik, V. *The Nature of Statistical Learning Theory*; Springer: New York, NY, USA, 1995.

40. Kampichler, C.; Wieland, R.; Calme, S.; Weissenberger, H.; Arriaga-Weiss, S. Classification in conservation biology: A comparison of five machine-learning methods. *Ecol. Inf.* **2010**, *5*, 441–450. [CrossRef]

41. Foody, G.M. Thematic map comparison: Evaluating the statistical significance of differences in classification accuracy. *Photogramm. Eng. Remote Sensing* **2004**, *70*, 623–633. [CrossRef]

42. Viscarra Rossel, R.; Behrens, T. Using data mining to model and interpret soil diffuse reflectance spectra. *Geoderma* **2010**, *158*, 46–54. [CrossRef]

43. Van Groenigen, J.W.; Mutters, C.S.; Horwath, W.R.; van Kessel, C. Nir and drift-mir spectrometry of soils for predicting soil and crop parameters in a flooded field. *Plant Soil* **2003**, *250*, 155–165. [CrossRef]

44. Vinterbo, S.A.; Kim, E.Y.; Ohno-Machado, L. Small, fuzzy and interpretable gene expression based classifiers. *Bioinformatics* **2005**, *21*, 1964–1970. [CrossRef] [PubMed]

45. Choe, E.; van der Meer, F.; van Ruitenbeek, F.; van der Werff, H.; de Smeth, B.; Kim, K.W. Mapping of heavy metal pollution in stream sediments using combined geochemistry, field spectroscopy, and hyperspectral remote sensing: A case study of the rodalquilar mining area, se spain. *Remote Sens. Environ.* **2008**, *112*, 3222–3233. [CrossRef]

46. Kemper, T.; Sommer, S. Use fo airborne hyperspectral data to estimate residual heavy metal contamination and acidification potential in the guadiamar floodplain andalusia, spain after the aznacollar mining accident. *Proc. SPIE* **2004**, *5574*, 224–234.

47. Stevens, A.; Udelhoven, T.; Denis, A.; Tychon, B. Measuring soil organic carbon in croplands at regional scale using airborne imaging spectroscopy. *Geoderma* **2010**, *158*, 32–45. [CrossRef]

sensors

MDPI

Article

Optical Sensing to Determine Tomato Plant Spacing for Precise Agrochemical Application: Two Scenarios

Jorge Martínez-Guanter [1], Miguel Garrido-Izard [2], Constantino Valero [2], David C. Slaughter [3] and Manuel Pérez-Ruiz [1,*]

1   Aerospace Engineering and Fluid Mechanics Department, University of Seville, 41013 Seville, Spain; martinezj@us.es
2   Laboratorio de Propiedades Físicas (LPF_TAGRALIA), Universidad Politécnica de Madrid (UPM), 28040 Madrid, Spain; miguel.garrido.izard@upm.es (M.G.-I.); constantino.valero@upm.es (C.V.)
3   Department of Biological and Agricultural Engineering, University of California, Davis, CA 95616, USA; dcslaughter@ucdavis.edu
*   Correspondence: manuelperez@us.es; Tel.: +34-954-481-389

Academic Editor: Dimitrios Moshou
Received: 31 March 2017; Accepted: 8 May 2017; Published: 11 May 2017

Abstract: The feasibility of automated individual crop plant care in vegetable crop fields has increased, resulting in improved efficiency and economic benefits. A systems-based approach is a key feature in the engineering design of mechanization that incorporates precision sensing techniques. The objective of this study was to design new sensing capabilities to measure crop plant spacing under different test conditions (California, USA and Andalucía, Spain). For this study, three different types of optical sensors were used: an optical light-beam sensor (880 nm), a Light Detection and Ranging (LiDAR) sensor (905 nm), and an RGB camera. Field trials were conducted on newly transplanted tomato plants, using an encoder as a local reference system. Test results achieved a 98% accuracy in detection using light-beam sensors while a 96% accuracy on plant detections was achieved in the best of replications using LiDAR. These results can contribute to the decision-making regarding the use of these sensors by machinery manufacturers. This could lead to an advance in the physical or chemical weed control on row crops, allowing significant reductions or even elimination of hand-weeding tasks.

Keywords: LiDAR; light-beam; plant localization; Kinect

1. Introduction

Precision agriculture requires accurate plant or seed distribution across a field. This distribution is to be optimized according to the size and shape of the area in which nutrients and light are provided to plant to obtain the maximum possible yield. These factors are controlled by the spacing between crop rows and the spacing of plants/seeds in a row [1]. For many crops, row spacing is determined as much by the physical characteristics of agricultural machinery used to work in the field as by the specific biological spacing requirements of the crop [2]. According to the crop and machinery used, the accuracy of planting by the precision transplanter/seeder to the desired square grid pattern must be adequate for the operation of agricultural machinery in both longitudinal and transverse crop directions.

The current designs of vegetable crop transplanters and seeders utilize several uncoordinated planting modules mounted to a common transport frame. These systems use sub-optimal open-loop methods that neglect the dynamic and kinematic effects of the mobile transport frame and of plant motion relative to the frame and the soil. The current designs also neglect to employ complete mechanical control of the transplant during the entire planting process, producing an error in the final planting position, due to the increased uncertainty of plant location as a result of natural variations in plant size, plant mass, soil traction and soil compaction [3].

Accurately locating the crop plant, in addition to allowing automatic control of weeds, allows individualized treatment of each plant (e.g., spraying, nutrients). Seeking to ensure minimum physical interaction with plants (i.e., non-contact), different remote sensing techniques have been used for the precise localization of plants in fields. For these localization methods, some authors have decided to address automatic weed control by localizing crop plants with centimetre accuracy during seed drilling [4] or transplanting [5,6] using a global positioning system in real time (RTK-GNSS). These studies, conducted at UC Davis, have shown differences between RTK-GNSS-based expected seed location versus actual plant position. The position uncertainly ranged from 3.0 to 3.8 cm for seeds, and tomato transplants, the mean system RMS was 2.67 cm in the along-track direction. Nakarmi and Tang used an image acquisition platform after planting to estimate the inter-plant distance along the crop rows [7]. This system could measure inter-plant distance with a minimum error of ±30 cm and a maximum error of ±60 cm.

Today, one of the biggest challenges to agricultural row crop production in industrialized countries is non-chemical control of intra-row (within the crop row) weed plants. Systems such as those developed by Pérez-Ruiz et al. [8] or the commercial platforms based on computer-controlled hoes developed by Dedousis et al. [9] are relevant examples of innovative mechanical weeding systems. However, the current effectiveness of mechanical weed removal is constrained by plant spacing, the proximity of the weeds to the plant, the plant height and the operation timing. Other methods for non-chemical weed control, such as the robotic platform developed by Blasco et al. [10] (capable of killing weeds using a 15-kV electrical discharge), the laser weeding system developed by Shah et al. [11] or the cross-flaming weed control machine designed for the RHEA project by Frasconi et al. [12], demonstrate that research to create a robust and efficient system is ongoing. A common feature of all these technological developments is the need for accurate measurement of the distance between plants.

Spatial distribution and plant spacing are considered key parameters for characterizing a crop. The current trend is towards the use of optical sensors or image-based devices for measurements, despite the possible limitations of such systems under uncontrolled conditions such as those in agricultural fields. These image-based tools aim to determine and accurately correlate several quantitative aspects of crops to enable plant phenotypes to be estimated [13,14].

Dworak et al. [15] categorized research studying inter-plant location measurements into two types: airborne and ground-based. Research on plant location and weed detection using airborne sensors has increased due to the increasing potential of unmanned aerial systems in agriculture, which have been used in multiple applications in recent years [16]. For ground-based research, one of the most widely accepted techniques for plant location and classification is the use of Light Detection and Ranging (LiDAR) sensors [17]. These sensors provide distance measurements along a line scan at a very fast scanning rate and have been widely used for various applications in agriculture, including 3D tree representation for precise chemical applications [18,19] or in-field plant location [20]. This research continues the approach developed in [21], in which a combination of LiDAR + IR sensors mounted on a mobile platform was used for the detection and classification of tree stems in nurseries.

Based on the premise that accurate localization of the plant is key for precision chemical or physical removal of weeds, we propose in this paper a new methodology to precisely estimate tomato plant spacing. In this work, non-invasive methods using optical sensors such as LiDAR, infrared (IR) light-beam sensors and RGB-D cameras have been employed. For this purpose, a platform was developed on which different sensor configurations have been tested in two scenarios: North America (UC Davis, CA, USA) and Europe (University of Seville, Andalucia, Spain). The specific objectives, given this approach, were the following:

- To design and evaluate the performance of multi-sensor platforms attached to a tractor (a UC Davis platform mounted on the rear of the tractor and a University of Seville platform mounted on the front of the tractor).
- To refine the data-processing algorithm to select the most reliable sensor for the detection and localization of each tomato plant.

2. Materials and Methods

To develop a new sensor platform to measure the space between plants in the same crop row accurately, laboratory and field tests were conducted in Andalucia (Spain) and in California (USA). This allowed researchers to obtain more data under different field conditions and to implement the system improvements required, considering the plant spacing objective. These tests are described below, characterizing the sensors used and the parameters measured.

2.1. Plant Location Sensors

2.1.1. Light-Beam Sensor Specifications

IR light-beam sensors (Banner SM31 EL/RL, Banner Engineering Co., Minneapolis, MN, USA) were used in two configurations: first as a light curtain (with three pairs of sensors set vertically, Figure 1 central and Figure 4) and later a simpler setup, using only one pair of sensors (Figure 2), which simplifies the system while still allowing the objective (plant spacing measurement) to be attained. In the light curtain, light-beam sensors were placed transversely in the middle of the platform to detect and discriminate the plant stem in a cross configuration to prevent crossing signals between adjacent sensors. Due to the short range and focus required in laboratory tests, it was necessary to reduce the field of view and the strength of the light signal by masking the emitter and receiver lens with a 3D-printed conical element. In laboratory tests, the height of the first emitter and receiver pair above the platform was 4 cm, and the height of 3D plants (artificial plants were used in laboratory tests; see Section 2.2) was 13 cm. In the field tests, the sensor was placed 12 cm from the soil (the average height measured manually for real plants in outdoor tests was 19.5 cm) to avoid obstacles in the field (e.g., dirt clods, slight surface undulations). In both cases, the receiver was set to obtain a TTL output pulse each time the IR light-beam was blocked by any part of the plant. The signals generated by the sensors were collected and time-stamped by a microcontroller in real time and stored for off-line analysis. Technical features of the IR light-beam sensors are presented in Table 1.

Table 1. IR light-beam sensor features.

Operational Voltage (V)	10–30 V
Detection range (m)	30 m
Response time (milliseconds)	1 ms
Sinking and sourcing outputs (mA)	150 mA

2.1.2. Laser Scanner

A LMS 111 LiDAR laser scanner (SICK AG, Waldkirch, Germany), was used in the laboratory and field testing platforms to generate a high-density point cloud on which to perform the localization measurements. Its main characteristics are summarized in Table 2. The basic operating principle of the LiDAR sensor is the projection of an optical signal onto the surface of an object at a certain angle and range. Processing the corresponding reflected signal allows the sensor to determine the distance to the plant. The LiDAR sensor was interfaced with a computer through an RJ 45 Ethernet port for data recording. Data resolution was greatly affected by the speed of the platform's movement; thus, maintenance of a constant speed was of key importance for accurate measurements. During data acquisition, two digital filters were activated for optimizing the measured distance values: a fog filter (becoming less sensitive in the near range (up to approximately 4 m)); and an N-pulse-to-1-pulse filter, which filters out the first reflected pulse in case that two pulses are reflected by two objects during a measurement [22]. Different LiDAR scan orientations were evaluated: scanning vertically with the sensor looking downwards (Figure 1), scanning with a 45° inclination (push-broom) and a lateral-scanning orientation (side-view).

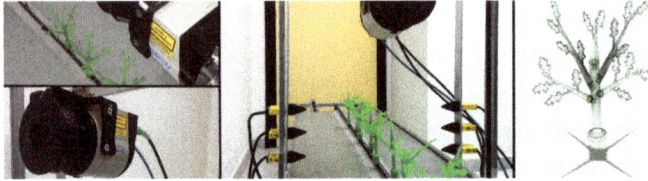

Figure 1. Details of the sensors on the laboratory platform (vertical LiDAR and Light-beam sensors) for the detection and structure of the modular 3D plant.

Table 2. LMS 111 technical data.

Operational Range	From 0.5 to 20 m
Scanning field of view	270°
Scanning Frequency	50 Hz
Angular resolution	0.5°
Light source	905 nm
Enclosure rating	IP 67

2.1.3. RGB-D Camera

A Kinect V2 commercial sensor (Microsoft, Redmond, WA, USA), originally designed for indoor video games, was mounted sideways on the research platform during field trials. This sensor captured RGB, NIR and depth images (based on time-of-flight) of tomato plants, although for further analysis, only RGB images were used for the validation of stick/tomato locations obtained from the LiDAR scans, as detailed in Section 2.4.3. Kinect RGB-captured images have a resolution of 1920 × 1080 pixels and a field of view (FOV) of 84.1 × 53.8°, resulting in an average of approximately 22 × 20 pixels per degree. NIR images and depth camera have a resolution of 512 × 424 pixels, with an FOV of 70 × 60° and a depth-sensing maximum distance of 4.5–5 m. Although systems such as the Kinect sensor were primarily designed for use under controlled light conditions, the second version of this sensor has higher RGB resolution (640 × 480 in v1) and its infrared sensing capabilities were also improved, enabling a more lighting-independent view and supporting its use outdoors under high-illumination conditions. Despite this improvement, we observed that the quality of the RGB images were somewhat affected by luminosity and direct incident light, and therefore, the image must be post-processed to obtain usable results. The images taken by the Kinect sensor were simultaneously acquired and synchronized with the LiDAR scans and the encoder pulses. Because the LabVIEW software (National Instruments, Austin, TX, USA) used for obtaining the scan data was developed to collect three items (the scans themselves, the encoder pulses and the timestamp), a specific Kinect recording software had to be developed to embed the timestamp value in the image data. With the same timestamp for the LiDAR and the image, the data could be matched and the images used to provide information about the forward movement of the platform.

2.2. Lab Platform Design and Tests

To maximize the accuracy of the distance measurements obtained by the sensors, an experimental platform was designed to avoid the seasonal limitations of testing outdoors. Instead of working in a laboratory with real plants, the team designed and created model plants (see Figure 1) using a 3D printer (Prusa I3, BQ, Madrid, Spain). These plants were mounted on a conveyor chain at a predetermined distance. This conveyor chain system, similar to that of a bicycle, was driven by a small electric motor able to move the belt at a constant speed of 1.35 km·h^{-1}. For the odometry system, the shaft of an incremental optical encoder (63R256, Grayhill Inc., Chicago, IL, USA) was mounted so that it was attached directly to the gear shaft and used to measure the distance travelled, thus serving as a localization reference system. Each channel in this encoder generates 256 pulses per revolution, providing a 3-mm resolution in the direction of travel. The data generated by the light-beam sensors and the cumulative odometer pulse count were collected using a low-cost open-hardware Arduino Leonardo microcontroller (Arduino Project, Ivrea, Italy) programmed in a simple integrated development environment (IDE). This device enabled recording of data that were stored in a text file for further computer analysis. Several repetitions of the tests were made on the platform to optimize the functions of both light-beam and LiDAR sensors. From the three possible LiDAR orientations, lateral scanning was selected for the field trials because it provided the best information on the structure of the plant, as concluded in [17]. In lab tests, two arrangements of light-beam sensors were assessed: one in a light curtain assembly with three sensor pairs at different heights and another using only one emitter-receiver pair.

2.3. Field Tests

The initial tests, performed in Davis, CA (USA), were used to assess the setup of the light-beam sensor system and detected only the stem of the plants rather than locating it within a local reference system. Once the tomato plants were placed in the field, tests were conducted at the Western Center for Agriculture Equipment (WCAE) at the University of California, Davis campus farm to evaluate the performance of the sensor platform for measuring row crop spacing. For this test, an implement was designed to house the sensors as follows. The same IR light-beam sensor and encoder, both described in Section 2.1, were used (Figure 2). The output signals of the sensors were connected to a bidirectional digital module (NI 9403, National Instruments Co., Austin, TX, USA), while the signal encoder was connected to a digital input module (NI 9411, National Instruments Co.). Both modules were integrated into an NI cRIO-9004 (NI 9411, National Instruments Co.), and all data were recorded using LabVIEW (National Instruments Co.). In these early field trials, the team worked on three lines of a small plot of land 20 m in length, where the methodology for detecting the plants within a crop line was tested.

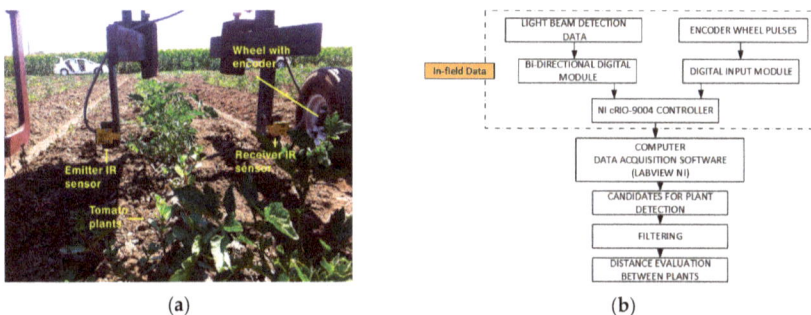

Figure 2. (**a**) Light-beam sensors mounted on the experimental platform designed for field trials at UC Davis, California; (**b**) Progressive monitoring flowchart using light-beam sensors.

To continue the study of plant localization in a different scenario, additional experiments were designed at the University of Seville, in which a refinement of the LiDAR sensors and data processing were performed. These tests were conducted on several lines of tomato plants manually transplanted from trays, with the plants placed with an approximate, though intentionally non-uniform, spacing of

30 cm. Two of these lines were analysed further, one with 55 tomato plants and the other with 51, and a line of 19 wooden sticks was also placed to provide an initial calibration of the instruments. Due to the initial test conditions, where tomato plants were recently transplanted and had a height of less than 20 cm, the team built an inverted U-shaped platform attached to the front of a small tractor (Boomer 35, New Holland, New Holland, PA, USA, Figure 3). The choice of the small tractor was motivated by the width of the track, as the wheels of the tractor needed to fit on the sides of the tomato bed, leaving the row of tomatoes clear for scanning and sensors.

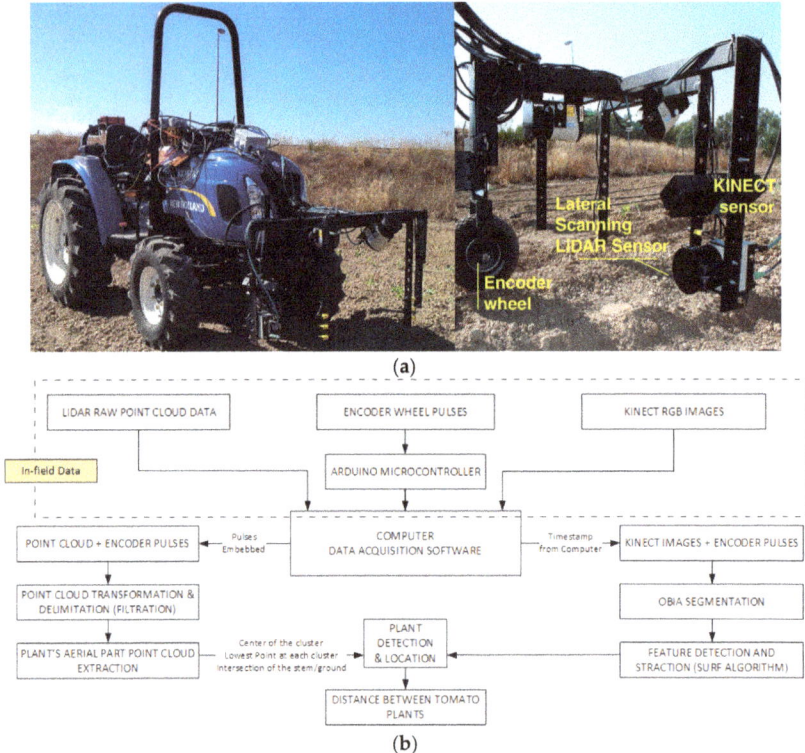

Figure 3. (a) Structure housing the sensors mounted on the tractor (left) and detail of the LiDAR and Kinect setup (right) for field trials at the University of Seville; (b) Progressive monitoring flowchart using LiDAR and Kinect sensors.

As was done in the laboratory platform, the encoder described in Section 2.1 was used as an odometric system, this time interfaced with an unpowered ground wheel, to determine the instantaneous location of the data along the row.

During the tests, the platform presented several key points that were addressed: (i) The encoder proved to be sensitive to vibrations and sudden movements, so it was integrated into the axis of rotation of an additional wheel, welded to the structure and dampened as much as possible from the vibrations generated by the tractor. In addition, the team had to reduce the slippage of the encoder wheel on the ground to avoid losing pulses; (ii) Correct orientation of the sensors was also key because the mounting angles of the LiDAR sensors would condition the subsequent analysis of the data obtained in the scans and the determination of which data contributed more valuable information; (iii) The speed of the tractor should be as low as possible remain uniform (during the test the average speed was 0.36 m/s) and maintain a steady course without steering wheel movements to follow a straight path.

2.4. Data-Processing Methodology

To detect precisely and determine properly the distances between plants in both laboratory and field tests, the data provided by the sensors were merged and analysed.

2.4.1. Plant Characterization Using Light-Beam Sensors

The methodology followed to analyse data obtained from the light-beam curtain (which was formed by three light-beam sensors in line) was similar to that described in [21]. The algorithm outputs the moment that the beam was interrupted and associates the beam with an encoder pulse. Because the 3D plants had a wider shape at the top (leaves) than the bottom (stem), and therefore more interruptions were received, the algorithm had to be adapted to each sensor pair and each height for plant detection. To discriminate correctly between plants for the light curtain case, the developed algorithm implemented a distance range, measured in pulses from the encoder, that allowed the verification of the presence or absence of a plant after the detection of the stem, inferring that interruptions received from the sensors placed at the middle and top heights before and after the stem corresponded to the leaves and the rest of the plant structure, respectively. For the analysis of data obtained from the single pair of IR light-beam sensors, a Matlab routine (MATLAB R2015b, MathWorks, Inc., Natick, MA, USA) was developed. System calibration was performed using 11 artificial plants in the laboratory test and 122 real tomato plants in the UC Davis field test. The methodology used for the detection of tomato plants was based on the following steps:

1. Selection of Values for the Variables Used by the Programme for Detection:

    a) *Pulse_distance_relation*: This variable allowed us to convert the pulses generated by the encoder into the distances travelled by the platforms. In laboratory trials, the encoder was coupled to the shaft that provided motion to the 3D plants, and in the field, it was coupled to a wheel installed inside the structure of the platform. The conversion factors used for the tests were 1.18 and 0.98 mm per pulse for the laboratory and the field, respectively.
    b) *Detection_filter*: To eliminate possible erroneous detections, especially during field trials due to the interaction of leaves, branches and even weeds, the detections were first filtered. We filtered every detection that corresponded to an along-track distance of less than 4 mm while the sensor was active (continuous detection).
    c) *Theoretical_plant_distance*: The value for the theoretical distance between plants in a crop line. The value set during testing was 100 mm and 380 mm for the laboratory and the field, respectively.
    d) *Expected_plant_distance*: Expected distance between plants in a crop line was defined as the theoretical plant distance plus an error of 20%.

2. Importing of raw data recorded by the sensors (encoder and existence "1" or absence "0" of detection by the IR sensors). The conversion factor (*pulse_distance_relation*) provided the distance in mm for each encoder value.
3. Data were filtered by removing all detections whose length or distance travelled, while the sensors were active, was less than the set value (*detection_filter*). Thus, potential candidates were selected by registering the following:

    a) The distance at the start of the detection;
    b) The distance at the end of the detection;
    c) The distance travelled during detection and (iv) the mean distance during the detection, which was considered the location of the stem of the plant.

4. Distance evaluation between the current candidate and the previous potential plant:

   a)   If the evaluated distance was greater than the value set (*expected_plant_distance*), we considered this candidate as a potential new plant, registering in a new matrix: the number of the plant, the detections that defined it, the midpoint location and the distance from the previous potential plant.

   b)   If the evaluated distance was less than the set value (*expected_plant_distance*), plant candidate data was added to the previous potential plant, recalculating all components for this potential plant. The new midpoint was considered the detection closest to the theoretical midpoint.

2.4.2. Plant Characterization Using a Side-View LiDAR

For the analysis of the data obtained from the LiDAR, it is important to mention the high complexity of its data, in both volume and format, compared with those data obtained by the light-beam. This is reflected in the following section, which explains the proposed methodology for obtaining both the aerial point clouds of the tomato rows referenced to the encoder sensor and the tomato plant identification. This is a prerequisite for tomato plant localization. For this purpose, it was necessary to pre-process the data, followed by a transformation and translation from the LiDAR sensor to the scanned point.

Pre-Processing of Data

(i)   Data pre-processing was performed at the LiDAR sensor.

An off-line Matlab process was used with the actual field data collected during the field experiments. Data were filtered to eliminate false positives or those that did not contribute relevant information, considering only those detections with a distance greater than 0.05 m. Later, the resulting detections were transformed from polar to Cartesian coordinates using a horizontal orientation coordinate system as a reference.

(ii) From the LiDAR sensor to the scanned point: transformations and data delimitation.

To transform the horizontal LiDAR coordinates to the actual LiDAR orientation (lateral in our case), the following steps were followed:

a)   The starting points were the Cartesian coordinates obtained using the horizontal orientation as a reference $(x'_{point}, y'_{point}, z'_{point})$.

b)   To integrate the scan from LiDAR into the platform coordinate system, a different transformation $(x_\varphi, y_\theta, z_\psi)$ was applied (see Equation (1)), considering the actual orientation of the LiDAR (see Table 3). Each LiDAR scanned point in the platform coordinate system $(x_{point}, y_{point}, z_{point})$ was obtained:

$$
\begin{bmatrix} x_{point} \\ y_{point} \\ z_{point} \\ 1 \end{bmatrix} = \begin{bmatrix} x'_{point} \\ y'_{point} \\ z'_{point} \\ 1 \end{bmatrix} \times \begin{bmatrix} 1 & 0 & 0 & 0 \\ 0 & \cos(x_\varphi) & \sin(x_\varphi) & 0 \\ 0 & -\sin(x_\varphi) & \cos(x_\varphi) & 0 \\ 0 & 0 & 0 & 1 \end{bmatrix} \times \begin{bmatrix} \cos(y_\theta) & 0 & -\sin(y_\theta) & 0 \\ 0 & 1 & 0 & 0 \\ \sin(y_\theta) & 0 & \cos(y_\theta) & 0 \\ 0 & 0 & 0 & 1 \end{bmatrix} \times
$$
$$
\begin{bmatrix} \cos(z_\psi) & \sin(z_\psi) & 0 & 0 \\ -\sin(z_\psi) & \cos(z_\psi) & 0 & 0 \\ 0 & 0 & 1 & 0 \\ 0 & 0 & 0 & 1 \end{bmatrix} \times \begin{bmatrix} 1 & 0 & 0 & 0 \\ 0 & 1 & 0 & 0 \\ 0 & 0 & 1 & 0 \\ Enc_t & 0 & 0 & 1 \end{bmatrix} \tag{1}
$$

Table 3. Transformation and translation values applied to LiDAR data with a lateral orientation.

Roll "φ" (°)	Pitch "θ" (°)	Yaw "ψ" (°)	x Translation (m)
0	−180	0	Enct

Once transformed, the x translation was applied to coordinates obtained for the actual LiDAR orientation. The encoder values recorded at each scan time were used to update the point cloud x coordinate related to the tractor advance. Additionally, the height values (z coordinate) were readjusted by subtracting the minimum obtained.

Plant Localization

The 3D point cloud processing was performed at each stick or tomato row. Thus, using manual distance delimitation, point clouds were limited to the data above the three seedbeds used during the tests.

(i)   Aerial Point Cloud Extraction

The aerial point data cloud was extracted using a succession of pre-filters. First, all points that did not provide new information were removed using a gridding filter, reducing the size of the point cloud. A fit plane function was then applied to distinguish the aerial points from the ground points. In detail, the applied pre-filters were:

a)   Gridding filter: Returns a downsampled point cloud using a box grid filter. GridStep specifies the size of a 3D box. Points within the same box are merged to a single point in the output (see Table 4).

b)   pcfitplane [23]: This Matlab function fits a plane to a point cloud using the M-estimator SAmple Consensus (MSAC) algorithm. The MSAC algorithm is a variant of the RANdom SAmple Consensus (RANSAC) algorithm. The function inputs were: the distance threshold value between a data point and a defined plane to determine whether a point is an inlier, the reference orientation constraint and the maximum absolute angular distance. To perform plane detection or soil detection and removal, the evaluations were conducted at every defined evaluation interval (see Table 4).

Table 4. Aerial point cloud extraction parameters selected.

Test	Grid Step (m³)	MSAC				
		Theoretical Distance Between Plants (m)	Evaluation Intervals (m)	Threshold (m)	Reference Vector	Maximum Absolute Angular Distance
Sticks		0.240	0.08			
Tomatoes 1	$(3 \times 3 \times 3) \times 10^{-9}$	0.290	0.096	0.04	[0,0,1]	5
Tomatoes 2						

Table 4 shows the parameter values chosen during the aerial point cloud extraction. The chosen values were selected by trial and error, selecting those that yielded better results without losing much useful information.

(ii)   Plant Identification and Localization

• Plant Clustering

A k-means clustering [24] was performed on the resulting aerial points to partition the point cloud data into individual plant point cloud data. The parameters used to perform the k-means clustering were as follows:

○ An initial number of clusters: Floor ((distance_travelled_mm/distance_between_plants_theoretical) + 1) × 2

○ The squared Euclidean distance for the centroid cluster. Each centroid is the mean of points in the cluster.

○ The squared Euclidean distance measure and the k-means++ algorithm were used for cluster centre initialization.

○ The clustering was repeated five times using the initial cluster centroid positions from the previous iteration.

○ Method for choosing initial cluster centroid positions: Select k seeds by implementing the k-means++ algorithm for cluster centre initialization.

A reduction in the number of clusters was determined directly by evaluating the cluster centroids. If pair of centroids were closer than the *min_distance_between_plants*, the process was repeated by reducing the number of clusters by one (Table 5).

Table 5. Plant identification and stem identification parameters.

Minimum Distance between Plants	Minimum Cluster Size	Histogram Jumps (mm)
Distance_between_plants_theoretical×0.2	5	4

A clustering size evaluation was performed, excluding clusters smaller than *min_cluster_size*.

(iii) Plant Location

Three different plant locations were considered:

• Centre of the cluster
• Location of the lowest point at each plant cluster
• Intersection of the estimated stem line and ground line

   ○ By dividing the aerial plant data in slices defined by "histogram jumps", the x limits on the maximal number of counts were obtained. For the z limits, data belonging to the bottom half of the aerial point cloud were considered.

   ○ The remaining data inside these limits were used to obtain a line of best fit, which was considered the stem line.

   ○ Plant location was defined as the intersection between the stem line and the corresponding ground line obtained previously from the MSAC function.

2.4.3. Validation of Plant Location Using RGB Kinect Images

To obtain the distance between the stems of two consecutive plants using the Kinect camera, it is necessary to characterize the plants correctly and then locate the stems with RGB images from the Kinect camera. This characterization and location of the stem was conducted as follows: a sequence of images of the entire path was obtained (~250 images in each repetition), where the camera's shooting frequency was established steadily in at 1-s intervals. Obtaining the string with the timestamp of each image was a key aspect of the routine developed in Matlab, as this string would later be used for integration with the LiDAR measurement. The relationship between the timestamp and its corresponding encoder value was used to spatially locate each image (x-axis, corresponding to the tractor advance).

Image processing techniques applied in the characterization of tomato plants from Kinect images generally followed these steps:

(i) According to [25], the first step in most works regarding image analysis is the pre-processing of the image. In this work, an automatic cropping of the original images was performed (Figure 4a), defining a ROI. Because the test was conducted under unstructured light conditions, the white balance and the saturation of the cropped image (Figure 4b) were modified;

(ii)   Next, the image segmentation step of an object-based image analysis (OBIA) was performed to generate boundaries around pixel groups based on their colour. In this analysis, only the green channel was evaluated to retain most of the pixels that define the plant, generating a mask that isolates them (Figure 4c) and classifying the pixels as plant or soil pixels. In addition, morphological image processing (erosion and rebuild actions) was performed to eliminate green pixels that were not part of the plant;

(iii) Each pair of consecutive images was processed by a routine developed in Matlab using the Computer Vision System Toolbox. This routine performs feature detection and extraction based on the Speeded-Up Robust Features "SURF" algorithm [26] to generate the key points and obtain the pairings between characteristics of the images.

Once the plant was identified in each image, the location of the stem of each plant was defined according to the following subroutine:

(1)   Calculating the distance in pixels between two consecutive images, as well as the encoder distance between these two images, the "forward speed" in mm/pixel was obtained for each image;

(2)   For each image, and considering that the value of the encoder corresponds to the centre of the image, the distance in pixels from the stem to the centre of the image was calculated. Considering whether it was to the left or to the right of the centre, this distance was designated positive or negative, respectively;

(3)   The stem location was calculated for each image using the relation shown in Equation (2) below;

(4)   Plant location was obtained for each image in which the plant appeared; the average value of these locations was used to calculate the distance between plants:

$$\text{Stem Location} = \text{Encoder Value} \pm \text{Distance (from stem to centre)} \times \text{motion relation}\left(\frac{mm}{pixel}\right) \quad (2)$$

Figure 4. (**a**) raw image captured by the Kinect sensor with the ROI indicated; (**b**) ROI with white balance and saturation adjustment; (**c**) OBIA resulting image with isolated plant pixels.

3. Results

3.1. Laboratory Test Results

Several laboratory tests were conducted to properly adjust and test the sensors. As an example, Figure 5 shows various profiles representing detections, comparing the pair of light-beam sensors placed on the bottom (stem detection) with those placed in the middle, detecting the aerial part of the

3D plant. Detections of the third pair of sensors, placed above the plants, were not considered relevant because most of their beam was above the plants. Figure 5 also shows that the detection of the stem using a single pair of light-beam located at a lower height with respect to the soil was more effective and robust than trying distinguish between different plants based on the detection of the aerial part of the plant. From our point of view and in agreement with the results in [21], this justifies the use of a single pair of light-beam sensors for the field tests rather than the use of the curtain mode (3 pairs as originally tested in the laboratory). The algorithm to analyze data from a single light-beam is also simpler and faster, being more adequate for real-time usage.

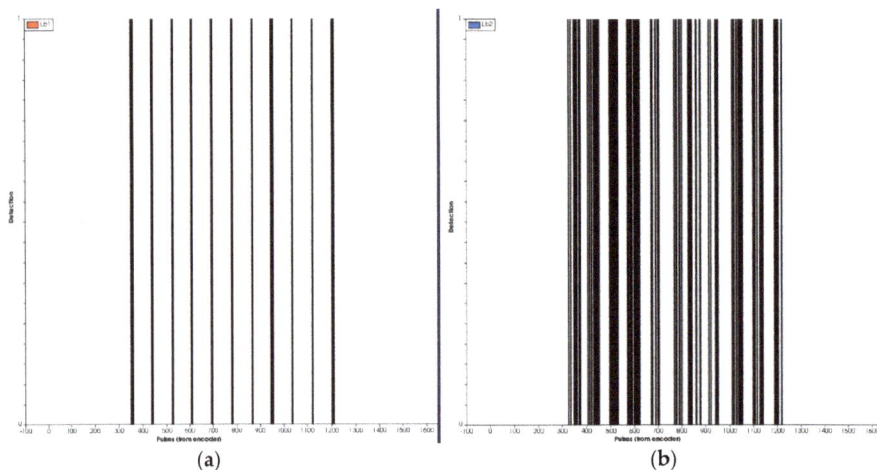

(a)   (b)

Figure 5. (**a**) Stem detection of the 11 artificial plants through the lower pair of light-beam sensors. (**b**) Detection of the aerial part (leaves and branches) of the 11 artificial plants through the centre pair of light-beam sensors.

Figure 6 shows the positions of the stem detections and the estimated distances between them. In all laboratory tests, 100% of the stems were detected using the light-beam sensor. Notably, under laboratory conditions, there were no obstacles in the simulated crop line, which does not accurately represent field conditions. Figure 6b shows the stem diameter measured from a test, with an estimated average diameter of 11.2 mm for an actual value of 10 mm.

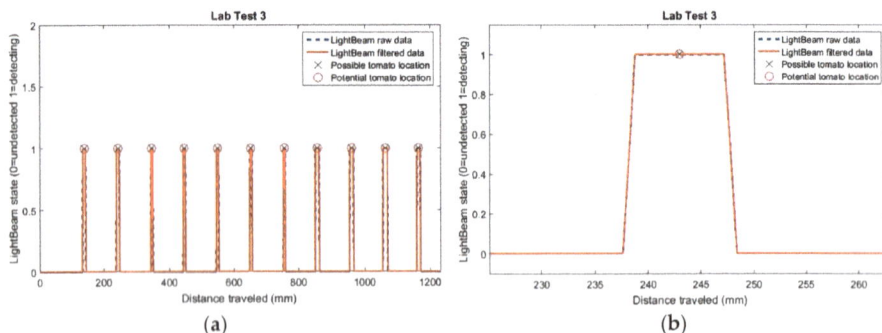

(a)   (b)

Figure 6. (**a**) Positions of the stems of the 11 plants in the laboratory test; (**b**) Average plant position for a distance of 11 mm.

Three laboratory tests corresponding to three different stages of the light-beam sensor adjustment process are shown in Figure 7.

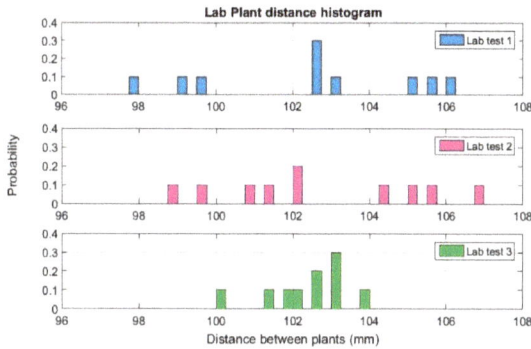

Figure 7. Histogram of measured distances between 3d plant stems during the adjusting process of sensors on the detection platform in the laboratory.

The histogram in this figure represents the estimated distances between plant stems, where the average distance is 102.5 mm when the real distance between plants was 100.0 mm. The average standard deviation for the three trials was 2.2 mm. For test 3 (last histogram in Figure 7), all distance values between plants were estimated between 100.1 mm and 103.8 mm.

For the results obtained by the LiDAR in the laboratory, as described in Section 2.2, multiple scanning configurations and orientations were tested. After a visual comparison of the results of these tests, lateral-scanning orientation gave the best point clouds, and its plant representation and spacing measurements were therefore more accurate. Figure 8 shows the point cloud representation of one of the laboratory scans made using the LiDAR sensor with lateral orientation.

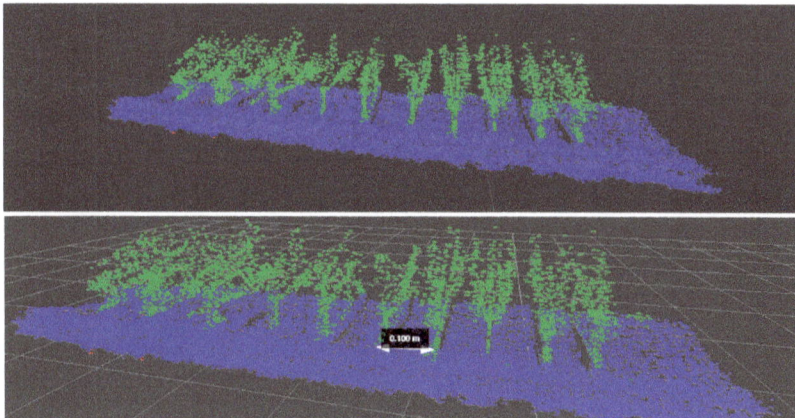

Figure 8. Point cloud representation of artificial 3D plants obtained using a lateral-scanning LiDAR sensor during laboratory tests.

3.2. Field Tests Results

Preliminary tests results using the light-beam sensors are presented in Figure 9a. Unlike laboratory detections, in the field data, the determined distances between stems were more variable. This variability is mainly due to crop plants missing from the line or the presence of a weed very close to the stem of the detected plant. Figure 9b reveals that several positions for potential plants (marked with a cross) were established, while only one plant was present (marked with a circle).

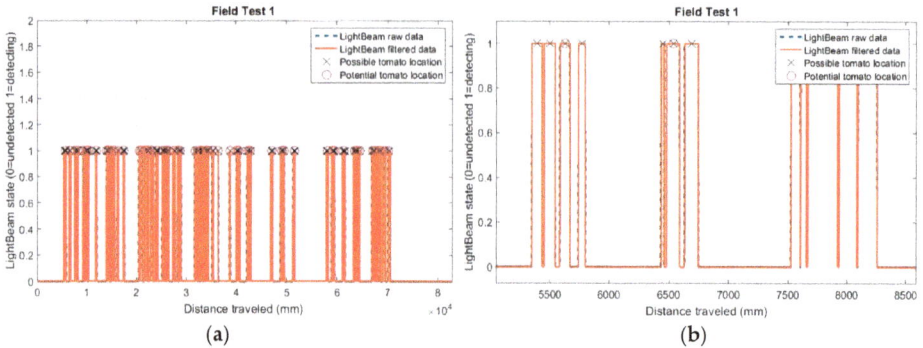

Figure 9. (a) Detections and positions of the 41 tomato plants in first field test with IR sensors; and (b) detail of the positions of potential plants.

During the first field test, 41 detections occurred when there were 32 real plants growing in the line; thus, for this trial, the error in the number of detections was 22% (Table 6). We speculate three possible causes of this error in the initial test: the plants were very close together, soil clods were detected between plants, or the evaluation of the distance between plants for the selection of potential plants was not optimal. However, in trial two, 32 stem detections occurred when 34 real plants were in the line, resulting in a 94% success rate. For test three, the accuracy was 98%.

Table 6. Plant detection ratio in field trials.

Test	Real Plants	Detected Plants	% Accuracy
First test	32	41	78
Second test	34	32	94
Third test	48	49	98

Regarding the experiments conducted to estimate the response of the LiDAR system in real field conditions, the detection system was used on three lines (one with wooden sticks and two with tomatoes). The system was intended to detect the presence of plant stems correctly and to distinguish between the foliage and the stem. In addition, this system was used to measure the distances between plants accurately, providing valuable information for future herbicide treatments.

Because the use of the laser scanner for detection implies the generation of a very dense cloud of points, in which not all data are relevant, an initial filtering was performed as explained above. Table 7 presents the number of original points obtained by eliminating the seedbed, and those considered to be representative of the aerial part of the plant, which comprised only 4.5% of the total.

Table 7. Point cloud reduction during plant point cloud extraction.

Test	Seedbed Delimit	Gridding	Plant Points
Stick: Test 1	44,708 (100%)	40,903 (91.5%)	1667 (3.7%)
Tomatoes 1: Test 1	374,963 (100%)	328,593 (87.6%)	14,720 (3.9%)
Tomatoes 1: Test 2	237,396 (100%)	220,714 (93%)	13,098 (5.5%)
Tomatoes 2: Test 2	3,807,858 (100%)	340,051 (89.3%)	18,706 (4.9%)

Regarding plant detection, Table 8 summarizes the results of some of the tests, implementing the methodology explained previously. Based on these data, it is observed that 100% of the sticks have been correctly detected, without obtaining false positives or negatives. When this analysis was performed on tomato plants, about 3–21% of false positive or negative detections were found. A plant detected by the LiDAR was defined as a false positive when the estimated plant centre was more than half of the plant spacing from the real plant centre.

Table 8. Plant and sticks detection results.

Test	Plant Location Method	Correctly Detected	False Positive	False Negative
Stick with filter Test 1	Centre of the cluster	19	0	0
	Lowest point	19	0	0
	Intersection of stem line and ground line	19	0	0
Tomatoes 1 with filter Test 1	Centre of the cluster	49	2	6
	Lowest point	48	3	7
	Intersection of stem line and ground line	46	5	9
Tomatoes 1 with filter Test 2	Centre of the cluster	53	6	2
	Lowest point	52	7	3
	Intersection of stem line and ground line	47	12	8
Tomatoes 2 with filter Test 2	Centre of the cluster	51	5	0
	Lowest point	48	8	3
	Intersection of stem line and ground line	42	14	9

Related to the data presented in Table 8, Figure 10 below shows stick and plant LiDAR detection results in three rows. Platform advance information (*x*-axis on the plot) provided by the encoder is given in mm. Each plant/stick detected is marked with different colour (note the overlapping on some plants, explaining false negative and false positive detections).

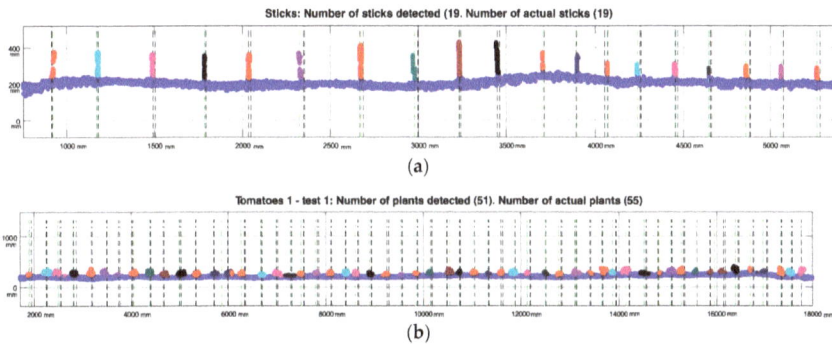

(a)

(b)

Figure 10. Detection results on three tested lines. Green dotted lines represent the cluster centre and black dotted lines the real plant interval obtained by the Kinect image (location ± Std.). (a) 19 stick detections using LiDAR; (b) 51 plants detected during Test 1 in row 1.

As explained in Section 2.4.2.2, the plant location method based on the point-to-ground intersection has been evaluated (in addition to the lowest-point and centre-of-cluster methods). Figure 11 shows an example of the insertion point obtained from the intersection of the two lines (aerial part and soil line) in tomato plants and sticks.

Plant locations obtained from the LiDAR and Kinect data are shown in Table 9. As explained in Section 2.4.3, when processing the Kinect images, a value of the encoder was automatically selected for each image. Mean values are obtained from the difference between the actual location and the location obtained with the LiDAR. The negative mean value obtained for Tomatoes 2 with filter Test 2 for the intersection of the stem and ground line method means that the LiDAR detected the plant at a distance greater than the actual distance (obtained from the Kinect). High standard deviation values can be explained due to the high variability found in the encoder values assigned to each plant or stick during the image processing.

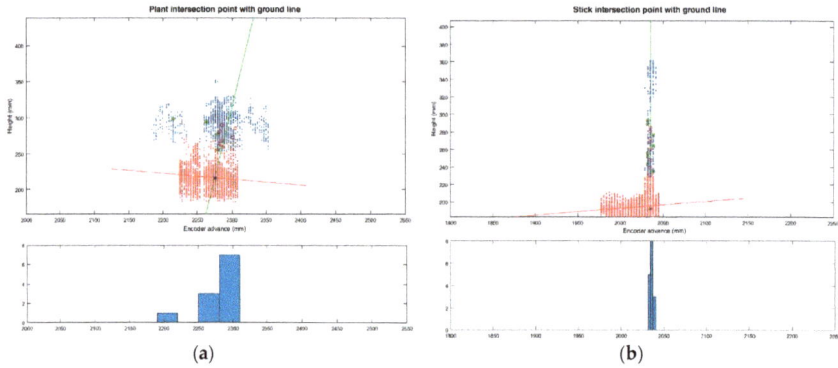

(a) (b)

Figure 11. Plant location method based on the intersection of stem line and ground line. (**a**) Intersection point between the average aerial part of the plant line (green line) and the average ground line (red line); (**b**) Intersection point between the stick aerial part line (green line) and the ground line (red line). Both histograms of aerial part of the plant points are shown on the bottom.

Table 9. Mean and standard deviation (Std.) of the plant and stick locations during the tests.

Test	Plant Location Method	Mean (mm)	Std. (mm)
Stick—Test 1	Centre of the cluster	8.32	10.09
	Lowest point	7.25	8.47
	Intersection of stem line and ground line	6.15	9.45
Tomatoes 1—Test 1	Centre of the cluster	20.74	40.37
	Lowest point	10.06	51.72
	Intersection of stem line and ground line	5.34	62.65
Tomatoes—Test 2	Centre of the cluster	30.42	37.03
	Lowest point	35.48	52.90
	Intersection of stem line and ground line	27.60	50.36
Tomatoes 2—Test 2	Centre of the cluster	1.33	41.49
	Lowest point	10.00	61.86
	Intersection of stem line and ground line	−5.95	53.32

Figure 12 shows the plant identifications in each line during the different tests. For each detection previously shown in Figure 10, tags for each of the localization methods are shown by coloured lines.

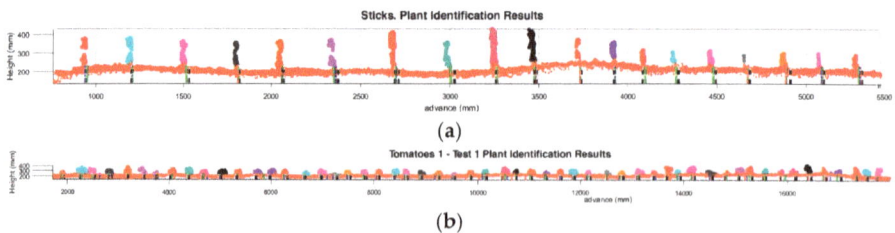

(a)

(b)

Figure 12. *Cont.*

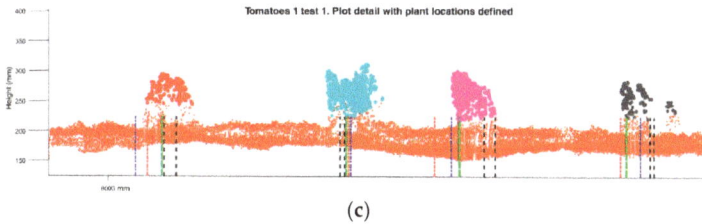

(c)

Figure 12. Plant and sticks location results obtained by the three different methods: Centre of the cluster; lowest point; and ground-plant intersection. (**a**) Sticks row data using LiDAR; (**b**) Tomato plants detected during Test 1 on row 1; (**c**) Detail of the aerial point cloud of tomato plants generated by the LiDAR during Test 1 on row 1. Each plant location method is marked with different coloured dotted line.

The centre of the cluster is marked in green, the lowest point in blue, and the intersection in red. Encoder stick/plant locations by the Kinect image (location ± Std.) are both marked in black lines, and their distances are represented by black doted.

4. Conclusions

A combination of optical sensors mounted on a frame on a tractor, which are capable of detecting and locating plants and utilizing ground-wheel odometry to determine a local reference system, was successfully developed and tested. Combining the use of affordable sensors (light-beam and Kinect) with more expensive ones (LiDAR), a laboratory setup platform and two field test platforms have been created. The following conclusions were drawn based upon the results of this research:

- The results obtained from the precise locations of the plants allow us to consider the viability of further methods capable of performing individualized treatments for each of the plants, or accurate agrochemical treatments to remove existing weeds between crop plants in the same row.
- A light-beam detection system was improved by reduction of the number of sensor pairs used. The reduction from three pairs to one pair had no effect on the desired plant detection results. In the field tests, 98% precision of detection was obtained, similar to that obtained in [21], showing that this is a robust technology and can be deployed in the field. Also, light-beam detection data allows faster processing than the LiDAR, so it could be used in real-time applications.
- Based on the methodology presented in the analysis of the data from the LiDAR and the Kinect sensors, different considerations can be established regarding the location of the same plant. According to the structure and morphology of the plant, it is assumed that the aerial part will not always be vertically in line with the stem (in fact, in the tests done this occurred frequently). For this reason, one of three proposed locations can be established as the location of the plant: the aerial part, the cluster centroid of the filtered point cloud or the insertion of the stem. Depending on the type of treatment to be made, one of these locations could be more interesting to evaluate than the others. For example, if a variable foliar herbicide treatment is to be applied (discriminating between weeds and crop plants), the distance between the aerial parts of the plants should be given greater weight in the application system (to avoid applications on the crop plants and maximize efficiency by ensuring the herbicide is applied to weeds). In the case of physical weed-removal systems, as proposed in [8], priority should be given to the location of stem insertion, and adjustments should be made to detect this element more precisely.
- The high volume of data generated by the more accurate sensors, such as the LiDAR used in this work, can be a hurdle for automatic weed detection machines when working in real time. However, it is important to emphasize the exponential growth of the processing algorithms available for the researcher, which can significantly reduce the time required for point cloud

data analysis. Reducing the density of information necessary while continuing to give accurate information is an interesting subject for future work.

- The precise locations of the plants were determined using an encoder. The use of this type of sensor is vital to implementation of low-cost referencing. Correct localization and integration to enable determination of reliable reference of the movement is of great importance.

- Agricultural use of affordable electronic components will lower costs in the near future, but the authors conclude that these systems must be robust and provide a rapid response and processing times.

Acknowledgments: This research was supported in part by the project P12-AGR-1227 funded by the Regional Government of Andalusia and the project AGL2013-46343-R funded by the Ministry of Economy, Industry and Competitiveness- Government of Spain. The authors thank Juan Agüera, Burt Vannucci and C.J. Gliever for technical assistance.

Author Contributions: Manuel Pérez-Ruiz and David C. Slaughter conceived and designed the experiments; Jorge Martínez-Guanter and Manuel Pérez-Ruiz performed the experiments; Jorge Martínez-Guanter and Miguel Garrido-Izard analyzed the data; Constantino Valero and David C. Slaughter contributed reagents/materials/analysis tools; Jorge Martínez-Guanter and Manuel Pérez-Ruiz wrote the paper.

Conflicts of Interest: The authors declare no conflict of interest.

References

1. Klenin, N.I.; Popov, I.F.; Sakun, V.A. *Agricultural Machines: Theory of Operation, Computation of Controlling Parameters, and the Conditions of Operatio*; Amerind: New Delhi, India, 1985.
2. Blas, A.; Barreireo, P.; Hernández, C.G.; Dias, G.; Griepentrog, H.W. Even-sowing pattern strategies for a low-input organic system in forage maize. *Agric. Eng. Int.* **2013**, *15*, 992.
3. Prasanna Kumar, G.V.; Raheman, H. Vegetable Transplanters for Use in Developing Countries—A Review. *Int. J. Veg. Sci.* **2008**, *14*, 232–255. [CrossRef]
4. Ehsani, M.R.; Upadhyaya, S.K.; Mattson, M.L. Seed location mapping using RTK GPS. *Trans. ASABE* **2004**, *47*, 909–914. [CrossRef]
5. Sun, H.; Slaughter, D.C.; Pérez-Ruiz, M.; Gliever, C.; Upadhyaya, S.K.; Smith, R.F. RTK GPS mapping of transplanted row crops. *Comput. Electron. Agric.* **2010**, *71*, 32–37. [CrossRef]
6. Pérez-Ruiz, M.; Slaughter, D.C.; Gliever, C.; Upadhyaya, S.K. Tractor-based Real-time Kinematic-Global Positioning System (RTK-GPS) guidance system for geospatial mapping of row crop transplant. *Biosyst. Eng.* **2012**, *111*, 64–71. [CrossRef]
7. Nakarmi, A.D.; Tang, L. Automatic inter-plant spacing sensing at early growth stages using a 3D vision sensor. *Comput. Electron. Agric.* **2012**, *82*, 23–31. [CrossRef]
8. Pérez-Ruiz, M.; Slaughter, D.C.; Fathallah, F.A.; Gliever, C.; Miller, B.J. Co-robotic intra-row weed control system. *Biosyst. Eng.* **2014**, *126*, 45–55. [CrossRef]
9. Dedousis, A.P.; Godwin, R.J.; O'Dogherty, M.J.; Tillett, N.D.; Grundy, A.C. Inter and intra-row mechanical weed control with rotating discs. *Precis. Agric.* **2007**, *7*, 493–498.
10. Blasco, J.; Aleixos, N.; Roger, J.; Rabatel, G.; Moltó, E. Robotic weed control using machine vision. *Biosyst. Eng.* **2002**, *83*, 149–157.
11. Shah, R.; Lee, W.S. An approach to a laser weeding system for elimination of in-row weeds. In Proceedings of the X European Conference on Precision Agriculture, Tel Aviv, Israel, 12–16 July 2015; pp. 309–312.
12. Frasconi, C.; Fontanelli, M.; Raffaelli, M.; Peruzzi, A. Design and full realization of physical weed control (PWC) automated machine within the RHEA project. In Proceedings of the International Conference of Agricultural Engineering, Zurich, Switzerland, 6–10 July 2014.
13. Li, L.; Zhang, Q.; Huang, D. A Review of Imaging Techniques for Plant Phenotyping. *Sensors* **2014**, *14*, 20078–20111. [CrossRef] [PubMed]
14. Fahlgren, N.; Gehan, M.A.; Baxter, I. Lights, camera, action: High-throughput plant phenotyping is ready for a close-up. *Curr. Opin. Plant Biol.* **2015**, *24*, 93–99. [CrossRef] [PubMed]
15. Dworak, V.; Selbeck, J.; Ehlert, D. Ranging sensor for vehicle-based measurement of crop stand and orchard parameters: A review. *Trans. ASABE* **2011**, *54*, 1497–1510. [CrossRef]

16. López-Granados, F. Weed detection for site-specific weed management: Mapping and real-time approaches. *Weed Res.* **2011**, *51*, 1–11. [CrossRef]

17. Garrido-Izard, M.; Paraforos, D.; Reiser, D.; Vázquez Arellano, M.; Griepentrog, H.; Valero, C. 3D Maize Plant Reconstruction Based on Georeferenced Overlapping LiDAR Point Clouds. *Remote Sens.* **2015**, *7*, 17077–17096. [CrossRef]

18. Rosell, J.R.; Llorens, J.; Sanz, R.; Arnó, J.; Ribes-Dasi, M.; Masip, J.; Escolà, A.; Camp, F.; Solanelles, F.; Gràcia, F.; et al. Obtaining the three-dimensional structure of tree orchards from remote 2D terrestrial LiDAR scanning. *Agric. For. Meteorol.* **2009**, *149*, 1505–1515. [CrossRef]

19. Garrido-Izard, M.; Méndez, V.; Valero, C.; Correa, C.; Torre, A.; Barreiro, P. Online dose optimization applied on tree volume through a laser device. In Proceedings of the First RHEA International Conference on Robotics and Associated High Technologies and Equipment for Agriculture, Pisa, Italy, 19–21 September 2012.

20. Shi, Y.; Wang, N.; Taylor, R.K.; Raun, W.R.; Hardin, J.A. Automatic corn plant location and spacing measurement using laser line-scan technique. *Precis. Agric.* **2013**, *14*, 478–494. [CrossRef]

21. Garrido-Izard, M.; Pérez-Ruiz, M.; Valero, C.; Gliever, C.J.; Hanson, B.D.; Slaughter, D.C. Active optical sensors for tree stem detection and classification in nurseries. *Sensors* **2014**, *14*, 10783–10803. [CrossRef] [PubMed]

22. Waldkirch, S.A. Laser Measurement Systems of the LMS100 Product Family: Operating Instructions. Available online: https://mysick.com/saqqara/im0031331.pdf (Accessed on 2 March 2017).

23. Torr, P.H.S.; Zisserman, A. MLESAC: A New Robust Estimator with Application to Estimating Image Geometry. *Comput. Vis. Image Underst.* **2000**, *78*, 138–156. [CrossRef]

24. Arthur, D.; Vassilvitskii, S. K-means++: The advantages of careful seeding. In Proceedings of the Eighteenth Annual ACM-SIAM Symposium on Discrete Algorithms, New Orleans, LA, USA, 7–9 January 2007; pp. 1027–1035.

25. Hamuda, E.; Glavin, M.; Jones, E. A survey of image processing techniques for plant extraction and segmentation in the field. *Comput. Electron. Agric.* **2016**, *125*, 184–199. [CrossRef]

26. Bay, H.; Ess, A.; Tuytelaars, T.; Van Gool, L. Speeded-up robust features (SURF). *Comput. Vis. Image Underst.* **2008**, *110*, 346–359. [CrossRef]

sensors

MDPI

Article

Leaf Area Index Estimation Using Chinese GF-1 Wide Field View Data in an Agriculture Region

Xiangqin Wei [1,2,3], Xingfa Gu [1,2,3,*], Qingyan Meng [1,3], Tao Yu [1,3], Xiang Zhou [1,3], Zheng Wei [3], Kun Jia [4,*] and Chunmei Wang [1,3]

[1] Institute of Remote Sensing and Digital Earth, Chinese Academy of Sciences, Beijing 100101, China; weixq@radi.ac.cn (X.W.); mengqy@radi.ac.cn (Q.M.); yutao@radi.ac.cn (T.Y.); zhouxiang@radi.ac.cn (X.Z.); wangcm@radi.ac.cn (C.W.)

[2] University of Chinese Academy of Sciences, Beijing 100049, China

[3] Application Technology Center of China High-Resolution Earth Observation System, Beijing 100101, China; weizheng@irsa.ac.cn

[4] State Key Laboratory of Remote Sensing Science, Faculty of Geographical Science, Beijing Normal University, Beijing 100875, China

* Correspondence: guxingfa@radi.ac.cn (X.G.); jiakun@bnu.edu.cn (K.J.); Tel.: +86-10-6488-9565 (X.G.); +86-10-5880-0152 (K.J.)

Received: 26 April 2017; Accepted: 5 July 2017; Published: 8 July 2017

Abstract: Leaf area index (LAI) is an important vegetation parameter that characterizes leaf density and canopy structure, and plays an important role in global change study, land surface process simulation and agriculture monitoring. The wide field view (WFV) sensor on board the Chinese GF-1 satellite can acquire multi-spectral data with decametric spatial resolution, high temporal resolution and wide coverage, which are valuable data sources for dynamic monitoring of LAI. Therefore, an automatic LAI estimation algorithm for GF-1 WFV data was developed based on the radiative transfer model and LAI estimation accuracy of the developed algorithm was assessed in an agriculture region with maize as the dominated crop type. The radiative transfer model was firstly used to simulate the physical relationship between canopy reflectance and LAI under different soil and vegetation conditions, and then the training sample dataset was formed. Then, neural networks (NNs) were used to develop the LAI estimation algorithm using the training sample dataset. Green, red and near-infrared band reflectances of GF-1 WFV data were used as the input variables of the NNs, as well as the corresponding LAI was the output variable. The validation results using field LAI measurements in the agriculture region indicated that the LAI estimation algorithm could achieve satisfactory results (such as $R^2 = 0.818$, RMSE = 0.50). In addition, the developed LAI estimation algorithm had potential to operationally generate LAI datasets using GF-1 WFV land surface reflectance data, which could provide high spatial and temporal resolution LAI data for agriculture, ecosystem and environmental management researches.

Keywords: leaf area index; radiative transfer model; neural networks; GF-1 satellite; wide field view

1. Introduction

Vegetation is the basic component of the terrestrial ecosystem and plays an important role in energy exchange, carbon cycling and hydrological cycling process on the earth surface [1–3]. Therefore, timely and accurate land surface vegetation information is of great significance for earth system science, ecological environment assessment and climate change related studies [4]. Leaf area index (LAI), generally defined as one half of the total green leaf area per unit of horizontal ground surface area [5], is an important parameter for characterizing land surface vegetation conditions [6–8]. LAI is an essential parameter that characterizes the density of leaves and canopy structure, which can reflect the vegetation's ability for biophysical processes such as photosynthesis, respiration and transpiration, and is also an important input variable for the carbon cycle models, crop growth models and water cycle models [9–13]. Therefore, accurate estimation of LAI on regional and global scales is of great significance for ecosystem modelling, biogeochemical cycle modelling, agriculture monitoring and related applications.

Remote sensing provides the only effective means to estimate LAI at the regional and global scales for its ability of continuous observation and to provide broad and impartial earth observation data [7,8,14,15]. Currently, several global LAI products have been generated using medium to low spatial resolution satellite data, such as the MODIS LAI products [6], GEOV1 LAI Products [7], MERIS LAI products [16], GLASS LAI products [17] and GLOBCARBON LAI products [18]. However, the currently existed LAI products generally have low spatial resolutions at the kilometric level, whereas decametric spatial resolution LAI data will be better suited for applications related to agriculture monitoring, as compared to kilometric resolution LAI data, which are usually larger than the typical scales of most croplands. The wide field view (WFV) sensor on board GF-1, the first satellite of the China High-resolution Earth Observation System, can acquire multi-spectral data with high spatial and temporal resolutions [19], which are valuable data sources for dynamic LAI monitoring at regional scale. Therefore, exploring the application potential of GF-1 WFV data on land surface LAI estimation and developing the specific LAI estimation algorithm for GF-1 WFV data are urgently needed.

The key of LAI estimation using remote sensing data is how to establish the relationship between LAI and land surface reflectance data according to the radiative transfer process of photon in the vegetation canopy and its spectral response characteristics [20–23]. Many algorithms have been developed to retrieve LAI using satellite remote sensing data and generally two types of algorithms can be distinguished including empirical methods and physical model based methods [6,24–27]. The empirical LAI estimation methods are based on the statistical relationships between LAI and vegetation indices (VIs), which are calibrated using field LAI measurements and remote sensing data or simulated data from canopy radiative transfer models [28,29]. The empirical methods simplify the complex radiative transfer process in the canopy, which are simple and can obtain satisfactory LAI estimation accuracy in small regions. Furthermore, when using VIs as independent variables, the VIs can highlight vegetation information and weaken the influences of canopy shadows, soil backgrounds, atmospheric conditions and angle effects [29]. However, the empirical methods only use the VIs calculated from several band reflectances, which cannot make full use of the multi-band spectral information of remote sensing data. Furthermore, the reduction of the VIs from multi-band reflectances to an index also reduces the LAI estimation constraints and increases the uncertainty of LAI estimation results, resulting in the empirical relationship between LAI and VIs changing with sensor types, vegetation types, time and geographical areas. Therefore, the application of empirical LAI estimation methods to a large scale is very difficult, due to the complexity of land surface.

The physical model based methods are mainly based on the simulation of radiative transfer in vegetation canopy and establishing the physical relationship between canopy reflectances and LAI, and then the canopy radiative transfer models are inversed [30,31]. The direct inversion of radiative transfer models is very difficult due to the complexity of the models, and iterative optimization (OPT) method [32], lookup table (LUT) method [6] and machine learning methods [26] are usually used for indirect inversion of physical models to achieve LAI estimation. The OPT method is based on the iterative minimization of a cost function which requires hundreds of runs of the canopy radiative transfer model for each pixel and therefore computationally too demanding. For practical applications, LUT method and machine learning methods are popular LAI estimation methods, which are based on the database simulated by physical models. The LUT method is conceptually the simplest technique by finding the solution for a given set of reflectance measurements, which consisted of selecting the closest cases in the database according to a cost function, and then based on extracting the corresponding set of LAI [6,33]. However, the LUT method usually requires a fixed number of input variables, unless there are very large lookup tables, which are difficult to manipulate. Machine learning methods can efficiently and accurately approximate the complex nonlinear functions, and train the algorithm parameters through training samples to realize the efficient and accurate estimation of LAI using remote sensing data [26,34]. Commonly used machine learning methods mainly include neural networks (NNs), as well as support vector machines and decision trees [35]. NNs trained over radiative transfer model simulations have been applied, with success, to estimate LAI from several sensors' data, leading to several operational LAI production algorithms, such as the CYCLOPES and MERIS LAI products [26,36]. Therefore, based on the reality of work in the field of LAI estimation using remote sensing data, the NNs inversion of physical models is a potential and reasonable choice for LAI estimation from GF-1WFV data.

Therefore, the objective of this study is to develop a general LAI estimation algorithm for GF-1 WFV reflectance data under various land surface conditions using NNs trained over radiative transfer model simulations. Meanwhile, the field LAI measurements in an agriculture region with maize as the dominated crop type are used to assess the performance of GF-1 WFV data on LAI estimation using the proposed algorithm.

2. Study Area

The Shenzhou county of China (Figure 1) is selected as the study area to investigate the performance of LAI estimation algorithm for GF-1 WFV data. The study area is located on the North China Plain (centred at 115°35′ E, 37°53′ N) covering approximately 20 km × 25 km. The study area belongs to the temperate climate zone and is a typical upland field agriculture area in the North China Plain. The annual average temperature is approximately 13.4 °C, and the annual average precipitation is approximately 486 mm, which is mainly concentrated in July and August. It is relatively flat farmland with an average altitude of about 20 m above the sea level, so that uncertainties of LAI estimation caused by topographical facts will be reduced to a minimum. The main autumn grain crops planted in the study area are maize. Maize season begins in mid-June and harvests in early October. Though the study area is not big, it has the representative characteristics of crop type distribution in the North China Plain.

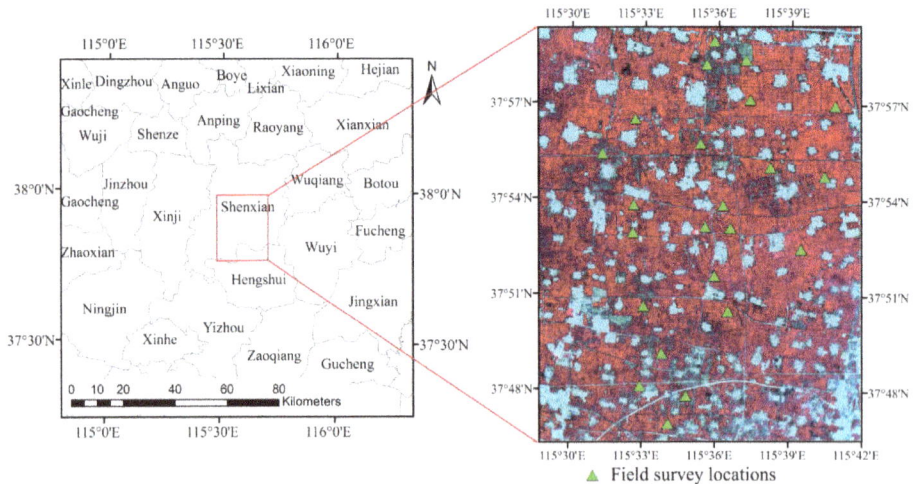

Figure 1. Square region in the left image shows the geo-location of the Shenzhou study area in Hebei Province, and the right image is the GF-1 WFV data acquired on 15 August 2014. Generally, the red patches are farmland in the right image and the blue patches are non-vegetated regions, such as residential areas and roads. The green triangles on the right image indicate the field survey locations.

3. Data and Pre-Processes

3.1. Field LAI Measurements

In order to validate the LAI estimation accuracy using GF-1 WFV data, the field campaigns were completed covering the main growing seasons of maize, and the specific field observation dates were 27 June, 21 July, 14 August and 5 September 2014. Twenty-three representative square sample sites (30 m × 30 m) within the study area (Figure 1) were selected based on the crop growth conditions and the high spatial resolution remote sensing data. The sample sites were located in relatively homogeneous regions with approximately 50 m around the sample sites having similar crop growth conditions, thus the uncertainty of LAI estimation caused by the pixel matching error between GF-1 WFV data and the sample site would be minimized. The center of each sample site was determined using the handheld global positioning system (GPS) receiver which had a positioning accuracy of approximately ±3 m. LAI was measured using an LAI-2000 plant canopy analyzer (Li-Cor, Inc., Lincoln, NE, USA), and there were three measuring plots in each sample site and the average LAI value of the three LAI measurements was regarded as the LAI of the sample site. The number of LAI measurement sites on 27 June was four because maize in this period was very small, and on 21 July, 14 August and 5 September were all twenty-three. These field LAI measurements were used to validate the LAI estimates using the GF-1 WFV data.

3.2. GF-1 WFV Data and Pre-Processes

In order to validate the proposed LAI estimation algorithm, GF-1 WFV data synchronized with the field LAI measurements were required. The spatial resolution of GF-1 WFV data was 16 m and the temporal resolution was four days for the four WFV sensors combined. According to the field survey dates and the image quality of GF-1 WFV data, the high-quality GF-1 WFV data close to the field survey dates were selected as much as possible, and finally 5 GF-1 WFV data were collected (Table 1). The GF-1 WFV data acquired on 24 June, 18 July and 15 August 2014 were close to the corresponding field survey dates (3 days or less), which could be approximated that LAI in this close time did not change and the field LAI measurements could be directly used to validate the LAI estimates using GF-1 WFV data. However, corresponding to the field LAI measurements on 5 September 2014, there were no suitable GF-1 WFV data close to the field survey date due to the effect of clouds, thus the GF-1 WFV data acquired on 24 August and 18 September 2014 were selected to estimate LAI, which would be interpolated to estimate LAI on 5 September using the spline interpolation method and then the LAI estimation accuracy would be validated using the filed LAI measurements on 5 September.

Table 1. The main characteristics of GF-1 wide field view (WFV) data used in this study.

WFV Sensor	Date (dd/mm/yy)	Matched Field Survey Date	Data Quality
WFV1	27 June 2014	27 June 2014	Good
WFV2	18 July 2014	21 July 2014	With cloud
WFV1	15 August 2014	14 August 2014	Good
WFV3	24 August 2014	5 September 2014	Good
WFV4	18 September 2014	5 September 2014	With cloud

The pre-process of GF-1 WFV data contained radiance calibration, atmospheric correction and geometric correction. The radiance calibration was converting the DN values of the raw data to radiances using the following equation:

$$L_e = \text{Gain} \times \text{DN} + \text{Offset}$$

where L_e was the radiance, and Gain and Offset were calibration coefficients obtained from the China Centre for Resources Satellite Data and Application, which were listed in Table 2.

Table 2. The calibration coefficients of GF-1 WFV data in 2014.

WFV Sensor	Bands	Gain	Offset
WFV1	Blue band (Band1)	0.2004	0
	Green band (Band2)	0.1648	0
	Red band (Band3)	0.1243	0
	NIR band (Band4)	0.1563	0
WFV2	Blue band (Band1)	0.1733	0
	Green band (Band2)	0.1383	0
	Red band (Band3)	0.1122	0
	NIR band (Band4)	0.1391	0
WFV3	Blue band (Band1)	0.1745	0
	Green band (Band2)	0.1514	0
	Red band (Band3)	0.1257	0
	NIR band (Band4)	0.1462	0
WFV4	Blue band (Band1)	0.1713	0
	Green band (Band2)	0.1600	0
	Red band (Band3)	0.1497	0
	NIR band (Band4)	0.1435	0

The Fast Line-of-sight Atmospheric Analysis of Spectral Hypercubes (FLAASH) model was used to atmospheric correction of GF-1 WFV data [37]. The input parameters for FLAASH model were determined based on the imaging time and imaging parameters of each GF-1 WFV data. After the atmospheric correction, the GF-1 WFV land surface reflectance data were obtained and the reflectances of green, red and near-infrared (NIR) bands would be used to estimate LAI. The geometric correction of GF-1 WFV data was conducted using two-order polynomial transformation method with bilinear interpolation resampling. High quality Landsat-8 Operational Land Imager data [38] were selected as the base map, and ground control points were selected from the images manually. The resulted geometric co-registration error for each GF-1 WFV data was less than one pixel of the Landsat data (30 m). Finally, the subset images of the atmospherically and geometrically corrected GF-1 WFV data covered the sample sites were extracted to further investigate the performance of the LAI estimation algorithm for GF-1 WFV data (Figure 1).

4. Methods

A flowchart of the LAI estimation algorithm development for GF-1 WFV data was presented in Figure 2. The training sample dataset was firstly generated using the spectral reflectance simulations based on the radiative transfer model and the relative spectral response profiles of GF-1 WFV sensors. Then, the NNs were trained over the training sample dataset to develop the LAI estimation algorithm for GF-1 WFV data. Finally, LAI could be estimated using the pre-processed GF-1 WFV land surface reflectance data using the trained NNs or set to zero when the pixel was non-vegetation which was judged by the normalized difference vegetation index (NDVI) value.

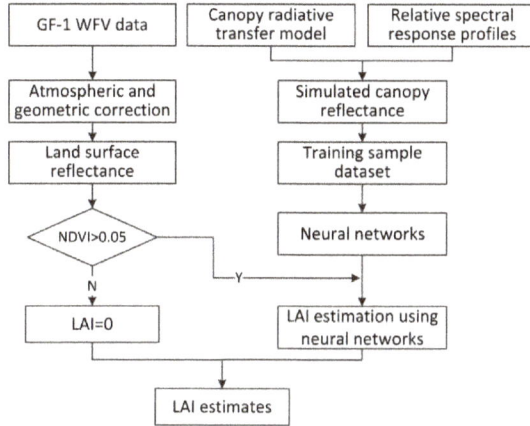

Figure 2. Flowchart of the leaf area index (LAI) estimation algorithm for GF-1 WFV data.

4.1. Generating LAI Training Sample Dataset from Radiative Transfer Model Simulations

The canopy radiative transfer model quantitatively described the physical relationship between LAI and canopy spectral reflectances. The widely used coupled PROSPECT and SAIL (PROSAIL) model, which was easy to use, had general robustness and also had consistent performance in validation practices [22,39,40] was selected to simulate the satellite observations of canopy reflectance based on the relative spectral response profiles of the GF-1 WFV sensors. The PROSPECT model was a plate model based radiative transfer model, which simulated the hemispherical reflectance and transmittance of leaves from spectral wavelength from 0.4 to 2.5 μm based on the leaf biochemical and biophysical parameters [41,42]. The input parameters for PROSPECT model included leaf chlorophyll a + b concentration (C_{ab}), water content (C_w), dry matter content (C_m), brown pigment content (C_{brown}), carotenoid content (C_{ar}) and leaf structure parameter (N), and the output parameters were hemispherical reflectance and transmittance of leaves, which were also the input parameters for SAIL model. The SAIL model with hot-spot correction, which assumed the canopy as a turbid medium, was selected in this study [22,26]. The canopy structure in the SAIL model was characterized by LAI, the average leaf angle inclination (ALA) assuming an ellipsoidal distribution and the hot-spot parameter.

The underlying soil reflectances were also input variables for the PROSAIL model. The soil reflectances from a globally distributed soil spectral library released by the International Soil Reference and Information Centre (access at: http://www.isric.org), which contained various soil types with various properties and had representative of various soil types [19,43], were selected for the inputs of PROSAIL model. To remove data redundancy generated by similar soil reflectances and to avoid huge computations in PROSAIL simulations, several representative soil reflectances should be determined from the original data. The spectral angle mapper method [44,45] was used to evaluate the similarity of different soil reflectances and further determined the representative soil reflectances for PROSAIL model. The soil reflectances having spectral angle value smaller than 0.05 would be considered as similar soil reflectances which would be averaged as a representative soil reflectance. Finally, 13 soil reflectances were determined to represent the possible range of soil reflectances in the PROSAIL model (Figure 3).

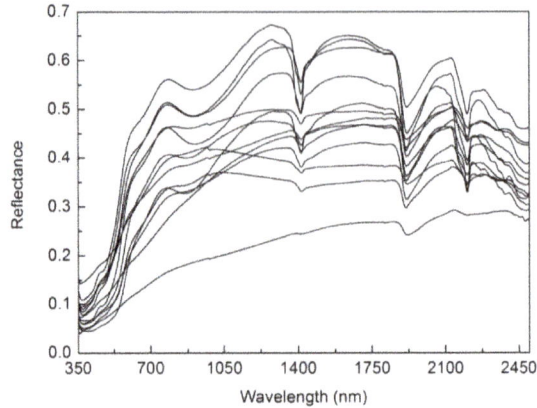

Figure 3. The 13 soil reflectances used to represent the possible range of spectral shapes for the PROSAIL model.

It had been demonstrated that reasonable error of the input variables of radiative transfer model was permitted which did not lead to obvious loss of parameter inversion accuracy, therefore some input variables in the PROSAIL model could be fixed in the simulations [46,47]. Considering the objective of this study was to develop a general LAI estimation algorithm using GF-1 WFV data under various land surface conditions, the input variables of PROSAIL model were given reasonable ranges or fixed values (Table 3) based on previous studies, such as the Leaf Optical Properties Experiment 93 and the algorithms of CYCLOPES LAI product [22,26,48,49]. Therefore, the PROSAIL model could simulate the vegetation canopy reflectances based on the vegetation physicochemical parameters, geometric parameters and soil reflectances, meanwhile the remote sensing data could also obtain vegetation canopy reflectances by atmospheric correction, thus the remote sensing reflectance data were associated with LAI through the physical process of radiation transmission.

Table 3. The input variables of PROSAIL model for LAI estimation algorithm development.

Parameters	Units	Value Range	Step
LAI	m^2/m^2	0–7	0.2
ALA	°	30–70	10
N	-	1–2	0.5
C_{ab}	$\mu g/cm^2$	30–60	10
C_m	g/cm^2	0.005–0.015	0.005
C_{ar}	$\mu g/cm^2$	0	-
C_w	cm	0.005–0.015	0.005
C_{brown}	-	0–0.5	0.5
Hot	-	0.1	-
Solar zenith angle	°	25–55	10

For any combination of the input variables, the canopy reflectance was computed for each wavelength from 0.4 to 2.5 μm at a spectral resolution of 1 nm, and then the canopy reflectance was resampled to simulate GF-1 WFV observations using the relative spectral response profiles. Because the relative spectral response profiles of the 4 WFV sensors on board GF-1 satellite were slightly different, the simulations of GF-1 WFV reflectance were separately generated for each WFV sensor. The canopy reflectance simulations using PROSAIL model resulted in 2,021,760 cases of matched reflectances and LAI values for each WFV sensor. To account for uncertainties in the satellite observations and model simulations, a white Gaussian noise with signal to noise ratio of 100 was added to the simulated reflectances, which were used as the learning dataset for the NNs.

4.2. Neural Networks

The NNs mimic human learning to build relationships between variables, which can approximate multivariate non-linear relationships and are robust to noisy data [50,51], thus have been widely used for estimating land surface parameters from remote sensing data [7,52,53]. The popular back propagation NNs (BPNNs) were selected for LAI estimation algorithm development using GF-1 WFV data [4,7]. The BPNNs could learn from the training sample dataset and built relationships between reflectances and LAI, then the trained BPNNs could generate the optimal LAI estimates based on the actual reflectances from GF-1 WFV data. The architecture of the BPNNs used for LAI estimation from GF-1 WFV data was shown in Figure 4. The input variables of the BPNNs included the reflectances of the green, red and NIR bands, and the output variable was the corresponding LAI. The number of nodes in the hidden layer was set to six. The BPNNs activation functions in the hidden layer and output nodes were set to "sigmoid" and "linear", respectively. The Levenberg–Marquardt minimization algorithm was used to calibrate the synaptic coefficients because of its efficient convergence capacity [19,54]. The training sample dataset made of pairs of reflectances and LAI, which were generated from the PROSAIL model simulations, was randomly split into two parts: 90% of the cases were used for training BPNNs, meanwhile the rest 10% of the cases were used to test the hyper-specialization during the training process. The BPNNs models were generated for each of the four GF-1 WFV sensors based on the specific training sample dataset for each WFV sensor.

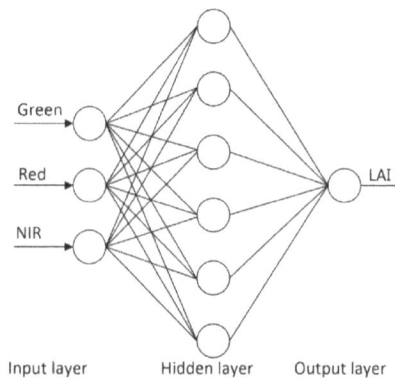

Figure 4. The architecture of the back propagation neural networks (BPNNs) used for LAI estimation from GF-1 WFV reflectance data.

4.3. LAI Estimating Procedure for GF-1 WFV Data

The NDVI generated from red and NIR band reflectances was an important indicator for vegetation growth conditions, and usually used to estimate LAI based on empirical relationships between LAI and NDVI [55–57]. Therefore, to avoid the presentation of abnormal NNs inversion values caused by the non-vegetation pixels, it was a good strategy to remove non-vegetation pixels using NDVI threshold value before LAI estimation using the NNs. In this study, the NDVI threshold value of 0.05 was used to distinguish vegetation and non-vegetation pixels. The pixels with NDVI value smaller than 0.05 was identified as non-vegetation pixels and LAI would be set to zero, whereas the LAI for other pixels would be estimated using the trained NNs, from which the GF-1 WFV data should be judged from that sensor and then adopt the corresponding trained NNs.

4.4. LAI Estimation Accuracy Assessment

Direct comparison of field LAI measurements with LAI estimates using GF-1 WFV land surface reflectance data is a reliable way to assess the accuracy of the proposed LAI estimation algorithm. Comparison between the field survey and predicted LAI will be performed by using five statistical indices: R^2 of linear regression, root mean square error (RMSE), relative RMSE (RRMSE), Nash–Sutcliffe model efficiency coefficient (EF) [58] and coefficient of residual mass (CRM) [59]. The RMSE values show how much the predicted LAI values under- or over-estimate the field observations. The RRMSE is calculated by dividing the RMSE by the mean field observations. The EF value compares the predicted LAI values to the average value of the field observations. A negative EF value indicates that the average value of the field observations gives a better estimate than the predicted values. The CRM is a measure of the tendency of the developed LAI estimation method to overestimate or underestimate the field observations. A positive CRM value indicates that the method underestimates the field observations and the negative value indicates a tendency to overestimate. For a perfect fit between field observed and predicted LAI data, values of R^2, RMSE, RRMSE, CRM, and EF should equal 1.0, 0.0, 0.0, 0.0 and 1.0, respectively.

5. Results

The LAI estimation results using the proposed algorithm and the pre-processed GF-1 WFV land surface reflectance data are shown in Figure 5. In the visual aspect, the spatial and temporal distributions of LAI estimates are reasonable. As for the spatial distribution, LAI estimates using GF-1 WFV data acquired on 27 June generally show small values, because maize at this period is at the early emergence stage with small LAI, while larger LAI estimates only present at an orchard on the upper left corner of the study area. After July, with the rapid growth of maize, the large LAI estimates are mainly distributed in the area of farmland, and the LAI estimates at other areas are smaller. Meanwhile, LAI estimates in the cloud and cloud shadow regions are all zero. Regarding the temporal variations, LAI estimates of maize show low values in late June and continuous increases in July and mid-August, reaching the peak LAI values in the late August and a downward trend in mid-September. The temporal variation characteristics of LAI estimates exactly reflect the growth processes of maize from germination, rapid growth to peak value, and then gradually decline, which is in accord with the growth characteristics of maize. Therefore, the visual observation of the spatial and temporal characteristics of LAI estimates using GF-1 WFV data could preliminarily indicate the reasonability of the proposed LAI estimation algorithm.

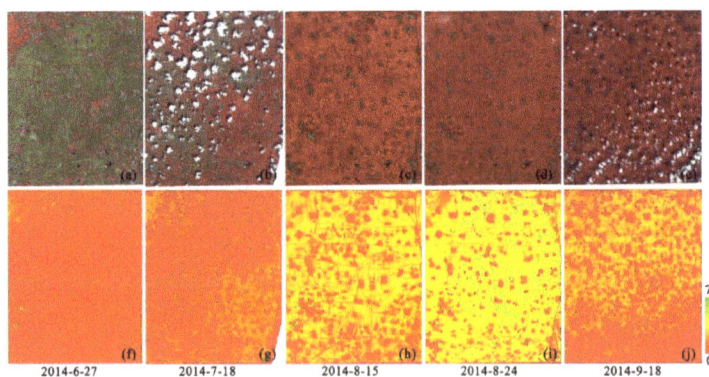

Figure 5. GF-1 WFV land surface reflectance data ((**a**) 27 June 2014, (**b**) 18 July 2014, (**c**) 15 August 2014, (**d**) 24 August 2014, and (**e**) 18 September 2014) and their corresponding LAI estimates ((**f**): 27 June 2014, (**g**) 18 July 2014, (**h**) 15 August 2014, (**i**) 24 August 2014, and (**j**) 18 September 2014).

The field LAI measurements located in the cloud and cloud shadow regions were deleted based on the visual observation, and the remaining field LAI measurements, which contained 43 LAI data, were used to directly validate the LAI estimates from GF-1 WFV data using the proposed algorithm (Figure 6). It could be seen that there was a good linear relationship between field LAI measurements and LAI estimates from GF-1 WFV data. The LAI estimation accuracy was satisfactory with values of R^2, RMSE, RRMSE, CRM, and EF equaling 0.818%, 0.5%, 19.0%, 0.04% and 0.797%, respectively. In addition, some larger differences between the field survey LAI and the predicted LAI were observed in the region of large LAI values. The main reason might be that the field LAI values measured on 5 September were matched with LAI estimates interpolated from LAI estimates on 24 August and 18 September, whereas maize in this period was close to the mature stage with great LAI change rate, thus the interpolated LAI results might have some uncertainties. The other reason might be that the clouds influenced the atmospheric conditions of the study area which further influenced the accuracy of obtaining GF-1 WFV land surface reflectances and caused the uncertainties of LAI estimates.

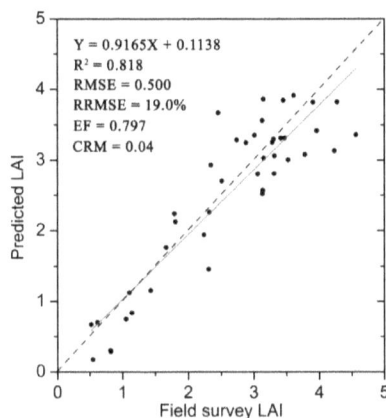

Figure 6. Scatter plots between field survey LAI and GF-1 WFV data predicated LAI.

In summary, the difference between the field LAI measurements and LAI estimates from GF-1 WFV data was small and the LAI estimation results were satisfactory. Therefore, the results indicated that the proposed LAI estimation algorithm for GF-1 WFV data was reliable and GF-1 WFV data could achieve acceptable performance on LAI estimation, which had the potential for providing high temporal and spatial resolution LAI dataset for related applications, such as agriculture monitoring.

6. Discussion and Conclusions

This study proposed a LAI estimation algorithm for GF-1 WFV land surface reflectance data based on BPNNs with training samples generated from the radiative transfer model simulations under different soil and vegetation conditions, thus the algorithm could adapt to a variety of different underlying conditions. The validation results using field LAI measurements in an agriculture region with maize as the dominated crop type showed that the proposed algorithm could achieve satisfactory LAI estimation accuracy (e.g., RMSE = 0.50), which indicated GF-1 WFV data had good performance on LAI estimation and the algorithm had potential to operationally estimate LAI from GF-1 WFV land surface reflectances. The LAI estimation algorithm was automatically operated without prior knowledge on land cover, and no human interaction and no empirical model parameters were needed. Therefore, the proposed LAI estimation algorithm for GF-1 WFV data could overcome the difficulties in determination of model parameters for empirical LAI estimation methods, which were generally changed with time, region and vegetation types. In addition, the studies of LAI estimation using

Sensors **2017**, *17*, 1593

remote sensing data based on NNs trained over radiative transfer model simulations in the past mainly focused on the kilometric spatial resolution remote sensing data, and several global LAI products were generated from SPOT VEGETATION, MERIS and MODIS data. Therefore, the developed LAI estimation algorithm for decametric spatial resolution remote sensing data using NNs based on physical model was a new try in this study, and LAI with decametric spatial resolution would be more useful for agriculture, ecosystem and environment management than kilometric spatial resolution data which were usually larger than the typical scales of most landscapes.

There were also some limitations about this study, even though it achieved satisfactory LAI estimation performance. Firstly, the field LAI measurements for validation were collected from an agriculture region covering the main growing season of maize, though they had temporal–spatial representativeness of cropland and the agriculture monitoring was the main application field of LAI, exploratory investigations should be strengthened using much more validation data from various vegetation types in the future work. Moreover, the pre-processes of GF-1 WFV data were conducted manually. If an automated data preprocessing approach for radiation correction and geometric correction could be developed, a streamlined LAI estimation workflow would be formed from the original DN values to LAI estimates, which would make the LAI estimation from GF-1 WFV data much simpler and quicker. Finally, the accuracy assessment of LAI estimation algorithm was difficult to achieve due to the difficulties in obtaining real LAI values from field measurements to match with the pixel level LAI estimates. Currently, using plant canopy analyzer to obtain field LAI measurements was the commonly used method, which could obtain relatively accurate LAI data. However, some errors might be made when matching the field LAI measurements and pixel scale LAI estimates due to the non-absolute uniformity of the land surface, though averaging the multi-point measurements could reduce this error to a certain extent. Perhaps using an unmanned aircraft system flying in the lower altitude to take photos from the nadir and developing corresponding LAI extraction algorithm from photos was a considerable way for real LAI data collection. Although the field LAI measurements had many uncertainties, they still played an important role in assessing LAI estimation algorithms using remote sensing data, and more effective real LAI obtaining approaches would be expected in the future.

In conclusion, the proposed LAI estimation algorithm for GF-1 WFV data was reliable and GF-1 WFV data were confirmed as having a good performance on LAI estimation, which could provide reliable high spatial and temporal resolution LAI data for related applications. Further work would focus on validating the proposed LAI estimation algorithm using field LAI measurements with less uncertainties under various land cover type conditions.

Acknowledgments: This study was partially supported by the Major Special Project of the China High-Resolution Earth Observation System and the Science and Technology Subsidies to Leading Scientists by National Administration of Surveying, Mapping and Geoinformation.

Author Contributions: X.W., X.G. and K.J. conceived and designed the experiments; X.W. performed the experiments; X.W. and K.J. analyzed the data; X.W. wrote the paper; and X.W., X.G., Q.M., T.Y., X.Z., Z.W., K.J. and C.W. contributed to revise the paper.

Conflicts of Interest: The authors declare no conflict of interest.

References

1. Jiang, B.; Liang, S.L.; Yuan, W.P. Observational evidence for impacts of vegetation change on local surface climate over northern China using the Granger causality test. *J. Geophys. Res. Biogeosci.* **2015**, *120*, 1–12. [CrossRef]
2. Jia, K.; Liang, S.; Liu, S.; Li, Y.; Xiao, Z.; Yao, Y.; Jiang, B.; Zhao, X.; Wang, X.; Xu, S.; et al. Global land surface fractional vegetation cover estimation using general regression neural networks from MODIS surface reflectance. *IEEE Trans. Geosci. Remote Sens.* **2015**, *53*, 4787–4796. [CrossRef]
3. Bonan, G.B. Forests and climate change: Forcings, feedbacks, and the climate benefits of forests. *Science* **2008**, *320*, 1444–1449. [CrossRef] [PubMed]

4. Liang, S. *Advances in Land Remote Sensing System, Modeling Inversion and Application*; Springer: Dordrecht, The Netherlands, 2008.

5. Chen, J.M.; Black, T.A. Defining leaf area index for non-flat leaves. *Plant Cell Environ.* **1992**, *15*, 421–429. [CrossRef]

6. Myneni, R.B.; Hoffman, S.; Knyazikhin, Y.; Privette, J.L.; Glassy, J.; Tian, Y.; Wang, Y.; Song, X.; Zhang, Y.; Smith, G.R.; et al. Global products of vegetation leaf area and fraction absorbed PAR from year one of MODIS data. *Remote Sens. Environ.* **2002**, *83*, 214–231. [CrossRef]

7. Baret, F.; Weiss, M.; Lacaze, R.; Camacho, F.; Makhmara, H.; Pacholcyzk, P.; Smets, B. GEOV1: LAI and FAPAR essential climate variables and FCOVER global time series capitalizing over existing products. Part 1: Principles of development and production. *Remote Sens. Environ.* **2013**, *137*, 299–309. [CrossRef]

8. Xiao, Z.Q.; Liang, S.L.; Wang, J.D.; Chen, P.; Yin, X.J.; Zhang, L.Q.; Song, J.L. Use of General Regression Neural Networks for Generating the GLASS Leaf Area Index Product from Time-Series MODIS Surface Reflectance. *IEEE Trans. Geosci. Remote Sens.* **2014**, *52*, 209–223. [CrossRef]

9. Myneni, R.B.; Ross, J.; Asrar, G. A review on the theory of photon transport in leaf canopies. *Agric. For. Meteorol.* **1989**, *45*, 1–153. [CrossRef]

10. Sellers, P.J.; Dickinson, R.E.; Randall, D.A.; Betts, A.K.; Hall, F.G.; Berry, J.A.; Collatz, G.J.; Denning, A.S.; Mooney, H.A.; Nobre, C.A.; et al. Modeling the exchanges of energy, water, and carbon between continents and the atmosphere. *Science* **1997**, *275*, 502–509. [CrossRef] [PubMed]

11. Arora, V. Modeling vegetation as a dynamic component in soil-vegetation-atmosphere transfer schemes and hydrological models. *Rev. Geophys.* **2002**, *40*. [CrossRef]

12. Bu, H.; Sharma, L.K.; Denton, A.; Franzen, D.W. Sugar Beet Yield and Quality Prediction at Multiple Harvest Dates Using Active-Optical Sensors. *Agron. J.* **2016**, *108*, 273–284. [CrossRef]

13. Bu, H.; Sharma, L.K.; Denton, A.; Franzen, D.W. Comparison of Satellite Imagery and Ground-Based Active Optical Sensors as Yield Predictors in Sugar Beet, Spring Wheat, Corn, and Sunflower. *Agron. J.* **2017**, *109*, 299–308. [CrossRef]

14. Sharma, L.K.; Bu, H.; Denton, A.; Franzen, D.W. Active-Optical Sensors Using Red NDVI Compared to Red Edge NDVI for Prediction of Corn Grain Yield in North Dakota, U.S.A. *Sensors* **2015**, *15*, 27832–27853. [CrossRef] [PubMed]

15. Sharma, L.K.; Bu, H.; Franzen, D.W. Comparison of two ground-based active-optical sensors for in-season estimation of corn (*Zea mays* L.) yield. *J. Plant Nutr.* **2016**, *39*, 957–966. [CrossRef]

16. Gobron, N.; Pinty, B.; Verstraete, M.; Govaerts, Y. The MERIS Global Vegetation Index (MGVI): Description and preliminary application. *Int. J. Remote Sens.* **1999**, *20*, 1917–1927. [CrossRef]

17. Liang, S.L.; Zhao, X.; Liu, S.H.; Yuan, W.P.; Cheng, X.; Xiao, Z.Q.; Zhang, X.T.; Liu, Q.; Cheng, J.; Tang, H.R.; et al. A long-term Global LAnd Surface Satellite (GLASS) data-set for environmental studies. *Int. J. Digit. Earth* **2013**, *6*, 5–33. [CrossRef]

18. Plummer, S.; Arino, O.; Simon, M.; Steffen, W. Establishing a earth observation product service for the terrestrial carbon community: The globcarbon initiative. *Mitig. Adapt. Strateg. Glob. Chang.* **2006**, *11*, 97–111. [CrossRef]

19. Jia, K.; Liang, S.; Gu, X.; Baret, F.; Wei, X.; Wang, X.; Yao, Y.; Yang, L.; Li, Y. Fractional vegetation cover estimation algorithm for Chinese GF-1 wide field view data. *Remote Sens. Environ.* **2016**, *177*, 184–191. [CrossRef]

20. Chen, J.M.; Leblanc, S.G. A four-scale bidirectional reflectance model based on canopy architecture. *IEEE Trans. Geosci. Remote Sens.* **1997**, *35*, 1316–1337. [CrossRef]

21. Verhoef, W. Light-scattering by leaf layers with application to canopy reflectance modeling: The SAIL model. *Remote Sens. Environ.* **1984**, *16*, 125–141. [CrossRef]

22. Jacquemoud, S.; Verhoef, W.; Baret, F.; Bacour, C.; Zarco-Tejada, P.J.; Asner, G.P.; Francois, C.; Ustin, S.L. PROSPECT + SAIL models: A review of use for vegetation characterization. *Remote Sens. Environ.* **2009**, *113*, S56–S66. [CrossRef]

23. Sharma, L.K.; Franzen, D.W. Use of corn height to improve the relationship between active optical sensor readings and yield estimates. *Precis. Agric.* **2014**, *15*, 331–345. [CrossRef]

24. Liang, S. Recent developments in estimating land surface biogeophysical variables from optical remote sensing. *Prog. Phys. Geogr.* **2007**, *31*, 501–516. [CrossRef]

25. Liang, S.; Li, X.; Wang, J. *Advanced Remote Sensing: Terrestrial Information Extraction and Applications*; Academic Press: Oxford, UK, 2012; p. 800.

26. Baret, F.; Hagolle, O.; Geiger, B.; Bicheron, P.; Miras, B.; Huc, M.; Berthelot, B.; Nino, F.; Weiss, M.; Samain, O.; et al. LAI, fAPAR and fCover CYCLOPES global products derived from VEGETATION—Part 1: Principles of the algorithm. *Remote Sens. Environ.* **2007**, *110*, 275–286. [CrossRef]

27. Sharma, L.; Bali, S.; Dwyer, J.; Plant, A.; Bhowmik, A. A Case Study of Improving Yield Prediction and Sulfur Deficiency Detection Using Optical Sensors and Relationship of Historical Potato Yield with Weather Data in Maine. *Sensors* **2017**, *17*, 1095. [CrossRef] [PubMed]

28. Wang, F.-M.; Huang, J.-F.; Tang, Y.-L.; Wang, X.-Z. New Vegetation Index and Its Application in Estimating Leaf Area Index of Rice. *Rice Sci.* **2007**, *14*, 195–203. [CrossRef]

29. Chen, J.M. Evaluation of Vegetation Indices and a Modified Simple Ratio for Boreal Applications. *Can. J. Remote Sens.* **1996**, *22*, 229–242. [CrossRef]

30. Kimes, D.S.; Knyazikhin, Y.; Privette, J.L.; Abuelgasim, A.A.; Gao, F. Inversion methods for physically-based models. *Remote Sens. Rev.* **2000**, *18*, 381–439. [CrossRef]

31. Gascon, F.; Gastellu-Etchegorry, J.P.; Lefevre-Fonollosa, M.J.; Dufrene, E. Retrieval of forest biophysical variables by inverting a 3-D radiative transfer model and using high and very high resolution imagery. *Int. J. Remote Sens.* **2004**, *25*, 5601–5616. [CrossRef]

32. Bicheron, P.; Leroy, M. A Method of Biophysical Parameter Retrieval at Global Scale by Inversion of a Vegetation Reflectance Model. *Remote Sens. Environ.* **1999**, *67*, 251–266. [CrossRef]

33. Weiss, M.; Baret, F.; Myneni, R.B.; Pragnère, A.; Knyazikhin, Y. Investigation of a model inversion technique to estimate canopy biophysical variables from spectral and directional reflectance data. *Agronomie* **2000**, *20*, 3–22. [CrossRef]

34. Roujean, J.L.; Lacaze, R. Global mapping of vegetation parameters from POLDER multiangular measurements for studies of surface-atmosphere interactions: A pragmatic method and its validation. *J. Geophys. Res. Atmos.* **2002**, *107*. [CrossRef]

35. Yang, L.; Jia, K.; Liang, S.; Liu, J.; Wang, X. Comparison of Four Machine Learning Methods for Generating the GLASS Fractional Vegetation Cover Product from MODIS Data. *Remote Sens.* **2016**, *8*, 682. [CrossRef]

36. García-Haro, F.J.; Camacho, F.; Meliá, J. Inter-comparison of SEVIRI/MSG and MERIS/ENVISAT biophysical products over Europe and Africa. In Proceedings of the 2nd MERIS/(A)ATSR User Workshop, ESA SP-666, Frascati, Italy, 22–26 September 2008.

37. Cooley, T.; Anderson, G.P.; Felde, G.W.; Hoke, M.L.; Ratkowski, A.J.; Chetwynd, J.H.; Gardner, J.A.; Adler-Golden, S.M.; Matthew, M.W.; Berk, A.; et al. FLAASH, a MODTRAN4-based atmospheric correction algorithm, its application and validation. In Proceedings of the 2002 IEEE International Geoscience and Remote Sensing Symposium & 24th Canadian Symposium on Remote Sensing, Toronto, Canada, 24–28 June 2002; pp. 1414–1418.

38. Jia, K.; Wei, X.Q.; Gu, X.F.; Yao, Y.J.; Xie, X.H.; Li, B. Land cover classification using Landsat 8 Operational Land Imager data in Beijing, China. *Geocarto Int.* **2014**, *29*, 941–951. [CrossRef]

39. Kuusk, A. The hot spot effect in plant canopy reflectance. In *Photon-Vegetation Interactions*; Myneni, R.B., Ross, J., Eds.; Springer: Berlin/Heidelberg, Germany, 1991; pp. 139–159.

40. Verhoef, W.; Jia, L.; Xiao, Q.; Su, Z. Unified optical-thermal four-stream radiative transfer theory for homogeneous vegetation canopies. *IEEE Trans. Geosci. Remote Sens.* **2007**, *45*, 1808–1822. [CrossRef]

41. Jacquemoud, S.; Baret, F. PROSPECT—A model of leaf optical-properties spectra. *Remote Sens. Environ.* **1990**, *34*, 75–91. [CrossRef]

42. Feret, J.-B.; François, C.; Asner, G.P.; Gitelson, A.A.; Martin, R.E.; Bidel, L.P.R.; Ustin, S.L.; le Maire, G.; Jacquemoud, S. PROSPECT-4 and 5: Advances in the leaf optical properties model separating photosynthetic pigments. *Remote Sens. Environ.* **2008**, *112*, 3030–3043. [CrossRef]

43. Shepherd, K.D.; Palm, C.A.; Gachengo, C.N.; Vanlauwe, B. Rapid characterization of organic resource quality for soil and livestock management in tropical agroecosystems using near-infrared spectroscopy. *Agron. J.* **2003**, *95*, 1314–1322. [CrossRef]

44. Dennison, P.E.; Halligan, K.Q.; Roberts, D.A. A comparison of error metrics and constraints for multiple endmember spectral mixture analysis and spectral angle mapper. *Remote Sens. Environ.* **2004**, *93*, 359–367. [CrossRef]

45. Jia, K.; Li, Q.Z.; Tian, Y.C.; Wu, B.F.; Zhang, F.F.; Meng, J.H. Accuracy Improvement of Spectral Classification of Crop Using Microwave Backscatter Data. *Spectrosc. Spectr. Anal.* **2011**, *31*, 483–487.

46. Goel, N.S.; Strebel, D.E. Inversion of vegetation canopy reflectance models for estimating agronomic variables. I. Problem definition and initial results using the suits model. *Remote Sens. Environ.* **1983**, *13*, 487–507. [CrossRef]

47. Qu, Y.H.; Zhang, Y.Z.; Wang, J.D. A dynamic Bayesian network data fusion algorithm for estimating leaf area index using time-series data from in situ measurement to remote sensing observations. *Int. J. Remote Sens.* **2012**, *33*, 1106–1125. [CrossRef]

48. Hosgood, B.; Jacquemoud, S.; Andreoli, G.; Verdebout, J.; Pedrini, G.; Schmuck, G. *Leaf Optical Properties EXperiment 90 (LOPEX93)*; Report EUR 16095 EN; European Commission, Joint Research Center, Institute for Remote Sensing Applications: Brussels, Italy, 1990.

49. Qu, Y.; Wang, J.; Wan, H.; Li, X.; Zhou, G. A Bayesian network algorithm for retrieving the characterization of land surface vegetation. *Remote Sens. Environ.* **2008**, *112*, 613–622. [CrossRef]

50. Murthy, C.S.; Raju, P.V.; Badrinath, K.V.S. Classification of wheat crop with multi-temporal images: Performance of maximum likelihood and artificial neural networks. *Int. J. Remote Sens.* **2003**, *24*, 4871–4890. [CrossRef]

51. Ahmad, S.; Kalra, A.; Stephen, H. Estimating soil moisture using remote sensing data: A machine learning approach. *Adv. Water Resour.* **2010**, *33*, 69–80. [CrossRef]

52. Verger, A.; Baret, F.; Camacho, F. Optimal modalities for radiative transfer-neural network estimation of canopy biophysical characteristics: Evaluation over an agricultural area with CHRIS/PROBA observations. *Remote Sens. Environ.* **2011**, *115*, 415–426. [CrossRef]

53. Jia, K.; Liang, S.; Wei, X.Q.; Li, Q.; Du, X.; Jiang, B.; Yao, Y.J.; Zhao, X.; Li, Y. Fractional forest cover changes in Northeast China from 1982 to 2011 and its relationship with climatic variations. *IEEE J. Sel. Top. Appl. Earth Obs. Remote Sens.* **2015**, *8*, 775–783. [CrossRef]

54. Ngia, L.S.H.; Sjoberg, J. Efficient training of neural nets for nonlinear adaptive filtering using a recursive Levenberg-Marquardt algorithm. *IEEE Trans. Signal Process.* **2000**, *48*, 1915–1927. [CrossRef]

55. Curran, P.J.; Dungan, J.L.; Gholz, H.L. Seasonal LAI in slash pine estimated with landsat TM. *Remote Sens. Environ.* **1992**, *39*, 3–13. [CrossRef]

56. Duan, H.; Yan, C.; Tsunekawa, A.; Song, X.; Li, S.; Xie, J. Assessing vegetation dynamics in the Three-North Shelter Forest region of China using AVHRR NDVI data. *Environ. Earth Sci.* **2011**, *64*, 1011–1020. [CrossRef]

57. Masson, V.; Champeaux, J.L.; Chauvin, F.; Meriguet, C.; Lacaze, R. A global database of land surface parameters at 1-km resolution in meteorological and climate models. *J. Clim.* **2003**, *16*, 1261–1282. [CrossRef]

58. Nash, J.E.; Sutcliffe, J.V. River flow forecasting through conceptual models part I—A discussion of principles. *J. Hydrol.* **1970**, *10*, 282–290. [CrossRef]

59. Loague, K.; Green, R.E. Statistical and graphical methods for evaluating solute transport models: Overview and application. *J. Contam. Hydrol.* **1991**, *7*, 51–73. [CrossRef]

![sensors logo] *sensors*

MDPI

Article

Phenoliner: A New Field Phenotyping Platform for Grapevine Research

Anna Kicherer [1,*], Katja Herzog [1], Nele Bendel [1], Hans-Christian Klück [2], Andreas Backhaus [2], Markus Wieland [3], Johann Christian Rose [3], Lasse Klingbeil [3], Thomas Läbe [4], Christian Hohl [5], Willi Petry [5], Heiner Kuhlmann [3], Udo Seiffert [2] and Reinhard Töpfer [1]

[1] Julius Kühn-Institut, Federal Research Centre of Cultivated Plants, Institute for Grapevine Breeding Geilweilerhof, 76833 Siebeldingen, Germany; katja.herzog@julius-kuehn.de. (K.H.); nele.bendel@julius-kuehn.de (N.B.); reinhard.toepfer@julius-kuehn.de (R.T.)
[2] Fraunhofer Institute for Factory Operation and Automation (IFF), Biosystems Engineering, Sandtorstr. 22, 39108 Magdeburg, Germany; hans-christian.klueck@iff.fraunhofer.de (H.-C.K.); andreas.backhaus@iff.fraunhofer.de (A.B.); udo.seiffert@iff.fraunhofer.de (U.S.)
[3] Institute of Geodesy and Geoinformation, Department of Geodesy, University of Bonn, Nussallee 17, 53115 Bonn, Germany; wieland@igg.uni-bonn.de (M.W.); rose@igg.uni-bonn.de (J.C.R.); klingbeil@igg.uni-bonn.de (L.K.); heiner.kuhlmann@uni-bonn.de (H.K.)
[4] Institute of Geodesy and Geoinformation, Department of Photogrammetry, University of Bonn, Nussallee 15, 53115 Bonn, Germany; laebe@ipb.uni-bonn.de
[5] ERO-Gerätebau GmbH, Simmerner Str. 20,55469 Niederkumbd, Germany; chohl@ERO-Weinbau.de (C.H.); wpetry@ERO-Weinbau.de (W.P.)
* Correspondence: anna.kicherer@julius-kuehn.de; Tel.: +49-063-454-1124

Received: 3 May 2017; Accepted: 11 July 2017; Published: 14 July 2017

Abstract: In grapevine research the acquisition of phenotypic data is largely restricted to the field due to its perennial nature and size. The methodologies used to assess morphological traits and phenology are mainly limited to visual scoring. Some measurements for biotic and abiotic stress, as well as for quality assessments, are done by invasive measures. The new evolving sensor technologies provide the opportunity to perform non-destructive evaluations of phenotypic traits using different field phenotyping platforms. One of the biggest technical challenges for field phenotyping of grapevines are the varying light conditions and the background. In the present study the Phenoliner is presented, which represents a novel type of a robust field phenotyping platform. The vehicle is based on a grape harvester following the concept of a moveable tunnel. The tunnel it is equipped with different sensor systems (RGB and NIR camera system, hyperspectral camera, RTK-GPS, orientation sensor) and an artificial broadband light source. It is independent from external light conditions and in combination with artificial background, the Phenoliner enables standardised acquisition of high-quality, geo-referenced sensor data.

Keywords: big data; geo-information; plant phenotyping; grapevine breeding; *Vitis vinifera*

1. Introduction

With new developments in electronics, software and sensor techniques, plant phenotyping has become a key technology in the agriculture sector. Platforms for the assessment of phenotypic data under controlled conditions are widespread [1–4]. These systems allow a very detailed assessment of plants under a controlled environment, genotype-environment interaction not taken into consideration. These systems are not however applicable for perennial crops, e.g., cultivated in trellis systems. Grapevine (*Vitis vinifera*), for example, is a large perennial liana that needs to be screened directly in the field for traits like plant architecture, yield, grape quality, abiotic and biotic stress. The application of non-invasive, sensor-to-plant methods facilitates the record of objective and repeatable phenotypic data of these traits. The lowest level of phenotyping applications are hand held devices applied by an operator. The Multiplex (Force A, Paris, France), a non-imaging fluorescence sensor was used to assess health status [5,6] or maturity [7,8] in grapevine. Some image-based hand-held prototypes have been used to evaluate yield parameters on grapes still attached to the vine using an artificial background and an RGB camera [9–11]. A completely different concept, being as distant as possible from the observed object, is the remote sensing approach using satellites [12], aircrafts [13], or unmanned aerial vehicles (UAVs) [14]. For most of these applications the focus is on plant health and water status, mostly expressed through vegetation indices, using mainly spectral sensors to validate the vigour of vineyards. In recent years, proximal sensing using moving vehicles as sensor carrier has been the biggest field of innovation in grapevine research. The simplest way is to attach sensors to tractors or smaller moving vehicles to monitor the crop for precision viticulture. This is for example done with commercial sensors like the GreenSeeker® (NTECH Industries Inc., Ukiah, CA, USA) [15,16] or the Cropcircle (Netherlands Scientific Inc., Lincoln, NE, USA) [17], which are based on multispectral sensors, providing vegetation indices correlating with leaf area index (LAI) and canopy density. The GapeSense (Lincoln Ventures Ltd, Hamilton, New Zealand) uses digital images of the canopy side to assess height and texture [18]. Another opportunity is the use of LiDAR sensors for 3D reconstruction to validate the canopy size [19] and LAI [20]. Other more prototype-like approaches are cameras and light units mounted on movable vehicles like the approach by Nuske, et al. [21] for automatic yield estimations. The most automated approaches are the so-called "phenomobiles" [22] or "agbots" [23]. They use automation and robotic technologies to control the movement speed and direction within the vineyard, based on GPS or proximity sensors. They are either equipped with several sensors to monitor the plant status along the row and to gain phenotypic information or they can be used for agronomic operations. Approaches like the VineRobot [24], VINBOT [25] and the Wall-ye [26] are robots designed to in the end monitor various parameters non-destructively such as yield, vigour, water stress and grape quality using sensors ranging from RGB, multispectral, fluorescence to thermal and infrared. Most of them are still in the prototype stage, looking already very promising. The PHENObot [27] was recently introduced for the specific application in grapevine breeding to screen single vines in the grapevine repository for different phenotypic traits. Prototypes build for an agronomic concept are the VineGuard [28] which was designed for foliar applications and aims at implementing a robotic arm for harvesting grapes. Besides monitoring the vineyard Wall-ye [26] is designed to carry out precise pruning whereas the Vision Robotics Cooperation has developed an optical pruning system based on 3D reconstruction in a tractor pulled "tunnel" [29]. The Vitirover [30] is a small robot developed to automatically cut the grass within the row up to a 2–3 cm distance to the base of the vine. The usage of RTK GPS systems for geo-referencing is crucial for all field phenotyping approaches. Nevertheless, precision viticulture is focused on a whole plots level, whereas grapevine research requires a single vines resolution.

All of these phenotyping platforms have some challenges in common. The variation of the trait, for example berry colour, size and shape could differ, especially with regard to the evaluation of genetic resources in a repository. Therefore, the algorithms used for image analysis have to be very flexible [31]. Depending on the training system and the viticulture practice the occlusion through leaves or other grapes is challenging and needs to be considered when validating the sensor results. Screening vine rows in the side view is very challenging at the early phase of vegetation, as plants in images show the same colour distribution. Using a 3D reconstruction approach can help to overcome this point [32,33]. By far the biggest challenge in grapevine field phenotyping is the variable light condition. Some efforts have been made using artificial light units [21,27,28] to overcome the sun and doing the data acquisition at dawn or during the night [21,27]. Bourgeon, et al. [16] used an umbrella to avoid the Sun's influence, whereas the Vision Robotics Cooperation [29] worked with a tractor pulled "chamber" to standardise the light conditions.

In the present study we developed a new, very robust field phenotyping platform, the Phenoliner, for applications in grapevine research and breeding. The biggest advantage of the Phenoliner compared to already used platforms is a standardized data acquisition, overcoming changing lighting conditions and changing background, by using a grape harvester as sensor carrier. We implemented two sensor systems, using RGB, NIR and hyperspectral imaging to classify different kinds of biological parameters optically and contact-free on the spot. Synchronic data acquisition and automated geo-referencing of high-resolute image data enable evaluation of several plant traits of whole breeding populations. Beside the Phenoliner setup, the accuracy of data acquisition and precision of geo-referencing was proved; the quality of sensor data was investigated on the example of bunch/berry detection and health status.

2. Materials and Methods

For improved, sensor-based field phenotyping in vineyards the Phenoliner was developed. It consists of an emptied grape harvester as base, a differential GPS system and two sensor systems with their respective artificial light sources. The technical setup of the Phenoliner will be explained in the following paragraph.

2.1. Plant Material

Field tests were conducted in October 2016 in an experimental vineyard plot at the JKI Geilweilerhof located in Siebeldingen, Germany (49°21.747′ N, 8°04.678′ E). Rows were planted in north-south direction and consisted of 20 (*Vitis vinifera* cv. 'Acolon'; experiment sensor A) and 24 (*Vitis vinifera* cv. 'Riesling'; experiment sensor B) individuals. Inter-row distance was 2 m, and grapevine spacing was 1 m. At the southern end, the last four vines were not treated with chemical plant protection.

2.2. Vehicle

The ERO-Grapeliner SF200 (ERO Gerätebau, Niderkumbd, Germany) was used as sensor carrier (Figure 1). All parts originally intended for the harvest, i.e., the shaking unit, destemmer, grape tank, and all parts used for the grape transport within the machine have been removed, including the hydraulic system needed for these parts. The emerging space was used for the sensor setup in the right part of the tunnel (minimal height: 2.05 m; maximum height: 2.80 m; width: 0.86 m). The space generated was used to integrate sensor systems and their peripheral components:

- In order to standardise the light conditions within the tunnel the base frame on the right hand side of the vehicle was extended and covered with metal plates. All slots in between were sealed and a curtain was installed in the back of the tunnel to avoid direct sun light interference
- Due to safety reasons on top of the machine the railing was enlarged where parts of the harvesting machine had been removed.
- The energy necessary for sensors, light units, and computer is provided by a generator driven by the vehicle. Two operating modes are possible: (1) diesel engine on and (2) diesel engine off. Due to the removal of the original harvesting hydraulics the free energy of the diesel engine can be used for powering a generator when the engine is on. Furthermore, it is possible to connect the vehicle to a regular power socket (230 V) when the engine is off. Two backup batteries (minimum 0.5 kWh) are bridging the time between turning off the engine and connecting the vehicle to the socket. This solution permits the transfer of the acquired data from the computers on the vehicle to the memory location without having the engine run for hours. There are 20 sockets available on the vehicle (cab: 2; front part: 9; back part: 9), provided with a suitable fuse through a distribution box.

2.3. RTK-GPS

All vines screened with the Phenoliner are surveyed using an real-time-kinematic (RTK)-GPS system (SPS852, Trimble®, Sunnyvale, CA, USA) with 2 cm accuracy. The geo-reference of each grapevine and the associated plant ID (unique for each plant) is stored in a database, Plant Location

Administration (PLA) [27]. The GPS antenna is placed on top on the vehicle (see Figure 1c) and the receiver provides NMEA strings for both sensor systems: Sensor A: GGA 20 Hz; baud rate 115,200; Sensor B: GGA 20 Hz; GST 20 Hz; RMC 1 Hz; baud rate 38,400). The consideration of the lever arm between GPS antenna and the sensors is described below.

Figure 1. Overview of the Phenoliner. (**a**) Phenoliner construction plan. (**b**) Scheme of the sensor layout in the tunnel (marked red above); right: Sensor A: consisting of RGB cameras 1–3 and 5, and a NIR camera 4, left: Sensor B consisting of two hyperspectral cameras; (**c**) Phenoliner in the vine row.

2.4. SensorA: Multicamerasystem (RGB, NIR)

Sensor A on the Phenoliner is a multi-camera system (MCS) consisting of four RGB cameras (DALSA Genie NanoC2590, Teledyne DALSA Inc., Waterloo, ON, Canada) and one near-infrared (NIR) camera (DALSA Genie Nano-M2590-NIR, Teledyne DALSA Inc., Waterloo, ON, Canada) arranged as shown in Figure 1b. Three of the RGB cameras (1–3, Figure 1a) are stacked vertically. Horizontally next to the lowest RGB camera 3, the NIR camera (4, Figure 1a) and afterwards the last RGB camera (5, Figure 1a) are positioned, with their protective cases touching each other. The cameras are equipped with 5.1 Megapixel sensors and 12 mm lenses. Given a distance of about 75 cm to the canopy, each camera covers an area of about 60 cm × 70 cm of the vine row, with a resolution of about 0.3 mm and a theoretical framerate of up to 51 frames per second. The illumination is realized using six 300 W halogen lamps (Hedler C12, Hedler Systemlicht, Runkel/Lahn, Germany), arranged around the camera system and pointing towards the canopy. In order to avoid hard shadows each lamp is equipped with a diffusor plate. All cameras are connected to a computer (Intel Core i7-860 with 2,8 GHz, 4 GB-DDR RAM, 2 × 480 GB SSD storage) via a GigE Interface. In order to enable the potentially high data rates, each camera is connected to a separate ethernet port and the images are stored on fast solid state disc (SSD) drives. For camera set up, camera control and synchrone image acquisition the IGG Geotagger 2.0 was developed in LabVIEW (National Instruments® GmbH, Munich, Germany). It is a further development based on privious versions [27,34], and also provides precise georeference information for every single image using the GPS receiver and 2-axis inclinometer (DOG2 MEMS-Series USB Rev.1; TE Connectivity Sensors Germany GmbH, Dortmund).

2.5. Sensor B: Hyperspectral Camera System

Parallel to Sensor A that provides the three channels red, green and blue, the Phenoliner is equipped with Sensor B providing a total of 416 spectral bands covering a spectrum from 400 nm to 2500 nm. Sensor B consists of two separate commercially available line scanning hyperspectral cameras (Norsk Elektro Optikk AS, Skedsmokorset, Norway) covering the visual-near infrared range (HySpex VNIR 1600) from 400 to 1000 nm providing 160 channels across a continuous visible part of light and the short-wave infrared range from 1000–2500 nm (HySpex SWIR 320m-e, Norsk Elektro Optikk AS, Skedsmokorset, Norway) equally distributed over 256 channels. With an achievable frame rate of 160 Hz (VIS) and 100 Hz (SWIR) the combination of both cameras and the high spectral sampling rate of 3.2 nm and 5.45 nm, respectively, allows for a continuous acquisition of 16 Bit digitized high resolution reflectance data. The available space within the tunnel is limited, resulting in a maximum distance of 1m between lens and the vine canopy. Therefore both cameras are equipped with lenses of 1 m fixed focal length. To match the focal length, the hyperspectral line cameras are setup alongside the driving direction with a rectangular mirror diverting the reflected light to a 90° angle (Figure 1b). The optical industry provides silver- and gold-coated VNIR- and SWIR-specific mirrors adjusted to a highly constant reflectance of >95% across their entire respective spectrum but also low-cost mirrors have proven suitable for the task. Additionally, an artificial illumination of two 300 W short-wave spotlights (Hedler C12, Hedler Systemlicht) with a broad power spectral density were installed. In order to measure reflectance a 1 × 1 m spectralon with certified reflectance values (Sphere Optics, Hersching, Germany) was set up in the background (left tunnel side). It is covered with a custom foil specifically designed for the purpose of reducing absorption and retaining spectral features as much as possible while at the same time providing suitable mechanical protection for the pad. A previously conducted spectral measurement of the foil-covered pad gave satisfactory results.

3. Application of the Phenoliner: Pilot Study

3.1. Sensor A

3.1.1. IGG Geotagger 2.0: Geo-Referencing of Images

The software enables accurate geotagging of all acquired images and the selection of images of single vines, when their coordinates are provided by the database (PLA). This is important with regard to automated data management. The coordinates of the vine stem from the database and the known offset direction between the stem and the area of interest (e.g., middle of the cane, bunch zone). This selection procedure enables significant reduction of storage space in the cases where not all images are needed. It also allows a direct association between the images and the database records of the grapevines.

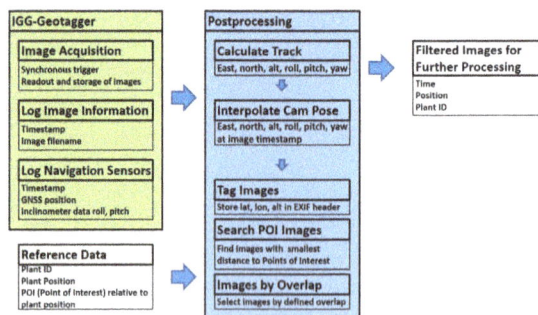

Figure 2. Tasks within IGG (Institute of Geodesy and Geoinformation) Geotagger 2.0. GNSS: Global Navigation Satellite System; Plant ID: Plant identification; EXIF: Exchangeable image file format.

There are several processes running on the image acquisition system and within the geotagging software (see Figure 2): During the motion of the vehicle through the vine row, the camera system is acquiring time synchronized images from every camera with a preconfigured frame rate and storing them to the SSD. If all cameras are used, the frame rate is limited to about 5 Hz, mainly due to the SSD writing speed. At the same time GPS positions (20 Hz) and inclinometer readings (roll and pitch angles, 20 Hz) are stored.

In a post processing step, a full 6D (position and orientation) trajectory of the system is calculated from these data. The missing third rotation angle (heading) is estimated based on the sequence of positions and the assumption, that the Phenoliners motion direction is restricted to its long axis. This trajectory is then interpolated to the time steps of image acquisition. Knowing the lever arm between the GPS antenna and the cameras and its orientation now allows the calculation of a coordinate for each image, which is then written into the metadata of the image file. It should be noted here, that the determination of the roll and pitch angles of the system are a crucial step, because the angles can be controlled by the driver while driving through the rows. This means, that the angle between the camera system and the ground cannot be assumed to be small and constant as it may be possible for other ground vehicles.

In a further preprocessing step an image filter can be applied to reduce the number of images based on the purpose of the application. For the 3D reconstruction of the full vine row (see below) a selection based on minimum overlap between neighbouring images can be applied. To select images of a certain point of interest (POI), such as the bunch zone of single vines, a distance between the coordinates of the camera image center and the POI is calculated. For every POI the image with the minimal distance is selected.

3.1.2. Validation

As mentioned above, the lever arm between the GPS antenna and the camera system has to be known precisely, in order to calculate the image coordinates from the position and orientation measurements. This lever arm has been measured with an accuracy of millimetres using a 3D terrestrial laser scanner (Leica P20; Leica Microsystems GmbH, Wetzlar, Germany). Figure 3 shows the scan of the Phenoliner used for the lever arm determination.

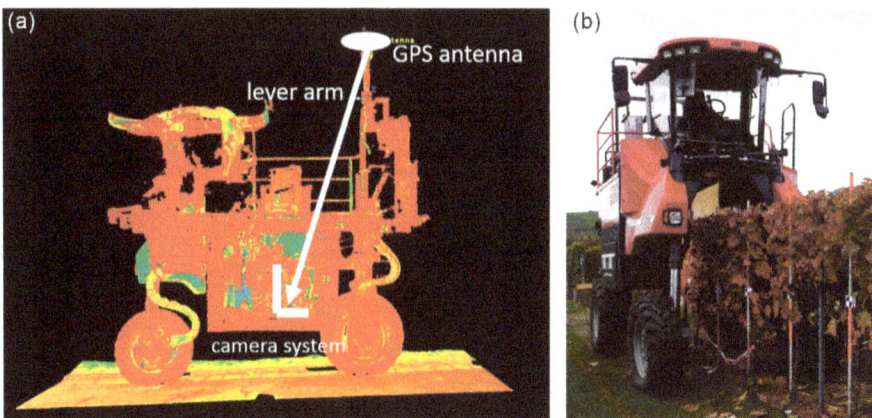

Figure 3. (**a**) Terrestrial laser scan of the Phenoliner to measure the lever arm between the GPS antenna and the camera system. (**b**) Evaluation measurement to test the georeferencing accuracy of the system. The coordinates of the black and white targets at the poles are known and can be compared with the target positions seen in the images.

The experiment is driven by the need to automatically take images of a certain POIs, such as the bunch zones of single vines, having this POI in the middle of the field of view of the camera. Within one vine row (*Vitis vinifera* cv. 'Acolon', 20 vines), black and white targets were attached to poles at the position of the vine stem to mark a POI (see Figure 4a). The exact position of these poles was surveyed using an RTK GPS receiver and their coordinates were given to the software as "Reference Data" (Figure 2). Then the vehicle was driving through the row, taking five images per second. The geo-reference of each image was determined by calculating the trajectory and using the determined lever arm and the tagger software selected one image for every POI as described above. Figure 4a shows one of these images. Here the distance of the target (POI) to the vertical central axis of the image is considered as the "deviation" of the measurement. Please note, that we only evaluate the accuracy of one dimension, which is the one in the driving direction. The other two dimension are not relevant in this particular application, since the vertical field of view of the camera system is big enough to cover the whole canopy and the distance to the canopy is more or less constant due to the given row geometry. There is also no reason to assume, that the other two dimensions are less accurate than the evaluated one, as the accuracy in driving direction is the most critical due to time synchronization effects and the limited framerate (see below). To ensure the functionality of the system calibration and the image georeferencing procedure, an evaluation measurement was conducted.

Figure 4. Experiment set up of POIs at the position of vine stems. (**a**) Image, selected as the one closest to an POI, with the deviation e as the error of the selection process. (**b**) Distribution of deviations from 10 runs at different speeds.

This experiment has been performed five times at two different vehicle speeds (0.2 m/s and 0.4 m/s) each of them driving in both directions of the row. The distribution of the deviations for the 20 POIs in all runs is shown in Figure 4b. The overall accuracy of this POI based selection process is shown to be in the order of a few centimetres. However, note that this accuracy contains the accuracy of the image tagging procedure (trajectory calculation, calibration), the accuracy of the target position determination (RTK GPS) and the minimum distance between two consecutive images (vehicle speed, frame rate). The two different speeds combined with the frame rate of 5 Hz lead to an image distance of 4 cm and 8 cm limiting the image selection resolution to 2 cm and 4 cm, respectively. Given that the maximum deviation in these experiments is about 6 cm, we can assume the accuracy of the image tagging process itself is in the order of 2–3 cm, which corresponds to the expected accuracy of the GPS receiver.

3.1.3. Application Example: 3D Reconstruction of the Full Vine Row

As shown in Figure 1b RGB cameras 1–3 are arranged vertically. These are used for a full 3D reconstruction of the vine row using multi-view stereo approaches [35]. These methods enable the reconstruction of 3D point clouds based on multiple overlapping images. This overlap is about 70–80% in both, the horizontal and vertical directions of the images. The vertical overlap is realized by the arrangement of the cameras. The horizontal overlap is realized by the motion of the Phenoliner. Here the combination of driving speed and image frame rate has to ensure a maximum distance of about 15 cm in order to achieve the required overlap. A vehicle speed of 0.2–0.3 m/s and an image frame rate of 5 Hz has shown to be a suitable parameter setup in practice.

Abraham, et al. [36] captured images with the PHENObot [27] and used them for three-dimensional (3D) reconstruction of vine rows. In this study images were accordingly acquired using the Phenoliner to test transferability of developed workflow. Figure 5 shows the principle of the image acquisition for point cloud reconstruction using the vertical MCS (a) and black and green grapes (b) from the front (upper images) and in profile (lower images), that have been reconstructed from the images using the Software Pix4D. The black grape varieties have been gathered deploying a heavy grey tarpaulin as background while the green grape varieties have been gathered deploying a white heavy blanket as background. The choice of the background color has a significant impact on the quality of the 3D-reconstruction and is discussed in Section 3.1.3. The geo-reference of the images has been incorporated in the processing procedure, so each point in the point cloud has also a metric coordinate, enabling measurements with correct scale within the data.

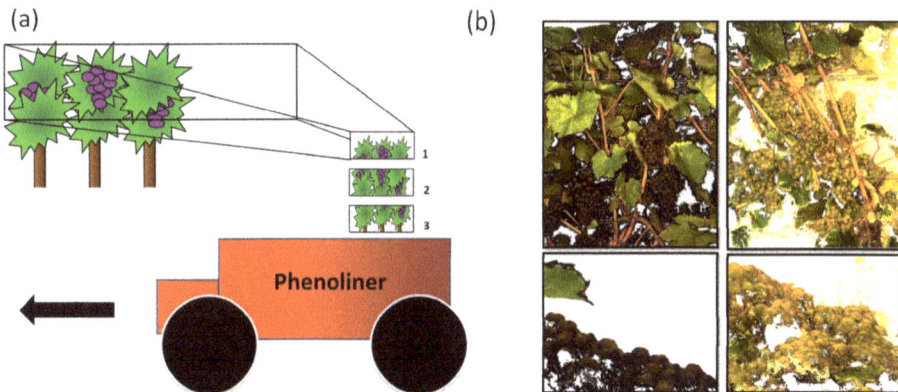

Figure 5. (**a**) Basic principle of the system setup for 3D reconstruction of the full vine row. The Phenoliners tunnel is driven over the vine rows. The MCS is oriented parallel to the rows, capturing images automatically while in motion. (**b**) Reconstructed point cloud of black (left side) and green grape (right side) varieties. The upper images show the rows from a frontal view and the lower images show grapes from a profile view. Single berry elevations of their spherical geometry are clearly distinguishable.

As described in Abraham, et al. [36] these point clouds now undergo further processing steps, where the data are segmented and classified in order to provide information about the number of grapes and berries. Individual berry elevations are clearly visible, highlighting the high level of geometric detail of the point cloud. This level of geometric detail is necessary to count single berries via geometric modelling.

3.1.4. Application Example: Depth Map Creation and Segmentation of Single Vines

Another application is the acquisition of a single image or stereo image pairs of single vines. While the application for the point cloud reconstruction needed images of the whole row with a sufficient overlap in horizontal and vertical direction, this application needs only selected images of a certain area of interest, for example the bunch zone. Calibrated cameras 3–5 are arranged based on the PHENObot experiences [27]. Camera calibration was performed by using a test field calibration according to [35] in order to determine the camera constant, principle point and camera lens distortions. No approximate values are necessary for this process. Afterwards, the internal parameters of the cameras are known enabling post processed image rectification. The rectified images then strictly follow the pinhole camera model with principle point in the centre, thus enabling increased precision for the subsequent steps.

The RGB image tools to detect berry size and colour as shown by [27,37] are currently adapted to Phenoliner images. First experiments regarding the computation of depth maps and segmentation of images, using stereo images acquired with the Phenoliner are showing promising result (see Figure 6). For these first tests the depth map has been calculated with the free software "pmvs2" [38] and is used to seperate foreground and background. Manually set thresholds for the colour channels of the RGB image are used to discriminate the other classes ("cane", "canopy", "grapes"). The artifical lightning may help to use a classifier with constant parameters over several data sets, because the brightness, constrast and colour temperature of the images will not change. Adjusting the published tools to the new sensors, the stereo system (cameras 3 and 5) can be used to segment an RGB image into the classes "cane", "canopy" and "background" in order to calculate phenotypic parameters in the same manner as by Kicherer, et al. [32] for pruning mass or Klodt, et al. [33] for leaf area. For a detailed explanation of the stereo system layout of cameras 3 and 5, please refer to Kicherer, et al. [32]. The NIR-camera 4 is meant to be used for plant disease detection purposes and to improve the colour segmentation of different classes like canopy, cane, and grape bunches.

| (a) | (b) | (c) |

Figure 6. (**a**) Rectified image acquired with Sensor A (camera (5), only overlapping part of stereo pair shown). (**b**) Depth map calculated with stereo pair from camera (3) and camera (5). The brightness indicates the distance to the cameras (white for near points, dark gray for far points). Black pixels indicate positions with no depth which can be assumed to be background. (**c**) First result of a test for classification with manually set thresholds. The RGB image from (**a**) and the depth map from (**b**) are used as input. Classes are: blue for "grapes", green for "canopy", brown for "cane", black for "background".

3.2. Sensor B

3.2.1. Hyperspectral Image Acquisition

Image recording was achieved using proprietary image acquisition software implemented by Fraunhofer Institute for Factory Operation and Automation IFF, which integrates both hyperspectral cameras and the Phenoliners GPS receiver in order to record georeferenced hyperspectral images.

Data pre-processing and analysis was done offline using Matlab 2013a (The MathWorks, Natick, MA, USA). Spectral data per image was clustered using a Neural Gas algorithm [39]. Spectra are grouped due to their similarity measured by the Euclidean distance to a number of prototype spectra, which are optimized to achieve minimal quantization error. The cluster or group that is representing plant material is selected and the segmentation mask is further processed using morphological eroding operations. Finally, a classifier Artificial Neural Network is trained on examples of leave spectra and background spectra. This classifier is applied to all images and achieves an automated classification of plant materials in all images. Region of interests in the image representing single vines are marked using the geo information recorded along the camera system. In order to obtain the dataset for the subsequent machine learning, a list of vines with the status "sprayed" and "not sprayed" was provided. Among all vines representing one of the two classes, 10,000 spectral pixels were sampled. Datasets are treated separately for the VNIR (160 features) and SWIR (256 features) camera. No pixel averaging was performed. In Figure 7 examples of the principal pre-processing steps are depicted.

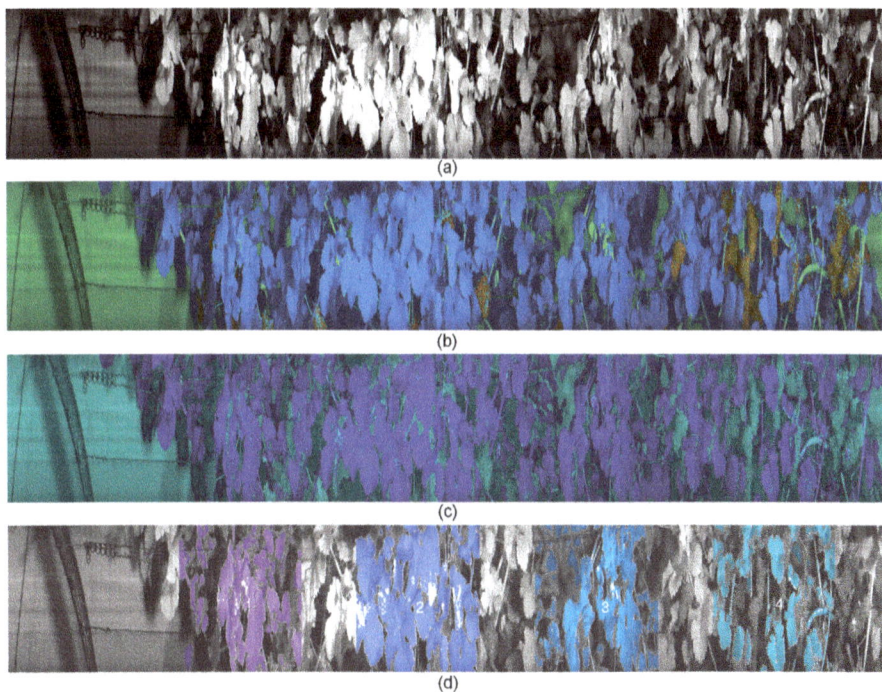

Figure 7. Pre-processing of hyperspectral image recording: (**a**) channel image at 1100nm, (**b**) clustering of spectral data, colour indicates pixel groups, (**c**) classification for foreground (leaves) vs. background, (**d**) image after removal of dark areas and marking of vine position from GPS.

The image segmentation can be performed in real-time on the vehicles computer system and will be the first processing component for the Phenoliners in-field detection capability of plant diseases based on leave spectral reflectance pattern. Since we image a geometrical complex scene, leaves can be overexposed or shadowed by other leaves. In order to decrease the variance in the spectral signal and to reduce noise, dark areas are segmented out (Figure 7d).

3.2.2. Validation

Field tests were conducted in October 2016 in one row of 24 individuals of *Vitis vinifera* cv. 'Riesling'. Since the cameras were located on the right side of the movable tunnel, the row was recorded from the right (in the direction of motion). The zone around the grapes was scanned from the west and east side, respectively. Scanning was performed at 0.1 m/s. Data was acquisitioned at different time points during the day: 10 a.m., 12 p.m., 2 p.m., 4 p.m., and 6 p.m. In Figure 8 the averaged normalized reflectance pattern for both spray classes, measured from east and west at 2 p.m. are depicted, (a) in the range of 400–1000 nm and (b) 1000–2500 nm.

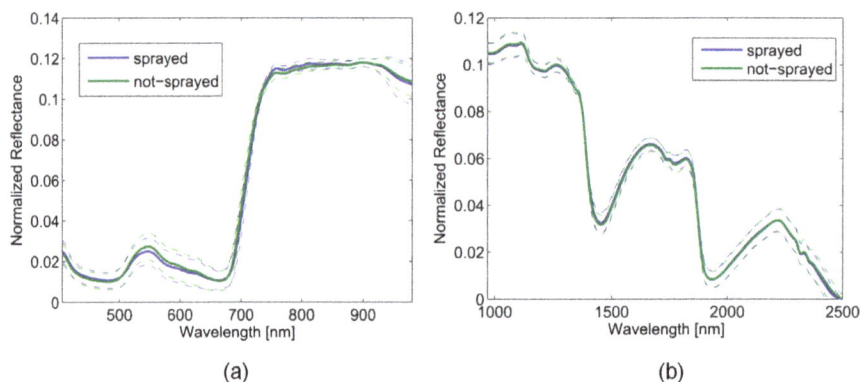

Figure 8. Normalized reflectance spectra for the VNIR 400–1000 nm (**a**) and SWIR 1000–2500 nm (**b**) range imaged at 2 p.m. and pooled over west and east side.

The final dataset of labelled spectral data per time point is then analyzed for discrimination of spray status (last four vines have not been sprayed) via a Linear Discriminant Analysis (LDA). An LDA finds a data projection that maximizes the class discrimination by maximizing between class distances and minimizing within class variance. Beyond a simple linear discrimination, a number of machine learning models were tested. On a dataset of labelled spectra, a Partially Least Square (PLS) model [40], a Radial Basis Function (RBF) network [41], a Multi-Layer Perceptron with linear output (MLP) [42] as well as soft-max output layer (PNET) [43] were performed (Table 1).

In order to evaluate generalization performance of the used machine learning algorithms, a five-fold cross validation was performed. For this purpose the data set is divided into five parts of equal sample size. Additionally the number of samples for both spray classes per fold is equalized. A model is then trained on four folds and tested on the fifth fold. All possible combinations are run and result in a mean accuracy for the test data over all folds. Mean accuracy and standard deviation are then used to determine the best machine-learning model. In Table 1, achieved prediction accuracy on the test folds are shown. A comparison of different machine learning approaches is worthwhile since as clearly indicated by the results, method can differ greatly in performance.

Figure 9 shows the achievable classification accuracy for a differentiation of sprayed vs. non-sprayed grapevine leaves measured from west and east (Figure 9a), only west side (Figure 9b) and only east side (Figure 9c) for the different times of day. Here we compare the best performing machine learning approach with the linear discrimination of the LDA. For these datasets, a machine learning approach just shows slight improvement over the linear discrimination method. Because the only condition both groups of grape plants are differing in is the spraying status with plant protection and along with the known fact of high infection pressure in this plot, detected changes in the spectral reflectance are probably determined by the counter reaction of the plants metabolism towards mildew infection. Across the results, the spectral reflectance in the VIS-NIR range seems to be the more robust predictor of the spray-status e.g., the suspected infection status. These initial results also indicate that the VIS-NIR range seems to be much less effected by time of day as well as the recording from west or east (for example SWIR at 12 p.m.). These initial results should be further investigated in the Phenoliner campaign 2017.

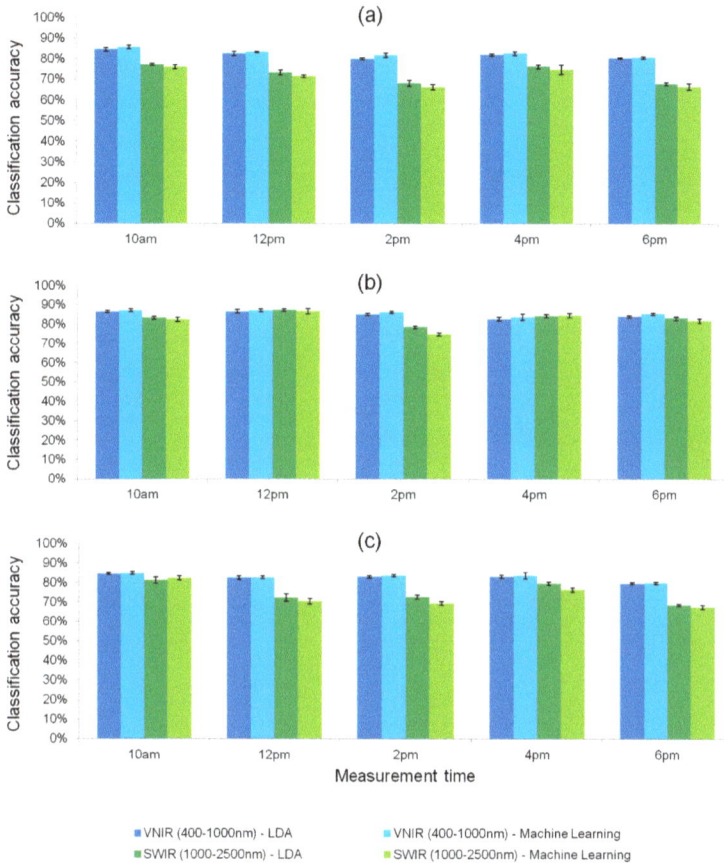

Figure 9. Classification accuracy for differentiation of sprayed vs. non-sprayed leaves measured from (**a**) west and east; (**b**) only west and (**c**) only east side. Across the results, the spectral reflectance data of the VIS-NIR range seems to be the more robust predictor for the spray status. Machine Learning and LDA approach are compared.

The correct classification rate is based on a per-pixel evaluation. For the evaluation of the whole plant, a voting of pixel results can be achieved to derive a per-plant prediction of the infection status. Furthermore, the machine learning approach will benefit from a broader database, since more confounding variations in the plant will be represented and aggregated by the machine learning model.

The machine learning approach RBF also calculates a relevance profile, which indicates what wavelength is informative for the task at hand. This technique will enable the overall goal of the project to find a set of wavelength that are informative for the pathogen detection while keeping the system flexible for new applications beyond the current scope. Figure 10 showing the relevance profile for the visual-near infrared range is depicted. The y-axis shows a percentage down- or up-regulation of wavebands according to the set tasks. As clearly visible, the profile differs greatly from an equal weighting of all wavebands. In future work we will test if this information can reduce the number of necessary wavelength.

Table 1. Accuracy of spray status prediction based on a single spectrum. Machine learning models tested: Soft-max output layer (PNET), Multi-Layer Perceptron with linear output (MLP), Radial Basis Function (RBF) and network Partially Least Square (PLS) model. The best results, also shown in Figure 9, are marked in bold. Depicted are average performances on the test sets in the 5-fold cross validation.

	VNIR					SWIR			
	West and east canopy side								
	PNET	MLP	RBF	PLS		PNET	MLP	RBF	PLS
10 a.m.	81.80%	79.40%	**86.85%**	86.55%	10 a.m.	70.95%	70.25%	77.65%	**77.80%**
12 p.m.	81.30%	81.40%	**83.90%**	83.05%	12 p.m.	66.65%	66.70%	72.25%	**73.95%**
2 p.m.	77.05%	77.65%	**83.40%**	80.95%	2 p.m.	60.30%	59.90%	67.95%	**69.25%**
4 p.m.	80.10%	79.75%	**84.00%**	82.45%	4 p.m.	67.30%	67.15%	**78.00%**	77.15%
6 p.m.	78.75%	78.95%	81.30%	**81.80%**	6 p.m.	63.25%	64.45%	**69.10%**	68.40%
	West side								
	PNET	MLP	RBF	PLS		PNET	MLP	RBF	PLS
10 a.m.	81.25%	83.05%	**87.95%**	87.65%	10 a.m.	73.35%	72.05%	**84.60%**	83.95%
12 p.m.	84.90%	85.20%	**88.05%**	87.80%	12 p.m.	81.20%	80.65%	88.00%	**88.25%**
2 p.m.	79.50%	80.70%	**87.05%**	85.60%	2 p.m.	68.55%	71.45%	75.30%	**77.50%**
4 p.m.	82.95%	81.85%	**84.70%**	83.90%	4 p.m.	70.60%	73.15%	**85.60%**	84.60%
6 p.m.	83.65%	82.15%	**86.30%**	85.80%	6 p.m.	72.45%	72.20%	83.95%	**84.10%**
	East side								
	PNET	MLP	RBF	PLS		PNET	MLP	RBF	PLS
10 a.m.	80.25%	82.05%	**85.60%**	85.15%	10 a.m.	76.20%	73.45%	**83.85%**	83.50%
12 p.m.	78.05%	79.00%	83.40%	**83.85%**	12 p.m.	64.20%	63.55%	71.75%	**73.60%**
2 p.m.	80.35%	78.55%	**84.55%**	83.45%	2 p.m.	64.90%	64.85%	70.45%	**73.90%**
4 p.m.	80.05%	79.45%	**85.55%**	83.95%	4 p.m.	71.10%	71.25%	77.95%	**80.75%**
6 p.m.	77.65%	76.70%	**80.45%**	80.15%	6 p.m.	63.55%	63.50%	68.90%	**69.50%**

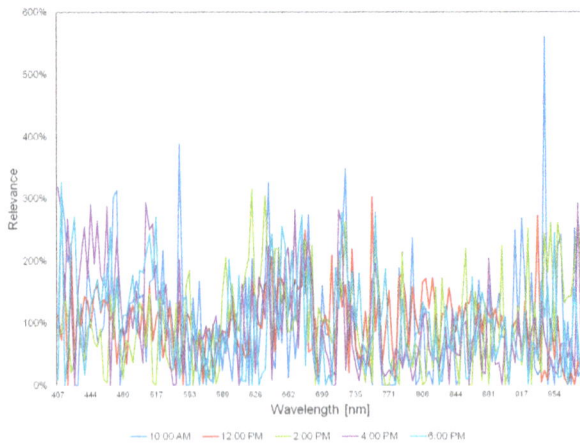

Figure 10. Relevance profile of the visual-near infrared range based on RBF performance at different recording times.

3.3. Discussion of One Year Experiences with the Phenoliner

The new, robust phenotyping platform has been successfully tested in season 2016. Two sensor systems (Sensors A and B) have been implemented on the Phenoliner and the sensor data have been connected to the geo-reference to successfully link this information for phenotypic trait evaluation. The idea of a moving tunnel has been shown to work very well to overcome changing light conditions, however some small adjustments have to be added to improve the standardization of the light conditions in the tunnel. Therefore, a second curtain will be installed in the front of the machine.

3.3.1. Sensor A

In season 2016 numerous rows were visually captured with the Phenoliner moving at a speed of 0.2–0.3 m/s, obtaining images with a frequency of 5 Hz. The error of image location was within a few centimetres or even below. Using this system, a near-to-complete high throughput screening of multiple vine rows is possible within one day. Future experiments need to determine the possibility to adapt the Phenoliner's operating speed to the usual operating speed of 10–15 km/h of field working machines. The speed limit is currently restricted by the possible frequency of image acquisition and storage. To reach the required image overlap of 70–80% at a speed of 15 km/h would require an imaging frequency of 26 Hz and a data storage speed of 1.15 Gb/s for the three vertical cameras. While a realisation is generally possible, this would require high investment costs. The second restriction is eventual plant movements through enhanced airflow at greater speeds. A solution might be a pre-preparation of the vine rows but at the costs of a higher work-load.

The illumination was standardized as good as possible by closing all major openings towards the tunnel and utilizing external illumination-units. Some improvements regarding the intensity of the external illumination and the background colour still have to be carried out to enhance the quality of the point cloud. Regarding point cloud reconstruction, three phenomena are connected with a high intensity illumination. The first phenomenon is the false reconstruction of the background at the border of foreground objects. The second phenomenon is the high intensity reflection of light at the background, which may lead to oversaturation in the images. Two background colours and materials were tested during our experiments. A heavy grey tarpaulin (Figure 5b left side, black berries) and a white blanket (Figure 5b right side, green berries) constituted two alternative backgrounds. Images taken with a white background, exhibit a high level of oversaturation derived from high intensity reflections of the white background. The resulting point cloud suffers from false colours and falsely reconstructed points at object borders in the point cloud, as can be seen in Figure 5b.

In contrast, the images taken with a grey background, exhibit correct colour assignments and sharper object borders. Still, high intensity reflections at the background are visible as well and may cause background points to be reconstructed at a wrong position. The third phenomenon is the high intensity reflection at berry arches (Figure 11).

In future experiments it is planned to address these problems in two ways. First, by the utilization of polarizing filters to reduce the high intensity reflections on berry arches and background. Secondly, the background will be painted in a matt colour to further minimize reflections. For an efficient background segmentation the colour will either be pitch black or of a neon-colour shade to distinguish it from the foreground.

Figure 11. High intensity reflections result in a wrong point positions at object borders (left image). At berry arches, they may corrupt the spherical geometry of the berries during reconstruction.

3.3.2. Sensor B

The integration of two optical sensor systems, computer hardware and light sources in a compartment while trying not to overuse the space, but getting good results for both sensors at the same time, was one of the biggest tasks to begin with. Despite earlier observations, Sensor B had to be rotated $90°$ since the distance between camera (fixed focal length) and plant canopy was still too short. A mirror capable of reflecting all relevant wavelengths needed to be introduced into the optical system. After that, the system was up and running while the Phenoliner was keeping a steady velocity of 0.1 m/s. During the season, the foil covering of the reflectance pad was exposed to some stress through plant contact and will be replaced with a thin glass window. The camera acquisition data as well as the GPS data stream was generally stable not counting the several improvements that needed to be done to the acquisition software. All in all the current hardware setup delivered promising results and its use will also be extended to the laboratory. Preliminary data taken from a repeated measurement of potentially infected vines showed that a machine learning model so far does not consistently outperformed a linear discrimination of the sensor data used so far. The VIS-NIR range seems to be the better and more stable indicator for the spray-status e.g., infection status while the shortwave-infrared range was influenced by day-time and recording direction. This fact has to be evaluated in the coming campaign with plants analysed for true infection. With a growing database of hyperspectral data samples it is to expect that the machine learning approach gains robustness against confounding variations in the data. The advantage (Sensor B) of the current system is the collection of reflectance data narrowly sampled across the wavelength range from 400 to 2500 nm which sets it apart from usual multispectral approaches whose wavelengths are chosen according to spectral indices known in advance to have a correlation with the target values (for example chlorophyll, biomass, nitrogen). Machine learning is the tool to cope with the high-dimensional data produced by the system and its potential non-linear relationship with the target value.In the long run, the hyperspectral imaging system is intented to be used on a commercial platform like a tractor in a productive environment. Some prerequisites must be met before, that is a higher framerate supported by better artifical lighting allowing lower exposure times. Also, if the evidence hardens, that VIS-NIR is the spectral range holding sufficient information for the detection of grapevine diseases, this would be beneficial for the realization of a multispectral system for the commercial application.

4. Conclusions

A new phenotyping platform for grapevine research has been successfully introduced. The Phenoliner was build based on a grape harvester and is equipped with two high-end visual camera systems using RGB, NIR and hyperspectral imaging to screen vines directly in the field. It has the potential to enable high-throughput phenotyping taking different phenotypic traits like yield parameters and health status into account. Compared to other phenotypic platforms it is independent of the surrounding light conditions. Furthermore the Phenoliner is very robust for field application and its functionality could be extended with additional sensors in the future.

Acknowledgments: We gratefully acknowledge the financial support of Projektträger Jülich and the German Federal Ministry of Education and Research (Bundesministerium für Bildung und Forschung (BMBF), Bonn, Gemany), as well as the Federal Office for Agriculture and Food (Bundesanstalt für Landwirtschaft und Ernährung (BLE), Bonn, Germany) and the Federal Ministry of Food and Agriculture (Bundesministerium für Ernährung und Landwirtschaft (BMBF), Bonn, Gemany).This work was funded by BMBF in the framework of the project novisys (FKZ 031A349), as well as by BMEL in the framework of the projects Big Grape (FZK 2815702515) and Vitismart (FZK 2815ERA05C).

Author Contributions: R.T., U.S. and H.K. designed the study, A.K., coordinated it. A.K., J.C.R., M.W., H.-C.K., A.B., K.H. and N.B. carried out the field trials. A.K., M.W., L.K., J.C.R., H.-C.K. and A.B. validated the results and drafted the manuscript. J.C.R. did the setup of sensor A. K.H. and T.L. did the rectification of Sensor A. H.-C.K. and A.B. did the setup of sensor B.M.W. developed the software IggGeotagger2.0. C.H. and W.P. coordinated the technical reconstruction work on the Phenoliner. K.H., N.B. and J.C.R helped to draft the manuscript. All authors read and approved the final manuscript.

Conflicts of Interest: The authors declare no conflict of interest.

References

1. Granier, C.; Aguirrezabal, L.; Chenu, K.; Cookson, S.J.; Dauzat, M.; Hamard, P.; Thioux, J.J.; Rolland, G.; Bouchier-Combaud, S.; Lebaudy, A.; et al. Phenopsis, an automated platform for reproducible phenotyping of plant responses to soil water deficit in arabidopsis thaliana permitted the identification of an accession with low sensitivity to soil water deficit. *New Phytol.* **2006**, *169*, 623–635. [CrossRef] [PubMed]
2. Walter, A.; Scharr, H.; Gilmer, F.; Zierer, R.; Nagel, K.A.; Ernst, M.; Wiese, A.; Virnich, O.; Christ, M.M.; Uhlig, B.; et al. Dynamics of seedling growth acclimation towards altered light conditions can be quantified via growscreen: A setup and procedure designed for rapid optical phenotyping of different plant species. *New Phytol.* **2007**, *174*, 447–455. [CrossRef] [PubMed]
3. Hartmann, A.; Czauderna, T.; Hoffmann, R.; Stein, N.; Schreiber, F. Htpheno: An image analysis pipeline for high-throughput plant phenotyping. *BMC Bioinform.* **2011**. [CrossRef] [PubMed]
4. Reuzeau, C.; Frankard, V.; Hatzfeld, Y.; Sanz, A.; Van Camp, W.; Lejeune, P.; De Wilde, C.; Lievens, K.; De Wolf, J.; Vranken, E.; et al. Traitmill™: A functional genomics platform for the phenotypic analysis of cereals. *Plant Genet. Resources* **2006**, *4*, 20–24. [CrossRef]
5. Lejealle, S.; Bailly, G.; Masdoumier, G.; Ayral, J.L.; Latouche, G.; Cerovic, Z. Pre-symptomatic detection of downy mildew using multiplex-330®. In Proceedings of the 10e Conférence Internationale sur les Maladies des Plantes, Tours, France, 3–5 December 2012; pp. 63–73.
6. Latouche, G.; Debord, C.; Raynal, M.; Milhade, C.; Cerovic, Z.G. First detection of the presence of naturally occurring grapevine downy mildew in the field by a fluorescence-based method. *Photochem. Photobiol. Sci.* **2015**, *14*, 1807–1813. [CrossRef] [PubMed]
7. Agati, G.; D'Onofrio, C.; Ducci, E.; Cuzzola, A.; Remorini, D.; Tuccio, L.; Lazzini, F.; Mattii, G. Potential of a multiparametric optical sensor for determining in situ the maturity components of red and white vitis vinifera wine grapes. *J. Agric. Food Chem.* **2013**, *61*, 12211–12218. [CrossRef] [PubMed]
8. Ghozlen, N.B.; Cerovic, Z.G.; Germain, C.; Toutain, S.; Latouche, G. Non-destructive optical monitoring of grape maturation by proximal sensing. *Sensors* **2010**, *10*, 10040. [CrossRef] [PubMed]
9. Diago, M.P.; Sanz-Garcia, A.; Millan, B.; Blasco, J.; Tardaguila, J. Assessment of flower number per inflorescence in grapevine by image analysis under field conditions. *J. Sci. Food Agric.* **2014**, *94*, 1981–1987. [CrossRef] [PubMed]
10. Grossetete, M.; Berthoumieu, Y.; Da Costa, J.P.; Germain, C.; Lavialle, O.; Grenier, G. Early estimation of vineyard yield: Site specific counting of berries by using a smartphone. In Proceedings of the International Conference of Agricultural Engineering on Infomation Technology, Automation and Precision Farming, Valencia, Spain, 8–12 July 2012.

11. Rabatel, G.; Guizard, C. Grape berry calibration by computer vision using elliptical model fitting. In Proceedings of the 6th European Conference on Precision Agriculture ECPA, Skiathos, Greece, 3–6 June 2007; pp. 581–587.

12. Johnson, L.F.; Roczen, D.E.; Youkhana, S.K.; Nemani, R.R.; Bosch, D.F. Mapping vineyard leaf area with multispectral satellite imagery. *Comput. Electron. Agric.* **2003**, *38*, 33–44. [CrossRef]

13. Losos, J.; Arnold, S.; Bejerano, G.; Brodie, E.I.; Hibbett, D. Evolutionary biology for the 21st century. *PLoS Biol.* **2013**, *11*, e1001466. [CrossRef] [PubMed]

14. Bellvert, J.; Zarco-Tejada, P.J.; Girona, J.; Fereres, E. Mapping crop water stress index in a 'pinot-noir' vineyard: Comparing ground measurements with thermal remote sensing imagery from an unmanned aerial vehicle. *Precis. Agric.* **2014**, *15*, 361–376. [CrossRef]

15. Mazzetto, F.; Calcante, A.; Mena, A.; Vercesi, A. Integration of optical and analogue sensors for monitoring canopy health and vigour in precision viticulture. *Precis. Agric.* **2010**, *11*, 636–649. [CrossRef]

16. Bourgeon, M.-A.; Paoli, J.-N.; Jones, G.; Villette, S.; Gée, C. Field radiometric calibration of a multispectral on-the-go sensor dedicated to the characterization of vineyard foliage. *Comput. Electron. Agric.* **2016**, *123*, 184–194. [CrossRef]

17. Bramley, R.G.V.; Trought, M.C.T.; Praat, J.P. Vineyard variability in marlborough, new zealand: Characterising variation in vineyard performance and options for the implementation of precision viticulture. *Aust. J. Grape Wine Res.* **2011**, *17*, 72–78. [CrossRef]

18. Bramley, R.; Kleinlagel, B.; Ouzman, J. A Protocol for the Construction of Yield Maps from Data Collected Using Commercially Available Grape Yield Monitors—Supplement No. 2. April 2008—Accounting for 'Convolution' in Grape Yield Mapping. Available online: http://www.cse.csiro.au/client_serv/resources/protocol_supp2.pdf (accessed on 10 January 2017).

19. Llorens, J.; Gil, E.; Llop, J.; Queralto, M. Georeferenced lidar 3d vine plantation map generation. *Sensors* **2011**, *11*, 6237–6256. [CrossRef] [PubMed]

20. Arnó, J.; Escolà, A.; Vallès, J.; Llorens, J.; Sanz, R.; Masip, J.; Palacín, J.; Rosell-Polo, J. Leaf area index estimation in vineyards using a ground-based lidar scanner. *Precis. Agric.* **2013**, *14*, 290–306. [CrossRef]

21. Nuske, S.; Wilshusen, K.; Achar, S.; Yoder, L.; Narasimhan, S.; Singh, S. Automated visual yield estimation in vineyards. *J. Field Robot.* **2014**, *31*, 837–860. [CrossRef]

22. Araus, J.L.; Cairns, J.E. Field high-throughput phenotyping: The new crop breeding frontier. *Trends Plant. Sci.* **2014**, *19*, 52–61. [CrossRef] [PubMed]

23. Matese, A.; Di Gennaro, S. Technology in precision viticulture: A state of the art review. *Int. J. Wine Res.* **2015**, *7*, 69–81. [CrossRef]

24. Vinerobot. Available online: http://www.vinerobot.eu/ (accessed on 1 February 2017).

25. Robotnik Vinbot Project—Robotnik. Available online: http://www.robotnik.eu/portfolio/robotnik-proyecto-vinbot/ (accessed on 1 February 2017).

26. Wall-ye. Available online: http://www.wall-ye.com/ (accessed on 1 February 2017).

27. Kicherer, A.; Herzog, K.; Pflanz, M.; Wieland, M.; Rüger, P.; Kecke, S.; Kuhlmann, H.; Töpfer, R. An automated phenotyping pipeline for application in grapevine research. *Sensors* **2015**, *15*, 4823–4836. [CrossRef] [PubMed]

28. Vineguard. Available online: http://robotics.bgu.ac.il/index.php/Development_of_an_Autonomous_vineyard_sprayer (accessed on 1 February 2017).

29. Vision Robotics Corporation. Available online: http://www.visionrobotics.com/vr-grapevine-pruner (accessed on 1 February 2017).

30. Vitirover. Available online: http://www.vitirover.com/fr/ (accessed on 1 February 2017).

31. Kicherer, A. *High-Throughput Phenotyping of Yield Parameters for Modern Grapevine Breeding*; University of Hohenheim, Julius Kühn-Institut, Federal Research Centre for Cultivated Plants: Quedlinburg, Germany, 2015.

32. Kicherer, A.; Klodt, M.; Sharifzadeh, S.; Cremers, D.; Töpfer, R.; Herzog, K. Automatic image-based determination of pruning mass as a determinant for yield potential in grapevine management and breeding. *Aust. J. Grape Wine Res.* **2016**. [CrossRef]

33. Klodt, M.; Herzog, K.; Töpfer, R.; Cremers, D. Field phenotyping of grapevine growth using dense stereo reconstruction. *BMC Bioinf.* **2015**. [CrossRef] [PubMed]

34. Herzog, K.; Roscher, R.; Wieland, M.; Kicherer, A.; Läbe, T.; Förstner, W.; Kuhlmann, H.; Töpfer, R. Initial steps for high-throughput phenotyping in vineyards. *Vitis* **2014**, *53*, 1–8.

35. Abraham, S.; Hau, T. Towards autonomous high-precision calibration of digital cameras. In Proceedings of the SPIE Annual Meeting, San Diego, CA, USA, 27 July 1997; pp. 82–93.
36. Rose, J.; Kicherer, A.; Wieland, M.; Klingbeil, L.; Töpfer, R.; Kuhlmann, H. Towards automated large-scale 3d phenotyping of vineyards under field conditions. *Sensors* **2016**, *16*, 2136. [CrossRef] [PubMed]
37. Roscher, R.; Herzog, K.; Kunkel, A.; Kicherer, A.; Töpfer, R.; Förstner, W. Automated image analysis framework for high-throughput determination of grapevine berry sizes using conditional random fields. *Comput. Electron. Agric.* **2014**, *100*, 148–158. [CrossRef]
38. Furukawa, Y.; Ponce, J. Accurate, dense, and robust multiview stereopsis. *IEEE Trans. Pattern Anal. Mach. Intell.* **2010**, *32*, 1362–1376. [CrossRef] [PubMed]
39. Martinetz, T.M.; Berkovich, S.G.; Schulten, K.J. 'neural-gas' network for vector quantization and its application to time-series prediction. *IEEE Trans. Neural Netw.* **1993**, *4*, 558–569. [CrossRef] [PubMed]
40. Wold, S.; Sjöström, M.; Eriksson, L. Pls-regression: A basic tool of chemometrics. *Chemometrics Intellig. Lab. Syst.* **2001**, *58*, 109–130. [CrossRef]
41. Moody, J.; Darken, C.J. Fast learning in networks of locally-tuned processing units. *Neural Comput.* **1989**, *1*, 281–294. [CrossRef]
42. Cybenko, G. Approximation by superpositions of a sigmoidal function. *Math. Control Signals Syst. (MCSS)* **1989**, *2*, 303–314. [CrossRef]
43. Bishop, C.M. *Pattern Recognition and Machine Learning*; Springer: New York, NY, USA, 2006; Volume 4, Chapter 4.

Article

Combining Multi-Agent Systems and Wireless Sensor Networks for Monitoring Crop Irrigation

Gabriel Villarrubia [1,*], Juan F. De Paz [1], Daniel H. De La Iglesia [1] and Javier Bajo [2]

[1]   Faculty of Science, University of Salamanca, Plaza de la Merced s/n, 37002 Salamanca, Spain; fcofds@usal.es (J.F.D.P.); danihiglesias@usal.es (D.H.D.L.I.)
[2]   Department of Artificial Intelligence, Universidad Politécnica de Madrid, Campus Montegancedo s/n, Boadilla del Monte, 28660 Madrid, Spain; jbajo@fi.upm.es
*   Correspondence: gvg@usal.es; Tel.: +34-923-29-44-00

Received: 10 July 2017; Accepted: 31 July 2017; Published: 2 August 2017

Abstract: Monitoring mechanisms that ensure efficient crop growth are essential on many farms, especially in certain areas of the planet where water is scarce. Most farmers must assume the high cost of the required equipment in order to be able to streamline natural resources on their farms. Considering that many farmers cannot afford to install this equipment, it is necessary to look for more effective solutions that would be cheaper to implement. The objective of this study is to build virtual organizations of agents that can communicate between each other while monitoring crops. A low cost sensor architecture allows farmers to monitor and optimize the growth of their crops by streamlining the amount of resources the crops need at every moment. Since the hardware has limited processing and communication capabilities, our approach uses the PANGEA architecture to overcome this limitation. Specifically, we will design a system that is capable of collecting heterogeneous information from its environment, using sensors for temperature, solar radiation, humidity, pH, moisture and wind. A major outcome of our approach is that our solution is able to merge heterogeneous data from sensors and produce a response adapted to the context. In order to validate the proposed system, we present a case study in which farmers are provided with a tool that allows us to monitor the condition of crops on a TV screen using a low cost device.

Keywords: ambient intelligence; wireless sensor; fuzzy logic; smart irrigation; virtual organizations of agents

1. Introduction

Drought, climate change and pollution subject our water resources to big changes, and as the situation gets worse with time, more people experience its negative effects; currently, four out of every 10 people in the world are affected by a lack of water. Our population continues to grow and with it our needs for more water grow, for both industrial and domestic purposes. Our work focuses on optimizing water management in the agricultural sector, being the largest economic sector in the world. It is estimated that the agricultural industry wastes 60% of the 2.500 billion litres of water used each year [1–3]. In comparison to the current crop irrigation systems, we seek a more economic and effective solution that incorporates intelligence and context-awareness. This is possible due to the remarkable progress that has been made in the field of electronics in the last decade, whereby the size of end devices has decreased and their production costs as well. As a result, a variety of low cost sensors and communication devices is now available, allowing us to propose new solutions that can solve many every day challenges in an economic way. The possibility of using sensor networks in the agricultural sector would allow us to acquire data and look for intelligent solutions, helping us to create a system that ensures proper crop growth and optimizes water usage. However, the current monitoring systems do not incorporate a minimum degree of intelligence and do not have the ability to adapt to the environment. Moreover, the implementation of industrial equipment for the control of crop irrigation is hindered by its high cost and complexity; the lack of such equipment on farms results in unnecessary water wastage.

A multi-agent system (MAS) [4,5] is the most suitable option for this solution. This is because a MAS includes an Intelligent Distributed Artificial System, which incorporates social aspects and human reasoning to solve problems. Specifically, this work proposes an architecture based on virtual organizations through the use of a multi-agent open source platform called PANGEA [6], which incorporates services that allow sensor networks to be interconnected. This design makes it possible for agent societies to include organizational aspects of human societies, improving self-regulation and self-organization. The use of autonomous agents embedded in devices with limited computing capabilities makes the PANGEA platform a perfect candidate for the planned system. Each agent may represent an autonomous entity that consumes and provides different services. Collaboration between agents from the same organization offers more distributed and complex functionalities. One of the challenges that must be faced in this work is the fusion of information that comes from heterogeneous sources; the aim is to design an agent type that fuses heterogeneous data and obtains the crops' growth patterns. The proposed system should guarantee growth and scalability of the platform.

The assignment of roles to virtual organizations of agents will ensure system flexibility and will provide us with the ability to add new functionalities, creating a completely transparent layer for the applications that are developed for the user. The communication between the different components of the system must be efficient and low power consuming, since it must be capable of operating without direct sunlight. The proposed system must be able to monitor a crop and automatically supply the amount of water needed, this will be done through a predictive model with input variables regarding the climate, humidity of the subsoil, the force of wind, sunlight, temperature and the time of the day. Fields generally have very vast areas of land and sensors often have to be installed at large distances from one another, this is why a wireless connection between them is necessary. To this end, a network of autonomous sensors has been designed, with manageable battery consumption, which coordinates and generates different events produced in the system. Wireless Sensor Networks (WSNs) [7] are used to analyse environmental behaviour and the existing interaction between different sensors, they make decisions automatically on a daily basis [8].

The key novelty of the system presented in this paper is the use of fuzzy-logic and its application in a multi-agent environment. Those agents interact autonomously giving the system greater flexibility and intelligence. We also describe how the open-hardware platform called "Open Garden" is used; it has been designed specifically for the collection of essential agricultural data.

In addition, new interfaces have been implemented. Among these interfaces, the described system is a pioneer in using television as a data representation interface. This is very important because it is one of the devices most commonly used by final users. We should also point out that it is a low cost system, this enables small and medium-sized farmers to access this technology without having to invest a large sum of money.

This article is organized as follows: Section 2 reviews the state of the art and related projects, Section 3 describes the proposed architecture, Section 4 describes the case study conducted to evaluate the system, and finally Section 5 presents the results obtained and our conclusions.

2. Background

Currently, a variety of automatic crop irrigation systems are available in the literature, [9,10] and this section provides a detailed analysis of them. Existing solution are limited by various factors: their design does not adapt to the particular requirements of different crop species; the quantity of water supplied by each manufacturer is different, making it difficult to determine the exact amount of water to be supplied to a crop. Moreover, irrigation systems are extremely expensive and are provided by their manufacturers with closed architectures that restrict customization or inter-compatibility with other devices. For example, it is not possible to interconnect sensors from different manufacturers or to integrate data in an application that could be controlled from a smart TV.

The current issue of wasting natural resources has called the European Union to action. The EU is now encouraging the development of solutions that ensure ecological efficiency. Some of the best known projects that have been funded by FP7 [9] include the following:

WATERBEE DA (REF: 283638 Funded FP7-SME): this system allows farmers to save water by watering only at the time and place required [11]. Financed by European Community funds, the project team developed a prototype of a sustainable irrigation system. The tests showed savings of 21%, registering peaks of up to 44%. The impact of irrigation was also reduced to 23%. This system features wireless communication and environmental sensors, providing intelligent, flexible, easy to use, affordable and accurate programming. Moreover, this system can be adapted to the specific requirements of each user, the humidity of the soil and environmental conditions, and to different agricultural management systems [9,12,13]. Related works such as [14] include a mobile application that has been used to manage irrigation with pivot in the state of Colorado (USA), while another work in Florida uses evapotranspiration-based irrigation controllers [15] to define schedules. Some works such as [16,17] operate with a Smart Irrigation Decision Support System; these systems include machine learning techniques such as artificial neural networks, fuzzy decision systems to analyze the water in the soil, or to establish previous irrigation patterns. However, these types of supervised systems require previous expert knowledge to train the algorithms.

OPTIFERT (REF: 2836772 Funded FP7-SME): is based on an innovative automatic irrigation system for medium and large scale agricultural holdings [18,19]. This system combines fertilization and irrigation, and reflects the increasingly widespread trend among farmers to use computers, making it easier to keep track of the consumption of water and fertilizer. The system is composed of a soil sensor, a data processor, a control and distribution unit that monitors fundamental parameters of soil, and plant requirements in real time. The control software is able to access databases containing information about crop growth and relate it to crop species and soil (type, structure and fertility) data, as well as economic data on costs and prices. It is also possible to get weather forecasts and insert them into the system. In addition, the user can add data, such as reports on crops and planting times. It can obtain data that determine the right amount of water and fertilizer for each stage of crop growth.

ENORASIS (REF: 282949 Funded FP7-SME): by using this system, farmers can install a network of wireless sensors on their farms to gather information on factors affecting the crops' need for water, including soil moisture, atmospheric temperature, insolation, wind speed and precipitation [20,21]. The system also has a set of valves to measure any increase in the amount of water. This solution saves water, prevents soil erosion and generates both environmental and economic benefits. This system also uses a weather forecasting model that combines satellite images from the fields and the information from sensors to create a specific meteorological prediction. The model offers such a fine resolution that predictions can focus on areas of up to two square kilometres. Moreover, crop data can be used to prepare a watering plan, allowing the farmers to decide if they need to add more water to the ground [21].

IRRIMAN LIFE (funded program Life+): granted in 2015, the project is based on an automated system. Using an algorithm [19,22,23], irrigation needs are determined according to the water contained in the soil, the plant, and the atmosphere, all of which are measured on a continual basis using different sensors in the endometrial system. The project ensures the efficient use of water resources, the improvement of the quantitative management of water, and the preservation of the high quality level of water, and avoiding the misuse and deterioration of water resources. This is a very interesting project which has recently started [24].

This section has provided clear initiatives [13,25] to implement solutions that combine different sensors for the purpose of using natural resources in a more efficient way e.g., rationing electric power employed by the irrigation equipment or rationing limited resources, such as water. While these systems are composed of different sensors, they use closed platforms and lack the capacity to interact with external agents. Moreover, they lack the intelligence that equips them with learning and adaptation capabilities. Consequently, we need an open and heterogeneous platform that allows us to merge information from all the sensors for subsequent analysis and study.

Having begun as recently as 2014–2015, these projects are still in the development phase. Their use on conventional farms requires a significant investment, making them appropriate only for large areas. Extrapolating these systems for use on smaller areas, such as a small vegetable garden or greenhouse, or using them simply to monitor a crop during a short period of time, would make the cost of acquiring the necessary equipment far too expensive for most farmers.

Nevertheless, a comparative study of commercial solutions has been carried out for small scale farms. The solutions that incorporate sensors do not include systems based on fuzzy logic which allow to establish the watering quantities in a precise way. Aifro WaterEco [26] considers climatology in order to lower or increase irrigation but it is focused on the definition of threshold values and does not include fuzzy logic or sensors, such as soil and land humidity. Blossom [27], encompasses crop irrigation and generation of calendars, depending on the climate these calendars can be edited manually, it has common functionalities but allows for remote management, it also does not include fuzzy logic in its behavior. BlueSpray [28] includes seasonal information to adjust irrigation as in the previous example, it does not include fuzzy logic based behavior. GreenIQ [29] and IrrigationCaddy [30] are conventional programs that can be managed remotely from mobile applications and include the feature of creating irrigation calendars. Lono [31] incorporates threshold values and seasonal information and reduces crop watering according to the thresholds, as in the previous cases it does not include fuzzy logic and does not have weather sensors.

On the other hand, the Orbit B-Hyve [32] system incorporates a control through smartphones that is able to change some parameters in order to edit the system schedule. The parameters that device takes into account when configuring the irrigation timer are: the slope of the site, the soil type, if it is in the sun or shade, history of rainfall in the area and the current weather. The Rachio Smart Sprinkler Controller [33] system also has a Wi-Fi connection and is able to send the data from the sensor to the user's smartphone. This device requires an initial configuration which is established by indicating the type of crop and the type of soil. In this way, the system can estimate the irrigation time required by the crop. The fuzzy system is not applied, nor are the flexible rules. Rainmachine [34] is another commercial system which incorporates an automatic irrigation program. It is capable of calculating the percentage of evaporation and transpiration of the soil, according to the weather conditions obtained from the data of the meteorological service. This system, like the others, does not include fuzzy knowledge. The Spruce irrigation [35] system combines the information obtained from all the temperature and humidity sensors and rainfall forecasts. Lastly, we list the Raincommander [36] system for its ease of use and its integration with mobile devices for remote irrigation control. However, this system lacks an intelligent configuration, it has no fuzzy logic rules, and only considers the schedule and the irrigation time that has been configured manually by the user.

After a careful review of the related literature, this work focuses on a novel design of an open architecture composed of virtual agent organizations. The proposed system is economic and can be customized to fit the needs of each farmer making it possible to monitor and automate the irrigation of any crop species. From an analytical point of view, it will be necessary to store the information of each sensor in a remote database, this will allow farmers to examine the effectiveness of the system. Finally, we can deliver these functionalities to the user as services; users will be able to control irrigation from a TV screen, using a remote. In conclusion, the major novelties of this work are: (a) the ability to estimate irrigation time through the use of multi-agent virtual organization technology that executes a fuzzy algorithm, (b) the deployment of agent models in devices with limited capabilities using the PANGEA architecture, (c) the monitoring and control of the irrigation system with a TV remote (thanks to the use of wireless sensors networks).

3. Proposed Architecture

In the field of computer science and artificial intelligence the use of multi-agent systems deals with the interconnectivity of intelligent agents that collaborate together to solve a complex problem. The use of a combination of agents in wireless sensor networks allows for the design of new platforms with advanced computing capabilities. The design of a multi-agent system based on virtual organizations allows one to monitor and control an irrigation system. The different algorithms that make up the case study should be embedded in embedded devices like sensors or small microcontrollers. To achieve this, we have chosen a multi-task architecture that makes it possible for virtual organizations to have a dialogue between them, this architecture makes up the case study since distributed processing techniques can also be used with it. The proposed architecture must be dynamic, have the ability to merge information from heterogeneous data sources, and contain advanced analysis and prediction capabilities. The dynamism that a multi-agent architecture offers allows us to add new sensors, adapting them to the requirements of the environment. One of the main innovations of this architecture is a design based on organizational theory, which can both imitate and collaborate with human organizations related to crop irrigation. This Section will present the design of an architecture that (1) allows for the creation of an open and self-organizing system, and (2) can handle different types of sensor networks, thus facilitating the dynamic addition of new protocols based on the emergence of new technologies such as Zigbee, RFID, Wi-Fi, and Bluetooth. We will explain the design of the architecture in detail, as well as the agents that make up each virtual organization, as shown in Figure 1.

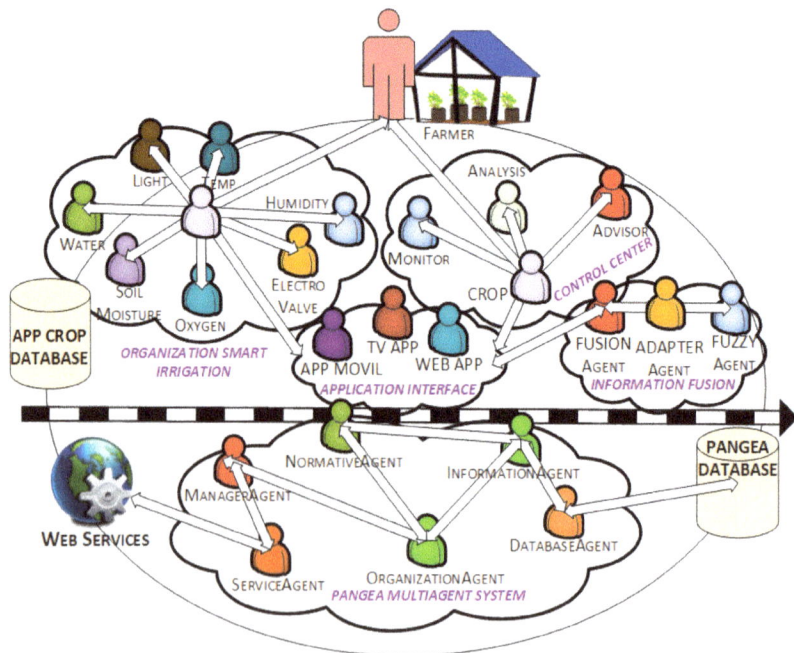

Figure 1. Agent organization flow chart.

The architecture is composed of two distinct parts: the bottom is formed by the minimum agents that make up the multi-agent PANGEA system; the top consists of different virtual organizations on which this case study is based, and whose operation is explained below:

Organization Information Fusion: This refers to an organization whose objective is to merge the information provided by the sensor networks (lower layers), which is then integrated with the virtual agent organizations (upper layers). In this organization, agents emulate the human behaviour of adding environmental information, thus making it possible to obtain far more advanced knowledge than what is generally provided by individual data. Also, the information formats controlled by each sensor are transformed to a common and manageable standard for all architecture. The internal message protocol chosen for the communication between agents of the platform is a messaging protocol of plain text that is based on the standard RFC1459 [37].

Organization Smart Irrigation: This refers to an organization that is in charge of extracting and collecting information from different sensors. Its main function is to transform the physical layer data so that they can be used by other agent organizations. Each agent communicates in a unidirectional way with a central officer who organizes and manages the communication. In this organization, there are two different roles: one held by officers, who obtain the values of the sensors; and another secondary role, in the coordination of tasks and communication with other organizations of the architecture. The different agents that form part of this organization are shown in Table 1. These agents are deployed in the nodes to extract information from the environment, the obtained data are sent to the central node which sends them to the main server.

Table 1. Monitoring variables.

Variable	I/O	Description
LightAgent	I	Obtains the brightness of the environment.
TempAgent	I	Responsible for measuring the environmental temperature.
HumidityAgent	I	Agents whose primary function is to measure the moisture in the air.
ElectroValveAgent	I	Responsible for increasing or decreasing the water flow.
OxygenAgent	I	Obtains the level of oxygen in the air.
SoilMostureAgent	I	Measures the degree of moisture in the subsoil
WaterAgent	I	Indicates the amount of water in the tank.
OrganizationAgent	I/O	Responsible for the communication between the different agents.

Organization Control Center: This organization is responsible for monitoring information obtained by agents, and belongs to the *Smart Irrigation Organization*. The most important task is the intelligent analysis of information and prediction based on the data collected from different sensors. The Crops agent is in charge of coordinating monitoring tasks, analysis and alerts, additionally this agent is responsible for managing the defined rules for each type of crop. In the case of an anomaly, an alerting situation, or a value outside of the usual range, this organization will be responsible for initializing the process of resolving the anomaly, which is then notified to the system administrator.

Organization Application Interface: This organization is in charge of adapting data from the other virtual organizations, and then representing this data in the application layer. As the organization is an interface, the applications inside the client can easily interact with the platform. For example, in the case of an external device that has to request a particular functionality from the system, or any application such as "Web Application or Smart TV", the data have to be adapted from the raw data to a standard format. This organization will develop an adaptation function, also known as the connector, for later use in any application. The presented case study has several connectors or gateways whose main function is to transform data from the architecture so that the data can then be represented on a smartphone, a Smart TV or a web application.

Pangea MultiAgent System: The decision to use PANGEA was based on its ability to create virtual organizations, which are characterized by their dynamic nature. This is the most singular feature, since other alternatives, such as THOMAS or JADE are not dynamic. PANGEA is a cost-free, multi-agent framework developed by the BISITE research group and anyone can use it. The PANGEA architecture can function with devices with limited computing capabilities, this feature is a big advantage because it enables us to deploy agents embedded in hardware. The fact that sensors are powered by sunlight makes this feature even more essential for the system. Moreover, limited computing capabilities are necessary for the algorithms responsible for data processing, as well as for efficient communication between the sensors in the system. The agents specialized in the management of virtual organizations are defined in [38], these agents are responsible for managing the agents inside the whole virtual organization. Below, we focus on the basic functions of the agents that manage the virtual organization executed within PANGEA.

- *DatabaseAgent*: This agent plays a storage role in the organization to provide persistence to the information in the organization. It is the only agent with database access privileges. Its objective is to perform backup tasks, as well as to ensure the correct consistency and storage of information. This agent communicates with the rest of the agents in the organization.
- *Information Agent*: This agent manages the services inside the virtual organization. It is also known as the "yellow pages" agent, as it allows other agents to publish the services provided, so that others can access them. When a new device or application uses the architecture for the first time, the corresponding agent should consult the specific services offered in the virtual organization.
- *Normative Agent*: One of the most important aspects in a virtual organization are the norms that govern the organization. This agent is responsible for the security when establishing communication between devices. When an application uses a specific functionality, this agent is in charge of checking whether it is authorized to do so, using a rules engine based on DROOLS [39,40].
- *Service Agent*: This agent distributes functionalities as web services. It is also a gateway to communicate external web services outside the system with the agents in the organization. To encourage greater abstraction, functionalities, and different capabilities offered by the architecture, some services are exposed; this mode favors greater integration independent of programming languages.
- *Manager Agent*: This agent is responsible for performing periodic system management, verifying if there is any overloaded functionality, and ensuring free of errors communication between people and organizations.
- *Organization Manager*: This agent plays a very important role in the architecture given that it is responsible for verifying the operation of all the virtual organizations, dealing with security, and balancing and providing encryption of the frames between the most important agents.

In the APP Crop Database different information is included, such as the irrigation rules liked to the type of crop. In these rule we include information on the geographic location, this data base is synchronized with a central server to ease the addition of new crops.

4. Monitoring Platform and Irrigation

This Section presents a case of study of a small crop environment combining a low cost hardware and multi-agent systems, which allows the fusion of information captured by different sensors. The chosen hardware platform is called the OpenGarden. Due to the wide variety of crops and the source that we can monitor, the architecture can be implemented in three different scenarios: indoors (houses and greenhouses), outdoors (gardens and fields), and with hydroponic agriculture (plants in water-based facilities).

The system must provide the ability to control the state of the plant through the detection of several parameters: moisture in the ground, humidity, brightness sensors for pH, conductivity, temperature, oxygen, and water level. The topology between different sensors is represented in Figure 2. There is a slave node for each type of crop or plant to be monitored, and a single central node that connects the cultivated area. There are two types of nodes: slave nodes, which send the information from the interconnected sensors; and a central or primary node, which acts as a gateway sending data to an agent that resides on a web server, using existing wireless technologies (Wi-Fi, GPRS, 3G).

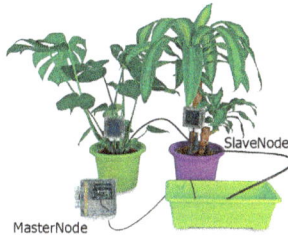

Figure 2. Architecture sensor network.

Below is a detailed explanation of the hardware used in this solution. The Gateway is powered by an Arduino-one controller [41]. The slave nodes send information to the central node via network, with a star topology for the transmission of information, using an Amplitude-Shift Keying (ASK) modulation. The selected band frequency is 433 MHz, due to the autonomy of the devices and the need for efficient communication, where the quantity identity of data shared between the different nodes is not very high. The gateway node is composed of an OpenGarden Shield [42]. The number of central nodes varies depending on the size of the farm, independent subzones can be established with different configurations. The distance between the central zones depends on the visibility of the environment. Using a 433 MHz range we attained interconnectivity between the nodes at a distance of 250 linear meters with total visibility. The functionality offered by each of the controller pins is shown in Figure 3.

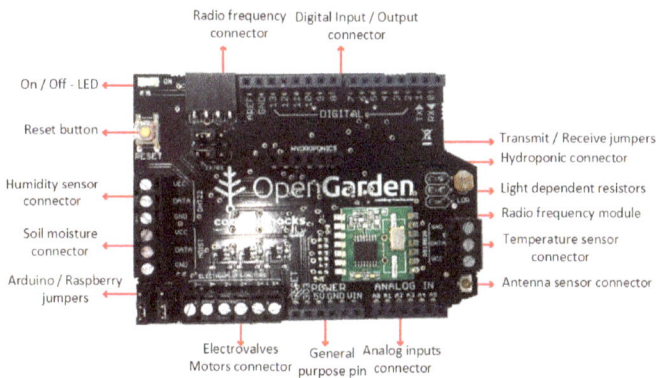

Figure 3. OpenGarden Gateway Node functionality board [42].

The shield of the master node allows us to technologically connect different types of sensors and to gather information from any sensor that is available on the market. In addition, it ensures interconnectivity with external hardware as Arduino or Raspberry by using a serial port. Of note, the shield incorporates a battery for autonomous operation which uses sunlight to power itself during the day. The controller is based on a DS1307 chip to time programming. It has an I2C interface that allows the interconnection of virtually any sensor currently on the market, and an accurate clock that will adjust to time changes. It can detect if there is a fault in the electrical circuit, and consumes less than 500 nA. The central node is capable of expanding its functionalities, providing us with the possibility of adding any type of sensor or functionality that we might need. This expansion port consists of 12 pins (analog and digital) which allow to, for example, adapt the system for activating monitoring systems, monitoring the condition of motors and pumps or if we want to use the system in greenhouses; to control the ventilation system, airflow and motorized doors.

The slave nodes are based on the use of OpenGarden Node Shield. As opposed to the central node, this board is simple, only in charge of connectivity with the different crop sensors to be monitored. Figure 4 shows a diagram of the connectivity board.

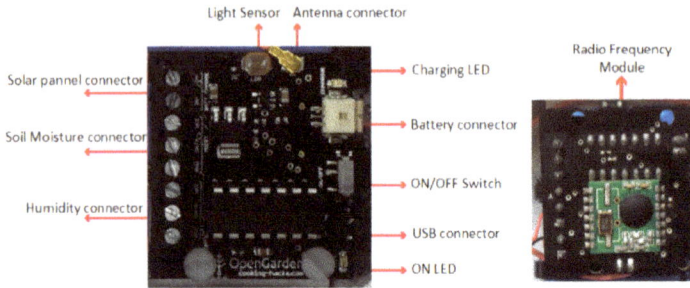

Figure 4. OpenGarden Slave Node functionality board [42].

As the functioning should be autonomous, this board also has a solar panel that is continually charging a lithium battery. The result of the complete assembly of the slave node and the central node are shown in Figure 5.

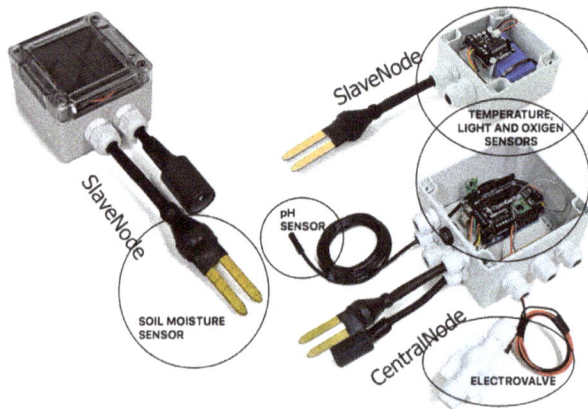

Figure 5. Look and feel of the devices.

One of the novelties of the system is the use of the light agents that are embedded in the nodes [24]. The light agents are especially designed for implementation in devices and sensors with limited resource constraints. In this case the sensors have limited resources and are therefore embedded in software agents that can communicate with the PANGEA architecture; to reduce computational costs, a simple communication protocol is used. The central node contains an agent that retrieves information for the agents in the Slave Node, this communication is made using the 433 MHz radio frequency. The central node sends the information to the server with REST and the information is made available to the other agent in the virtual organization so that it can be displayed by different devices.

4.1. Irrigation System Based on Fuzzy Logic

As mentioned in Section 3, the virtual *Information Fusion* organization aims to adapt and process information from each of the sensors. This organization merges the information collected from each of the individual sensors, and estimates the flow of necessary irrigation at each moment. For the fusion of information from the sensors and the establishing of the volume of water for irrigation, fuzzy logic is used as explained in this section. The reason for using fuzzy logic as opposed to other alternatives, such as Bayes, is because we want to establish a continuous irrigation level and not by categories [43]. A diagram of the flowchart detailing the procedures that take place in the Information Fusion organization in provided in Figure 6. Readers may check [3] for further information.

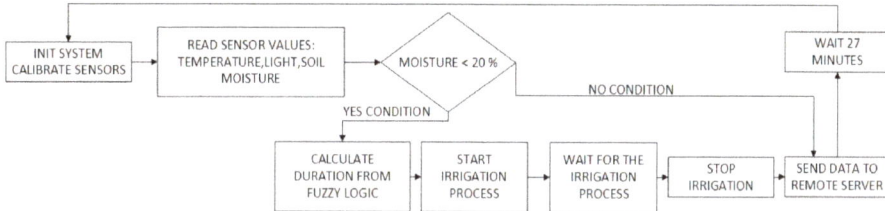

Figure 6. Process workflow: the main loop introduces a time delay in the whole process. Once the sensors are calibrated and measures obtained, the moisture level is checked. If the humidity crosses a given threshold, data is sent to the server. Otherwise, the input values of the sensors are sent to the fuzzy logic algorithm and the irrigation process is triggered.

The workflow of the irrigation process is described in the following paragraphs:

After the initial installation and activation carried out by the farmer, our system begins an auto-evaluation process where it verifies the condition of the installed sensors. If some type of interconnection error occurs, it is reported to the user through an alert. The nodes have an initial connection time of 10 s to connect to the master node. Once the sensors are connected to the WSN and the link with the central node is established via radio, they are ready to collect measurement data.

When the sensors are launched correctly, each one collects data according to the sampling frequency established by the user. Each of the slave nodes is in charge of collecting different measurements from the sensors, converting them into a format that can be read by a human and transmitting them to the central node. Each measure received by the central node is compared with the previous measures and the state of the humidity sensor is analyzed. If the value of the sensors is above 20%, all the information is sent to the central server with the aim of visualizing these data in the developed applications. However, if the humidity sensor displays a value that is below 20%, although all the necessary irrigation conditions are supplied (temperature, radiation, light, humidity) the required irrigation time will also have to be determined apart from sending the data. The sensors' measures are used as input variables for the fuzzy logic system which measures the exact irrigation time. The empirical rules used by the fuzzy logic system, have been established by a farmer who is an experienced tomato cultivator. These rules can be seen in Figure 7.

Moreover, the server continually stores data in the database; the values collected by the sensors as well as the decisions taken by fuzzy logic, enabling the user to access all this information remotely through the application designed in Section 4.2.

As shown in Figure 6, when the crops are being irrigated, sensor readings cannot be taken until 27 min after the irrigation started, this is due to the effects of transpiration. If soil is watered under conditions of extreme heat, the water will evaporate and the ground will not dampen immediately, resulting in an incorrect reading. The 27 min period allows the sensor to retrieve the correct value for subsoil humidity. This time window is fixed and was calculated by performing evaporation tests during the month of July in the town of Salamanca, Spain. It is possible to find literature on how to calculate time dynamically [44,45], however it is not the focus of this work.

The goal of the fuzzy logic based algorithm [3,46] is to determine the volume of water and the duration of irrigation (opening of electric valve) required in each case. Knowledge rules are established for the humidity sensor in three situations: when the sensor is wet, when the sensor is partly wet and finally when the sensor is dry. Table 2 shows the irrigation time for each case.

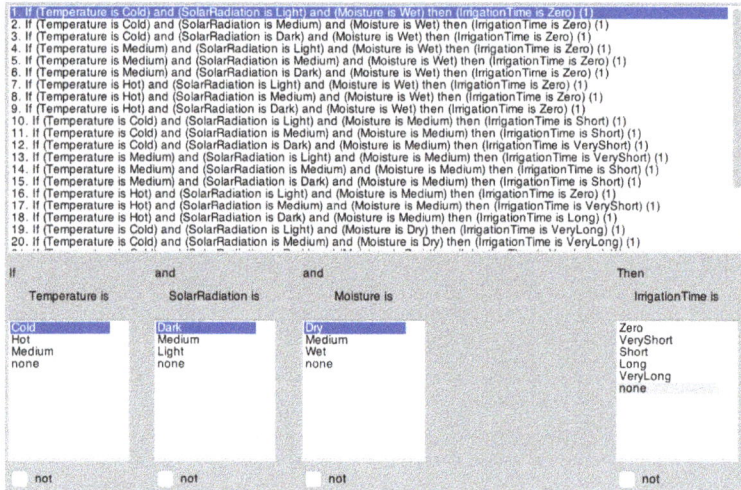

1. If (Temperature is Cold) and (SolarRadiation is Light) and (Moisture is Wet) then (IrrigationTime is Zero) (1)
2. If (Temperature is Cold) and (SolarRadiation is Medium) and (Moisture is Wet) then (IrrigationTime is Zero) (1)
3. If (Temperature is Cold) and (SolarRadiation is Dark) and (Moisture is Wet) then (IrrigationTime is Zero) (1)
4. If (Temperature is Medium) and (SolarRadiation is Light) and (Moisture is Wet) then (IrrigationTime is Zero) (1)
5. If (Temperature is Medium) and (SolarRadiation is Medium) and (Moisture is Wet) then (IrrigationTime is Zero) (1)
6. If (Temperature is Medium) and (SolarRadiation is Dark) and (Moisture is Wet) then (IrrigationTime is Zero) (1)
7. If (Temperature is Hot) and (SolarRadiation is Light) and (Moisture is Wet) then (IrrigationTime is Zero) (1)
8. If (Temperature is Hot) and (SolarRadiation is Medium) and (Moisture is Wet) then (IrrigationTime is Zero) (1)
9. If (Temperature is Hot) and (SolarRadiation is Dark) and (Moisture is Wet) then (IrrigationTime is Zero) (1)
10. If (Temperature is Cold) and (SolarRadiation is Light) and (Moisture is Medium) then (IrrigationTime is Short) (1)
11. If (Temperature is Cold) and (SolarRadiation is Medium) and (Moisture is Medium) then (IrrigationTime is Short) (1)
12. If (Temperature is Cold) and (SolarRadiation is Dark) and (Moisture is Medium) then (IrrigationTime is VeryShort) (1)
13. If (Temperature is Medium) and (SolarRadiation is Light) and (Moisture is Medium) then (IrrigationTime is VeryShort) (1)
14. If (Temperature is Medium) and (SolarRadiation is Medium) and (Moisture is Medium) then (IrrigationTime is Short) (1)
15. If (Temperature is Medium) and (SolarRadiation is Dark) and (Moisture is Medium) then (IrrigationTime is Short) (1)
16. If (Temperature is Hot) and (SolarRadiation is Light) and (Moisture is Medium) then (IrrigationTime is Short) (1)
17. If (Temperature is Hot) and (SolarRadiation is Medium) and (Moisture is Medium) then (IrrigationTime is VeryShort) (1)
18. If (Temperature is Hot) and (SolarRadiation is Dark) and (Moisture is Medium) then (IrrigationTime is Long) (1)
19. If (Temperature is Cold) and (SolarRadiation is Light) and (Moisture is Dry) then (IrrigationTime is VeryLong) (1)
20. If (Temperature is Cold) and (SolarRadiation is Medium) and (Moisture is Dry) then (IrrigationTime is VeryLong) (1)

If	and	and		Then
Temperature is	SolarRadiation is	Moisture is		IrrigationTime is
Cold	Dark	Dry		Zero
Hot	Medium	Medium		VeryShort
Medium	Light	Wet		Short
none	none	none		Long
				VeryLong
				none
not	not	not		not

Figure 7. Rules provided by a human expert.

Table 2. Rules for irrigation estimation time.

	Solar Radiation (Lux)			
Humidity	Light	Medium	Dark	Temperature
	No Water	No Water	No Water	Cold
Wet	No Water	No Water	No Water	Medium
	No Water	No Water	No Water	Hot
	Short	Short	Very Short	Cold
Half-Wet	Very Short	Short	Short	Medium
	No Water	Very Short	Long	Hot
	Very Long	Very Long	Very Long	Cold
Dry	Very Short	Long	Long	Medium
	No Water	Very Short	Very Long	Hot

To determine the reduction of uncertainty levels that comes with the inclusion of these three variables, an analysis of irrigation estimations is carried out through the Bayes application, on the basis of the use of these variables. The accuracy percentages obtained are listed in Table 3. As can be seen, when the three variables are used the accuracy increases. From this we can conclude that the three variables used are important in reducing the uncertainty when estimating the level of irrigation. We should also highlight that when using other classifiers, based on decision trees, such as J48 accuracy rises to 100%, however Bayes has been used given that it is the alternative to fuzzy logic listed in [43].

The figures below provide information regarding the membership functions that have been used by the fuzzy sets. Figures 8–10 represent the inference rules for temperature, solar radiation and soil moisture, respectively.

Table 3. Accuracy percentage (calculated with Bayes) depending on the irrigation levels, according to the variables indicated in Table 2.

Sensors	Accuracy
Humidity	62.96%
Light	40.74%
Temperature	40.74%
Humidity/light	62.96%
Humidity/temperature	74.07%
Light/temperature	40.74%
Humidity/light/temperature	81.48%

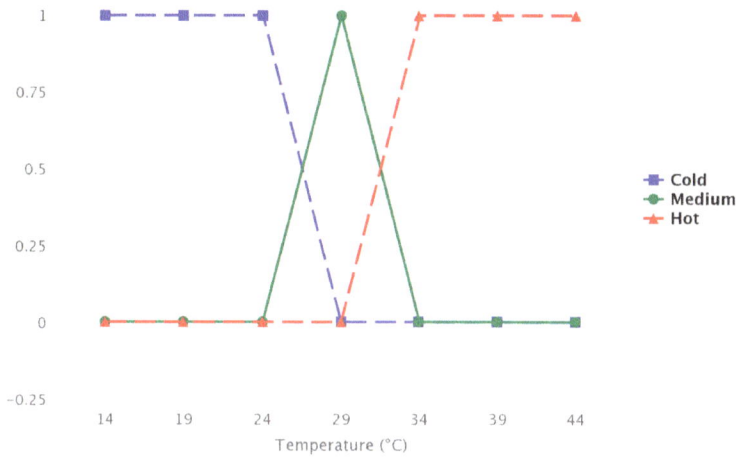

Figure 8. Temperature inference rule.

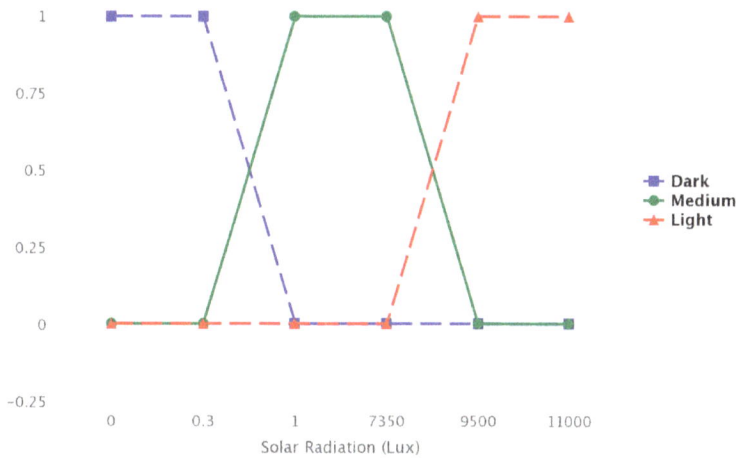

Figure 9. Solar radiation inference rule.

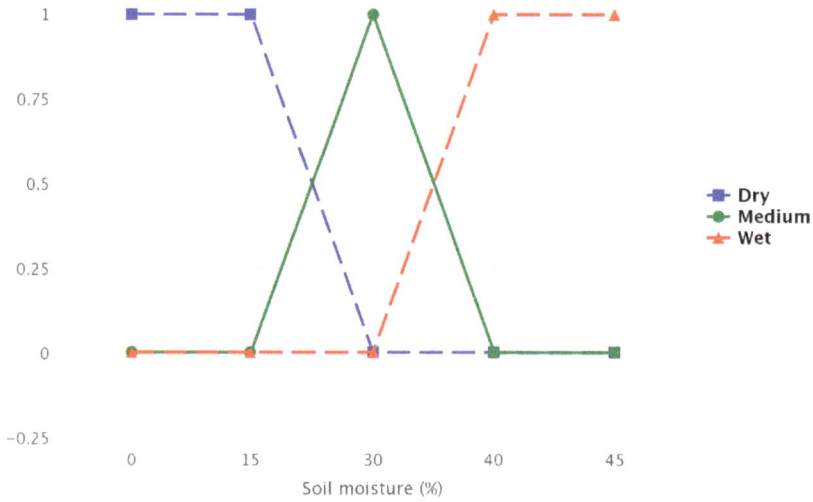

Figure 10. Soil inference rule.

Finally, Figure 11 combines the inference rules, showing the estimated irrigation time.

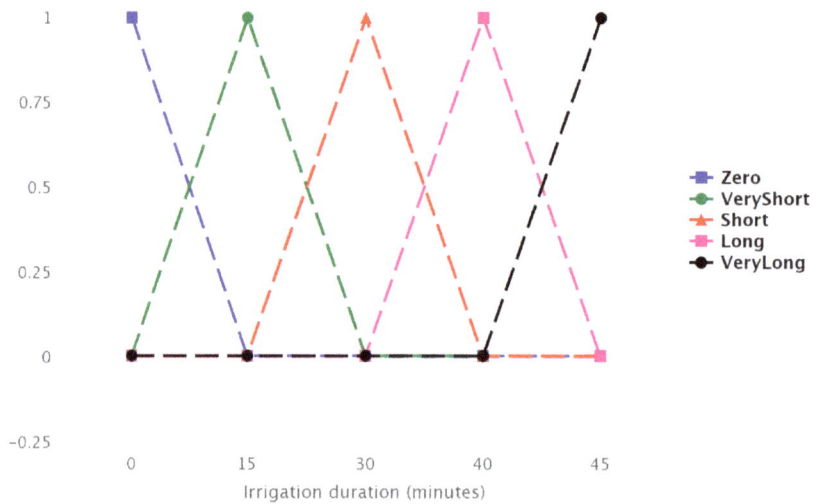

Figure 11. Irrigation time estimate.

The algorithm chosen in this work is based on a Mamdani system [3], in which the membership functions are trapezoidal. The reason for using this fuzzy system is that the library [47], which allows one to develop applications with fuzzy logic in microcontrollers based on the ATmega328p chipset, is the only one that possesses the Mamdani fuzzy system. In addition, the Takagi-Sugeno method is less intuitive and more computationally complex. While the defuzzification process can be done using different methods, the centroid technique method [48,49] was selected in this case. The fuzzy logic system was designed with MATLAB software. Figure 12 shows the general scheme.

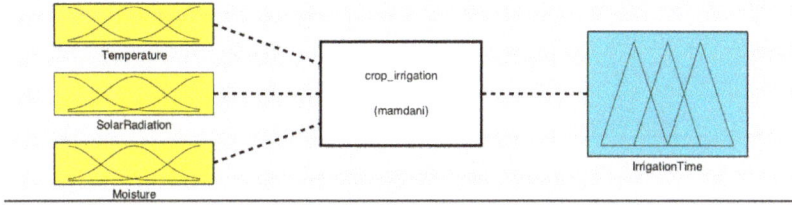

Figure 12. Inputs/output.

Figure 13 shows the output produced by the fuzzy system, when the value of temperature is 30 °C, solar radiation is 3000 lux and the percentage of the subsoil moisture sensors is 20%.

Figure 13. Output simulation.

As shown in Figure 14, when the temperature is high, irrigation time is completely determined by it, this helps to avoid water evaporation. In addition, Figure 15 shows how irrigation time increases as the humidity sensor approaches dry values and brightness.

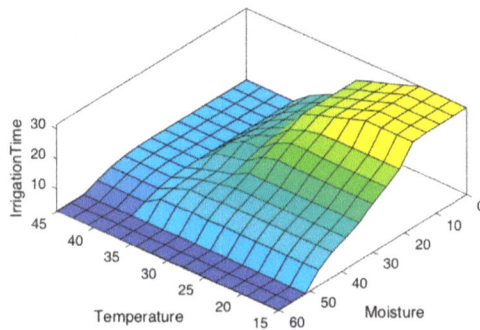

Figure 14. Relation between temperature and moisture.

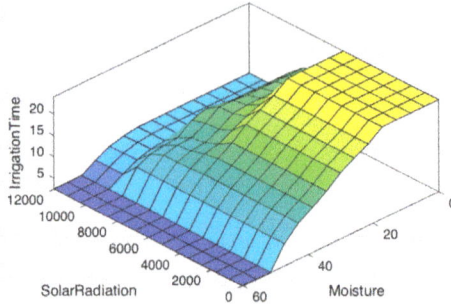

Figure 15. Relation between solar radiation and moisture.

Once the defuzzification process has been carried out, the following preliminary conclusions were obtained:

- The *subsoil moisture sensor* provides the most important system information; it measures the moisture of the subsoil and indicates when it is necessary to activate the irrigation mechanism. In addition, it estimates the amount of water needed.
- The *outdoor temperature sensor* measures the outside temperature. If the temperature is high, this sensor prevents the watering process that, if activated, would simply result in water evaporation and unnecessary water wastage.
- The *solar radiation sensor* is as necessary as the outside temperature sensor, since sunlight causes water to evaporate.

4.2. Platform Display

All irrigation systems must be controlled and monitored remotely. This section describes the physical connectivity of the system. The developed system can be deployed in any geographical location, provided that there is a data connection (Wi-Fi/3 G GSM) allowing the data from the sensors to be sent to a platform that resides on a central server. All wireless sensors have batteries that are continuously charged by solar energy. The objective of this section is to describe how individuals who are not familiar with technology could use the proposed system to check the state of their plants in real time. The overall architecture, including the screen display, is shown in Figure 16.

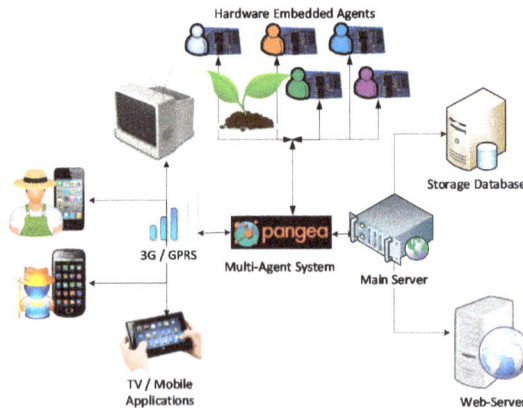

Figure 16. Scheme platform.

The agents embedded in the sensor network send the collected data periodically to organizations of agents located on the central server. This data can be kept and subsequently displayed. The communication between agents is done via RESTful web services, which allows for minimal battery consumption and high speed. Data exchange is done through JSON frames. This format was chosen because the data can be parsed by agents that are embedded in limited computing devices and very little time is required to plot the information. In Figure 17 below we give an example of the information structure.

```
{
  "sensors": [
    {
      "sensor_id": 1,
      "name": "Temperature",
      "value": 34,
      "date": "20/07/2015 16:56:01"
    },
    {
      "sensor_id": 2,
      "name": "Light",
      "value": 54000,
      "date": "20/07/2015 16:56:02"
    },
    {
      "sensor_id": "DEVICE_IDENTIFICATION",
      "name": "NAME_OF_THE_DEVICE",
      "value": "VALUE_OF_SENSOR",
      "date": "DATE_OF_MESURE"
    }
  ]
}
```

Figure 17. Message structure.

The most innovative feature presented in this Section is the use of a display agent installed on a Raspberry PI device, which allows us to connect to any type of Smart TV browser that has an HDMI adapter. The goal is to provide all users, particularly elderly farmers, with a visualization agent which will allow them to view the condition of their crops easily and from their own home. To do this, the architectural design was implemented, as shown in the figure below (Figure 18).

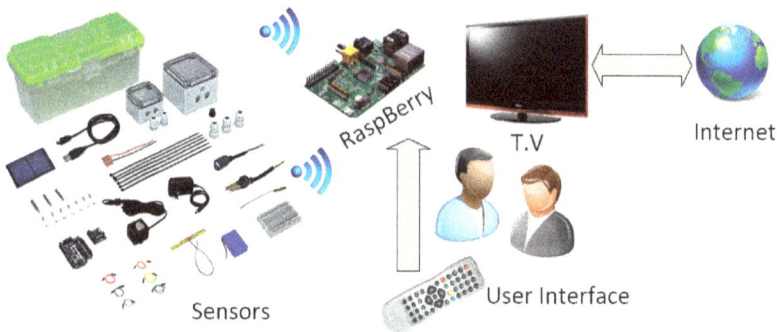

Figure 18. Remote control platform of the irrigation system.

The display agent allows the user to interact with the system via the TV remote control. The first time that the farmer opens the application, he has to carry out an initial configuration, in which he chooses the type of crop and its geographic location. In this way, we preload the initial irrigation configuration which the user can modify according to their preferences. Using an infrared sensor, the user will be able to monitor and control the state of the different sensors in a web environment. In addition, the display agent alarms the user if one of the system sensors fails, even if the user is watching TV, a warning will display on the screen. Figures 19 and 20 show a general view of the user interface of the proposed system, which uses a normal television to check the condition of plants.

In Figure 20, we can see the place where the system has been implemented, the farm has a size of 250 m^2 and was loaned to us by a farmer for the purpose of this case study. The farm in this case study does not have large dimensions, this is because we wanted to avoid economic loss if the result happened to be negative and growth would be affected.

For now, the cameras are simply used to provide the user with a snapshot of the system at a given moment; in future works, however, we are planning to add a camera-based monitoring system to our architecture.

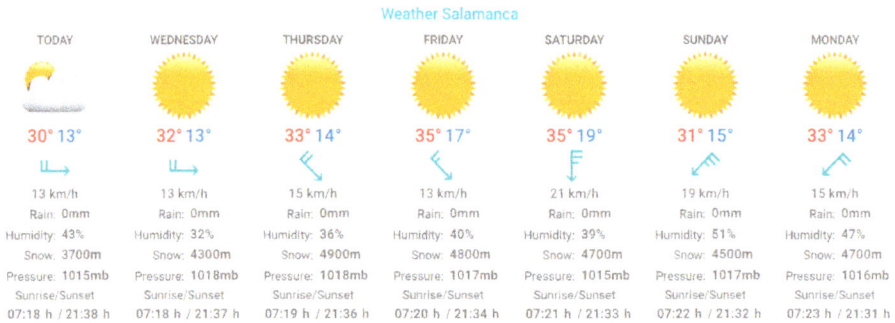

Figure 19. Smart TV application that shows the weather forecast.

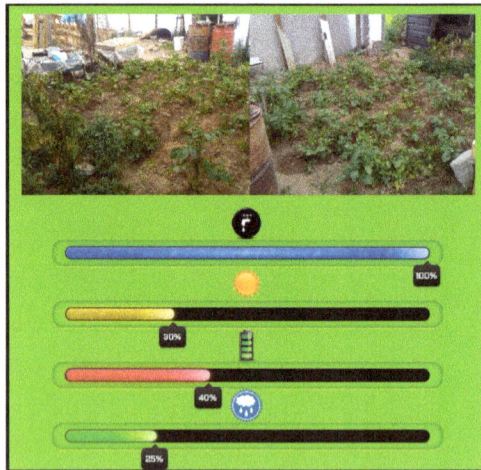

Figure 20. User interface screen capture with information from the different sensors displayed on the TV platform.

5. Results

This work presented the development of an intelligent system based on WSN that monitors and automates crop irrigation in an easy and economical way. The multi-agent architecture chosen to develop the case study is based on PANGEA, due to its ease of use and the ability to deploy agent drivers on computationally limited devices. The low cost of the proposed system ($100 €/250 m^2$) is a key factor, it makes it an accessible tool to the majority of farmers who cannot afford to implement existing solutions.

The location chosen to test and validate the system was a rural garden located in the outskirts of Salamanca, in the town of Roblija de Cojos. The tomato crop in the garden had a WSN composed of various sensors which measured soil moisture, soil temperature, external temperature, light, rain and wind. The nodes were evenly distributed, with one node placed every 5 m^2. Since the garden has an area of 250 m^2, slave nodes and a central node, which coordinate communication, were also installed. The main characteristics of the field included: clay soil, no crop yield in the last five years, fallow land, no presence of nematodes.

Table 4 provides a comparison between the costs of the commercial systems that have been described in the Background Section 2 and the system proposed in this work. The price of these devices has been calculated on the basis of the configuration that they would require for the case study conducted in this work; a field of 250 m^2. The calculated costs do not include additional teleoperator expenses for 3G/GPRS connection. The chosen systems do not have any installation costs since they are self-installing systems and the procedure can be carried out by the user. 250 m^2 is established as the baseline size, which is the minimum field range within which the system is useful and its measurements are conclusive. The implementation costs of the other systems are also calculated for a similar area, between 150 and 250 m^2. As can be seen, the proposed system has a smaller cost in comparison to the rest of commercial devices, even though the sensitizing areas are broader.

Table 4. Comparative between the proposed solution and the main commercial solutions currently available on the market.

NAME	INCLUDED SENSORS	MAXIMUM ZONES	CONNECTIVITY	SCALABLE (PRICE)	SYSTEM PRICE
AIFRO WATERECO [26]	Humidity and temperature	16	Wi-Fi	No	180 $
BLOSSOM [27]	Humidity and temperature	12	Wi-Fi	No	200 $
BLUESPRAY [28]	Humidity, light, oxygen and water	16	Wi-Fi & Ethernet	Yes (80 $)	250 $
GREENIQ [29]	Humidity and light	6	Wi-Fi & Ethernet	No	212 $
IRRIGATIONCADDY [30]	Humidity	10	Wi-Fi & Ethernet	No	175 $
LONO [31]	Humidity, light and oxygen	20	Wi-Fi & Bluetooth	No	250 $
ORBIT B-HYVE [32]	Humidity	12	Wi-Fi	No	130 $
RACHIO SMART SPRINKLER CONTROLLER [33]	Humidity, temperature and water	8	Wi-Fi	Yes (60 $)	200 $
RAINMACHINE [34]	Humidity, soil moisture, temperature, water	16	Wi-Fi	Yes (80–110 $)	300 $
SPRUCE IRRIGATION [35]	Humidity, soil moisture and temperature	16	-	No	170 $
RAINCOMMANDER [36]	Humidity and light	12	Wi-Fi	No	250 $
PROPOSED SYSTEM	Humidity, soil moisture, temperature, light, water and oxygen	256	Wi-Fi, GPRS, Ethernet & RF	Yes (15 $)	100 $

The type of tomato chosen for the testing process was the Pyros tomato. The Pyros tomato is a productive variety, of indeterminate growth, with a similar precocity to the Montfavet variety, resistant to cracking, with an average weight of 130 g, eye-catching green color, and resistant to *Verticillium*. Surface drip irrigation, 4 L/h of flow drip with a planting framework of 80 cm between rows and 25 cm between plants, with a plant density of 40 plants/100 m^2, in a single line of cultivation.

Above we present the diverse results obtained in the case study. Figure 21 shows the different temperature and radiation measurements taken at different times of the day. Figure 22 presents the relationship between temperature and duration of irrigation. We see that at times of extreme temperatures, the irrigation system was not activated in order to avoid water evaporation.

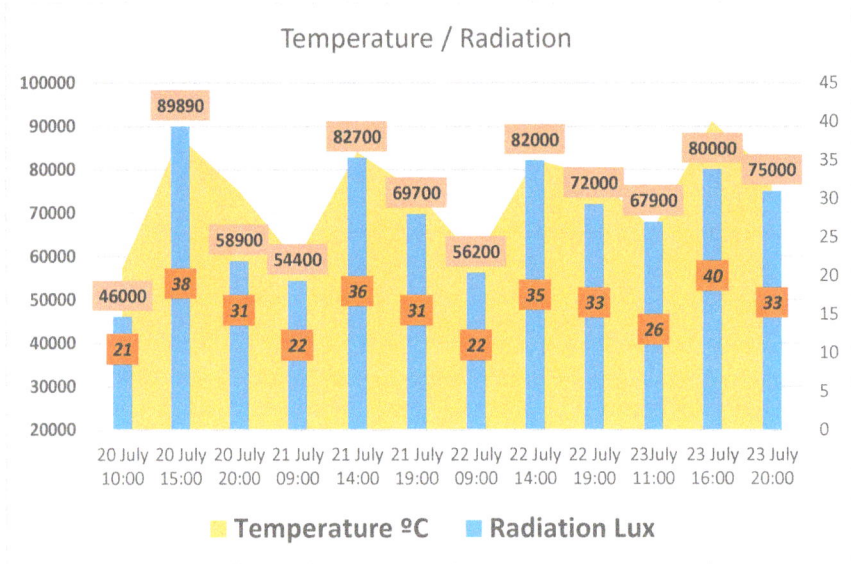

Figure 21. Relationship between temperature and solar radiation at different times of day.

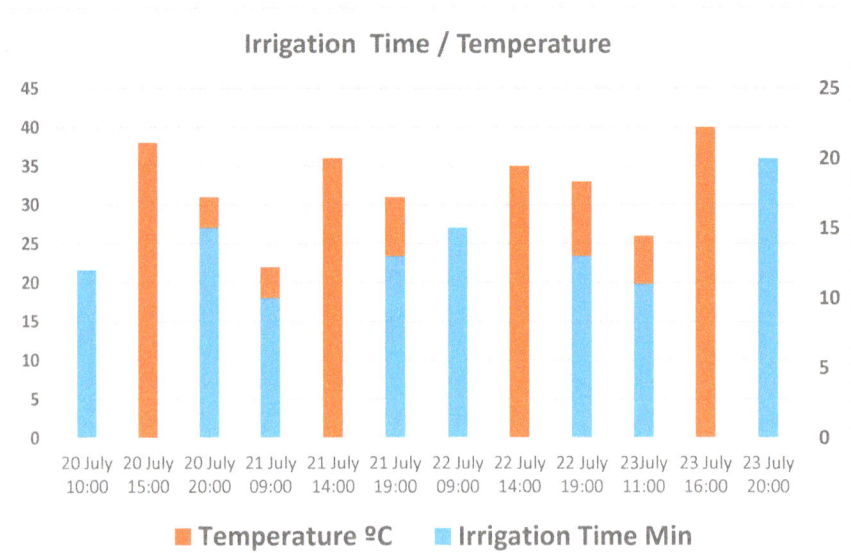

Figure 22. Relationship between the watering time and temperature.

Figure 23 shows how humidity drops to a minimum value at the hottest moments of the day. After irrigation is begun, we can immediately see that humidity increases. Figure 24 displays water consumption levels for an area of tomato crop measuring 50 m^2, using conventional programmed irrigation compared with the system proposed in this article.

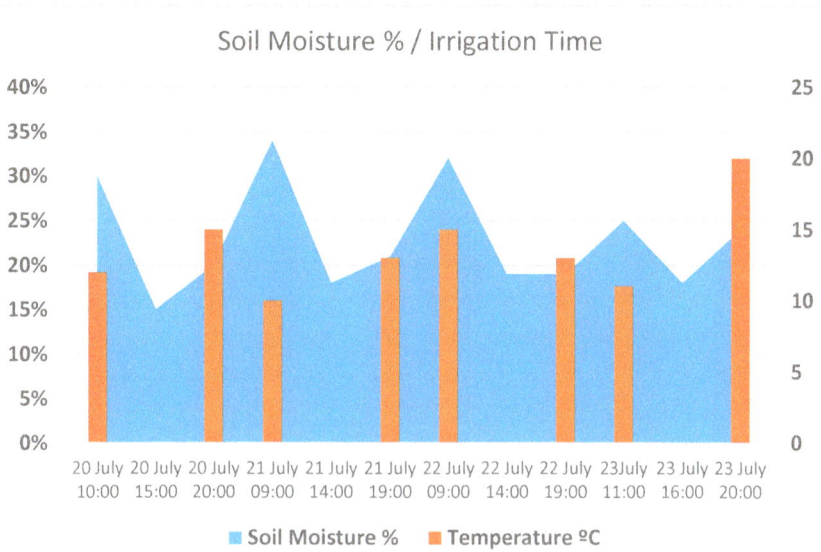

Figure 23. Relationship between moisture and watering time.

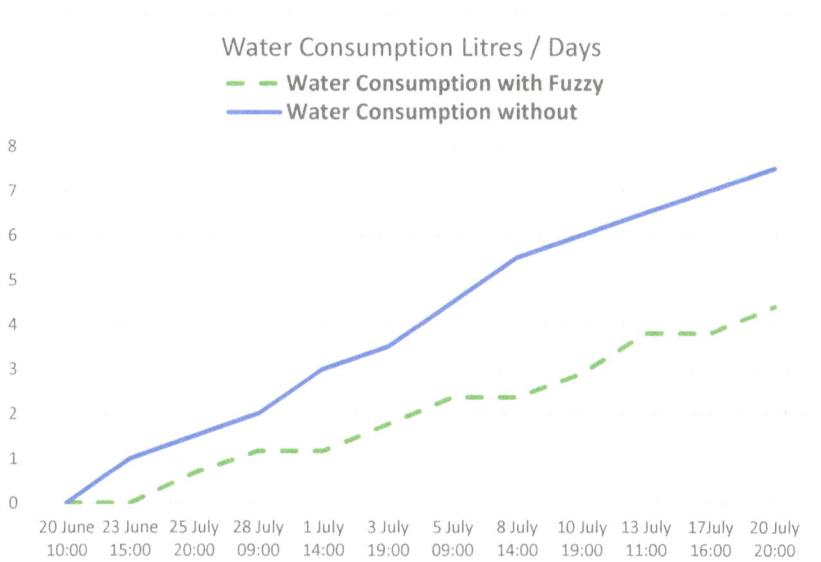

Figure 24. Total water consumption in the system using the fuzzy logic system.

The results of the case study have been compared with the traditional automatically programmed irrigation system. Concretely, an irrigation programming device has been used, it is called Orbit B-Hyve [32] with a cost of 130 $. A description and an image are included below (Figure 25).

Figure 25. Commercial device used for comparison with the conducted case study.

The crop was always irrigated at dawn, using 1 L of water, and in the evening, using 0.5 L of water. As shown in the image, water consumption in a traditional system is linear, and does not consider any external factors, meaning that the amount of water used for irrigation is always the same and occurs at the same hours of the day. However, the use of the proposed architecture guarantees that the precise amount of water is used, depending on the sensors values and the weather. Both systems were evaluated during 30 days, in comparison to the traditional system, 37% were achieved with the traditional system.

As mentioned before, the conventional system was applied to an area of 10 m^2 while the new system was used on an area of 50 m^2. The location was contiguous and there was no difference between the crops. Although less water was used with the proposed system, crop production per square meter is very similar; the proposed system 4.73 kg/m^2 tomatoes were collected as opposed to the 4.65 kg/m^2 of the commercial one. Production is not very high, given that it follows a normal cycle; crops are planted in January and harvested in summer. In other regions, where crops are grown in greenhouses, up to 10 kg/m^2 can be obtained.

Acknowledgments: This work has been supported by the European Commission H2020 MSCA-RISE-2014: Marie Skłodowska-Curie project DREAM-GO Enabling Demand Response for short and real-time Efficient And Market Based Smart Grid Operation—An intelligent and real-time simulation approach ref 641794.

Author Contributions: Gabriel Villarrubia and Juan F. De Paz have developed the system; they made the test and elaborate the review of the state of the art. Javier Bajo and Daniel Hernández formalized the problem, the algorithms and reviewed the work. All the authors contributed in the redaction of the paper.

Conflicts of Interest: The authors declare no conflict of interest.

References

1. Aqeel-ur-Rehman; Abbasi, A.Z.; Islam, N.; Shaikh, Z.A. A review of wireless sensors and networks' applications in agriculture. Comput. *Stand. Interfaces* **2014**, *36*, 263–270. [CrossRef]
2. Hedley, C.B.; Knox, J.W.; Raine, S.R.; Smith, R. Water: Advanced Irrigation Technologies. In *Encyclopedia of Agriculture and Food Systems*; Academic Press: Cambridge, MA, USA, 2014; pp. 378–406.
3. Touati, F.; Al-Hitmi, M.; Benhmed, K.; Tabish, R. A fuzzy logic based irrigation system enhanced with wireless data logging applied to the state of Qatar. *Comput. Electron. Agric.* **2013**, *98*, 233–241. [CrossRef]

4. Lim, C.H.; Anthony, P.; Fan, L.C. Applying Multi-Agent System in a Context Aware Smart Home. *Learning* **2009**, *24*, 53–64.

5. Rogers, A.; Corkill, D.D.; Jennings, N.R. Agent Technologies for Sensor Networks. *IEEE Intell. Syst.* **2009**, *24*, 13–17. [CrossRef]

6. Zato, C.; Villarrubia, G.; Sánchez, A.; Barri, I.; Rubión, E.; Fernández, A.; Rebate, C.; Cabo, J.A.; Álamos, T.; Sanz, J.; et al. PANGEA—Platform for automatic coNstruction of orGanizations of intElligent agents. In *Advances in Intelligent and Soft Computing*; Springer: Berlin/Heidelberg, Germany, 2012; Volume 151, pp. 229–239.

7. Gast, M. *802.11 Wireless Networks: The Definitive Guide*; O'Reilly: Sebastopol, CA, USA, 2005.

8. Ahmed, A.; Ali, J.; Raza, A.; Abbas, G. Wired vs Wireless Deployment Support for Wireless Sensor Networks. In Proceedings of the Tencon 2006—2006 IEEE Region 10 Conference, Hong Kong, China, 14–17 November 2006; pp. 1–3.

9. Giannakis, E.; Bruggeman, A. The highly variable economic performance of European agriculture. *Land Use Policy* **2015**, *45*, 26–35. [CrossRef]

10. Migliaccio, K.W.; Morgan, K.T.; Fraisse, C.; Vellidis, G.; Andreis, J.H. Performance evaluation of urban turf irrigation smartphone app. *Comput. Electron. Agric.* **2015**, *118*, 136–142. [CrossRef]

11. Zhang, K. Parameter Identification for Root Growth based on Soil Water Potential Measurements—An Inverse Modeling Approach. *Procedia Environ. Sci.* **2013**, *19*, 574–579.

12. Zhang, K.; Hilton, H.W.; Greenwood, D.J.; Thompson, A.J. A rigorous approach of determining FAO56 dual crop coefficient using soil sensor measurements and inverse modeling techniques. *Agric. Water Manag.* **2011**, *98*, 1081–1090. [CrossRef]

13. Vellidis, G.; Tucker, M.; Perry, C.; Kvien, C.; Bednarz, C. A real-time wireless smart sensor array for scheduling irrigation. *Comput. Electron. Agric.* **2008**, *61*, 44–50. [CrossRef]

14. Bartlett, A.C.; Andales, A.A.; Arabi, M.; Bauder, T.A. A smartphone app to extend use of a cloud-based irrigation scheduling tool. *Comput. Electron. Agric.* **2015**, *111*, 127–130. [CrossRef]

15. Davis, S.L.; Dukes, M.D. Irrigation scheduling performance by evapotranspiration-based controllers. *Agric. Water Manag.* **2010**, *98*, 19–28. [CrossRef]

16. Navarro-Hellín, H.; Martínez-del-Rincon, J.; Domingo-Miguel, R.; Soto-Valles, F.; Torres-Sánchez, R. A decision support system for managing irrigation in agriculture. *Comput. Electron. Agric.* **2016**, *124*, 121–131. [CrossRef]

17. Giusti, E.; Marsili-Libelli, S. A Fuzzy Decision Support System for irrigation and water conservation in agriculture. *Environ. Model. Softw.* **2015**, *63*, 73–86. [CrossRef]

18. Smolka, M.; Puchberger-enengl, D.; Bipoun, M.; Fercher, G.; Klasa, A.; Krutzler, C.; Keplinger, F.; Vellekoop, M.J. A new injection method for soil nutrient analysis in capillary electrophoresis. *Proc. SPIE* **2013**, *8763*, 1–7.

19. Falagán, N.; Artés, F.; Gómez, P.A.; Artés-Hernández, F.; Pérez-Pastor, A.; De La Rosa, J.M.; Aguayo, E. Combined effects of deficit irrigation and fresh-cut processing on quality and bioactive compounds of nectarines. *Hortic. Sci.* **2015**, *42*, 125–131. [CrossRef]

20. Kioutsioukis, I.; de Meij, A.; Jakobs, H.; Katragkou, E.; Vinuesa, J.-F.; Kazantzidis, A. High resolution WRF ensemble forecasting for irrigation: Multi-variable evaluation. *Atmos. Res.* **2016**, *167*, 156–174. [CrossRef]

21. Chatzikostas, G.; Boskidis, I.; Symeonidis, P.; Tekes, S.; Pekakis, P. Enorasis. *Procedia Technol.* **2013**, *8*, 516–519. [CrossRef]

22. Conesa, M.R.; Torres, R.; Domingo, R.; Navarro, H.; Soto, F.; Pérez-Pastor, A. Maximum daily trunk shrinkage and stem water potential reference equations for irrigation scheduling in table grapes. *Agric. Water Manag.* **2016**, *172*, 51–61. [CrossRef]

23. Conesa, M.R.; de la Rosa, J.M.; Domingo, R.; Bañon, S.; Pérez-Pastor, A. Changes induced by water ambient agents: Embedded agents for remote control and monitoring using the PANGEA platform grown in pots. *Sci. Hortic.* **2016**, *202*, 9–16. [CrossRef]

24. Albiac Murillo, J.; Kahil, M.T.; Dinar, A.; Esteban Gracia, E.; García, M.; Avella, L. El debate sobre la gestión sostenible de los recursos hídricos: Evidencia empírica de la Cuenca del Júcar. *Boletín Inter Cuencas* **2015**, *44*, 12.

25. Nikolidakis, S.A.; Kandris, D.; Vergados, D.D.; Douligeris, C. Energy efficient automated control of irrigation in agriculture by using wireless sensor networks. *Comput. Electron. Agric.* **2015**, *113*, 154–163. [CrossRef]

26. Aifro WaterEco. Available online: https://www.aifro.com/aifro-watereco.html (accessed on 19 June 2017).
27. Blossom Makes Your Sprinkler System Smarter. Available online: http://myblossom.com/ (accessed on 19 June 2017).
28. BlueSpray. Available online: https://www.bluespray.net/ (accessed on 19 June 2017).
29. Greeniq. Available online: http://greeniq.co/ (accessed on 19 June 2017).
30. Irrigationcaddy. Available online: http://irrigationcaddy.com/ (accessed on 19 June 2017).
31. Lono. Available online: https://lono.io/ (accessed on 19 June 2017).
32. Orbit B-Hyve. Available online: https://www.orbitonline.com/products/sprinkler-systems/timers/timers/outdoor-wifi-timer/b-hyve-12-station-smart-wifi-sprinkler-timer-3455 (accessed on 19 June 2017).
33. Rachio Smart Sprinkler Controller. Available online: https://rachio.com/ (accessed on 19 June 2017).
34. Rainmachine. Available online: http://www.rainmachine.com/ (accessed on 19 June 2017).
35. Spruce Irrigation. Available online: http://www.spruceirrigation.com/ (accessed on 19 June 2017).
36. Raincommander. Available online: http://www.raincommander.com (accessed on 19 June 2017).
37. Oikarinen, J.; Reed, D. RFC1459. *Netw. Work. Gr.* **1993**, *28*, 65.
38. Villarrubia, G.; De Paz, J.F.; Bajo, J.; Corchado, J.M. Ambient agents: embedded agents for remote control and monitoring using the PANGEA platform. *Sensors* **2014**, *14*, 13955–13979. [CrossRef] [PubMed]
39. Raeth, P.G. *Expert Systems: A Software Methodology for Modern Applications*; IEEE Computer Society Press: Washington, DC, USA, 1990.
40. Cemus, K.; Cerny, T.; Donahoo, M.J. Automated Business Rules Transformation into a Persistence Layer. *Procedia Comput. Sci.* **2015**, *62*, 312–318. [CrossRef]
41. Arduino—Open Source Products for Electronic Projects. Available online: http://www.arduino.org/ (accessed on 18 January 2017).
42. Libelium, C.-H. Open Garden. Available online: https://www.cooking-hacks.com/open-garden-outdoor-1node-1gw (accessed on 18 January 2017).
43. El Faouzi, N.-E.; Klein, L.A. Data Fusion for ITS: Techniques and Research Needs. *Transportation Research Procedia* **2016**, *15*, 495–512. [CrossRef]
44. Mousa, A.K.; Croock, M.S.; Abdullah, M.N. Fuzzy based Decision Support Model for Irrigation System Management. *Int. J. Comput. Appl.* **2014**, *104*, 14–20.
45. Güçlü, Y.S.; Subyani, A.M.; Şen, Z. Regional fuzzy chain model for evapotranspiration estimation. *J. Hydrol.* **2017**, *544*, 233–241. [CrossRef]
46. Bahat, M.; Inbar, G.; Yaniv, O.; Schneider, M. A fuzzy irrigation controller system. *Eng. Appl. Artif. Intell.* **2000**, *13*, 137–145. [CrossRef]
47. GitHub-zerokol/eFLL. Available online: https://github.com/zerokol/eFLL (accessed on 20 July 2017).
48. Patil, P.; Desai, B.L. Intelligent Irrigation Control System by Employing Wireless Sensor Networks. *Int. J. Comput. Appl.* **2013**, *79*, 33–40. [CrossRef]
49. Khan, F.; Shabbir, F.; Tahir, Z. A Fuzzy Approach for Water Security in Irrigation System Using Wireless Sensor Network. *Sci. Int.* **2014**, *26*, 1065–1070.

sensors

Communication

A True-Color Sensor and Suitable Evaluation Algorithm for Plant Recognition

Oliver Schmittmann * and Peter Schulze Lammers

Institute of Agricultural Engineering, University Bonn, 53115 Bonn, Germany; lammers@uni-bonn.de
* Correspondence: o.schmittmann@uni-bonn.de; Tel.: +49-228-73-3054

Received: 12 May 2017; Accepted: 4 August 2017; Published: 8 August 2017

Abstract: Plant-specific herbicide application requires sensor systems for plant recognition and differentiation. A literature review reveals a lack of sensor systems capable of recognizing small weeds in early stages of development (in the two- or four-leaf stage) and crop plants, of making spraying decisions in real time and, in addition, are that are inexpensive and ready for practical use in sprayers. The system described in this work is based on free cascadable and programmable true-color sensors for real-time recognition and identification of individual weed and crop plants. The application of this type of sensor is suitable for municipal areas and farmland with and without crops to perform the site-specific application of herbicides. Initially, databases with reflection properties of plants, natural and artificial backgrounds were created. Crop and weed plants should be recognized by the use of mathematical algorithms and decision models based on these data. They include the characteristic color spectrum, as well as the reflectance characteristics of unvegetated areas and areas with organic material. The CIE-Lab color-space was chosen for color matching because it contains information not only about coloration (a- and b-channel), but also about luminance (L-channel), thus increasing accuracy. Four different decision making algorithms based on different parameters are explained: (i) color similarity (ΔE); (ii) color similarity split in ΔL, Δa and Δb; (iii) a virtual channel 'd' and (iv) statistical distribution of the differences of reflection backgrounds and plants. Afterwards, the detection success of the recognition system is described. Furthermore, the minimum weed/plant coverage of the measuring spot was calculated by a mathematical model. Plants with a size of 1–5% of the spot can be recognized, and weeds in the two-leaf stage can be identified with a measuring spot size of 5 cm. By choosing a decision model previously, the detection quality can be increased. Depending on the characteristics of the background, different models are suitable. Finally, the results of field trials on municipal areas (with models of plants), winter wheat fields (with artificial plants) and grassland (with dock) are shown. In each experimental variant, objects and weeds could be recognized.

Keywords: CIE-Lab; precision plant protection; optical sensor; weed control

---

1. Introduction

The use of pesticides in agriculture and green areas is regarded critically. The possible impact of residues on human health and the environment causes decreasing societal acceptance. High costs of pesticides, as well as political regulations call for a reduction of herbicide application in agriculture [1,2]. For the chemical industry, it has become more expensive to develop new agents, but alternative weeding methods have deficiencies, e.g., mechanical methods do not allow weed control on the total crop area and thermal and applications with bio-herbicides are uneconomic [3].

One approach to reduce the amount of pesticides is spot application by decentralized injection of agents into the individual nozzles of a sprayer boom [4–10]. These treatments can be applied with total herbicides, corresponding selective herbicides, with bio-herbicides or alternative mechanical or thermal measures [11]. However, a precondition is the availability a high-resolution plant recognition system with real-time capability for triggering the actuators. Imaging methods for this purpose do not have this ability yet. The assignment of plants is difficult because plant contours can overlap, and the elapse and response times for real-time processing are not applicable [12].

Conventional RGB color sensors consisting of optical components are real-time capable, but do not have sufficient color accuracy. IR-sensors are not able to distinguish between plant types. High-precision spectrometers are too expensive for agricultural use and are not real-time capable.

The aim of the research is to develop a programmable sensor system, consisting of different single sensors, for the identification of individual crop plants or weeds in municipal and agricultural land and to initiate site-specific treatments.

Each sensor should cover a small spot of about 20 cm^2, analyze it and make a decision for or against spraying. The sensor should use the algorithm to develop and switch a valve next to one nozzle in real time. It is expected that small plants, for example in the two-leaf stage, can be detected. The advantage is that small plants can be eliminated with a small dosage very effectively, so that costs and impacts on the environment are reduced.

As a first step, each plant has to be detected. This is sufficient if no differentiation is needed and when unselective treatments are intended (e.g., with glyphosate or bio-herbicides). The second step is to distinguish between crop plants, weeds and plants which can be tolerated.

Smart elements are stated in the fact that each sensor can be programed individually with different decision models for different tasks. Previously, before starting the application, each sensor has to be adjusted. Databases with reflection properties are uploaded on each sensor. Finally, the appropriate decision model and algorithm will be selected and uploaded, as well.

- Positive recognition:

Spraying is performed if any plant is detected. The reflection properties are within a specified range. Application areas are weed destruction on municipal land or the use as an alternative to glyphosate.

- Negative recognition:

No spraying is performed if a crop plant is detected. The reflection properties are out of a specified range. Areas of application are weed destruction in row crops like sugar beet or maize.

- Recognition and differentiation between crops and weeds:

The range of reflection properties of different plants is compared with the database. Plants are assigned as crop, weed and harmless weed. Application areas are weed destruction on farmland. For example:

○ Arable land with crops like wheat or rape
○ Green areas with grass and broadleaf dock or lawn/golf courses with clover

For this purpose, databases with reflection properties (average and range of values) of different backgrounds (e.g., gravel, stones, soil, grassland) are determined and compiled. Different shades of green, characterizing the spectrum of plant colors, are selected and used for developing algorithms and decision models. These models are based on the color similarity of mixed areas in regard to ΔE, which includes all reflection values, different color and luminance similarity (ΔL, Δa and Δb) and a virtual color channel. The modeling results of the recognition of the greens on different backgrounds are presented.

Additionally, the results of field trials with plants and artificial plants are presented. Exemplarily, the results of the application for plant detection on stones, in winter wheat (*Triticum aestivum* L.) and dock (*Rumex acetosa* L.) in grassland by the use of those algorithms are shown. Finally, the suitability of true-color sensor systems for plant recognition will be evaluated.

2. State of the Art

Site-specific plant-protection has been an important field of research over the last few years. Different sensors and methods are used to detect and recognize plants to make a decision for the use of herbicides. In this section, an overview of the state of the art of real-time feasible sensors is presented. It is focused on the suitability for the use in sprayers. Airborne methods or systems that do not refer to single plants or small spots are not taken into account in this overview. The systems can be divided into opto-electronic sensors, imaging techniques and contour-based systems.

2.1. Opto-Electronic Sensors

DetectSpray, Weed-Seeker and Green-Seeker are systems for the detection of herbaceous plants on bare soil in reflection mode [13]. The detection principle is based on the fact that green plants absorb red light in the wavelength range between 630 and 660 nm and are highly reflective in the NIR range between 750 and 1200 nm. Basically, in all systems, two monochromatic diodes are used for the R- and the IR-range. For this basic plant identification, the ratio of the R channel to the IR channel is used as a decision criterion [14,15]. When exceeding a threshold value, the existence of plants can be concluded. In contrast to DetectSpray, the Green- and Weed-Seeker [16] devices use an active light source [14].

Approaches about the assignment of indices to plant groups are reported in the literature [17]. For the differentiation of plants, Biller [14] used five sensitive photodiodes with different wavebands to get a 'spectral fingerprint' for specific plant types.

Studies on the response accuracy of monochromatic sensors in comparison to real crop plants with weed population showed correlation coefficients of 0.6–0.9 [18]. These systems are described by the Alberta Farm Machinery Research Centre to be negatively affected by sunlight, preventing correct detections. Further, shadowing can hamper the application, especially in row crops [19].

Weed-IT is an Australian system using a sensor scanning a strip of 1 m [20]. It contains an NIR sensor with a light source. According to the manufacturer, it is applicable at high speeds up to 25 km·h^{-1} on stubble fields [12,21].

Crop-Cycle (Fa. Holland Scientific, Lincoln, NE, USA) uses reflections in three different wavelengths (670, 730 nm and NIR) for determining the nitrogen supply of plants. Using a calculation model with various indices including a preliminary calibration, the green color can be determined, as well. The manufacturer specifies the size of the measuring area as 20 cm in diameter [22].

Finally, traded under the name 'AmaSpot', a sensor-nozzle unit, which is based on the Weed-IT system, was awarded as a novelty in 2015 [23].

In these mentioned commercial systems, a 30 × 30 cm^2 area is scanned, and at least 3% of the scanned surface has to be green for a successful recognition. A scanning area of 50 × 50 cm^2 needs weeds with a size more than 75 cm^2. Consequently, either weeds have to be well developed (extended phenotype), or a high degree of coverage by weeds supports a successful recognition and herbicide application [24]. The distinction of plant type and the variation of the scan area size is not possible. Additionally, the system is not 'open' to the user.

Kluge [25] has stated that existing opto-electronic systems are not capable of the distinction between plants, but only generate information on the existence of plants. Therefore, the application of these sensor systems in agriculture is restricted to the period before crops emerge only.

In laboratory experiments, positive results for the determination of plant species by spectrometers are mentioned. Feyaert [26] stated that the differentiation of the reflection characteristics of plant species refers to their physical differences: in the red wavelength by chlorophyll content, which, however, depends on external factors (diseases, water and nutrients), and in the NIR wavelength of the internal structure of the plant, such as cell size and cell wall texture, waxes and trichomes.

2.2. Imaging Techniques

Imaging techniques are based on camera systems (CCD camera, bi-, multi- or hyper-spectral) with appropriate optical components and post-processing software. Plant contours can be detected if plants are freestanding [27]. The use of IR-channel shows good results to differentiate between soil and plants [28–30]. Imaging techniques are highly sensitive to varying external conditions. Extended computing power is required for weed detection based on shape factors [31–34]. Overlapping of plant parts makes recognition and plant type differentiation more difficult. 3D camera systems (time-of-flight cameras) actively emit light with a defined wavelength and receive the reflection of the object. They generate real-time images of all three dimensions and additionally a grey-scale image [35]. Time-of-flight cameras are a technical advancement and improve the quality of plant identification, but due to their low resolution and high costs, they are less suitable for practical use.

2.3. Plant Identification through Plant Contours without a Camera

Various sensors for plant phenotyping are known. Light grid sensors, consisting of horizontally-cascaded light barriers, were successfully tested [36]. In the described trial, these sensors have been mounted on a carrier vehicle, and measurements have been conducted in a maize field with approximately 20 cm-high plants. Plant identification with light grid sensors in row crops only works for large, non-herbaceous weeds under undisturbed conditions. These sensors are not suitable for narrow-spaced crop recognition.

Other methods are based on distance sensors (laser and ultrasonic), which determine the contour of the crop plant [37]. Due to the measurement speed, dynamic oscillation of the carrier vehicle and the sensitivity to small changes in distance, this method is not effective [38].

In conclusion, it can be pointed out that the mentioned optoelectronic sensors were well evaluated in former times, but they are not able to detect small plants in early leaf stages (smaller than 3% of the measuring spot). Differentiation of plants seems to be very difficult. Imaging technologies are more complicated, need more computer performance and are too expensive for practical use. Such systems do measure the reflection, but not the real coloration of the object. The identification by the use of plant contours is complicated, as movements of the sprayer or overlapping plants interfere with the measurement of crops. The use of true-color sensors in combination with algorithms is a further development of the presented optoelectronic sensors.

3. Material and Methods

3.1. Materials

3.1.1. Sensor Technology

Our sensor development is based on the true-color PR0126C sensor of Premosys GmbH (Wiesbaum, Germany). Compared to other sensor systems, true-color sensors represent a compromise between expensive spectrometers with high color accuracy and low cost, but imprecise RGB sensors. The velocity of true-color sensors is much higher than the velocity of spectrometers [39]. The PR0126C sensor can be equipped with different lenses, which determine the spot size of measurement (spot sizes).

For true-color determination and technical implementation of color standards, true-color sensors are coated with interference filters. Because of this filter characteristics, they are highly capable of color measuring and more sensitive than human eyes (standardized according to the German Institute for Standardization DIN 5033 norm). The sensitivity of the filters is related to a defined spectral wavelength. After normalizing the sensor, the color values are assigned to XYZ coordinates. The XYZ space provides the basis for the conversion into other color spaces.

The obtained color information then is converted into the CIE-Lab color space. L characterizes the luminance (L: 0 = black, 100 = white). Channels a and b refer to the coloration (a: −128 = green,

127 = red; b: −128 = blue, 127 = yellow). The color space was introduced in 1976 by the International Commission Internationale de l'Eclairage (CIE) and is frequently used by 3D color systems. CIE-Lab is device independent. For plant recognition, the wide green range is a major advantage.

The true-color sensor consists of 19-diode hexagon color ICs (integrated circuits, Figure 1) supplied by Mazet GmbH (Jena, Germany) [40]. Each diode has three segments with interference filters for the colors red, green and blue. The components of the sensor are the color-IC, a trans-impedance amplifier, a light-emitting diode with a defined wavelength, a fiber optic for emitting light and receiving reflection and an optical lens. The dimension of the lens (focal length range, measuring spot size and form (point or rod lens)) influences the characteristics of the sensor system in regard to the resolution.

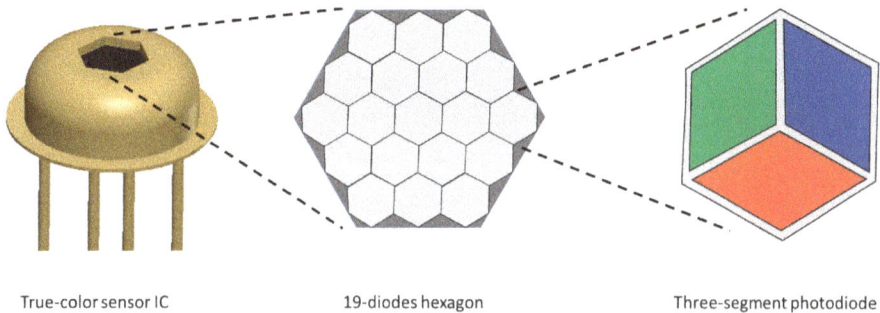

True-color sensor IC   19-diodes hexagon   Three-segment photodiode

Figure 1. Nineteen-diode hexagon sensor IC with three segment photodiodes for color detection.

The sensor was equipped with a double concave lens with a screen diameter of 50 mm (Figure 2). The spot size is about 20 cm^2.

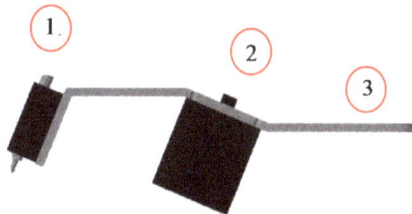

Figure 2. Design of the sensor-lens unit prepared for use in the experiments. Components: 1. True-color sensor with PC interface (RS232), power supply, switching outputs and two optical fibers (emitter and receiver); 2. Lens (double concave, diameter 5 cm, angle 22.5° connector for optical fiber); 3. Space for valve and nozzle.

3.1.2. Objects and Backgrounds

For mobile measurements, the backgrounds were divided into anthropogenic and natural. The artificial backgrounds include gravel, concrete slab or paving stones. Natural backgrounds are arable land without vegetation, with stubble or mulch. Green areas were also assigned to this category (grassland with and without dew). The backgrounds are displayed in Figure 3. To characterize different plants by color, four different green color cards from bright to dark green were used.

Figure 3. Representation of backgrounds. (**a**) Gray paving stones; (**b**) Stone steps; (**c**) Asphalt; (**d**) Fine chippings; (**e**) Gravel; (**f**) Red paving stones; (**g**) Sandy path; (**h**) Arable land with shadows; (**i**) Grassland; (**j**) Grassland with dew.

3.1.3. Test Facilities

(a)   Stationary Carrier:

Dynamic measurements on anthropogenic backgrounds were carried out by means of a driven rail system (Figure 4). The sensor was positioned at a predefined height and moved along a distance of 150 cm at a constant forward speed of 0.1 m s^{-1}. Different objects, backgrounds and mixed areas were placed below the sensor line. The measuring frequency was 10 Hz (recording of ten color values per second).

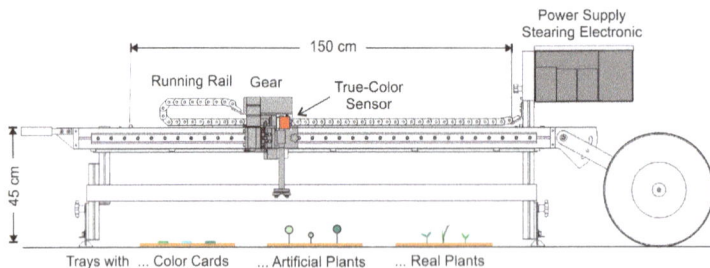

Figure 4. Stationary carrier for dynamic sensor tests.

(b)   Mobile Field Carrier:

For measurements in the field, a mobile carrier was built (Figure 5) for spray application. It can be trailed manually or by a vehicle. Time- and distance-based recordings of color values were performed.

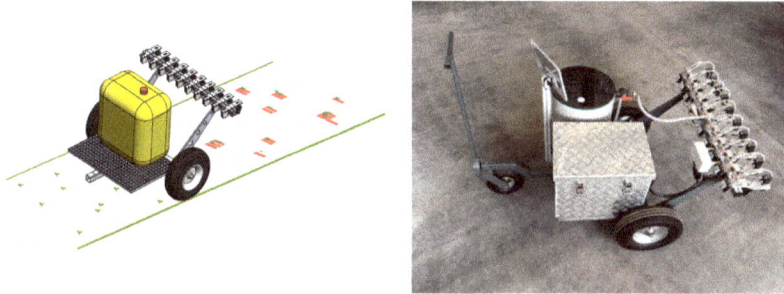

Figure 5. Concept (**Left**) and prototype (**Right**) of the mobile field carrier with spraying device.

3.2. Methods

3.2.1. Characterization of Backgrounds

In all experiments, objects (plants characterized by cards) and backgrounds (artificial and natural soils with and without vegetation) are characterized by Lab values and their variation (=noise). For this purpose, statistical indicators were used and presented by means of average and frequency distributions. Each background description contains more than 500 values. The backgrounds are divided into human (anthropogenic) and natural vegetation influenced (Figure 5).

3.2.2. Data Management and Processing

Data were collected by means of private domain software. The color values of objects and backgrounds were compiled and implemented into a database, which will be used for decision models and (real-time) plant identification. The setup, justification of the sensors, acquisition, visualization and recording of the data were realized by using this private domain software. However, the software is also important for other aspects, like course-controlled data acquisition, recording of data in a defined format, visualization of measured values and testing of algorithms.

3.2.3. Decision Making Based on Analyzing the Color Similarity of Mixed Areas

Mixed areas are defined as backgrounds covered with a large or small proportion of objects (plants). The database serves to estimate the potential of plant recognition and identification on different backgrounds. ΔE describes the color distance between two color values in the Lab-color space. According to ISO 12647 [41] and ISO 13655 [42], ΔE for this study is calculated by Equation (1):

$$\Delta E_{BG,MA} = \sqrt{(L_{BG} - L_{MA})^2 + (a_{BG} - a_{MA})^2 + (b_{BG} - b_{MA})^2} \tag{1}$$

$\Delta E_{BG,MA}$ = color similarity of background and mixed area (background with objects)
L_{BG} = luminance value of background
L_{MA} = luminance value of mixed area
a_{BG} = color value of background for Channel a
a_{MA} = color value of mixed area for Channel a
b_{BG} = color value of background for Channel b
b_{MAj} = color value of mixed area for Channel b

It describes the color similarity between the background without plants and the mixed area: background with plants (Table 1). The range of ΔE is used as a trigger for the decision procedure.

Table 1. Interpretation and evaluation of the ΔE values ([43], translated).

ΔE	Evaluation Categories
0.0...0.5	no to almost no difference
0.5...1.0	difference may be noticeable to the trained eye
1.0...2.0	weak perceptible color difference
2.0...4.0	perceived color difference
4.0...5.0	substantial difference in color, which is rarely tolerated
above 5.0	high difference defined as a different color

To get more information about color similarity, each channel was analyzed individually. The formulas are given in Equations (2)–(4):

$$\Delta L_{BG,MA} = \sqrt{\left(L_{BG} - L_{MA}\right)^2} \tag{2}$$

$$\Delta a_{BG,MA} = \sqrt{\left(a_{BG} - a_{MA}\right)^2} \tag{3}$$

$$\Delta b_{BG,MA} = \sqrt{\left(b_{BG} - b_{MA}\right)^2} \tag{4}$$

3.2.4. Decision Making Based on Modeling

A statistical model is based on the assumption that the difference of the color values of the background and mixed area should be bigger than the standard deviation of the background without objects and the standard deviation of the object (the sum of both standard deviations (Formula (5)). For decision making, a difference in only one channel may be sufficient. The information about the differences in each channel could be used to classify the plants.

An object exists if:

$$\begin{aligned} L_{MA} - L_{BG} &\geq \sigma_{LBG} + \sigma_{LObj} \\ \vee\, a_{MA} - a_{BG} &\geq \sigma_{aBG} + \sigma_{aObj} \\ \vee\, b_{MA} - b_{BG} &\geq \sigma_{bBG} + \sigma_{bObj} \end{aligned} \tag{5}$$

The required relative coverage area A is calculated as follows:

$$A_L = \frac{100 \cdot \left(\sigma_{LBG} + \sigma_{LObj}\right)}{\sqrt{\left(L_{Obj} - L_{BG}\right)^2}} \tag{6}$$

$$A_a = \frac{100 \cdot \left(\sigma_{aBG} + \sigma_{aObj}\right)}{\sqrt{\left(a_{Obj} - a_{BG}\right)^2}} \tag{7}$$

$$A_b = \frac{100 \cdot \left(\sigma_{bBG} + \sigma_{bObj}\right)}{\sqrt{\left(b_{Obj} - b_{BG}\right)^2}} \tag{8}$$

A_L, A_a, A_b = minimal relative area with plants covered for Channels L, a and b
$L_{Obj}, a_{Obj}, b_{Obj}$ = color values of the object for Channels L, a and b
L_{BG}, a_{BG}, b_{BG} = color values of the background for Channels L, a and b
$\sigma_{LObj}, \sigma_{aObj}, \sigma_{bObj}$ = standard deviation of the object for Channels L, a and b

σ_{LBG}, σ_{aBG}, σ_{bBG} = standard deviation of the background for Channels L, a and b

As an additional identification unit, a virtual 'd channel' is defined as the difference between the a and b value.

$$d = \sqrt{(a-b)^2} \tag{9}$$

4. Results and Discussion

4.1. Database to Characterize Backgrounds

In Table 2, the reflection values of the anthropogenic and natural backgrounds are listed and characterized by means Ø and standard deviations σ.

- Luminance:

The L-channel in the anthropogenic surfaces is located in a range from 6.39 to 19.31 (Table 2 and Appendix Figure A1). In comparison, the L-channel values of natural backgrounds are in a range from 7.57 to 12.03. Anthropogenic and natural backgrounds are in the same range in this channel. Also important is the variation of these values: it is very conspicuous that the arable land has the smallest standard deviation (0.13). It could be an indicator that even small differences in this channel may be sufficient to detect plants. The biggest deviations are in gravel (1.05, dark and bright stones), red paving stones (1.28, red stone and grey joint) and grassland with dew (2.23).

Table 2. Statistical description of reflection values of selected anthropogenic and natural backgrounds by the mean Ø and standard deviation σ.

	L				a				b			
	Ø	σ	Ø−σ	Ø+σ	Ø	σ	Ø−σ	Ø+σ	Ø	σ	Ø−σ	Ø+σ
Anthropogenic Backgrounds												
Grey paving stones	18.69	0.62	18.07	19.31	2.80	1.00	1.80	3.80	20.81	0.54	20.27	21.35
Asphalt	12.91	0.34	12.57	13.25	2.16	1.71	0.45	3.87	13.53	1.12	12.41	14.65
Gravel	7.44	1.05	6.39	8.49	3.15	1.00	2.15	4.15	12.24	1.46	10.78	13.70
Stone steps	13.55	0.50	14.05	14.05	2.34	0.52	1.82	2.86	15.44	0.76	14.68	16.20
Fine chippings	15.55	0.55	15.00	16.10	3.90	0.50	3.40	4.40	18.52	0.84	17.68	19.36
Red paving stones	15.28	1.28	14.00	16.56	13.61	2.89	10.72	16.50	24.12	0.91	23.21	25.03
Sandy path	11.45	0.41	11.04	11.86	4.90	1.62	3.28	6.52	19.72	1.18	18.54	20.90
Natural Backgrounds												
Arable land	11.32	0.13	11.19	11.45	4.81	0.70	4.11	5.51	20.10	0.47	19.93	20.57
Grassland	10.35	0.96	9.39	11.31	−2.37	2.40	−4.77	0.03	22.74	2.06	20.68	24.80
Grassland with dew	9.80	2.23	7.57	12.03	−6.32	2.54	−8.86	−3.78	19.47	3.32	16.15	22.79

- Coloration:

Channel a and Channel b give information about the coloration of the backgrounds. Excluding the red paving stones, all tested anthropogenic backgrounds in Channel a are in a range of 0.45–6.52. The range of natural backgrounds is from −8.86 to 5.51. The negative values are caused by the green color of grassland and can be an indicator for the presence of plants. It can be concluded that there is an overlapping of both kinds of background.

The b-channel is located in the paved surfaces in a range from 10.78 to 25.03 and in the natural backgrounds from 16.16 to 24.80. The range of natural backgrounds is within the range of anthropogenic surfaces.

4.2. Characterization of Different Green Tones in Regard to Plant Recognition

In Table 3, the Lab-values for four different green tones are exemplarily listed:

- Luminance:

The variation of the selected greens is higher than the variation of the backgrounds. The range of Channel L is between 29 and 45. In comparison to the backgrounds, the luminance is an outstanding criterion to detect plants. Furthermore, the magnitude of L could be a criterion to differentiate plants and weed.

- Coloration:

Channel a values are between −15 and −62 and differ in greens significantly from the chosen backgrounds. This wide range of values supports the idea of differentiating different plants by the use of true-color sensors and the CIE-Lab color space. It would be sufficient as the sole criterion for triggering a further evaluation step. In conclusion, the different shades of green can be clearly distinguished by the Lab channels.

Table 3. Description of the reflection properties of different selected greens.

	Green 1			Green 2			Green 3			Green 4		
	L	a	b	L	a	b	L	a	b	L	a	b
Median	44.78	−15.21	44.47	40.45	−26.27	51.75	33.95	−49.86	33.39	29.16	−61.65	−4.33
Mean	44.80	−15.24	44.39	40.45	−26.18	51.80	33.95	−49.95	33.54	29.15	−61.51	−4.22
Minimum	44.71	−14.85	43.79	40.39	−16.83	50.82	32.95	−49.61	32.95	28.98	−63.01	−3.25
Maximum	44.84	−15.73	44.69	40.55	−25.60	52.91	34.47	−50.53	34.47	29.32	−60.35	−4.96

4.3. Object Identification by Analyzing Color Similarity

The following figures display the described delta values depending on the background characteristics/scattering. For discrimination, a threshold has to be determined. In Figure 6, the delta-signals for solid background are shown exemplarily. This design is the most promising task for weed recognition.

The four objects (r = 1 cm) are detected accurately. The ΔE and ΔL are highly responsive to colored points. By setting a threshold of five, all green objects can be filtered out by the use of ΔE and ΔL. The threshold for Δa and Δb is two. If a measured value is higher, an object/plant exists. To determine the color (kind of plant), the relation of each delta value has to be assessed. The relation between ΔL, Δa and Δb suits the characterization and identification of objects.

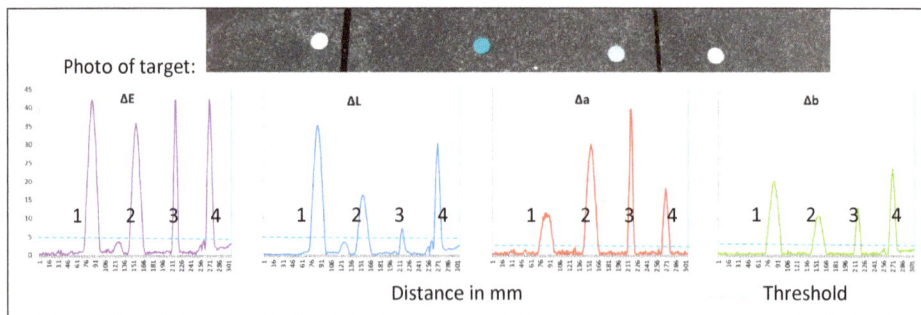

Figure 6. Reflection of paved ground with different colored cards, ΔE, ΔL, Δa, Δb and the defined threshold for detection.

The data in Figure 7 are recorded in a young wheat stand (BBCH 13 [44]; Figure 6). Herbicide application at this point in time is common. It is a good example for plant recognition and site-specific spraying in existing crops with selective herbicides. In contrast to the literature [25], it can be shown that the differentiation of plant and weeds is possible by the use of a threshold. With a threshold of 12, all six objects can be identified correctly.

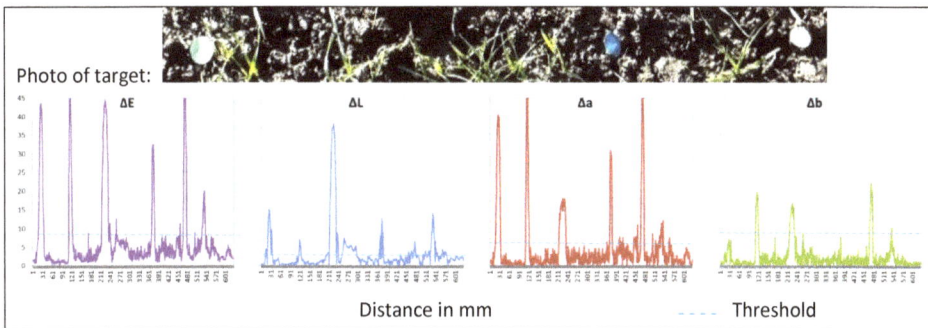

Figure 7. Reflection of a winter wheat field with different colored cards, ΔE, ΔL, Δa, Δb and the defined threshold for detection.

Figure 8 shows the results of the detection of broadleaf dock on grassland (dock plants are toxic for some animals [45]). A special task is to detect green plants on green areas. ΔE and ΔL are suitable to recognize dock by the use of a threshold of about seven. Δa provides no significant signals. The green color of both plants is quite similar. The reason why the differences in ΔL are higher is caused by the leaf position. A horizontal arrangement of leaves causes higher reflection intensity as the vertical arrangement of leaves from grass.

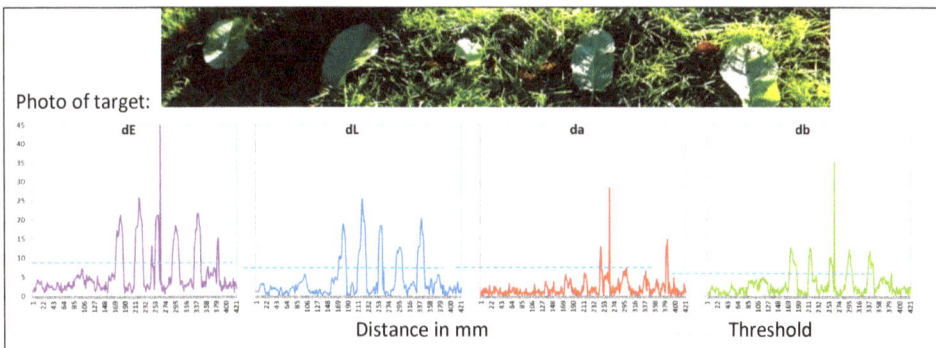

Figure 8. Reflection of grassland with dock, ΔE, ΔL, Δa, Δb and the defined threshold for detection.

4.4. Object Identification by Modeling

The result of the decision-based mathematical model (Section 3.2.4) is displayed in Table 4. For each background, the relative object (plant) size of each of the four greens is calculated. Exemplarily, the valuation is displayed for lens diameters of five (spot size 20 cm^2) and ten centimeters (spot size ~80 cm^2).

According to Table 4, it can be concluded that the quality of recognition for all channels is different. As described previously, the influence of the background is relatively low. Green 1 can be identified by the use of the luminance very well. A coverage of 0.5% on the field should be enough for detection. Channel a is also quite suitable, except on red paving stones.

For Channel L, the sufficient cover of the measurement spot is between 0.5% on fields (Green 1) and up to 13.3% on grassland with dew (Green 4). It is evident that soil without vegetation has the best detection success. The virtual channel d (the difference between a and b) does not show a big advantage. d is more suitable than b, but has no advantage in comparison to a.

Table 4. Required relative coverage of different shades of green on backgrounds for Channels L, a, b and Virtual Channel d.

	Green 1				Green 2				Green 3				Green 4			
	L	a	b	d	L	a	b	d	L	a	b	d	L	a	b	d
Gray pavings	2.5	7.1	3.5	3.0	3.6	7.7	8.5	7.1	5.9	3.6	16.4	5.2	9.2	5.7	6.8	12.2
Asphalt cover	1.2	11.5	4.6	4.4	1.8	10.3	8.4	7.7	2.9	5.0	13.3	5.9	4.2	6.8	12.8	12.3
Gravel	2.9	7.0	5.4	3.5	3.7	7.6	9.0	6.9	5.0	3.6	14.1	5.3	6.4	5.6	15.9	11.0
Flagged floor	1.8	4.6	3.6	3.4	2.5	6.1	7.8	7.1	3.8	2.8	12.7	5.3	5.4	5.0	9.8	11.6
Concrete	2.1	4.1	4.3	2.5	2.9	5.7	8.8	6.6	4.5	2.6	15.9	4.8	6.5	4.8	8.8	10.9
Red pavings	4.5	18.0	5.9	5.6	5.7	15.4	10.9	8.6	8.4	9.2	26.2	6.7	11.7	10.0	7.3	13.4
Sandy path	1.4	9.5	5.9	4.9	2.0	9.1	10.2	8.2	3.1	4.6	19.8	6.3	4.2	6.4	9.8	13.5
Field	0.5	4.9	3.1	2.5	1.0	6.2	8.1	6.6	1.8	3.0	15.0	4.8	2.6	5.1	6.7	8.0
Grassland	2.9	21.0	10.8	8.8	3.7	15.2	14.3	11.5	5.3	7.0	33.6	8.9	6.9	8.5	11.9	20.4
Grassland+dew	6.5	31.8	14.4	15.2	7.8	18.9	16.8	15.6	10.4	7.9	34.8	12.7	13.3	9.4	18.8	25.2

	Spot diameter 5 cm (~20 cm^2)				...10 cm (~80 cm^2)				
0–3.0		less:	0.6	cm^2		less:	2.4	cm^2	plant shape size
3.1–5.0			1	cm^2			4	cm^2	
5.1–7.5			1.5	cm^2			6	cm^2	
7.6–10			2	cm^2			8	cm^2	

5. Summary and Conclusions

True-color sensors are a good compromise between inexpensive RGB sensors and spectrometers with high spectral resolution. The objective was to study the suitability of this kind of sensor for the detection and differentiation of crop and weed plants in agricultural and municipal areas.

For plant-specific spraying, the boom of a sprayer can be equipped with those sensors. The sensor, valve and nozzle together make up an independent sensor-valve-nozzle unit. Depending on sensor spot and spraying angle of the nozzles, the distances between each nozzle can be up to 50 cm. Each true-color sensor is able to control one valve for one nozzle. If spot diameter and nozzle distance are not the same, more sensors (=sensor array) can control the same valve. The possible detection success and identification of small plants is calculated and tested by the use of different algorithms. The spot sizes should be between 20 cm^2 and 80 cm^2 to detect plants in the two-leaf stage. The presented detection method contains the following steps:

1. Normalization and adjusting of the sensor in the field. On a place without weeds, the background properties are determined, and a threshold for discrimination will be defined.
2. Selection of the mode: positive or negative detection.
3. Selection and upload of the algorithms and database with reflection properties to each sensor-valve-nozzle unit.
4. Running the plant recognition and differentiation process.
5. Control of the sprayer valve.

The used CIE-Lab color space is suitable for plant recognition, due to the distinction between luminance and coloration. The a-channel (green-red) of the color space is very sensitive for the discrimination of green-colored plants. Databases with reflection properties are assigned. It was shown that backgrounds and green objects are quite different especially in luminance and Channel a.

Four methods for detection are presented:

1. The detection based on color similarity ΔE,
2. The splitting color similarity into ΔL, Δa and Δb,
3. A Virtual Channel d and
4. A modeling algorithm based on statistics.

The suitability of the methods are introduced and have been applied for chosen true backgrounds and different green tones. Methodically, the detection quality and the potential of the sensor were tested under defined conditions on fields with and without vegetation. Based on the experiments carried out, the following statements can be made:

- Due to the built-in single sensor controller, actuators can be addressed and activated in real time.
- A specified evaluation algorithm for plant identification, the decision model and the current calibration values have to be updated to a central computer.
- Minimal differences in coloration (a- and b-channel) and reflectivity can be detected with the sensor. Coloration properties are applicable for plant identification.
- The differentiation of plants in crop and weed appears to be possible by a multistage model. The procedure is as follows:

    1. Extraction of suspicious points by the consideration of ΔE.
    2. Targeted analysis of these signals by the use of the individual Channels L, a and b and
    3. Comparison of all four channels (including Virtual Channel d) relative to each other to decrease the influence of the object size.

- The relative weed/plant coverage of the measuring spot was calculated by a mathematical model. By choosing a decision model previously, the detection quality can be increased. Depending on the background characteristics, different models are more suitable.
- Plants with a size of 1–5% of the measuring spot can be recognized. Weeds in the two-leaf stage can be identified.
- The detection success of the recognition system is displayed and described in field tests. Field trials on municipal areas (with models of plants), winter wheat fields (with models of plants) and grassland (with dock) are shown. In the experiment variants, objects and weeds can be recognized.

It can be stated that true-color sensors are able to detect small differences in luminance and the coloration of objects. They are real-time capable, easy to use and inexpensive. The sensor system is open for the user and can be adapted to the individual condition. In combination with the presented algorithms, it was proven that the sensor has the potential to differentiate between crop and weed.

In comparison to the existing optoelectronic systems presented in this paper, true-color sensors are further developed. Plants can be detected and discriminated. Even a discrimination of green plants is possible in some cases. An important next step is to evaluate this sensor system under a 'real field condition' in different crops. The amounts of savings of herbicides will be the most convincing evaluation parameter.

Acknowledgments: This project was funded by the Deutsche Bundesumweltstiftung (DBU). We thank Matthias Kuhl from the Premosys company for the cooperation and providing the CIE-Lab sensors and technical device.

Author Contributions: Oliver Schmittmann designed the sensor system and performed the experiments; Peter Schulze Lammers supported the cooperation with the Premosys company and writing of the paper.

Conflicts of Interest: The authors declare no conflict of interest.

Appendix A

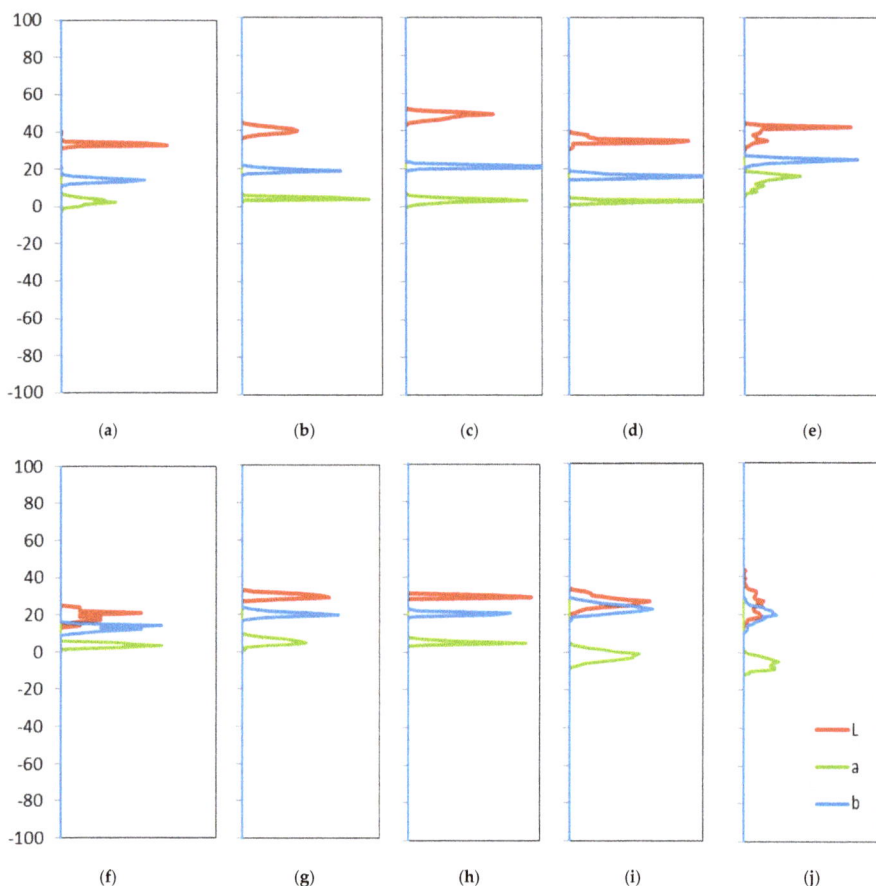

Figure A1. Signal distribution of backgrounds for Channels L, a and b. (**a**) Asphalt; (**b**) Fine chippings; (**c**) Red paving stones; (**d**) Stone steps; (**e**) Gray paving stones; (**f**) Gravel; (**g**) Sandy path; (**h**) Arable land; (**i**) Grazing land; (**j**) Grassland with dew.

References

1. Bundesministerium für Ernährung und Landwirtschaft. Nationaler Aktionsplan zur Nachhaltigen Anwendung von Pflanzen-Schutzmitteln. Federal Ministry for Consumer Protection, Food and Agriculture of Germany, 2013. Available online: http://www.bmel.de/SharedDocs/Downloads/Broschueren/NationalerAktionsplanPflanzenschutz.pdf?__blob=publicationFile (accessed on 5 May 2017).
2. International Institute for Sustainable Development. FAO, WHO Join Forces to Reduce Risks Posed by Pesticides. 2017. Available online: http://sdg.iisd.org/news/fao-who-join-forces-to-reduce-risks-posed-by-pesticides/ (accessed on 18 June 2017).
3. Schmittmann, O.; Kam, H.; Schulze Lammers, P. Position steered sowing of sugar beet Technology and precision. In Proceedings of the 2nd International Conference on Machine Control & Guidance—MCG, Bonn, Germany, 9–11 March 2010.
4. Hloben, P. Study on the Response Time of Direct Injection Systems for Variable Rate Application of Herbicides (VDI-MEG Schriftenreihe 459). Ph.D. Thesis, University of Bonn, Bonn, Germany, 2007.

5.   Vondricka, J. Study on the Process of Direct Nozzle Injection for Real-Time Site Specific Pesticide Application (VDI-MEG Schriftenreihe 465). Ph.D. Thesis, University of Bonn, Bonn, Germany, 2008.

6.   Walgenbach, M. Aufbau und Untersuchung Eines Versuchsträgers zur Direktein-Speisung an der Düse (VDI-MEG Schriftenreihe 533). Ph.D. Thesis, University of Bonn, Bonn, Germany, 2014.

7.   Vondricka, J.; Schulze Lammers, P. Evaluation of a carrier control valve for a direct nozzle injection system. *Biosyst. Eng.* **2009**, *103*, 43–46. [CrossRef]

8.   Rockwell, A.D.; Ayers, P.D. A variable rate, direct nozzle injection field sprayer. *Appl. Eng. Agric.* **1996**, *12*, 531–538. [CrossRef]

9.   Paice, M.E.R.; Miller, P.C.H.; Bodle, J.D. An experimental sprayer for the spatially selective application of herbicides. *J. Agric. Eng. Res.* **1995**, *60*, 107–116. [CrossRef]

10. Downey, D.; Crowe, T.G.; Giles, D.K.; Slaughter, D.C. Direct Nozzle Injection of Pesticide Concentrate into Continuous Flow for Intermittent Spray Applications. *Trans. ASABE* **2006**, *49*, 865–873. [CrossRef]

11. Cordeau, S.; Triolet, M.; Wayman, S.; Steinberg, C.; Guillemin, J.-P. Bioherbicides. Dead in the water? A review of the existing products for integrated weed management. *Crop Prot.* **2016**, *87*, 44–49. [CrossRef]

12. Kempenaar, C.; Groeneveld, R.M.W.; Uffing, A.J.M. *Evaluation of Weed IT Model 2006 MKII. Spray Volume and Dose Response Tests*; Plant Research International Wageningen UR: Wageningen, The Netherlands, 2006.

13. Knittel, G.; Stahli, W. *Spritz- und Sprühverfahren in Pflanzenschutz und Flüssigdüngung bei Flächenkulturen*; Books on Demand: Berlin, Germany, 2001.

14. Biller, R. Optoelektronik zur Unkrauterkennung—Erste Erfahrungen beim Test unter simulierten Bedingungen und beim Einsatz auf Versuchsflächen Innovative Verfahren zur Unkrauterkennung. *KTBL Arb.* **1996**, *236*, 75–85.

15. Gibson, P.; Power, C. *Introductory Remote Sensing: Digital Image Processing and Applications*; Routledge: London, UK, 2000.

16. Trimble. 2017. Available online: http://www.trimble.com/agriculture/weedseeker.aspx (accessed on 24 April 2017).

17. Vrindts, E.; de Baerdemaeker, J. Optical discrimination of crop, weed and soil for on-line weed detection. In Proceedings of the First European Conference on Precision Agriculture, Warwick University Conference Centre, Coventry, UK, 7–10 September 1997.

18. Wartenberg, G.; Langner, H.-R.; Böttger, H.; Schmidt, H.; Ruckelshausen, A. *Messsystem zur Bewertung des Unkrautvorkommens*; Bornimer Agrartechnische Berichte: Potsdam, Germany, 2005.

19. Hanks, J.E.; Beck, J.L. Sensor-controlled hooded sprayer for row crops. *Weed Technol.* **1998**, *12*, 308–314.

20. WeedIT. 2017. Available online: www.weedit.com.au (accessed on 24 April 2017).

21. Visser, R.; Timmermans, A.J.M. Weed-It: A new selective weed control system. In Proceedings of the SPIE 2907, Optics in Agriculture, Forestry, and Biological Processing II. 120, Boston, MA, USA, 18 November 1996.

22. Holland Scientific. Crop-Cycle ACS 470 Manual. 2014. Available online: http://hollandscientific.com/wp-content/uploads/files/ACS430_Manual.pdf (accessed on 15 December 2014).

23. Köller, K.H. Landtechnische Innovationen auf der Agritechnica 2015. Available online: http://www.dlg.org/aktuell_landwirtschaft.html?detail/2015dlg/1/1/7922 (accessed on 18 December 2015).

24. Department of Agriculture and Rural Development, Alberta, Canada, 2013. The Detectspray Spraying System. Available online: http://www1.agric.gov.ab.ca/$department/deptdocs.nsf/all/eng7995 (accessed on 24 February 2015).

25. Kluge, A. Methoden zur Automatischen Unkrauterkennung für die Prozesssteuerung von Herbizidmaßnahmen. Ph.D. Thesis, University of Stuttgart, Stuttgart, Germany, 2011.

26. Feyaert, F.; Pollet, P.; van Gool, L.; Wambact, P. Vision system for weed detection using hyper-spectral imaging, structural field information and unsupervised training sample collection. In Proceedings of the British Crop Protection Conference, Brighton, UK, 15–18 November 1999; pp. 607–614.

27. Choi, K.H.; Han, A.H.; Han, S.H.; Park, K.-H.; Kim, K.S.; Kim, S. Morphology-based guidance line extraction for an autonomous weeding robot in paddy fields. *Comput. Electron. Agric.* **2015**, *113*, 266–274. [CrossRef]

28. Brivot, R.; Marchant, J.A. Segmentation of plants and weeds for a precision crop protection robot using infrared images. *IEE Proc.-Vis. Image Signal Process.* **1996**, *143*, 118–124. [CrossRef]

29. Dammer, K.-H.; Kim, D.-S. Real-time variable-rate herbicide application for weed control in carrots. *Weed Res.* **2016**, *56*, 237–246. [CrossRef]

30. Gerhards, R.; Christensen, S. Real-time weed detection, decision making and patch spraying in maize, sugarbeet, winter wheat and winter barley. *Weed Res.* **2003**, *43*, 385–392. [CrossRef]

31. Gerhards, R.; Oebel, H. Practical experiences with a system for site-specific weed control in arable crops using real-time image analysis and GPS-controlled patch spraying. *Weed Res.* **2006**, *46*, 185–193. [CrossRef]

32. Laursen, M.S.; Jørgensen, R.N.; Midtiby, H.S.; Jensen, K.; Christiansen, M.P.; Giselsson, T.M.; Mortensen, A.K.; Jensen, P.K. Dicotyledon Weed Quantification Algorithm for Selective Herbicide Application in Maize Crops. *Sensors* **2016**, *16*, 1848. [CrossRef] [PubMed]

33. Sökefeld, M.; Gerhards, R.; Oebel, H.; Therburg, R.-D. Image acquisition for weed detection and identification by digital image analysis. In *Precision Agriculture '07*; Stafford, J.V., Ed.; Academic Publishers: Wageningen, The Netherlands, 2007; pp. 523–528.

34. Sökefeld, M.; Gerhards, R.; Kühbauch, W. Teilschlagspezifische Unkrautkontrolle—Von der Unkrauterfassung bis zur Herbizidapplikation. *J. Plant Dis. Prot.* **2000**, *17*, 227–233.

35. Klose, R.; Penlington, J.; Ruckelshausen, A. Usability study of 3D Time-of-Flight cameras for automatic plant phenotyping. In Proceedings of the 1st International Workshop on Computer Image Analysis in Agriculture, Potsdam, Germany, 27–28 August 2009.

36. Fender, F.; Hanneken, M.; Linz, A.; Ruckelshausen, A.; Spicer, M. Messende Lichtgitter und Multispektralkameras als bildgebende Systeme zur Pflanzenerkennung. *Bornimer Agrartech. Ber.* **2005**, *40*, 7–16.

37. Lee, W.S.; Alchanatis, V.; Yang, C.; Hirafuji, M.; Moshou, D.; Li, C. Sensing technologies for precision specialty crop production. *Comput. Electron. Agric.* **2010**, *74*, 2–33. [CrossRef]

38. Schmittmann, O. Teilflächenspezifische Ertragsmessung von Zuckerrüben in Echtzeit unter besonderer Berücksichtigung der Einzelrübenmasse (VDI-MEG Schriftenreihe 401). Ph.D. Thesis, University of Bonn, Bonn, Germany, 2002.

39. Jensen, K.; Nimz, T. Welche Genauigkeit Erreicht Man Mit Farbsensoren und Mini-Spektrometern? (2015). Available online: http://www.mazet.de/de/downloads/produkt-kundeninformationen/white-paper/item/download/39_9302d1d2d74f08ece546cc7c63cf0386.html (accessed on 24 April 2017).

40. Mazet. 2017. Available online: www.mazet.de (accessed on 24 April 2017).

41. *ISO 12647. Graphic Technology—Process Control for the Production of Half-Tone Colour Separations, Proof and Production Prints*; International Organization for Standardization: Geneva, Switzerland, 2013.

42. *ISO 13655. Graphic Technology—Spectral Measurement and Colorimetric Computation for Graphic Arts Images*; International Organization for Standardization: Geneva, Switzerland, 2009.

43. EN ISO 11664-4, 1976. Colorimetry—Part 4: CIE 1976 L*a*b* Colour Space. Available online: https://de.wikipedia.org/wiki/Delta_E (accessed on 17 July 2017).

44. Meier, U.; Bleiholder, H.; Buhr, L.; Feller, C.; Hack, H.; Heß, M.; Lancashire, P.D.; Schnock, U.; Stau, R.; van den Boom, T.; et al. The BBCH system to coding the phenological growth stages of plants—History and publications. *J. Kulturpflanzen* **2009**, *61*, 41–52.

45. Van Evert, F.K.; Samsom, J.; Polder, G.; Vijn, M.; van Dooren, H.-J.; Lamaker, A. A robot to detect and control broad-leaved dock (*Rumex obtusifolius* L.) in grassland. *J. Field Robot.* **2011**, *28*, 264–277. [CrossRef]

sensors

MDPI

Article

Discrimination of Transgenic Maize Kernel Using NIR Hyperspectral Imaging and Multivariate Data Analysis

Xuping Feng [1], Yiying Zhao [1], Chu Zhang [1], Peng Cheng [2] and Yong He [1,*]

[1]   College of Biosystems Engineering and Food Science, Zhejiang University, Hangzhou 310058, China;
      pimmmx@163.com (X.F.); zhaoyy@zju.edu.cn (Y.Z.); chuzh@zju.edu.cn (C.Z.)
[2]   Institute of Quality and Standard for Agro-Products, Zhejiang Academy of Agricultural Sciences,
      Hangzhou 310021, China; pc_phm@163.com
*   Correspondence: yhe@zju.edu.cn; Tel.: +86-571-8898-2143

Received: 3 July 2017; Accepted: 13 August 2017; Published: 17 August 2017

Abstract: There are possible environmental risks related to gene flow from genetically engineered organisms. It is important to find accurate, fast, and inexpensive methods to detect and monitor the presence of genetically modified (GM) organisms in crops and derived crop products. In the present study, GM maize kernels containing both *cry1Ab/cry2Aj-G10evo* proteins and their non-GM parents were examined by using hyperspectral imaging in the near-infrared (NIR) range (874.41–1733.91 nm) combined with chemometric data analysis. The hypercubes data were analyzed by applying principal component analysis (PCA) for exploratory purposes, and support vector machine (SVM) and partial least squares discriminant analysis (PLS–DA) to build the discriminant models to class the GM maize kernels from their contrast. The results indicate that clear differences between GM and non-GM maize kernels can be easily visualized with a nondestructive determination method developed in this study, and excellent classification could be achieved, with calculation and prediction accuracy of almost 100%. This study also demonstrates that SVM and PLS–DA models can obtain good performance with 54 wavelengths, selected by the competitive adaptive reweighted sampling method (CARS), making the classification processing for online application more rapid. Finally, GM maize kernels were visually identified on the prediction maps by predicting the features of each pixel on individual hyperspectral images. It was concluded that hyperspectral imaging together with chemometric data analysis is a promising technique to identify GM maize kernels, since it overcomes some disadvantages of the traditional analytical methods, such as complex and monotonous sampling.

Keywords: classification; NIR hyperspectral imaging; chemometrics analysis

1. Introduction

Maize (*Zea mays* L.) is one of the most important agricultural commodities in the world, and also serves as a key ingredient in feed for livestock. It is used extensively in industrial products all over the world, including the production of renewable fuel [1]. The application of genetic transformation to maize has made rapid strides in the past decades to meet some specific requirements. Some agronomic traits, including enhancement of disease and insect pest tolerance [2], quality improvement [3], and increasing nutritional value [4], have been introduced into maize. In recent years, genetically-modified (GM) crop cultivation has been following the trend of combining two or more agronomical traits by transgenic breeding, referred to as "stacked" events [5]. The first binary transgenic event in GM maize production was mainly dominated by GM plants containing insect protection through endotoxin genes, conferred by *Bacillus thuringiensis* (*Bt*) as well as herbicide tolerance characteristics [5]. However, it has been argued that the use of GM techniques could possibly result in unpredictable adverse effects on food and environment safety.

These unintended effects include the transfer of an uncontrollable escape of exogenous genes into neighboring wild plants by pollen, the formation of toxins associated with GM food, and modification of the biodiversity of the host plant by changing the expression of the existing genes [6]. The introduction of genetically-modified organisms (GMOs) in agro-food markets should be accompanied by a regulatory need to monitor and verify the presence and amount of GM varieties to guarantee consumer safety. Consequently, there is a need for GMO detection methods that are accurate, fast, and inexpensive. Currently, there are several analytical methods proposed for the determination, characterization, and authentication of GMOs in crops and derived crop products, such as polymerase chain reaction (PCR)/restriction enzyme assay [7], enzyme-linked immunosorbent assays [8], lateral flow strip [9], and microarray [10]. As a whole, the DNA- and protein-based methods for the identification of GMOs are versatile, sensitive, and accurate. However, there are also some disadvantages—they are destructive, laborious, expensive, time-consuming, and require highly-skilled operators; thus, they are unsuitable for online process control [11]. As non-destructive, synchronous, and coherent detection tools, spectroscopic techniques are environmentally friendly, fast, and easy to operate without complex sample pretreatments.

The application of a method involving near-infrared (NIR) combined with chemometrics for the identification of GMOs in the agro-food market is feasible [12–14]. NIR is the region of the electromagnetic spectrum between 750 nm and 2500 nm, and NIR spectroscopy is often used to gather information on the relative proportions of C–H, N–H, and O–H bonds in organic molecules [11]. The basis of this technology for the detection of mutants, mediated by transgenic technology, is that it can identify phenotypic changes caused by genotypic changes, which ultimately bring about changes of organic molecular bonds [11]. Liu et al. (2014) distinguished GM rice seeds from their counterparts by using visible/near-infrared spectroscopy (VIS-NIR) spectroscopy combined with a chemometric tool with classification accuracy up to 100% with the least squares-support vector machine (LS-SVM) model [15]. Garcíamolina et al. (2016) applied NIR spectroscopy technology to discriminate GM wheat gain and flour from non-GM wheat lines [14]. Guo et al. (2014) also demonstrated that clear differences between GM and non-GM tomatoes could be identified by using VIS-NIR together with discriminant partial least squares regression with excellent classification accuracy of up to 100% [13]. However, conventional NIR—widely used for transgenic foods identification—lacks spatial dimension information. In contrast, NIR hyperspectral imaging combines traditional optical imaging and the spectral method which is capable of capturing images over broad contiguous wavelengths in the NIR region, and has received much attention in cereal science [16–18]. These images form a three-dimensional structure (x, y, λ) of multivariate data for processing and analysis, where x and y are the spatial dimensions (the number of rows and columns in pixels), and λ represents the number of wavelengths [19,20]. NIR hyperspectral imaging is a powerful spectroscopic tool for seed classification, quality discrimination, and detection of an object by obtaining visual information about the samples [21]. The benefits of using NIR hyperspectral imaging for cereal science are numerous, including disease and pest diagnoses [18,22], kernel density classification [16,17], seed moisture determination [23], and rice cultivar identification [24]. Currently, limited research has used this technique to distinguish GM from non-GM. Prior to this study, no research had mapped the spatial heterogeneity between GMOs from non-GM controls based on their different spectral signatures.

The purpose of this study was to investigate four goals: (1) to examine the feasibility of using NIR hyperspectral imaging techniques to identify GM maize kernels mediated by *Agrobacterium tumefaciens* and detect spatial heterogeneity in spectral variability; (2) to identify important wavelengths that identify the differences between GM and non-GM maize kernels; (3) to build an optimal discrimination model based on these important wavelengths to simplify the prediction model and to speed up the operation; and (4) to visualize the number and locations of GM maize kernel by developing imaging processing algorithms.

2. Materials and Methods

2.1. Maize Samples

The GM maize kernels used in this study (containing insecticidal and herbicide tolerant traits, *cry1Ab/cry2Aj-G10evo* genes) and their non-GM control were provided by the Institute of Insect Sciences, Zhejiang University, China. For the test maize, variety zhengdan958 was used as the GM acceptor line. Glyphosate tolerance of maize was obtained by expression of a mutant 5-enolpyruvylshikimate-3-phosphate synthase (EPSPS) enzyme. Insect resistance of the maize was obtained by expression of a *Bacillus thuringiensis* delta endotoxin protein. The transgenic maize was created by an *Agrobacterium tumefaciens*-mediated transformation system (Figure 1). There were no other differences between the transgenic maize and the non-transgenic control kernels. The GM and non-GM maize crops were grown in the same field to eliminate any environmental effects.

Intact samples of 1050 transgenic maize kernels and 1050 non-transgenic maize kernels were used for image acquisition. In total, 1050 samples of each genotype were randomly selected to form the calibration and prediction sets in a ratio of 2:1. Thus, there were 700 samples used for the calibration set and 350 samples used for the prediction set. Samples were classified according to the genetic background by classification model, which preferably should be approximate to the values assigned. In this study, the spectral data from GM maize kernels were assigned 1, and those of non-GM maize kernels were assigned 2.

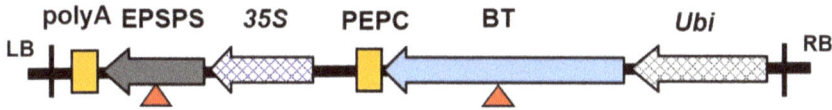

Figure 1. Structure of the plant expression vector containing coding regions of the *cry1Ab/cry2Aj-G10evo* genes. LB is left border; RB is the right border; poly A is a terminator; PEPC is a terminator; 35S is a promoter; Ubi is a promoter; EPSPS denotes the herbicide-resistant genes; BT denotes the insect-resistant genes; EPSPS and BT are marked with a red triangle.

2.2. Near-Infrared Hyperspectral Imaging

A ground hyperspectral imaging system was used to acquire NIR hyperspectral images. This system's equipment mainly consists of the following devices: a N17E-QE imaging spectrograph (Spectral Imaging Ltd., Oulu, Finland), two 150 W tungsten halogen lamps (Fiber-Lite DC950 Illuminator; Dolan Jenner Industries Inc., Boxborough, MA, USA) for illumination, a high-performance CCD camera (Hamamatsu, Hamamatsu City, Japan) coupled with a C-mount imaging lens (OLES22; Specim, Spectral Imaging Ltd., Oulu, Finland), a displacement platform driven by a stepper motor (Isuzu Optics Corp., Zhubei, Taiwan) to move the samples, and a computer. The hyperspectral imaging system acquires spectra in the form of pixels from the range of 874–1734 nm with a spectral resolution of 5 nm intervals. Maize kernel samples were positioned on the conveyer belt. The exposure time was set to 3 milliseconds, and the distance between the lens of the CCD camera and the sample was set to 258 mm. Maize kernels were placed on the conveyor stage and moved with a speed of 19 mm/s to be scanned.

Before spectral data and image processing, the acquired raw images must be corrected, and the calibrated image R was calculated using the following equation:

$$R = \frac{I_{raw} - I_{dark}}{I_{white} - I_{dark}} \qquad (1)$$

where I_{raw} is the raw hyperspectral image; R is the calibrated hyperspectral image; I_{dark} is the dark reference image by turning off the light source with reflectance close to 0; and I_{white} is the white reference image by using a white Teflon tile with 100% reflectance.

2.3. Spectral Collection and Pretreatment

To extract spectral data, the whole maize kernel was segmented from the background and the region of interest (ROI) was defined. The spectral mean of all the pixels of the ROI was taken as the average spectrum of the relative sample. For the purpose of eliminating the noise of the spectral data and to improve the predictive ability of the samples, three typical pre-processing methods were used—namely, wavelet transformation (WT) [25], standard normal variate (SNV) [26], and multiplicative scatter correction (MSC) [26]. The raw spectra were subjected to noise suppression by wavelet transformation using Daubechies 8 with decomposition scale 3, which was conducted by a series of MATLAB programs. SNV and MSC pre-processing was implemented using the Unscrambler software version 10.1 (CAMO PROCESS AS, Oslo, Norway).

2.4. Multivariate Chemometrics Analysis

Multivariate analyses including principal component analysis (PCA), partial least squares discriminant analysis (PLS-DA), and support vector machine (SVM) were used in the present study to classify and screen the GM and non-GM maize kernels. Exploratory classification was carried out by PCA analysis in order to find possible clustering by their average spectral characters. The contiguous spectral bands in hyperspectral image data are highly correlated, and thus the high dimensionality results are redundant information. It is essential to extract feature components to augment both efficiency and effectiveness. Next, competitive adaptive reweighted sampling (CARS) [27] was applied to select the important wavelengths. In the next stage, the PLS-DA and SVM discriminant analysis models were established based on the raw average spectral datasets (200 wavelengths) and optimal spectra (54 wavelengths) of all test samples. Finally, a prediction map was developed by applying the CARS-PLS-DA model based on each pixel at the optimal wavelengths. Image visualization helped to present the distribution of different features between different genotypes. In general, the prediction map is presented in a pseudo-color map, and the colors represent the corresponding feature values. The hyperspectral image processing procedure is illustrated in Figure 2, and includes spectral data extraction, optimal wavelengths selection, the development of discrimination models, and the building of a prediction map.

PCA is an effective algorithm for reducing the dimensionality of data into a set of principal components (PCs), for solving the problem of multicollinearity and handling any potential co-linearity between variables [28]. The PCA algorithm transforms multiple variables into a smaller number of PCs. First, exploratory classification was carried out by PCA analysis to identify clusters into the genetic background classes—GM and non-GM—based on their average spectral data. Because the PCs are orthogonal, we can view the possible distinction between different samples by plotting the PCs. PCA score images of the first three scores were conducted by combing all-pixel spectral information and then the score information in the next step. Anomalies in the interpretation of PCA score images between different genotypes would most likely be due to chemical components in heterogeneity [29]. According to Wold et al. (1996), discriminant analysis models established on optimal wavelengths might have the same or better results than those established with full spectra [30]. Moreover, the reduced number of wavelengths makes the model easier to apply and is sufficient to determine if classification works [31]. CARS is a promising procedure for variable selection and was applied in this work. The number of Monte Carlo sampling runs was set to 50, and 10-fold cross validation was used to evaluate the effectiveness of each subset of variables. The CARS method was implemented in MATLAB with open script code which is available at http://cn.mathworks. com/matlabcentral/fileexchange/64154-cars-algorithm-for-feature-variable-selecting. A detailed description of the CARS procedure can be found in Li et al. (2009) [27].

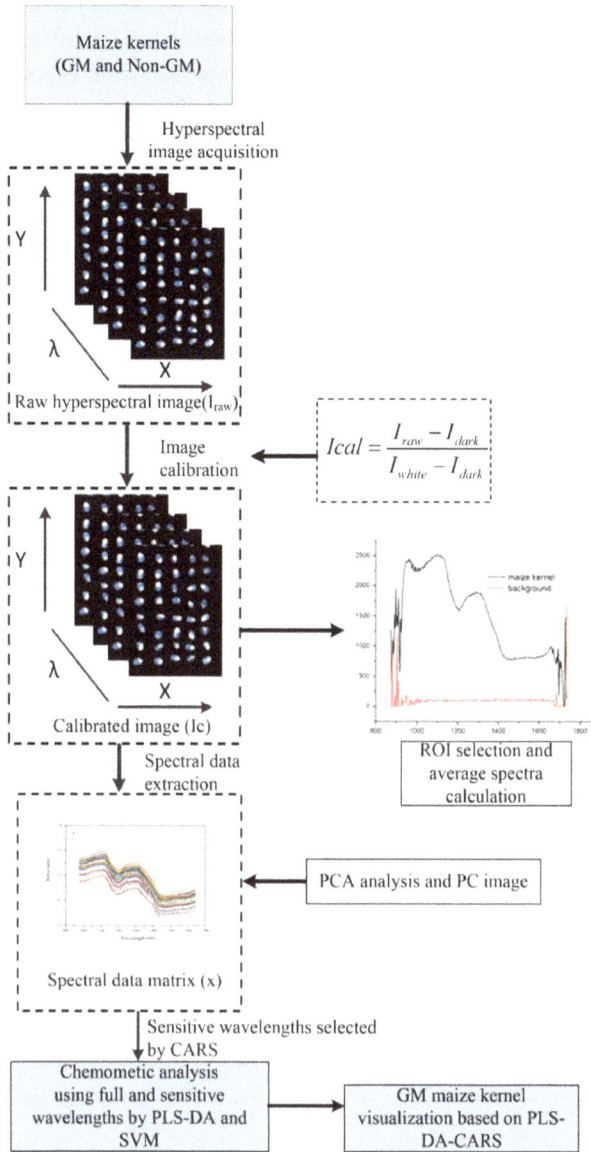

Figure 2. Flowchart of image processing and data analysis for discrimination of genetically-modified (GM) maize kernels. CARS: competitive adaptive reweighted sampling; PLS-DA: partial least squares discriminant analysis; ROI: region of interest; SVM: support vector machine.

PLS-DA is a supervised method used for classification purposes to explain the maximum discrimination between defined samples groups [32]. PLS-DA linearly models the relevant sources of data into new variables called latent variables (LVs), and the first few LVs carry the most useful information. In this case, the PLS-DA discrimination model was built by assigning reference values for all the samples. The GM maize kernel would be considered to be correctly evaluated if the value was between 0.5 and 1.5. A sample was considered non-GM if the value was between 1.5 and 2.5.

Otherwise, the samples were considered as incorrectly classified. The PLS-DA model was built using leave-one-out cross validation, and the number of optimal LVs was determined. The accuracy of the classification procedure is expressed as the fraction of correctly classified samples to the total samples for both the calibration and prediction sets.

SVM is a supervised learning model based on structured risk minimization that analyzes data used to perform multivariate function estimation or a non-probabilistic binary linear classification [33]. Compared to other machine learning methods, this method develops a model with less training samples, and overcomes the local minimum required for a neural network. SVM has been widely used for supervised pattern recognition. Detailed information about this popular model can be found in the literature [34]. For this study, SVM with the radial basis function (RBF) as the kernel function was used, and different penalty parameters (c) and kernel function parameters (g) were chosen to achieve the highest recognition rate. The best c and g were obtained by a grid-search procedure in the range of 2^{-8}–2^8 with the kernel function of RBF.

2.5. Software Tools

Images were analyzed by using Evince version 4.6 Hyperspectral image analysis soft package (ITT, Visual Information Solutions, Boulder, CO, USA) and MATLAB version R2010b (The Math-Works, Natick, MA, USA). In addition, origin Pro 7.0SR0 (Origin Lab Corporation, Northampton, MA, USA) software was used to design graphs. The model performance was evaluated by the classification accuracy of the calibration set and the prediction set.

3. Results and Discussion

3.1. Spectroscopic Analysis

The spectra were collected over the range of 874–1733 nm. Only the spectra of 971.66–1642.43 nm were used for analyses, as the front and rear parts of the spectra showed high noise levels caused by the optical equipment and the ambient environment. Figure 3A shows the extracted spectra of the ROI, and Figure 3B represents the average spectra of 1050 transgenic and 1050 non-transgenic maize kernel samples. The differences in spectra reflectance were observed, noting that the trends of most spectra were similar. The average reflectance of the non-GM samples was always higher than those of the GM samples, which reflects the differences in the hundreds of physical and chemical components between the genotypes. These differences might result from metabolites in the transgenic samples. It was hard to discriminate GM samples from their non-GM control based on the NIR spectral reflectance only. Therefore, chemometrics methods in combination with NIR spectra were introduced to build the discriminant analysis models for classification.

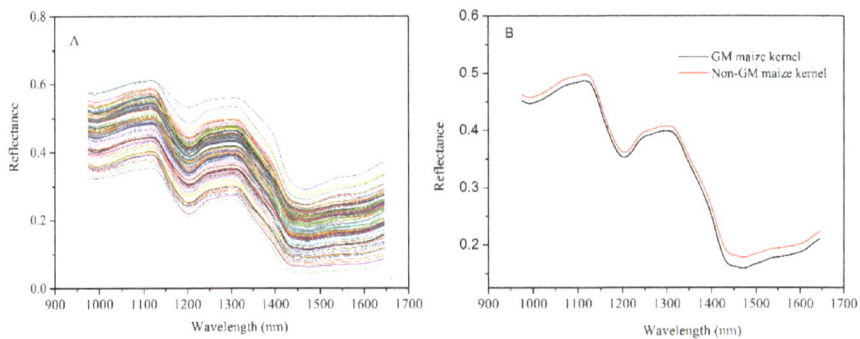

Figure 3. The profiles for raw spectra (**A**) and average spectra reflectance (**B**) from the near-infrared (NIR) multispectral images of GM and non-GM maize kernels.

3.2. Spectral Analysis by Principal Component Analysis

Spectra data were pre-processed to eliminated the systematic noise and highlight the differences between the samples. PLS-DA was applied using leave-one-out cross-validation for the original raw spectral data and the pre-processed spectra to test the different pre-treatment strategies. Table 1 summarizes the results acquired for raw spectra and the different pre-processing methods. In all cases, the optimal number of LVs for establishing the calibration set was nine. Discrimination performance of the calibrations can be improved by each pre-processing treatment, but the performance of the prediction model was only improved by WT pre-processing. From the different pre-treatments evaluated, WT correction was the most efficient pre-treatment. In order to establish a robust prediction model, WT was applied as the pre-treatment method in the next step.

Table 1. Comparison of discrimination performance obtained by partial least squares discriminant analysis (PLS-DA) with different preprocessing methods using full wavelengths.

Methods	PLS-DA		
	Parameter [1]	Calibration Set	Prediction Set
Raw	9	98.50%	97.86%
WT	9	99.43%	98.71%
SNV	9	99.50%	95.00%
MSC	9	99.36%	94.57%

[1] Model parameters means the optimal number of LVs for establishing the calibration model of PLS-DA; Raw means raw spectra; WT is wavelet transformation using Daubechies 8 with decomposition scale 3; SNV is standard normal variate; and MSC is multiplicative scatter correction.

After WT was applied, PCA programs were first developed to examine the qualitative difference of GM and non-GM maize kernels in PC space. All spectra of the 1050 GM and non-GM maize kernels were analyzed for PCA. The three-dimensional (3D) PC score plot of the samples is illustrated in Figure 4A. The first three PCs explained the most spectral variations, at a total of 99.02%, including 94.04%, 4.79%, and 0.20% for PC1, PC2, and PC3, respectively. It was evident that the two classes were well-separated along the third PC, which indicated that the spectral fingerprints carry discriminant information. The suitability of PCA for distinguishing *Bacillus thuringiensis*-mediated transgenic rice seeds from NIR has been previously demonstrated [15].

Since hyperspectral imaging possesses all-pixel spectral information, PCA visualization analysis on hyperspectral reflectance images was also introduced, instead of using the average spectrum of each sample. Score images (Figure 5) were investigated to identify and visualize the patterns detected on the score plots. The score plot of PC1 and PC2 did not show clear classification differences between genotypes, as these PCs were associated with maize kernel composition and anatomy [16,17]. The introduced foreign genes did not change the anatomical properties of the kernel and major kernel dominant traits, such as protein, fat, and starch concentration. Maize kernel mainly consists of two types of endosperm texture. In the vitreous endosperm, starch granules are polygonal-shaped and tightly compacted without air spaces. The floury endosperm comprises spherical starch granules that are covered with a protein matrix and air spaces [29]. The main source of spectral variation was explained by PC1. The germ region and pedicle of the maize kernel is composed of a floury endosperm, while the other pericarp of the kernel is composed of a glass endosperm [35]. As illustrated in Figure 5, the positive PC1 scores (shown in red color) were associated with floury endosperm in the germ and pedicle region, while negative PC1 scores (shown in blue color) were associated with the glass endosperm. The score image of PC2 showed different features linked to the pedicle and hull of maize's histological characteristics, as earlier described by Williams (2016) [17]. Similar findings regarding morphological classes including vitreous and soft endosperm were also reported by other researchers [16,17,28]. With the score image of PC3, the first visualization of a difference between GM and non-GM maize kernels was observed (Figure 5). The GM samples were largely characterized by positive scores (shown by the colors in the warm range) on the surface of the kernels, while non-GM samples were mainly covered by cool colors. The differences observed were the same as in the PCA 3D plot using the spectral data.

The value of the PCA loadings reflects the degree of correlation between the PCs and the raw wavelength variable; therefore, the variation observed in the PCA score plots and images can be explained by studying the accompanying loading. The variation is explained by the loading line plot of PC3 (Figure 4B). The absorption bands around 1206 nm are related to the second overtone of C–H stretching vibration of various functional groups: $-CH_2$, $-CH_3$, and $-CH=CH-$ [36]. The peak near 1311 nm is due to the first overtone of the OH stretch and OCO bending [37]. The remarkable peak centered around 1365 nm is related to the $C-H_3$ stretch and deformation overtone [38]. The band around at 1473 nm represents OH, CH, and CH_2 deformations [39].

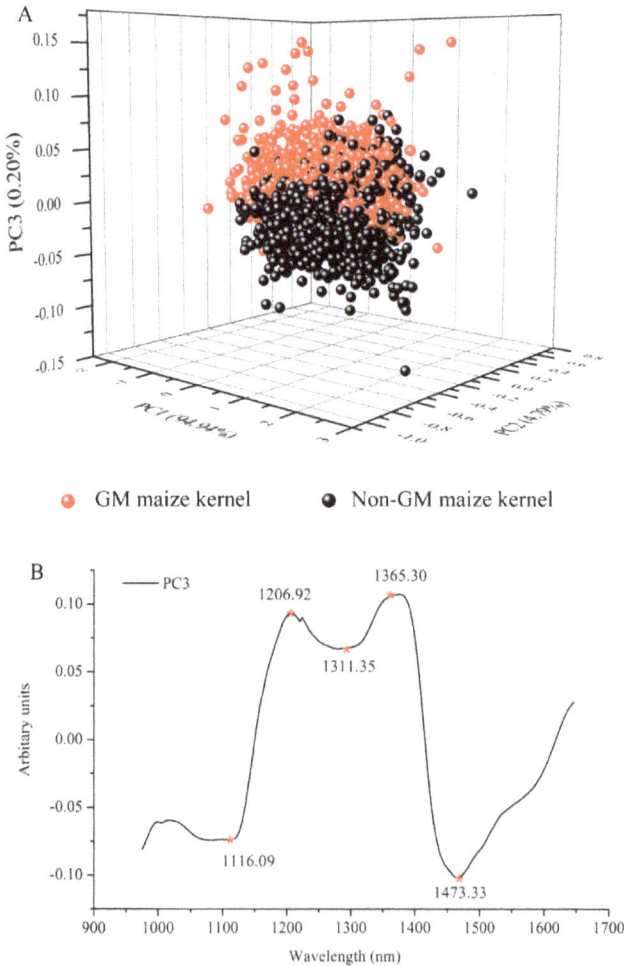

Figure 4. Three-dimensional principal component analysis (PCA) scores scatter plot of the first three principle components (PCs) for the GM and non-GM maize kernels with raw spectra (**A**) after pre-processing and (**B**) main peaks of PC3 loadings that are indicative of the differences between GM and non-GM maize kernels.

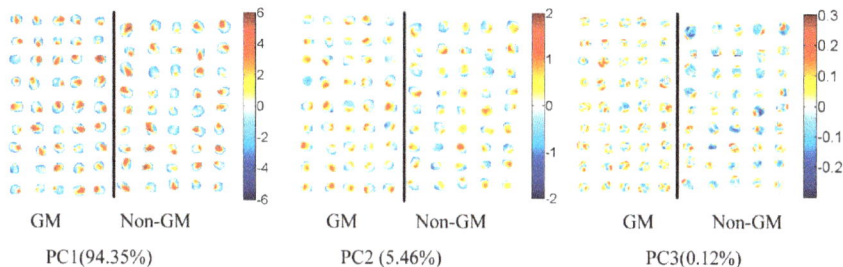

GM   Non-GM   GM   Non-GM   GM   Non-GM

PC1(94.35%)   PC2 (5.46%)   PC3(0.12%)

Figure 5. Score images of the first three principal components (**PC1–PC3**) of the images of maize kernels.

3.3. Selection of Optimal Wavelengths

Hyperspectral imaging data contain redundant information, which affects the prediction performance of the model. Variable selection was carried out using CARS election-based techniques to reduce the effect of non-related variables and speed up the classification. As shown in Figures 6 and 7, 54 optimal wavelengths were selected. The wavelength number was decreased by 73% ($\frac{200-54}{200} = 94\%$) after preprocessing all the wavelengths by CARS.

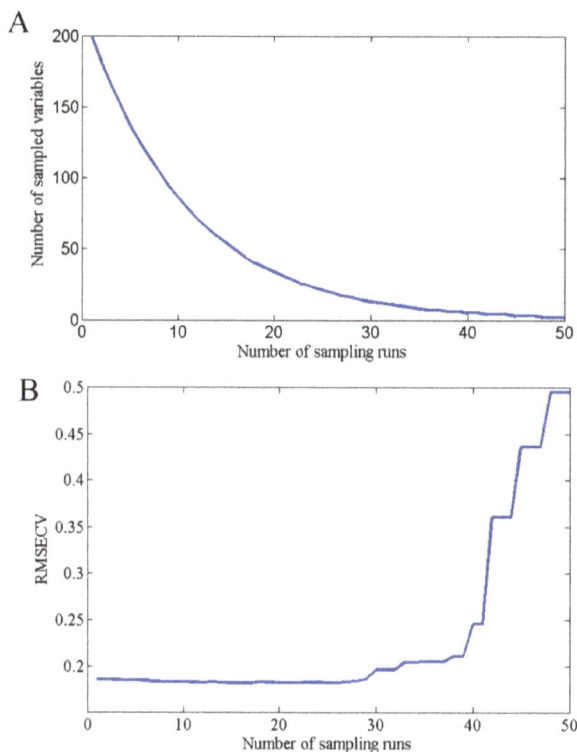

Figure 6. *Cont.*

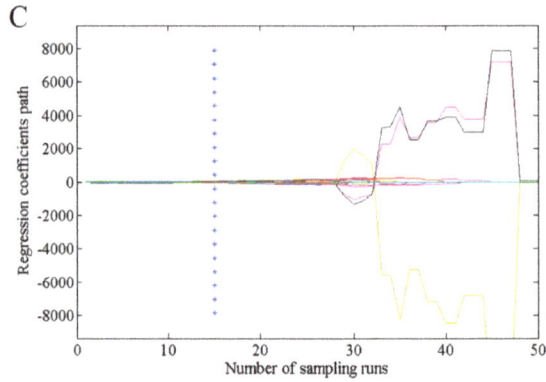

Figure 6. Selected optimal wavelengths by competitive adaptive reweighted sampling (CARS). (**A**) The number of variables as the function of iterations; (**B**) ten-fold root mean squared errors (RMSECV) values; and (**C**) regression coefficients of each variable with the number of sampling runs. Each line denotes the path of a variable.

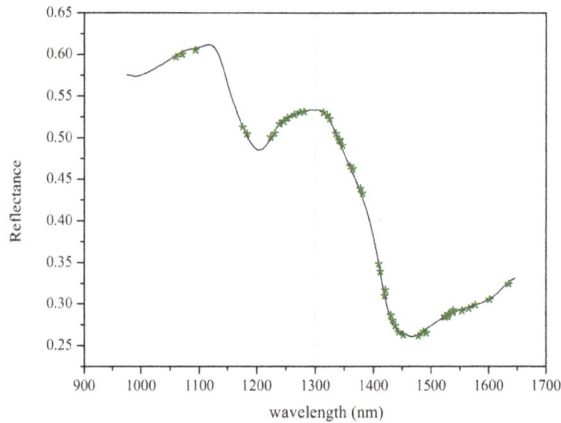

Figure 7. The distribution of optimal wavelengths selected by CARS.

The bands found between 1250–1350 nm were due to the combination between the first overtone of Amide B with the fundamental Amide III vibrations [40]. The spectral region (1410–1480 nm) was assigned for protein as a result of the first overtone of the N–H stretching vibration [40]. The bands at 1520–1600 nm were related to the N–H stretching vibrations [41]. Based on the above interpretations and observations, it is reasonable to assume that the change in conformation and composition status of the GM maize is due to the pleiotropic effect caused by the insertion of *cry1Ab/cry2Aj-G10evo* foreign genes into the parent genome, influencing the NIR spectra and causing variation between the different genotype backgrounds.

3.4. Classification Analysis by the Discrimination Model

In the next stage, spectra collected from the images of the kernel samples were used to build a model capable of discriminating the GM maize kernels based on their hyperspectral fingerprint. Calibration model measurements were conducted on the full spectrum of 200 wavelengths, and 54 optimum wavelengths were selected. The recognition effect of different discrimination models, developed with full and selected feature wavelengths, are compared in Table 2. The recognition accuracies obtained from the calibration and prediction sets were summarized. The SVM and PLS-DA models all achieved good recognition results with large sample size. The classification ability of the PLS-DA model was higher than that of SVM when all spectrum regions were used. The calibration set was 99.43% accurate for PLS-DA and 98.5% for SVM. The prediction set was 98.71% accurate for PLS-DA and 97% for SVM. The CARS algorithm was used to select optimal wavelengths from NIR hyperspectral imaging. The number of effective wavelengths decreased to 27% after using this algorithm. The variable selection made the modeling procedure faster. The discrimination ability of the calibration set from the PLS-DA model, based on optical wavelengths, was slightly worse than that obtained from all the wavelengths, but was still rated as acceptable. The discrimination ability of the prediction set from the PLS-DA model increased from 98.71% to 99.00%. The SVM model established on selected wavelengths performance improved, since it was 99.14% accurate with the calibration set and 98.29% for the prediction set. The reason for this might be that some wavelengths carrying useless interference were eliminated. Comparison of the results showed that CARS-PLS-DA performs better than CARS-SVM, since it made the prediction more robust and accurate. The overall results indicated that it was feasible to discriminate GM maize kernel by using hyperspectral imaging, and that the PLS-DA recognition model based on optimal wavelengths is a reliable and robust model.

Table 2. Comparison of discrimination ability obtained by different PLS-DA and SVM models, with the total spectral data and optimal wavelengths selected by the competitive adaptive reweighted sampling (CARS) method.

Methods	PLS-DA			SVM		
	Parameter [1]	Calibration Set	Prediction Set	Parameter [1]	Calibration Set	Prediction Set
Full spectra	9	99.43%	98.71%	(256, 0.0625)	98.5%	97.00%
Optimal wavelengths	8	99.35%	99.00%	(256, 16)	99.14%	98.29%

[1] Model parameters of the differentiating models; i.e., the optimal number of LVs for establishing the calibration model of partial least squares-discrimination analysis (PLS-DA), different penalty parameters (c) and kernel function parameters (g) for support vector machine (SVM).

3.5. Transgenic Maize Kernel Visualization

In addition to verifying the reliability of the proposed method, the classification of genotypes was visualized on prediction maps by predicting the features of each pixel on individual hyperspectral images. Accordingly, the PLS-DA model—computed using optimal wavelengths selected by CARS—was applied to every single pixel in the image to predict the class of kernels (GM and non-GM maize) for all surfaces of the sample. For creating a classification map, a binary code with a dummy variable was used to classify samples, with GM samples assigned as one and non-GM samples assigned as two. The result is shown in Figure 8 with pixels in prediction map colored according to the predicted category with the same dimension as the original hyperspectral image. Although it was difficult to determine the difference between the two classes from sample to sample and from point to point with the naked eye (Figure 8), GM maize kernels were obviously identified from the final chemical image. Green represents the non-GM maize kernel, and the red the GM maize kernel. Notably, some kernels on the classification map were misidentified based on the CARS-PLA-DA model. The morphological characteristics of the kernels in the classification map were altered due to the low resolution of the NIR imaging system and the image segmentation algorithm. However, the main shape of the kernels and their locations were clear on the prediction map. This approach is important because it facilitates the progress for rapid and high throughput detection of GM maize kernels and could be implemented as an online visualization system for iscrimination purposes.

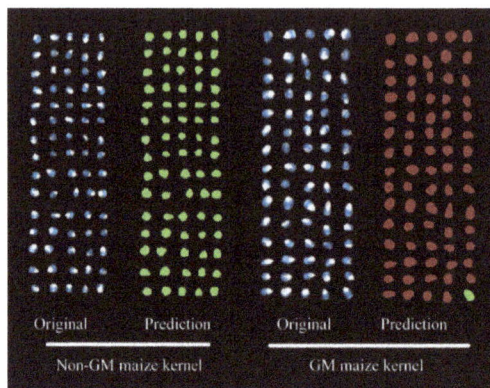

Figure 8. Visualization of independent maize kernels by the SVM classification process using optimal wavelengths in the hyperspectral images. Green denotes non-GM maize kernels, and red identifies GM maize kernels.

4. Conclusions

The above results demonstrate that it is possible to differentiate a stacked commercial maize hybrid containing both herbicide-tolerant and insect-resistant traits from a single transgenic event by coupling the hyperspectral imaging technique in the NIR region (1975.01–1645.82 nm) with chemometric processing. Both PCA and classification models were suitable for GM maize kernel variety identification. From the perspective of the pixel spectra combined with the spatial distribution of the maize kernel, a principal component pseudo-color map was drawn and the differences were intuitively displayed. High-dimensional hyperspectral image data were reduced by CARS to extract the characteristic spectrum. The classification models built by PLS-DA and SVM using full wavelengths had a predictive accuracy near 100%. Additionally, it was demonstrated that the PLS-DA model established with a reduced set of only 54 wavelengths resulted in excellent accuracy, with 99.35% for the calibration set and 99.00% for the prediction set. This last outcome was fairly promising, as it could significantly speed up the data processing, which could facilitate online detection in the future. The main benefit of the hyperspectral imaging technique is its ability to visualize the identification of GM maize kernel in a pixel-based manner that cannot be obtained with either common spectroscopy or imaging. Finally, the GM maize kernel could be identified on the prediction maps by using the features of each pixel on individual hyperspectral images obtained by the CARS-PLS-DA model. We conclude that it is feasible to use hyperspectral imaging to differentiate GM maize kernels from their non-GM parents. The experiment material used for classification were obtained from the same year. Further research is expected to use the seeds or kernel samples from different years, regions, and transgenic events involved to improve the reliability and adaptability of the discrimination model.

Acknowledgments: This work was supported financially by 863 National High-Tech Research and Development Plan (Project No: 2013AA10030401) and State Key Laboratory Breeding Base for Zhejiang Sustainable Pest and Disease Control (No. 2010DS700124-KF1712).

Author Contributions: X.F. and Y.Z. performed the measurements. X.F. and P.C. wrote the manuscript. X.F., Y.H. and C.Z. designed the experiment and reviewed the manuscript. Y.H. reviewed the initial design of the experiments and made guidance for the writing of the manuscript. All authors reviewed the manuscript.

Conflicts of Interest: The authors declare no conflict of interest.

Sensors **2017**, *17*, 1894

References

1. Shiferaw, B.; Prasanna, B.M.; Hellin, J.; Bänziger, M. Crops that feed the world 6. Past successes and future challenges to the role played by maize in global food security. *Food Secur.* **2011**, *3*, 307–327. [CrossRef]
2. Rensburg, J.B.J.V. First report of field resistance by the stem borer, Busseola fusca (Fuller) to Bt-transgenic maize. *S. Afr. J. Plant Soil* **2007**, *24*, 147–151. [CrossRef]
3. Vermerris, W.; Saballos, A.; Ejeta, G.; Mosier, N.S.; Ladisch, M.R.; Carpita, N.C.; Albrecht, B.; Bernardo, R.; Godshalk, E.B. Molecular breeding to enhance ethanol production from corn and sorghum stover. *Crop Sci.* **2007**, *47* (Suppl. 3), S142. [CrossRef]
4. Aluru, M. Generation of transgenic maize with enhanced provitamin A content. *J. Exp. Bot.* **2015**, *23*, 1–11. [CrossRef] [PubMed]
5. Agapito-Tenfen, S.Z.; Vilperte, V.; Benevenuto, R.F.; Rover, C.M.; Traavik, T.I.; Nodari, R.O. Effect of stacking insecticidal cry and herbicide tolerance epsps transgenes on transgenic maize proteome. *BMC Plant Biol.* **2014**, *14*, 1–19. [CrossRef] [PubMed]
6. Nap, J.P.; Metz, P.L.J.; Escaler, M.; Conner, A.J. The release of genetically modified crops into the environment. *Plant J. Cell Mol. Biol.* **2003**, *33*, 19–46. [CrossRef]
7. Taverniers, I.; Van, B.E.; De, L.M. Cloned plasmid DNA fragments as calibrators for controlling GMOs: Different real-time duplex quantitative PCR methods. *Anal. Bioanal. Chem.* **2004**, *378*, 1198–1207. [CrossRef] [PubMed]
8. Brunnert, H.J.; Spener, F.; Börchers, T. PCR-ELISA for the CaMV-35S promoter as a screening method for genetically modified Roundup Ready soybeans. *Eur. Food Res. Technol.* **2001**, *213*, 366–371. [CrossRef]
9. Huang, X.; Zhai, C.; You, Q.; Chen, H. Potential of cross-priming amplification and DNA-based lateral-flow strip biosensor for rapid on-site GMO screening. *Anal. Bioanal. Chem.* **2014**, *406*, 4243–4249. [CrossRef] [PubMed]
10. Tengs, T.; Kristoffersen, A.B.; Berdal, K.G.; Thorstensen, T.; Butenko, M.A.; Nesvold, H.; Holstjensen, A. Microarray-based method for detection of unknown genetic modifications. *BMC Biotechnol.* **2007**, *7*, 1–8. [CrossRef] [PubMed]
11. Alishahi, A.; Farahmand, H.; Prieto, N.; Cozzolino, D. Identification of transgenic foods using NIR spectroscopy: A review. *Spectrochim. Acta Part A Mol. Biomol. Spectrosc.* **2010**, *75*, 1–7. [CrossRef] [PubMed]
12. Xie, L.; Ying, Y.; Ying, T.; Yu, H.; Fu, X. Discrimination of transgenic tomatoes based on visible/near-infrared spectra. *Anal. Chim. Acta* **2007**, *584*, 379–384. [CrossRef] [PubMed]
13. Guo, H.; Chen, J.; Pan, T.; Wang, J.; Cao, G. Vis-NIR wavelength selection for non-destructive discriminant analysis of breed screening of transgenic sugarcane. *Anal. Methods* **2014**, *6*, 8810–8816. [CrossRef]
14. Dolores, G.M.M.; Juan, G.O.; Francisco, B. Effective Identification of Low-Gliadin Wheat Lines by Near Infrared Spectroscopy (NIRS): Implications for the Development and Analysis of Foodstuffs Suitable for Celiac Patients. *PLoS ONE* **2016**, *11*, e0152292.
15. Liu, C.; Wei, L.; Lu, X.; Wei, C.; Yang, J.; Lei, Z. Nondestructive determination of transgenic Bacillus thuringiensis rice seeds (*Oryza sativa* L.) using multispectral imaging and chemometric methods. *Food Chem.* **2014**, *153*, 87–93. [CrossRef] [PubMed]
16. Williams, P.; Geladi, P.; Fox, G.; Manley, M. Maize kernel hardness classification by near infrared (NIR) hyperspectral imaging and multivariate data analysis. *Anal. Chim. Acta* **2009**, *653*, 121–130. [CrossRef] [PubMed]
17. Williams, P.J.; Kucheryavskiy, S. Classification of maize kernels using NIR hyperspectral imaging. *Food Chem.* **2016**, *209*, 131–138. [CrossRef] [PubMed]
18. Williams, P.J.; Geladi, P.; Britz, T.J.; Manley, M. Investigation of fungal development in maize kernels using NIR hyperspectral imaging and multivariate data analysis. *J. Cereal Sci.* **2012**, *55*, 272–278. [CrossRef]
19. Burger, J.; Geladi, P. Hyperspectral NIR imaging for calibration and prediction: A comparison between image and spectrometer data for studying organic and biological samples. *Analyst* **2006**, *131*, 1152–1160. [CrossRef] [PubMed]
20. Grusche, S. Basic slit spectroscope reveals three-dimensional scenes through diagonal slices of hyperspectral cubes. *Appl. Opt.* **2014**, *53*, 4594–4603. [CrossRef] [PubMed]
21. Wang, L.; Liu, D.; Pu, H.; Sun, D.W.; Gao, W.; Xiong, Z. Use of Hyperspectral Imaging to Discriminate the Variety and Quality of Rice. *Food Anal. Methods* **2015**, *8*, 515–523. [CrossRef]

22. Min, H.; Wan, X.; Min, Z.; Zhu, Q. Detection of insect-damaged vegetable soybeans using hyperspectral transmittance image. *J. Food Eng.* **2013**, *116*, 45–49.

23. Sun, J.; Lu, X.; Mao, H.; Wu, X.; Gao, H. Quantitative Determination of Rice Moisture Based on Hyperspectral Imaging Technology and BCC-LS-SVR Algorithm. *J. Food Process Eng.* **2016**, *40*, e12446. [CrossRef]

24. Kong, W.; Zhang, C.; Liu, F.; Nie, P.; He, Y. Rice seed cultivar identification using near-infrared hyperspectral imaging and multivariate data analysis. *Sensors* **2013**, *13*, 8916–8927. [CrossRef] [PubMed]

25. Shao, X.; Zhuang, Y. Determination of chlorogenic acid in plant samples by using near-infrared spectrum with wavelet transform preprocessing. *Anal. Sci.* **2004**, *20*, 451–454. [CrossRef] [PubMed]

26. Rinnan, Å.; Berg, F.V.D.; Engelsen, S.B. Review of the most common pre-processing techniques for near-infrared spectra. *TrAC Trends Anal. Chem.* **2009**, *28*, 1201–1222. [CrossRef]

27. Li, H.; Liang, Y.; Xu, Q.; Cao, D. Key wavelengths screening using competitive adaptive reweighted sampling method for multivariate calibration. *Anal. Chim. Acta* **2009**, *648*, 77–84. [CrossRef] [PubMed]

28. Lee, K.M.; Herrman, T.J.; Lingenfelser, J.; Jackson, D.S. Classification and prediction of maize hardness-associated properties using multivariate statistical analyses. *J. Cereal Sci.* **2005**, *41*, 85–93. [CrossRef]

29. Manley, M.; Mcgoverin, C.M.; Engelbrecht, P.; Geladi, P. Influence of grain topography on near infrared hyperspectral images. *Talanta* **2012**, *89*, 223–230. [CrossRef] [PubMed]

30. Wold, J.P.; Jakobsen, T.; Krane, L. Atlantic Salmon Average Fat Content Estimated by Near-Infrared Transmittance Spectroscopy. *J. Food Sci.* **1996**, *61*, 74–77. [CrossRef]

31. Rodríguez-Pulido, F.J.; Barbin, D.F.; Sun, D.W.; Gordillo, B.; González-Miret, M.L.; Heredia, F.J. Grape seed characterization by NIR hyperspectral imaging. *Postharvest Biol. Technol.* **2013**, *76*, 74–82. [CrossRef]

32. Almeida, M.R.; Fidelis, C.H.; Barata, L.E.; Poppi, R.J. Classification of Amazonian rosewood essential oil by Raman spectroscopy and PLS-DA with reliability estimation. *Talanta* **2013**, *117*, 305–311. [CrossRef] [PubMed]

33. Chapelle, O.; Haffner, P.; Vapnik, V.N. Support vector machines for histogram-based image classification. *IEEE Trans. Neural Netw.* **1999**, *10*, 1055–1064. [CrossRef] [PubMed]

34. Burges, C.J.C. A Tutorial on Support Vector Machines for Pattern Recognition. *Data Min. Knowl. Discov.* **1998**, *2*, 121–167. [CrossRef]

35. Gwirtz, J.A.; Garcia-Casal, M.N. Processing maize flour and corn meal food products. *Ann. N. Y. Acad. Sci.* **2014**, *1312*, 66–75. [CrossRef] [PubMed]

36. Osborne, B.G.; Fearn, T.; Hindle, P.H. *Practical Nir Spectroscopy with Applications in Food & Beverage Analysis*; Longman Scientific & Technical: Harlow, UK, 1993.

37. Workman, J.; Weyer, L. *Practical Guide to Interpretive Near-Infrared Spectroscopy*; CRC Press, Inc.: Boca Raton, FL, USA, 2007.

38. Ishikawa, D.; Murayama, K.; Awa, K.; Genkawa, T.; Komiyama, M.; Kazarian, S.G.; Ozaki, Y. Application of a newly developed portable NIR imaging device to monitor the dissolution process of tablets. *Anal. Bioanal. Chem.* **2013**, *405*, 9401–9409. [CrossRef] [PubMed]

39. Woodcock, T.; Downey, G.; Kelly, J.D.; O'Donnell, C. Geographical classification of honey samples by near-infrared spectroscopy: A feasibility study. *J. Agric. Food Chem.* **2007**, *55*, 9128–9134. [CrossRef] [PubMed]

40. Daszykowski, M.; Wrobel, M.S.; Czarnik-Matusewicz, H.; Walczak, B. Near-infrared reflectance spectroscopy and multivariate calibration techniques applied to modelling the crude protein, fibre and fat content in rapeseed meal. *Analyst* **2008**, *133*, 1523–1531. [CrossRef] [PubMed]

41. Cséfalvayová, L.; Pelikan, M.; Kralj, C.I.; Kolar, J.; Strlic, M. Use of genetic algorithms with multivariate regression for determination of gelatine in historic papers based on FT-IR and NIR spectral data. *Talanta* **2010**, *82*, 1784–1790. [CrossRef] [PubMed]

sensors

MDPI

Article

Novelty Detection Classifiers in Weed Mapping: *Silybum marianum* Detection on UAV Multispectral Images

Thomas K. Alexandridis [1,*], Afroditi Alexandra Tamouridou [1,2], Xanthoula Eirini Pantazi [2], Anastasia L. Lagopodi [3], Javid Kashefi [4], Georgios Ovakoglou [1], Vassilios Polychronos [5] and Dimitrios Moshou [2]

[1] Laboratory of Remote Sensing and GIS, Faculty of Agriculture, Aristotle University of Thessaloniki, 54124 Thessaloniki, Greece; tamouridoualex@gmail.com (A.A.T.); georobak@hotmail.com (G.O.)
[2] Agricultural Engineering Laboratory, Faculty of Agriculture, Aristotle University of Thessaloniki, 54124 Thessaloniki, Greece; renepantazi@gmail.com (X.E.P.); dmoshou@auth.gr (D.M.)
[3] Plant Pathology Laboratory, Faculty of Agriculture, Aristotle University of Thessaloniki, 54124 Thessaloniki, Greece; lagopodi@agro.auth.gr
[4] USDA-ARS-European Biological Control Laboratory, Tsimiski 43, 7th floor, 54623 Thessaloniki, Greece; jkashefi@ars-ebcl.org
[5] Geosense S.A., Filikis Etairias 15-17, Pylaia, 55535 Thessaloniki, Greece; vpoly@geosense.gr
* Correspondence: thalex@agro.auth.gr; Tel.: +30-2310-991777; Fax: +30-2310-991778

Received: 28 July 2017; Accepted: 28 August 2017; Published: 1 September 2017

Abstract: In the present study, the detection and mapping of *Silybum marianum (L.) Gaertn.* weed using novelty detection classifiers is reported. A multispectral camera (green-red-NIR) on board a fixed wing unmanned aerial vehicle (UAV) was employed for obtaining high-resolution images. Four novelty detection classifiers were used to identify *S. marianum* between other vegetation in a field. The classifiers were One Class Support Vector Machine (OC-SVM), One Class Self-Organizing Maps (OC-SOM), Autoencoders and One Class Principal Component Analysis (OC-PCA). As input features to the novelty detection classifiers, the three spectral bands and texture were used. The *S. marianum* identification accuracy using OC-SVM reached an overall accuracy of 96%. The results show the feasibility of effective *S. marianum* mapping by means of novelty detection classifiers acting on multispectral UAV imagery.

Keywords: weeds; UAS; RPAS; one-class; machine learning; remote sensing; geoinformatics

1. Introduction

Mapping weed patches is an important aspect of effective site-specific application of herbicides. Remote sensing is a source of data that offers full field coverage at various spatial, spectral and temporal resolutions [1]. However, successful automated detection of weeds on remotely sensed images may be hindered by spectral, textural and shape similarities with the crop. Zhang et al. [2] produced a ground-level, hyperspectral weed mapping method for black nightshade and pigweed in tomato fields. Bayesian classifiers were developed for each season that reached high levels of accuracy with cross-validation species pixel classification (92%). Cross-season validation of single season-based classifiers proved that accuracy was affected by changes in the NIR. A machine learning algorithm was produced that utilised three artificial intelligence (AI) season "expert" classifiers in a multi-season, multi-classifier method. This approach achieved a recognition rate of 95.8%, slightly higher than a global calibration approach with multi-season species discrimination accuracies from 90% to 92.7%.

Weed identification methods are based on a variety of image processing techniques, including morphology, spectral features and visual shape. Åstrand and Baerveldt [3] worked on robot vision-based perception for weed identification and control. Their device carried a grey-scale camera with a near-infrared filter for high-contrast images for row detection and a colour camera to identify crop plants. The lateral offset error was ± 1 cm and the error measured with the downward-looking camera at the weeding tool position was ± 0.5 cm. The colour camera efficiently recognised crops via image segmentation and discriminated weeds from crops.

Techniques from the field of machine learning have been effectively applied in land cover classification with remote sensing [4]; meanwhile, during the last years the precision crop protection community has used the advantages of machine learning [5]. Support Vector Machines (SVMs) [6] are among the most prominent machine learning techniques which are utilised for pattern recognition and regression and can be characterised as unique and faster than other computational intelligence techniques due to the fact that they are always capable of converging to a global minimum, and have a simple geometric explanation. SVMs have the ability to effectively handle large volumes of data or attributes. Thus, they can be useful in processing very high-resolution UAV images.

Sluiter [7] studied a wide range of vegetation classification techniques using remote sensing imagery in the Mediterranean region. It was proven that random forests as well as support vector machines demonstrated better performance among traditional classification methods. Regarding feature extraction, Sluiter [7] utilised the spatial domain, more precisely, the spectral information per-pixel and of neighbouring pixels to classify natural vegetation types from remote sensing imagery. The contextual technique SPAtial Reclassification Kernel (SPARK) was implemented, and successfully detected vegetation types, which other per-pixel-based methods failed to detect. Im and Jensen [8] applied a three-channel neighbourhood correlation image model for monitoring vegetation changes using the pixel relations, and their contextual neighbours achieved similar results. More recently Dihkan et al. [9] utilised satellite imagery for mapping tea plantations based on support vector machines. In this study, a three-step approach was proposed and implemented on a test area with high slope, in order to discriminate tea plantations from other types of vegetation. The overall accuracies reached over 90% for land use/land cover mapping.

Novelty detection and machine learning techniques can be collectively recruited to detect abnormal situations. Crupi et al. [10] proposed a novel neural network based technique for the validation of vibration signatures and the indication of a fault occurrence. The model was trained using the data set consisting only of normal examples. Abnormalities, corresponding to fault occurrences, were detected as noteworthy deviations that varied from the normality description. In particular, one class classifiers are characterised by the following behaviours [11]: (i) the only data available correspond to the target class and not to the outlier class; (ii) data coming from the target class are the only means of estimating the separating threshold between the two classes; and (iii) describing a threshold encircling the target class is a challenge.

Silybum marianum (L.) Gaertn. is a weed that is hard to eradicate. It is found in cultivated areas and pastures [12]. The leaves can cause symptoms of toxicity to livestock due to the build-up of nitrates [13]. Khan et al. [14] have demonstrated the allelopathic consequences of *S. marianum* on various cultivated species. Herbicides are expensive and they pollute the ecosystems, rendering it necessary to map the spatial distribution of *S. marianum* and define appropriate management strategies.

Previous efforts to map *S. marianum* with UAV images include Tamouridou et al. [15], who identified the optimum resolution for mapping *S. marianum* patches with UAV images, Pantazi et al. [16] who tested three hierarchical self-organising map classifiers, namely, the supervised Kohonen network, counter-propagation artificial neural network, and XY-Fusion network. Efforts to map other weeds using UAV images include Peña et al. [17], who used UAV images acquired at six regions of the visible and near-infrared spectrum and identified weeds from wheat field with Object Based Image Analysis (OBIA) with an overall accuracy of 86%. Higher accuracy rates were achieved when mapping weeds in a sunflower field with RGB images taken by a UAV [18]. Various spatial resolution levels taken from different altitudes using two sensors on board a vertical take-off and landing UAV, were tested in relation to the needs of different weed management applications [19].

The aim of this work is to evaluate the proposed novelty detection classifiers in recognising *S. marianum* weed patches as a target class and detecting all other vegetation as outliers. The algorithms were applied on multispectral images acquired by a camera on board a fixed wing UAV. The features that were employed were constructed by combining spectral and textural information.

2. Materials and Methods

2.1. Experimental Study Location

The experimental study took place at a 10.1 ha field located in Thessaloniki, Greece (Figure 1). The topography of the field is almost level and the elevation is 75 m. The field was cultivated with cereals until 1990, and now graminaceous weeds have colonised it, together with sizeable patches of *S. marianum*. Other dominant weeds include *Avena sterilis* L., *Bromus sterilis* L., *Solanum elaeagnifolium* Cav., *Conium maculatum* L., *Cardaria draba* L. and *Rumex* sp. L.

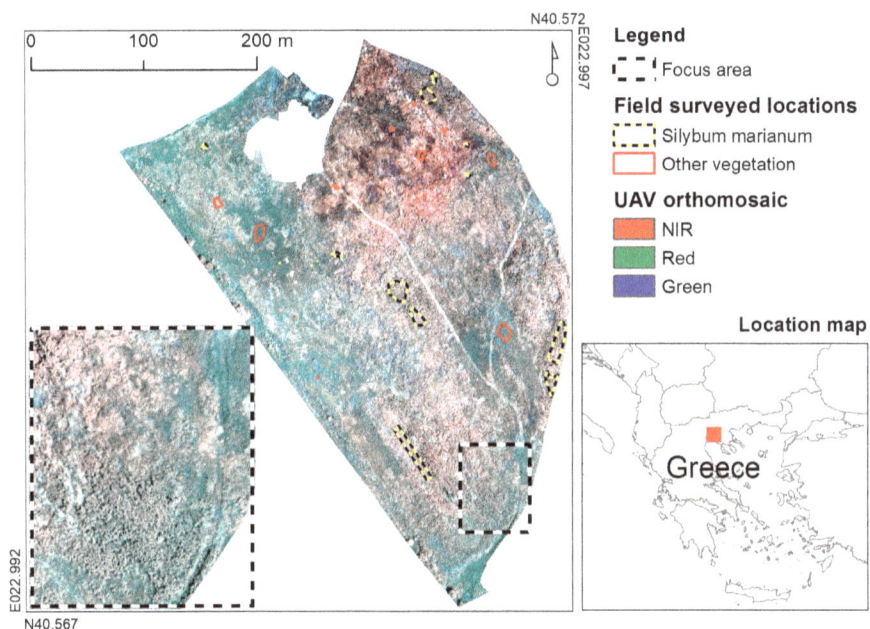

Figure 1. Study area, UAV orthomosaic, focus area, and field surveyed locations.

2.2. Datasets

Imagery was acquired on 19.05.2015, a sunny day with a light breeze not more than 3 m·s^{-1} during the flight. A fixed wing UAV was used (senseFly eBee) with camera payload Canon S110 NIR that captures spectral bands that correspond to green (560 nm, Full-Width Half-Maximum (FWHM): 50 nm), red (625 nm, FWHM: 90 nm) and near-infrared (850 nm, FWHM: 100 nm). The acquisition time frame was between 11:00 am and 12:00 pm to minimise ground shadows and maximise the reflectance effect. The camera was set to shutter priority at 1:2000 s shutter speed. To adapt to potential ground light variations and reflections, we let the camera decide the correct aperture and sensor sensitivity.

Flight was set at a height of 115 m above take-off (relative height), in flight paths that achieved 75% overlap and 70% sidelap. The above setup gave ground sampling distance up to 0.04 m and each image footprint was 160 m × 120 m. The total number of captured images was 55; all of them stored in raw format to avoid radiometric distortion due to file compression. During capture, the drone autopilot was registering coordinates for each image by its GPS.

After landing, the autopilot log file was matched with the acquired images. In order to provide precise geolocation, we measured six targets that were laid down on the ground prior to the flight with a dual frequency Spectra Precision SP80 GNSS receiver. Each target was marked with an absolute accuracy down to 2 cm, using a real time kinematic (RTK) method. Photogrammetric processing was done with Pix4D mapper Pro, photogrammetry software that produced an orthomosaic, a 2.5D digital surface model (DSM) with a typical accuracy of 2 × ground sampling distance (GSD) horizontal and 3 × GSD vertical. The orhomosaic and the DSM were further resampled to a final resolution of 0.5 m that was adequate for mapping the weed patches, while avoiding noise introduced to the classifiers from finer resolutions [15]. In addition to the three spectral bands, the texture of NIR band was created using the local variance algorithm with a moving window of 7 × 7 pixels, and added to the orthomosaic as inputs to the classification algorithms.

The location of patches of *S. marianum* and of other prevailing vegetation types were recorded on a Trimble GeoXH GPS (accuracy ~0.3 m), in order to help to build the training sets that were needed for image classification. As *S. marianum* patches were impenetrable, their outline was marked on the GPS. To ensure equal class representation, a calibration dataset was created consisting of two sets of 1434 pixels each for classes "*S. marianum*" and "other vegetation" (2868 pixels in total). From this calibration dataset, a random subset of 70% (2008 pixels for both classes) was used for algorithm training and the rest 30% (860 pixels for both classes) for testing the result. This dataset was common to all classification algorithms.

For the evaluation of weed recognition, spectral signatures were acquired from *S. marianum* plants, and also from randomly chosen plants of other species (Figure 2) on 29 May 2015 using a UniSpec-DC spectrometer (PP Systems, Inc., Amesbury, MA, USA). It is a handheld dual channel spectrometer, simultaneously measuring incident and reflected light. The spectrometer's wavelength range is 310–1100 nm (visible and near-infrared) and its spectral resolution is less than 10 nm.

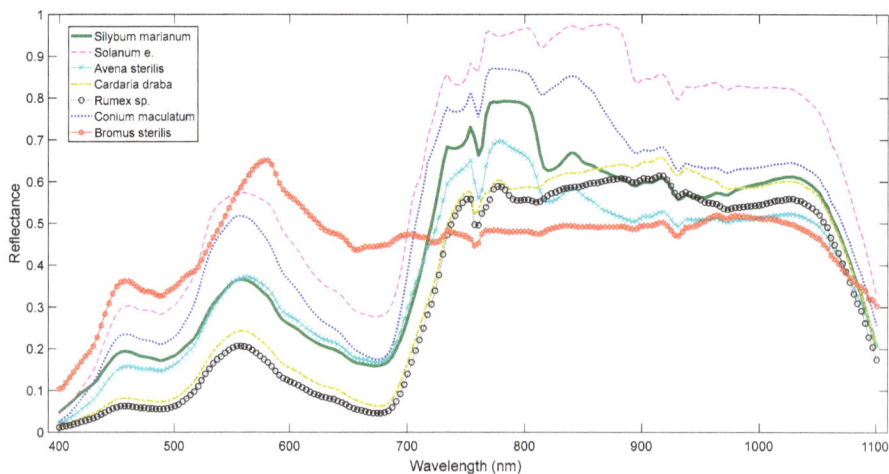

Figure 2. Spectra corresponding to each vegetation type.

It is illustrated in Figure 2 that various plant species demonstrate similar spectral behaviour, rendering the discrimination a challenge. For instance, *S. marianum* and *A. sterilis* presented very strong similarities in the visible range of the spectrum, in contrast to the NIR area.

2.3. Novelty Detector Classifiers

2.3.1. OC-SVM

To achieve One Class SVM Classification, a suitable description of SVM as a model to describe only target data was introduced by Tax and Duin [20] in the form of Support Vector Data Description (SVDD). Given a data set of targets $\{x_i\}, i = 1, 2, 3, \ldots, N$ used for training, the goal of SVDD is to implement by learning a decision function to decide whether an example is either a target or an outlier. SVDD is based on the hypothesis that target data are encircled by a close boundary in the feature space. The simplest type of closed boundary is a supersphere, which is simple to describe with only two parameters: the centre a and the radius R. A supersphere is asked to cover the given target data set members but with the smallest radius R. This optimization problem is constructed as follows:

$$\min R^2 + C \sum_{i=1}^{N} \xi_i$$
$$s.t. ||x_i - a||^2 \leq R^2 + \xi_i, \; i = 1, 2, \ldots, N \tag{1}$$

where C assigns a penalty to the loss term and ξ_i are the slack variables, the value of C is determined by the expected upper function bound v on misclassified targets:

$$C = \frac{1}{N \cdot v} \tag{2}$$

α is computed:

$$\alpha = \sum_{i=1}^{N} \alpha_i x_i, \quad 0 \leq \alpha_i \leq C \tag{3}$$

The value of α_i can be delivered into three categories:

$$\alpha_i = 0 \Rightarrow ||x_i - \alpha|| < R^2,$$

$$0 < \alpha_i < C \Rightarrow ||x_i - \alpha||^2 = R^2,$$

$$\alpha_i = C \Rightarrow ||x_i - \alpha||^2 < R^2.$$

To predict an example v, the distance between v and α is computed:

$$||v - \alpha||^2 \leq R^2 \Rightarrow v \text{ is target}$$
$$||v - \alpha||^2 > R^2 \Rightarrow v \text{ is outlier} \tag{4}$$

The OC-SVM develops a model by being trained in using normal data conforming to the SVDD description. At the second stage, it allocates test data based on the ocurring deviation from normal calibration data as being either normal or outlier [21]. The effect of the Radial Basis Function's (RBF) spreading parameter in $K(\mathbf{x}, \mathbf{z}) = \exp\{-||\mathbf{x} - \mathbf{z}||^2/\sigma^2\}$ can be determined by considering that a sizeable spread indicates a linear class of target data while on the other hand, many support vectors joint with a small spread indicate a highly nonlinear case as is illustarted in Figure 3. A spread parameter equivalent to 2.5 yielded the best results in the *S. marianum* detection presented here. The threshold for accepting outliers was set at 5%.

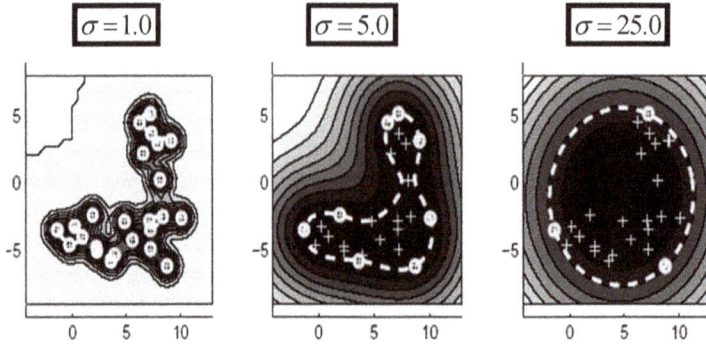

Figure 3. The influence of the RBF spreading parameter on the behaviour of the one class SVM [20].

2.3.2. One Class Self-Organizing Map (OC-SOM)

An OC-SOM is trained using normal operation data. Consequently, the feature vector that matches to a new measurement is inspected in order to evaluate its similarity to the weight vectors of each other map unit. If the smallest distance surpasses a predetermined threshold, it is expected that the process denotes a fault situation. This result stems from the assumption that quantisation faults exceeding a certain value are associated with operation points external to the training data region. Depending on the magnitude of deviation from the normal state, a degradation index can be designed. The OC-SOM [11] creates a model from healthy data and successively classifies new data conferring to its deviation from the healthy baseline condition. During novelty recognition, novel instances from feature combinations (in the current case the features from the multispectral image) of not definable state are used to form the input to the network, while the SOM algorithm chooses the best matching unit (BMU). In Saunders and Gero [22], if the quantisation error that is the outcome from the appraisal between the new exemplar data (x^{NEW}) and BMU is larger than a pre-specified threshold (d) then the example is considered as novel. Equation (5) represents the minimum distance for the BMU and inspects it against the threshold.

$$\min\left(\sum_{j=0}^{n-1} (x_j^{NEW} - m_i)^2\right) > d, \; i \in M \tag{5}$$

where M denotes the SOM grid of neurons.

There are several heuristics to define a threshold based on the effectiveness of the threshold and the precise structure of the data set. A way to define a threshold (d) depends on the resemblance between the SOM centroid vectors and the baseline indicating training vectors selected as BMUs which determines the quantisation error. These distances are calculated from Equation (6):

$$\text{distances} = \min\left(\sum_{k=0}^{N-1} (x_k^{TARGET} - m_i)^2\right), \; i \in M \tag{6}$$

The threshold is programmed in pseudo-code which is presented below:

```
Data_distances_sorted = sort(distances);
Fraction = round(fraction_targets × length(target_set));
Threshold = (Data_distances_sorted(fraction) + Data _distances_sorted(fraction + 1))/2;
```

By the threshold's selection designed to represent a fraction of distances from the whole training set, an assumption is to get distance values in order to represent the codebooks data vectors more faithfully, especially when the distances are formed. Hence, the fraction error represents a subset of distances that can isolate outlier values following a specific distribution. For instance, if the 99% fraction of the distances that separate data and codebooks is chosen to belong to the dataset, the determination of a descriptive hypersphere is direct with a radius that covers 99% of the data. The remaining 1% matches to outliers since they are located exterior to the target set description area. In Figure 4, it is shown that different areas are defined by the threshold relative to the BMUs. The BMUs define Voronoi polygons representing OCSOM neuron domains. It is clear that points selected as residing inside a domain of neuron are situated externally to the threshold-defined polygon which demarcates the border between target and novel data (for illustration purposes, actual data are of high dimension so direct visualisation is not possible). In this paper, an 8 × 8 rectangular OC-SOM gave the best results while 3 × 3 to 20 × 20 were tested. The threshold for accepting outliers was set at 5%.

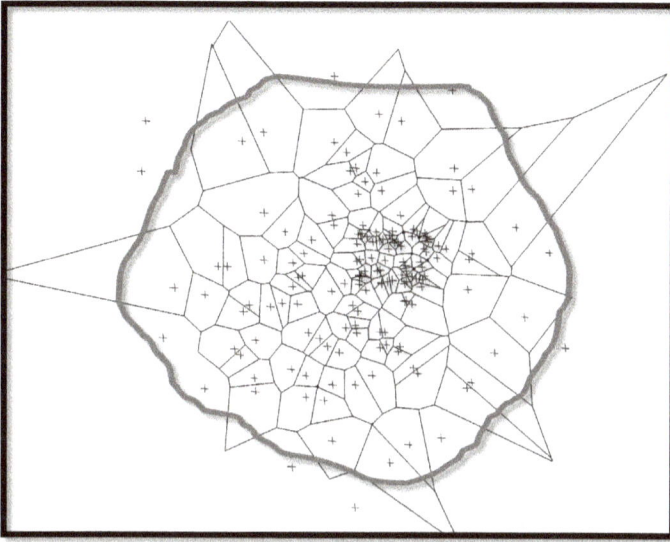

Figure 4. Target dataset regions and Voronoi polygons and the threshold perimeter for OCSOM. Target data defined by the threshold are resident inside the grey border line.

2.3.3. Auto-Encoders

Auto-encoders are neural networks able to learn a data representation [23–25]. They are trained to reproduce input patterns in a mirroring sense as an output (they perform an identity operation). In the auto-encoder architecture, the structure uses only one hidden layer with a number of hidden units which are related to input features. Sigmoidal transfer functions are used.

The auto-encoder networks will successfully reconstruct target objects but with a smaller error compared to outlier objects. With one hidden layer, a (linear) principal component type of solution is the outcome [26]. The auto-encoder network will tend to find a data description similar to the PCA description. To obtain a small reconstruction error from the target set, the network is forced to construct a compact mapping starting from the data into the subspace and structurally coded inside these hidden neurons. The total number of hidden units provides the dimensionality of this subspace. The non-linear transfer functions of the neurons in other layers deem this subspace non-linear.

Auto-encoder methods are flexible, but they inherit several problems from multilayer perceptrons because they are very similar to them [11]. These problems concern sensitivity to improper training parameter selection by the user and predefinition of the Auto-encoder architecture. In this paper, the inputs were the three spectral features and the texture, while 8 hidden units provided the best result ranging from a number of trials in which 3 to 20 were utilised.

2.3.4. OC-PCA

Principal Component Analysis (PCA) (or the Karhunen-Loeve transform) [27] is used for data belonging to a linear subspace. The PCA mapping discovers the orthonormal subspace which represents the variance of the data (in the squared error sense). The optimisation procedure uses eigenvalue decomposition to find the eigenvectors of the covariance matrix of target data. The eigenvectors that are associated to the largest eigenvalues are the principal axis of the d-dimensional data while their direction corresponds to the largest variance. These vectors form an orthonormal basis vectors for the data. The number of basis vectors M is adjusted to a user-defined fraction of the variance in the data. The basis vectors W become a d×M matrix. Since they form an orthonormal basis, the total free parameters in the PCA are determined by:

$$n_{freePCA} = \binom{d-1}{M} \tag{7}$$

When computing the PCA, a zero mean dataset is often assumed. If the data has to be estimated, this operation will add d more free parameters to $n_{freePCA}$.

The reconstruction error of an object z [11] is based on the squared distance from the original object and the corresponding mapped version:

$$d_{PCA}(z) = \|z - (W(W^TW)^{-1}W^T)z\|^2 = \|z - (WW^T)z\|^2 \tag{8}$$

where the second step is feasible since the basis W is orthonormal.

The PCA achieves good accuracy when a clear linear subspace is the case. For very low sample sizes, the data is located in a subspace. If the intrinsic data dimensionality is smaller than the feature size, the PCA can generalize from the low sample size data. If the available data has variance in all feature directions, then in some cases it would be impossible to reduce the dimensionality without reducing the explained variance too much. When data are in separate subspaces, the PCA yields an average subspace representing data in each subspace very poorly. Scaling alters the large variance directions and thus the PCA basis. When data directions are amplified, it will improve the PCA description. On the other hand, when the noise is amplified, it will damage the characterisation. PCA only focuses on the variance of the target set, a drawback is that PCA is not able to include negative examples in the training set. In this paper, a threshold of 10% was set for accepting outliers which gave the best performance.

3. Results and Discussion

The performance of the novelty detector classifiers into identifying *S. marianum* and other vegetation types are presented in Table 1. Accuracy assessment was achieved by computing the confusion tables on the validation dataset (860 pixels), with 441 belonging to *S. marianum* and 419 to other vegetation types [28]. All classifiers were characterised by very high overall accuracy, exceeding 90%. OC-SVM performed better, reaching 96% overall accuracy, and equally high user's and producer's accuracy. OC-SOM had marginally lower overall performance (94.65%) which is explained by the underestimation of 8.39% of *S. marianum* actual observations (omission errors). Auto-encoder had also slightly lower overall accuracy (94.3%), which is attributed to 5.7% of pixels confused between the two categories, leading to a uniformly lower user's and producer's accuracy (94%). OC-PCA had lower overall accuracy (90%) which is explained by the underestimation of 11.56% of *S. marianum* actual observations (omission errors), and the overestimation of 8.35% of other vegetation image pixels (commission errors).

Table 1. Contingency table and optimal parameters of each classifier tested for the identification of *S. marianum* against other vegetation.

		Network Prediction			
Classifier (Overall Accuracy %)	Actual Observations	*S. marianum* (Pixels)	Other Vegetation (Pixels)	User's Accuracy (%)	Producer's Accuracy (%)
OC-SVM σ = 2.5	*S. marianum*	416	25	97.88	94.33
(96.05)	Other vegetation	9	410	94.25	97.85
OC-SOM, 8 × 8	*S. marianum*	404	37	97.82	91.61
(94.65)	Other vegetation	9	410	91.72	97.85
Autoencoder,	*S. marianum*	416	25	94.55	94.33
8 hidden (94.30)	Other vegetation	24	395	94.05	94.27
OC-PCA (90.00)	*S. marianum*	390	51	91.76	88.44
	Other vegetation	35	384	88.28	91.65

The trained OC-SVM, OC-SOM, Autoencoder and OC-PCA models were provided with feature vectors originating from the whole UAV mosaic with 782,838 samples. Each of the input vectors produced a decision corresponding to 1 for *S. marianum* and 2 denoting other vegetation (Figure 5). The weeds map derived from OC-SVM classifier shows large patches of *S. marianum* along the eastern border of the study area, and a high concentration of patches in the central section of the study area, while only few patches are identified in the remaining western part. In the focus area, a large contiguous patch is evident in the southern part, and several densely vegetated patches towards the northwest corner. A slight underestimation of *S. marianum* is evident in the OC-SOM map as compared to OC-SVM, this is most notable in the central north and southeast parts of the study area. This is evident in the focus area, where several gaps have appeared in the large contiguous patch that was mapped with the OC-SVM classifier. In the Autoencoder map, patches of *S. marianum* are missing in the central north part of the study area, while other areas have been incorrectly classified as *S. marianum*. Similar patterns of misclassification appear in the OC-PCA map, where the omission of *S. marianum* patches is more prominent. In the focus area, the large contiguous patch at the southern side is almost entirely missing, while patches at the northwest corner appear denser than those mapped with the other three classifiers.

The vegetation surrounding the *S. marianum* patches consists of graminaceous weeds, forming an even surface with a very smooth texture, in contrast to the *S. marianum* patches, which consist of large individuals with high uneven texture. Thus, texture was a feature contributing to the novelty detection of all classifiers.

The autoencoder and OC-PCA classifiers displayed a notable over-estimation of the *S. marianum* category in the northern part of the study area, which was in fact dominated by *A. sterilis*. This over-estimation can be observed in Figure 5a, and may be due to the fact that *A. sterilis* plants remained in a healthy condition despite the season (late spring), when the rest of the area's vegetation was in senescence. As a result, the spectral signatures of the two species are highly similar (Figure 2) resulting in the aforementioned confusion. The autoencoder and OC-PCA were especially susceptible to this confusion, as they simply rotate data to new multidimensional axes, while OC-SVM and OC-SOM use support vectors and local neurons, respectively.

All of the examined novelty detection classifiers demonstrated a very high level of performance, with overall accuracies ranging from 90% to 96.05%. Compared to other classifiers used with the same dataset, the parametric maximum likelihood classifier achieved lower overall accuracy of 87% [15], while all hierarchical self-organising map classifiers achieved higher accuracies up to 98.87% [16].

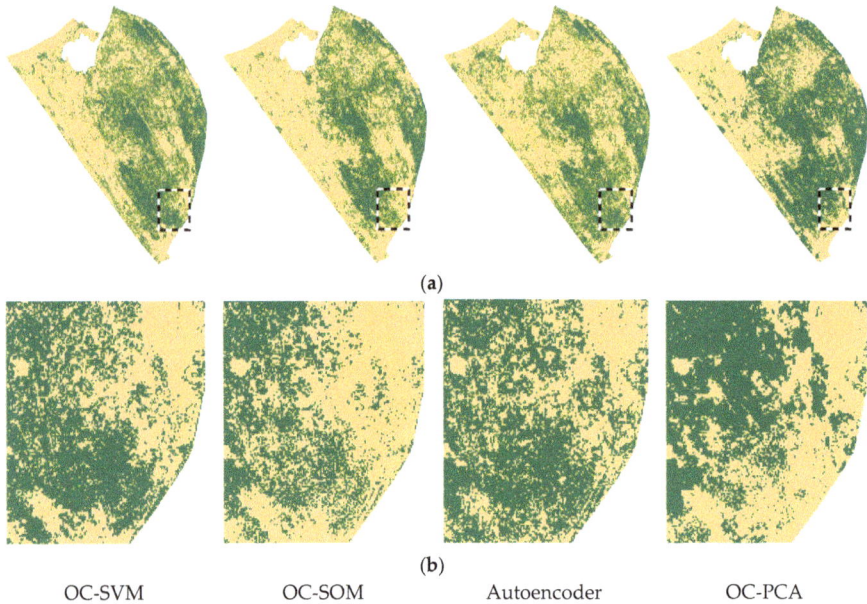

Figure 5. *Silybum marianum* coverage based on four novelty classifiers prediction in the study area (**a**); and in the focus area (**b**) *S. marianum* is green colour and other vegetation is yellow.

Aiming to employ site specific weed management, classification accuracy is of vital importance since the output maps will consequently be utilised as treatment maps. Furthermore, those thematic output maps can be utilised to monitor the spatial distribution of weeds (in this case *S. marianum*), as well as its variation in time, in the case of weed eradication applications.

Coverage maps on a national and regional level can be created by means of relatively low-cost technology like satellite remote sensing [29]. The aim of utilising satellite imagery is to map out the spatial distribution of larger-scale infestations.

On the other hand, when the aim is to apply treatment practices, high accuracy in detection and classification is imperative, even in cases of low-density infestations. Accordingly, multispectral or hyperspectral imagery is the necessary approach, and the suggested method of acquisition is by employing UAV technology. The acquired data and results can be utilised to detect areas susceptible to weed growth due to favourable conditions, to identify problematic areas where the weeds could spread rapidly, and lastly, to design new and evaluate existing management practices.

4. Conclusions

In the current study, it was proven that it is possible to locate *S. marianum* patches with the aid of four novelty detection classifiers. The classifiers were OC-SVM, OC-SOM, OC-PCA and Autoencoder. For the image acquisition, a multispectral camera (green-red-NIR) mounted on a UAV was utilised. As input features for the classifiers, three spectral bands from the camera and the texture, calculated as local variance of the NIR band, were applied. The highest rates of overall accuracy were achieved using OC-SVM network (96.05%), while slightly lower accuracies were scored by OC-SOM (94.65%), OC-PCA (90%) and Autoencoder (94.30%). The results advocate that using one class novelty detectors can be applied operationally for mapping *S. marianum* with UAV for several operations encompassing weed eradication programs.

Sensors **2017**, *17*, 2007

Acknowledgments: We are thankful to the American Farm School of Thessaloniki for providing access to the study area, and to Vasilios Kaprinis and Konstantinos Georgiadis for their assistance in the field surveys.

Author Contributions: T.K.A., A.L.L. and J.K. conceived and designed the experiments; T.K.A., A.A.T. and G.O. developed the algorithms; X.E.P. and D.M. coded and debugged the algorithms; A.A.T. performed the experiments; T.K.A., A.A.T., G.O. and V.P. analysed the data; J.K. provided field data; V.P. contributed field surveying tools; all authors wrote the paper.

Conflicts of Interest: The authors declare no conflict of interest.

References

1. Thorp, K.; Tian, L. A review on remote sensing of weeds in agriculture. *Precis. Agric.* **2004**, *5*, 477–508. [CrossRef]
2. Zhang, Y.; Slaughter, D.C.; Staab, E.S. Robust hyperspectral vision-based classification for multi-season weed mapping. *ISPRS J. Photogramm. Remote Sens.* **2012**, *69*, 65–73. [CrossRef]
3. Åstrand, B.; Baerveldt, A.-J. An agricultural mobile robot with vision-based perception for mechanical weed control. *Auton. Robots* **2002**, *13*, 21–35. [CrossRef]
4. Waske, B.; van der Linden, S.; Benediktsson, J.A.; Rabe, A.; Hostert, P. Sensitivity of support vector machines to random feature selection in classification of hyperspectral data. *IEEE Trans. Geosci. Remote Sens.* **2010**, *48*, 2880–2889. [CrossRef]
5. Ahmed, F.; Al-Mamun, H.A.; Bari, A.H.; Hossain, E.; Kwan, P. Classification of crops and weeds from digital images: A support vector machine approach. *Crop Prot.* **2012**, *40*, 98–104. [CrossRef]
6. Vapnik, V. *Statistical Learning Theory*; John Wiley & Sons, Inc.: New York, NY, USA, 1998.
7. Sluiter, R. *Mediterranean Land Cover Change: Modelling and Monitoring Natural Vegetation Using Gis and Remote Sensing*; Utrecht University: Utrecht, The Netherlands, 2005.
8. Im, J.; Jensen, J.R. A change detection model based on neighborhood correlation image analysis and decision tree classification. *Remote Sens. Environ.* **2005**, *99*, 326–340. [CrossRef]
9. Dihkan, M.; Guneroglu, N.; Karsli, F.; Guneroglu, A. Remote sensing of tea plantations using an svm classifier and pattern-based accuracy assessment technique. *Int. J. Remote Sens.* **2013**, *34*, 8549–8565. [CrossRef]
10. Crupi, V.; Guglielmino, E.; Milazzo, G. Neural-network-based system for novel fault detection in rotating machinery. *Modal Anal.* **2004**, *10*, 1137–1150. [CrossRef]
11. Tax, D.M.J. One-Class Classification. Ph.D. Thesis, Delft University of Technology, Delft, The Netherlands, 2001.
12. Parsons, W.T.; Cuthbertson, E. *Noxious Weeds of Australia*; CSIRO Publishing: Melbourne, Victoria, Australia, 2001.
13. Tucker, J.M.; Cordy, D.R.; Berry, L.J.; Harvey, W.A.; Fuller, T.C. *Nitrate Poisoning in Livestock*; University of California: Oakland, CA, USA, 1961.
14. Khan, M.A.; Kalsoom, U.; Khan, M.I.; Khan, R.; Khan, S.A. Screening the allelopathic potential of various weeds. *Pak. J. Weed Sci. Res.* **2011**, *11*, 73–81.
15. Tamouridou, A.; Alexandridis, T.; Pantazi, X.; Lagopodi, A.; Kashefi, J.; Moshou, D. Evaluation of uav imagery for mapping silybum marianum weed patches. *Int. J. Remote Sens.* **2017**, *38*, 2246–2259.
16. Pantazi, X.; Tamouridou, A.; Alexandridis, T.; Lagopodi, A.; Kashefi, J.; Moshou, D. Evaluation of hierarchical self-organising maps for weed mapping using uas multispectral imagery. *Comput. Electron. Agric.* **2017**, *139*, 224–230. [CrossRef]
17. Peña, J.M.; Torres-Sánchez, J.; de Castro, A.I.; Kelly, M.; López-Granados, F. Weed mapping in early-season maize fields using object-based analysis of unmanned aerial vehicle (UAV) images. *PLoS ONE* **2013**, *8*, e77151. [CrossRef] [PubMed]
18. Pérez-Ortiz, M.; Gutiérrez, P.A.; Peña, J.M.; Torres-Sánchez, J.; Hervás-Martínez, C.; López-Granados, F. An experimental comparison for the identification of weeds in sunflower crops via unmanned aerial vehicles and object-based analysis. In Proceedings of the 13th International Work-Conference on Artificial Neural Networks (IWANN), Palma de Mallorca, Spain, 6 June 2015; Rojas, I., Joya, G., Catala, A., Eds.; Springer: Palma de Mallorca, Spain, 2015; pp. 252–262.
19. Torres-Sánchez, J.; López-Granados, F.; De Castro, A.I.; Peña-Barragán, J.M. Configuration and specifications of an unmanned aerial vehicle (UAV) for early site specific weed management. *PLoS ONE* **2013**, *8*, e58210. [CrossRef] [PubMed]

20. Tax, D.M.; Duin, R.P. Support vector data description. *Mach. Learn.* **2004**, *54*, 45–66. [CrossRef]
21. Scholkopf, B.; Smola, A.J. *Learning with Kernels: Support Vector Machines, Regularization, Optimization, and Beyond*; MIT Press: Cambridge, MA, USA, 2002.
22. Saunders, R.; Gero, J.S. A curious design agent: A computational model of novelty-seeking behaviour in design. In Proceedings of the Sixth Conference on Computer-Aided Architectural Design Research in Asia (CAADRIA 2001), Sydney, Australia, 19–21 April 2001; pp. 345–350.
23. Japkowicz, N.; Myers, C.; Gluck, M. *A Novelty Detection Approach to Classification*; International Joint Conference on Artificial Intelligence (IJCAI 95): Montreal, QC, Canada, 1995; pp. 518–523.
24. Hertz, J.A.; Krogh, A.S.; Palmer, R.G. *Introduction to the Theory of Neural Computation*; Addison-Wesley Longman Publishing Co., Inc.: Boston, MA, USA, 1991.
25. Baldi, P.; Hornik, K. Neural networks and principal component analysis: Learning from examples without local minima. *Neural Netw.* **1989**, *2*, 53–58. [CrossRef]
26. Bourlard, H.; Kamp, Y. Auto-association by multilayer perceptrons and singular value decomposition. *Biol. Cybern.* **1988**, *59*, 291–294. [CrossRef] [PubMed]
27. Bishop, C.M. *Neural Networks for Pattern Recognition*; Oxford University Press: New York, NY, USA, 1995.
28. Congalton, R.G. A review of assessing the accuracy of classifications of remotely sensed data. *Remote Sens. Environ.* **1991**, *37*, 35–46. [CrossRef]
29. Metternicht, G. *Geospatial Technologies and the Management of Noxious Weeds in Agricultural and Rangelands Areas of Australia*; University of South Australia: Mawson Lakes, SA, Australia, 2007.

sensors

MDPI

Article

A Robust Deep-Learning-Based Detector for Real-Time Tomato Plant Diseases and Pests Recognition

Alvaro Fuentes [1], Sook Yoon [2,3], Sang Cheol Kim [4] and Dong Sun Park [5,*]

[1]   Department of Electronics Engineering, Chonbuk National University, Jeonbuk 54896, Korea;
      afuentes@jbnu.ac.kr
[2]   Research Institute of Realistic Media and Technology, Mokpo National University, Jeonnam 534-729, Korea;
      syoon@mokpo.ac.kr
[3]   Department of Computer Engineering, Mokpo National University, Jeonnam 534-729, Korea
[4]   National Institute of Agricultural Sciences, Suwon 441-707, Korea; sckim@rda.go.kr
[5]   IT Convergence Research Center, Chonbuk National University, Jeonbuk 54896, Korea
*   Correspondence: dspark@jbnu.ac.kr; Tel.: +82-63-270-2475

Received: 10 July 2017; Accepted: 28 August 2017; Published: 4 September 2017

Abstract: Plant Diseases and Pests are a major challenge in the agriculture sector. An accurate and a faster detection of diseases and pests in plants could help to develop an early treatment technique while substantially reducing economic losses. Recent developments in Deep Neural Networks have allowed researchers to drastically improve the accuracy of object detection and recognition systems. In this paper, we present a deep-learning-based approach to detect diseases and pests in tomato plants using images captured in-place by camera devices with various resolutions. Our goal is to find the more suitable deep-learning architecture for our task. Therefore, we consider three main families of detectors: Faster Region-based Convolutional Neural Network (Faster R-CNN), Region-based Fully Convolutional Network (R-FCN), and Single Shot Multibox Detector (SSD), which for the purpose of this work are called "deep learning meta-architectures". We combine each of these meta-architectures with "deep feature extractors" such as VGG net and Residual Network (ResNet). We demonstrate the performance of deep meta-architectures and feature extractors, and additionally propose a method for local and global class annotation and data augmentation to increase the accuracy and reduce the number of false positives during training. We train and test our systems end-to-end on our large Tomato Diseases and Pests Dataset, which contains challenging images with diseases and pests, including several inter- and extra-class variations, such as infection status and location in the plant. Experimental results show that our proposed system can effectively recognize nine different types of diseases and pests, with the ability to deal with complex scenarios from a plant's surrounding area.

Keywords: plant disease; pest; deep convolutional neural networks; real-time processing; detection

---

1. Introduction

Crops are affected by a wide variety of diseases and pests, especially in tropical, subtropical, and temperate regions of the world [1]. Plant diseases involve complex interactions between the host plant, the virus, and its vector [2]. The context of this problem is sometimes related to the effects of the climate change in the atmosphere and how it alters an ecosystem. Climate change basically affects regional climate variables, such as humidity, temperature, and precipitation, that consequently serve as a vector in which pathogens, virus, and plagues can destroy a crop, and thus cause direct impacts on the population, such as economic, health, and livelihood impacts [3].

Diseases in plants have been largely studied in the scientific area, mainly focusing on the biological characteristics of diseases [4]. For instance, studies on potato [5] and tomato [6,7] show how susceptible a plant is to be affected by diseases. The problem of plant diseases is a worldwide issue also related to food security [8]. Regardless of frontiers, media, or technology, the effects of diseases in plants cause significant losses to farmers [9]. An earlier identification of disease is nowadays a challenging approach and needs to be treated with special attention [10].

In our approach, we focus on the identification and recognition of diseases and pests that affect tomato plants. Tomato is economically the most important vegetable crop worldwide, and its production has been substantially increased through the years [11]. The worldwide cultivation of tomato exposes the crop to a wide range of new pathogens. Many pathogens have found this crop to be highly susceptible and essentially defenseless [6]. Moreover, viruses infecting tomato have been described, while new viral diseases keep emerging [12].

Several techniques have been recently applied to apparently identify plant diseases [13]. These include using direct methods closely related to the chemical analysis of the infected area of the plant [14–16], and indirect methods employing physical techniques, such as imaging and spectroscopy [17,18], to determine plant properties and stress-based disease detection. However, the advantages of our approach compared to most of the traditionally used techniques are based on the following facts:

- Our system uses images of plant diseases and pests taken in-place, thus we avoid the process of collecting samples and analyzing them in the laboratory.
- It considers the possibility that a plant can be simultaneously affected by more than one disease or pest in the same sample.
- Our approach uses input images captured by different camera devices with various resolutions, such as cell phone and other digital cameras.
- It can efficiently deal with different illumination conditions, the size of objects, and background variations, etc., contained in the surrounding area of the plant.
- It provides a practical real-time application that can be used in the field without employing any expensive and complex technology.

Plant diseases visibly show a variety of shapes, forms, colors, etc. [10]. Understanding this interaction is essential to design more robust control strategies to reduce crop damage [2]. Moreover, the challenging part of our approach is not only in disease identification but also in estimating how precise it is and the infection status that it presents. At this point, it is necessary to clarify the differences between the notions of image classification and object detection. Classification estimates if an image contains any instances of an object class (what), unlike a detection approach, which deals with the class and location instances of any particular object in the image (what and where). As shown in Figure 1, our system is able to estimate the class based on the probability of a disease and its location in the image shown as a bounding box containing the infected area of the plant.

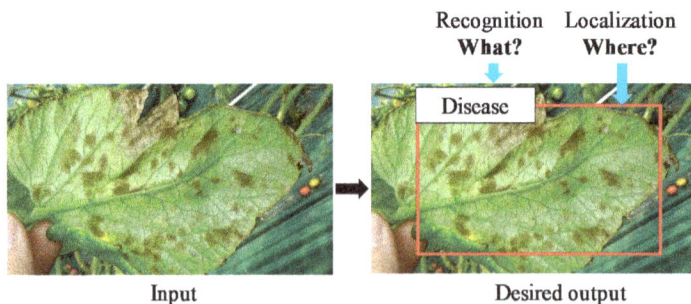

Figure 1. Recognition and localization of plant diseases and pests: problem formulation. Our system aims to detect both class (what) and location (where) of the affected areas in the image.

Recent advances in hardware technology have allowed the evolution of Deep Convolutional Neural Networks and their large number of applications, including complex tasks such as object recognition and image classification. Since the success of AlexNet [19] in the ImageNet Large Scale Visual Recognition Challenge [20] 2012 (ILSVRC), deeper and deeper networks [21–26] have been proposed and achieved state-of-the-art performance on ImageNet and other benchmark datasets [27]. Thus, these results evidence the need to study the depth and width, as deeper and wider networks generate better results [28].

In this paper, we address disease and pest identification by introducing the application of deep meta-architectures [29] and feature extractors. Instead of using traditionally employed methods, we basically develop a system that successfully recognizes different diseases and pests in images collected in real scenarios. Furthermore, our system is able to deal with complex tasks, such as infection status (e.g., earlier, last), location in the plant (e.g., leaves, steam), sides of leaves (e.g., front, back), and different background conditions, among others.

Following previous approaches [30–32], we aim to use meta-architectures based on deep detectors to identify Regions of Interest (ROI) in the image, which correspond to infected areas of the plant. Each ROI is then classified as containing or not containing a disease or pest compared to the ground-truth annotated data. Using deep feature extractors, our meta-architecture can efficiently learn complex variations among diseases and pests found in different parts of the plant and deal with different sizes of candidates in the image.

The contributions of this paper are as follows: we propose a robust deep-learning-based detector for real-time tomato diseases and pests recognition. The system introduces a practical and applicable solution for detecting the class and location of diseases in tomato plants, which in fact represents a main comparable difference with traditional methods for plant diseases classification. Our detector uses images captured in-place by various camera devices that are processed by a real-time hardware and software system using graphical processing units (GPUs), rather than using the process of collecting physical samples (leaves, plants) and analyzing them in the laboratory. Furthermore, it can efficiently deal with different task complexities, such as illumination conditions, the size of objects, and background variations contained in the surrounding area of the plant. A detailed review of traditional methods for anomaly detection in plants and deep-learning techniques is presented in Section 2. Our proposed deep-learning-based system and the process for detecting diseases and pests is detailed in Section 3. In Section 4, we show the experimental results to demonstrate how our detector is able to successfully recognize nine different diseases and pests and their location in the images while providing robust real-time results. Moreover, we found out that using a technique-based data annotation and augmentation method results in better performance. In the last section, we study some of the detection failures and conclude that, although the system shows outstanding performance when dealing with all complex scenarios, there is still room for prediction improvements as our dataset becomes larger and includes more classes.

2. Related Works

2.1. Anomaly Detection in Plants

Plant diseases identification is a critical topic that has been studied through the years, and is motivated by the need to produce healthy food. However, some desirable elements to take into account should be cost-effectiveness, user-friendliness, sensitiveness, and accuracy [33]. In the last decade, several works have proposed some nondestructive techniques to overcome those facts. In [34], hyperspectral proximal sensing techniques were used to evaluate plant stress to environmental conditions. Optical technologies are practical tools considered for monitoring plant health; for example, in [35], thermal and fluorescence imaging methods were introduced for estimating plant stress produced mainly by increased gases, radiation, water status, and insect attack, among others. Another important area includes the study of plant defense in response to the presence of pathogens.

For that effect, in [36], chemical elements were applied to leaves in order to estimate their defense capabilities against pathogens. To study plant robustness against nutritional facts, in [37], potato plants were cultivated in the presence of several nutritional elements to evaluate their effects in the crop.

As mentioned earlier, the area of plant anomaly detection has been dealt with by different media. Although previous methods show outstanding performance in the evaluated scenarios, they do not provide yet a highly accurate solution for estimating diseases and pests in a real-time manner. Instead, their experiments are mainly conducted in a laboratory or using expensive techniques. Therefore, our approach is focused on a cost-effective technique that uses images collected in situ as our source of information, including variations of the scenario in place. Before Deep Learning became popular in the Computer Vision field, several handcrafted feature-based methods had been widely applied specifically for image recognition. A handcrafted method is called so because of all the human knowledge implied in the development of the algorithm itself and the complex parameters that are included in the process. Some disadvantages of these methods are also the high computational cost and time consumption due to the complex preprocessing, feature extracting, and classifying. Some of the best-known handcrafted feature methods are the Histogram of Oriented Gradients (HOG) [38] and Scale-Invariant Feature Transform (SIFT) [39], which are usually combined with classifiers such as Adaptive Boosting (AdaBoost) [40] or Support Vector Machines (SVM) [41].

The facilities of Deep Learning have allowed researchers to design systems that can be trained and tested end-to-end (all included in the same process), unlike when using handcrafted-based methods that use separate processes. Due to the outstanding performance of Convolutional Neural Networks (CNNs) as a feature extractor in image recognition tasks, the idea has been also extended to different areas, such as in agriculture, automation, and robotics. Some of the applications for agriculture utilize Computer Vision and CNNs to solve complex tasks, such as plant recognition. For instance, in [42], it is shown how a CNN-based method outperforms local feature descriptors and bag of visual words techniques when recognizing 10 types of plants. In [43], the authors found that using a fusion of deep representations and handcrafted features leads to a higher accuracy of leaf plant classification. They applied a CNN for leaf segmentation, extracted handcrafted features with image processing techniques, trained an SVM with feature vectors, and used an SVM with a CNN to identify species among 57 varieties of trees.

Subsequently, due to the recent advance in Machine Learning, the principle of CNN has been applied to plant diseases recognition in different crops, such as [44] using a CNN-based LeNet and image processing to recognize two leaf diseases out of healthy ones. In [45], an image processing and statistical inference approach was introduced to identify three types of leaf diseases in wheat. In [46], the authors developed a method to discriminate good and bad condition images which contain seven types of diseases out of healthy ones in cucumber leaves. For that effect, they used an image-processing technique and a four-layer CNN, which showed an average of 82.3% accuracy under a 4-fold cross-validation strategy. Another approach for cucumber leaf diseases, [47], used a three-layer CNN to train images containing two diseases out of healthy ones. To support the application of machine learning, [48] proposed to use a method called Color and Oriented FAST and Rotated BRIEF (ORB) to extract features and tree classifiers (Linear Support Vector Classifier (SVC), K-Nearest Neighbor, Extremely Randomized Trees) to recognize four types of diseases in cassava. As a result, they present a smartphone-based system that uses the classification model that has learned to do real-time prediction of the state of health of a farmer's garden.

Other works that use deep convolutional neural networks for diseases recognition have been also proposed, showing good performance on different crops. For instance, [49] developed a CNN-based system to identify 13 types of diseases out of healthy ones in five crops using images downloaded from the internet. The performance of that approach shows a top-1 success of 96.3% and top-5 success of 99.99%. In [50], the authors evaluate two CNN approaches based on AlexNet [19] and GoogleNet [23], to distinguish 26 diseases included in 14 crops using the Plant Village Dataset [51]. Another work in the same dataset shows a test accuracy of 90.4% using a VGG-16 model trained with transfer learning [52]. However, the Plant Village Dataset contains only images of leaves that are previously cropped in the field and captured by a camera in the laboratory. This is unlike the images in our Tomato Diseases and Pest Dataset, which are directly taken in-place by different cameras with various resolutions, including not only leaves infected by specific pathogens at different infection stages but also other infected parts of the plant, such as fruits and stems. Furthermore, the challenging part of our dataset is to deal with background variations mainly caused by the surrounding areas or the place itself (greenhouse).

Although the works mentioned above show outstanding performance on leaf diseases recognition, the challenges, such as pattern variation, infection status, different diseases or pests and their location in the image, and surrounding objects, among others, are still difficult to overcome. Therefore, we consider a technique that not only recognizes the disease in the image but also identifies its location for the posterior development of a real-time system.

2.2. Deep Meta-Architectures for Object Detection

Convolutional Neural Networks are considered nowadays as the leading method for object detection. As hardware technology has been improved through the years, deeper networks with better performance have been also proposed. Among them, we mention some state-of-the-art methods for object recognition and classification. In our paper, we focus principally on three recent architectures: Faster Region-Based Convolutional Neural Network (Faster R-CNN) [30], Single Shot Multibox Detector (SSD) [31], and Region-based Fully Convolutional Networks (R-FCN) [32]. As proposed in [29], while these meta-architectures were initially proposed with a particular feature extractor (VGG, Residual Networks ResNet, etc.), we now apply different feature extractors for the architectures. Thus, each architecture should be able to be merged with any feature extractor depending on the application or need.

2.2.1. Faster Region-based Convolutional Neural Network (Faster R-CNN)

In Faster R-CNN, the detection process is carried out in two stages. In the first stage, a Region Proposal Network (RPN) takes an image as input and processes it by a feature extractor [30]. Features at an intermediate level are used to predict object proposals, each with a score. For training the RPNs, the system considers anchors containing an object or not, based on the Intersection-over-Union (IoU) between the object proposals and the ground-truth. In the second stage, the box proposals previously generated are used to crop features from the same feature map. Those cropped features are consequently fed into the remaining layers of the feature extractor in order to predict the class probability and bounding box for each region proposal. The entire process happens on a single unified network, which allows the system to share full-image convolutional features with the detection network, thus enabling nearly cost-free region proposals.

Since the Faster R-CNN was proposed, it has influenced several applications due to its outstanding performance on complex object recognition and classification.

2.2.2. Single Shot Detector (SSD)

The SSD meta-architecture [31] handles the problem of object recognition by using a feed-forward convolutional network that produces a fixed-size collection of bounding boxes and scores for the presence of an object class in each box. This network is able to deal with objects of various sizes by combining predictions from multiple feature maps with different resolutions. Furthermore, SSD encapsulates the process into a single network, avoiding proposal generation and thus saving computational time.

2.2.3. Region-based Fully Convolutional Network (R-FCN)

The R-FCN framework [32] proposes to use position-sensitive maps to address the problem of translation invariance. This method is similar to Faster R-CNN, but instead of cropping features from the same layer where region proposals are predicted, features (regions with a higher probability of containing an object or being part of it) are cropped from the last layer of features prior to prediction [29]. By the application of that technique, this method minimizes the amount of memory utilized in region computation. In the original paper [32], they show that using a ResNet-101 as feature extractor can generate competitive performance compared to Faster R-CNN.

2.3. Feature Extractors

In each meta-architecture, the main part of the system is the "feature extractor" or deep architecture. As mentioned in the previous section, year by year different deep architectures have been proposed and their application drastically depends on the complexity of problem itself. There are some conditions that should be taken into consideration when choosing a deep architecture, such as the type or number of layers, as a higher number of parameters increases the complexity of the system and directly influences the memory computation, speed, and results of the system.

Although each network has been designed with specific characteristics, all share the same goal, which is to increase accuracy while reducing computational complexity. In Table 1, some of the feature extractors used in this work are mentioned, including their number of parameters and performance achieved in the Image Net Challenge. We select some of the recent deep architectures because of their outstanding performance and applicability to our system.

Table 1. Properties of the deep feature extractors used in this work and their performance on the ImageNet Challenge.

Feature Extractor	Parameters (M)	Number of Layers	Top-5 Error
AlexNet [19]	61	8	15.3
ZFNet	-	8	14.8
VGG-16 [22]	138	16	7.40
GoogLeNet [23]	6.9	22	6.66
ResNet-50 [24]	25	50	3.57
ResNet-101 [24]	42.6	101	-
ResNetXt-101 [26]	42.6	101	3.03

As shown in Figure 2, our system proposes to treat the deep meta-architecture as an open system on which different feature extractors can be adapted to perform on our task. The input image captured by a camera device with different resolutions and scales is fed into our system, which after processing by our deep network (feature extractor and classifier) results in the class and localization of the infected area of the plant in the image. Thus, we can provide a nondestructive local solution only where the damage is presented, and therefore avoid the disease's expansion to the whole crop and reduce the excessive use of chemical solutions to treat them.

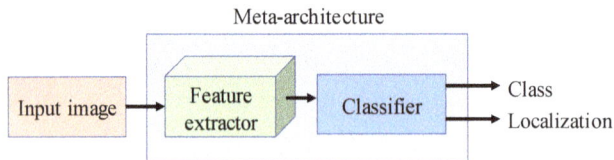

Figure 2. Flow chart of the deep meta-architecture approach used in this work. Our system proposes to treat a deep meta-architecture as an open system on which different feature extractors can be adapted to perform on our task. The system is trained and tested end-to-end using images captured in-place. The outputs are the class and localization of the infected area in the image.

3. Deep Meta-Architectures-Based Plant Diseases and Pest Recognition

3.1. System Background

Tomato plants are susceptible to several disorders and attacks caused by diseases and pests. There are several reasons that can be attributable to the effects on the crops: (1) abiotic disorders due to the environmental conditions, such as temperature, humidity, nutritional excess (fertilizer), light, and species of plant; (2) some pest that spread the disease from plant to plant, such as whiteflies, leaf miners, worms, bugs, etc; and (3) the most common diseases that include bacterial, virus, and fungal diseases. Those diseases and pests along with the plant may present different physical characteristics, such as a variety of shapes, colors, forms, etc. Therefore, due to similar patterns, those variations are difficult to be distinguished, which furthermore makes their recognition a challenge, and an earlier detection and treatment can avoid several losses in the whole crop.

Based on the facts above mentioned, we consider the following characteristics for our analysis:

- Infection status: A plant shows different patterns along with their infection status according to the life cycle of the diseases.
- Location of the symptom: It considers that diseases not only affect leaves, but also other parts of the plant such as stem or fruits.
- Patterns of the leaf: Symptoms of the diseases show visible variations either on the front side or the back side of the leaves.
- Type of fungus: Identifying the type of fungus can be an easy way to visibly differentiate between some diseases.
- Color and shape: Depending on the disease, the plant may show different colors or shapes at different infection stages.

In Figure 3, we show a representation of the diseases and pests under different conditions and variations identified in our work. A detailed study of each disease's and pest's symptoms is described in [10].

Figure 3. A representation of diseases and pests that affect tomato plants. (**a**) Gray mold, (**b**) Canker, (**c**) Leaf mold, (**d**) Plague, (**e**) Leaf miner, (**f**) Whitefly, (**g**) Low temperature, (**h**) Nutritional excess or deficiency, (**i**) Powdery mildew. The images are collected under different variations and environmental conditions. The patterns help to distinguish some proper characteristics of each disease and pest.

3.2. System Overview

Our work aims to identify nine classes of diseases and pests that affect tomato plants using Deep Learning as the main body of the system. A general overview of the system is presented in Figure 4. Following we describe in detail each component of the proposed approach.

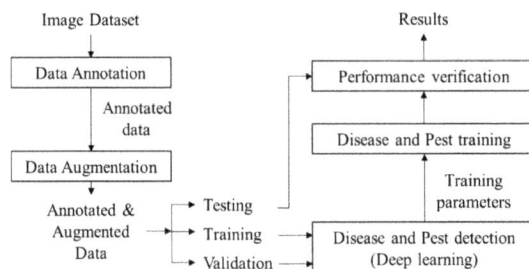

Figure 4. System overview of the proposed deep-learning-based approach for plant diseases and pest recognition. Our deep meta-architecture approach consists of several steps that use input images as a source of information, and provide detection results in terms of class and location of the infected area of the plant in the image.

3.3. Data Collection

Our dataset contains images with several diseases and pests in tomato plants. Using simple camera devices, the images were collected under several conditions depending on the time (e.g., illumination), season (e.g., temperature, humidity), and place where they were taken (e.g., greenhouse). For that purpose, we have visited several tomato farms located in Korea and fed our dataset with various types of data, including:

- Images with various resolutions.
- Samples at early, medium, and last infection status.
- Images containing different infected areas in the plant (e.g., stem, leaves, fruits, etc.).
- Different plant sizes.
- Objects surrounding the plant in the greenhouse, etc.

These conditions help to estimate the infection process and determine how a plant is affected by the disease or pest (origin or possible developing cause).

3.4. Data Annotation

Starting with the dataset of images, we manually annotate the areas of every image containing the disease or pest with a bounding box and class. Some diseases might look similar depending on the infection status that the present; therefore, the knowledge for identifying the type of disease or pest has been provided by experts in the area. That has helped us to visibly identify the categories in the images and infected areas of the plant.

This annotation process aims to label the class and location of the infected areas in the image. The outputs of this step are the coordinates of the bounding boxes of different sizes with their corresponding class of disease and pest, which consequently will be evaluated as the Intersection-over-Union (IoU) with the predicted results of the network during testing. To make it more clear, an example of an annotated bounding box can be visualized in Figure 1. The red box shows the infected areas of the plant, and parts of the background.

Since our images are collected in the field, many areas corresponding to the background could be included in the image, making the problem more challenging. Therefore, when collecting the images, we find out that the best way to get more precise information is to capture the samples containing the

ROIs as the main part of the image. As previously presented in Figure 1, the problem formulation of recognition and localization of the infected part of the plant makes our system different from others that are basically focused only on classification.

3.5. Data Augmentation

Although Deep Neural Network systems have shown outstanding performance compared to traditional machine learning or computer vision algorithms, the drawback of these systems is the overfitting problem. Overfitting is often referred to the hyper-parameters selection, system regularization, or a number of images used for training. Following [19], data augmentation is necessary to pursue when the number of images in the dataset is not enough. We use several techniques that basically increase the number of images of our dataset. These techniques consist of geometrical transformations (resizing, crop, rotation, horizontal flipping) and intensity transformations (contrast and brightness enhancement, color, noise).

3.6. Disease and Pest Detection

We now describe our main method for detecting diseases and pests. Our goal is to detect and recognize the class and location of disease and pest candidates in the image. To detect our target, we need to accurately localize the box containing the object, as well as identify the class to which it belongs.

As shown in Figure 2, our proposed solution aims to overcome such a complex problem by a simple and accurate form. We extend the idea of a meta-architecture-based object detection framework to adapt it with different feature extractors that detect diseases and pests and localize their position in the image. For that purpose, we have considered three meta-architectures due to their high performance in object detection. In the following, we explain in detail each meta-architecture and feature extractor.

3.6.1. Faster R-CNN

We extend the application of Faster R-CNN [30] for object recognition and its Region Proposal Network (RPN) to estimate the class and location of object proposals that may contain a target candidate. The RPN is used to generate the object proposals, including their class and box coordinates. Then, for each object proposal, we extract the features with an RoI Pooling layer and perform object classification and bounding-box regression to obtain the estimated targets.

3.6.2. SSD

We follow the methodology described in [31]. SSD generates anchors that select the top most convolutional feature maps and a higher resolution feature map at a lower resolution. Then, a sequence of the convolutional layer containing each detection per class is added with spatial resolution used for prediction. Thus, SSD is able to deal with objects of various sizes contained in the images. A Non-Maximum Suppression method is used to compare the estimated results with the ground-truth.

3.6.3. R-FCN

We follow the implementation of R-FCN [32] as another meta-architecture to perform our approach. Similar to Faster R-CNN, R-FCN uses a Region Proposal Network to generate object proposals, but instead of cropping features using the RoI pooling layer it crops them from the last layer prior to prediction. We used batch normalization for each feature extractor, and train end-to-end using an ImageNet Pretrained Network.

We have selected the feature extractors based on their performance and number of parameters from Table 1. These are VGG-16, ResNet 50-152, and ResNeXt-50 for Faster R-CNN, ResNet-50 for SSD, and ResNet-50 for R-FCN. To perform the experiments, we have adapted the feature extractors to the conditions of each meta-architecture. For instance, in Faster R-CNN, each feature extractor includes the RPN and features are extracted from the "conv5" layer of VGG-16, the last layer of the "conv4" block in ResNet 50-152, as well as from the "conv4" block in ReNeXt-50. In SSD, in contrast to the original work, we use ResNet-50 as its basis feature extractor. In R-FCN, ResNet-50 is used as its feature extractor and the features are extracted from the "conv4" block.

Our training objective is to reduce the losses between the ground-truth and estimated results, as well as to reduce the presence of false positives in the final results, by Non-Maximum Suppression (NMS) of each meta-architecture, which selects only candidates only with an IoU > 0.5 compared to their initial annotated ground-truth. The loss functions and bounding-box encoding used in this work are presented in Table 2.

Table 2. Details of Deep Learning Meta-architectures and Feature Extractors.

Meta-Architecture	Feature Extractor	Bounding Box	Loss Function
Faster R-CNN	VGG-16 ResNet-50 ResNet-101 ResNet-152 ResNeXt-50	$\left[\frac{x_c}{w_a}, \frac{y_c}{h_a}, \log w, \log h\right]$	SmoothL_1
SSD	ResNet-50	$\left[\frac{x_c}{w_a}, \frac{y_c}{h_a}, \log w, \log h\right]$	SmoothL_1
R-FCN	ResNet-50	$\left[\frac{x_c}{w_a}, \frac{y_c}{h_a}, \log w, \log h\right]$	SmoothL_1

Faster R-CNN: faster region-based convolutional neural network; SSD: single shot detector; R-FCN: region-based fully convolutional network.

4. Experimental Results

4.1. Tomato Diseases and Pests Dataset

Our dataset consists of about 5000 images collected from farms located in different areas of the Korean Peninsula. The images were taken under different conditions and scenarios. They include diseases that can develop depending on the season and variables such as temperature and humidity. Since not all diseases can be found all year round, but rather in seasons, the number of images corresponding to each class is different. The categories and the number of annotated samples used in our system can be seen in Table 3. The number of annotated samples corresponds to the number of bounding boxes labeled in the images after data augmentation. Every image contains more than one annotated sample depending on the infection areas of the plant, and the background class is collected as a transversal category (hard negatives) that is annotated in most of the images. The background class has been called so because it contains areas of the image that correspond to healthy parts of the plant and from the background itself, such as the structure of the greenhouse.

Table 3. List of Categories included in Our Tomato Diseases and Pests Dataset and their Annotated Samples.

Class	Number of Images in the Dataset [1]	Number of Annotated Samples (Bounding Boxes) [2]	Percentage of Bounding Box Samples (%)
Leaf mold	1350	11,922	27.47
Gray mold	335	2768	6.37
Canker	309	2648	6.10
Plague	296	2570	5.92
Miner	339	2946	6.78
Low temperature	55	477	1.09
Powdery mildew	40	338	0.77
Whitefly	49	404	0.93
Nutritional excess	50	426	0.98
Background [3]	2177	18,899	43.54
Total	5000	43,398	100

[1] Number of images in the dataset; [2] Number of annotated samples after data augmentation; [3] Transversal category included in every image.

4.2. Experimental Setup

We perform experiments on our Tomato Diseases and Pests dataset that includes nine annotated diseases and pest categories. As explained in the previous section, since the number of images in our dataset is still small and in order to avoid overfitting, we apply extensive data augmentation, including the techniques mentioned in Section 3.4. To perform the experiments, our dataset has been divided into 80% training set, 10% validation set, and 10% testing set. The training is proceeded on the training set, after that the evaluation is performed on the validation set, and when the experiments seem to achieve the expected results, the final evaluation is done on the testing set (unknown data). As in the Pascal Visual Object Classes (VOC) Challenge [53], the validation set is a technique used for minimizing overfitting and is a typical way to stop the network from learning. We use the training and validation sets to perform the training process and parameter selection, respectively, and the testing set for evaluating the results on unknown data. Our proposed system has been trained and tested end-to-end with an Intel Core I7 3.5 GHz Processor on two NVidia GeForce Titan X GPUs. Figure 5 illustrates the resultant loss curve for a number of two hundred thousand iterations, which demonstrates that our network efficiently learns the data while achieving a lower error rate at about one hundred thousand iterations.

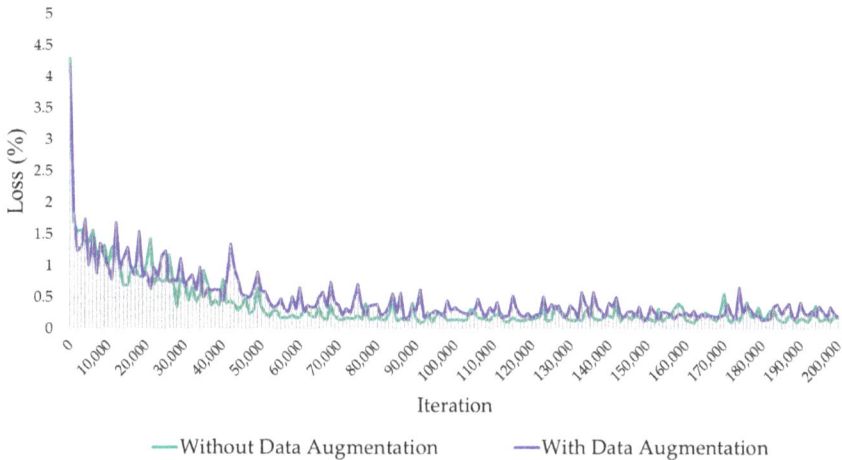

Figure 5. Training loss curve of our proposed approach. The comparison includes results with and without data augmentation. Our network efficiently learns the data while achieving a lower error rate at about one hundred thousand iterations.

4.3. Quantitative Results

Our proposed system implements meta-architectures and different feature extractors to deal with the detection and recognition of complex diseases and pests in the images. The performance of our system is evaluated first of all in terms of the Intersection-over-Union (IoU), and the Average Precision (AP) that is introduced in the Pascal VOC Challenge [53].

$$IoU(A, B) = \left| \frac{A \cap B}{A \cup B} \right| \qquad (1)$$

where A represents the ground-truth box collected in the annotation, and B represents the predicted result of the network. If the estimated IoU outperforms a threshold value, the predicted result is considered as a true positive, TP, or if not as a false positive, FP. TP is the number of true positives generated by the network, and FP corresponds to the number of false positives. Ideally, the number of

FP should be small and determines how accurate is the network to deal with each case. The *IoU* is a widely used method for evaluating the accuracy of an object detector.

The Average Precision is the area under the Precision-Recall curve for the detection task. As in the Pascal VOC Challenge, the AP is computed by averaging the precision over a set of spaced recall levels $[0, 0.1, \ldots, 1]$, and the mAP is the AP computed over all classes in our task.

$$AP = \frac{1}{11} \sum_{r \in \{0,\, 0.1, \ldots,\, 1\}} p_{interp}(r) \tag{2}$$

$$p_{interp}(r) = \max_{\tilde{r}:\tilde{r} \geq r} p(\tilde{r}) \tag{3}$$

where $p(\tilde{r})$ is the measure precision at recall $\tilde{r}$. Next, we compute the mAP averaged for an $IoU = 0.5$ (due to the complexity of the scenarios). The detection results are shown in Table 4.

Table 4. Detection Results of Our Proposed System using Deep-Learning Meta-architectures and Feature Extractors.

Class/Feature Extractor	Meta-Architectures					R-FCN	SSD
	Faster R-CNN						
	VGG-16	ResNet-50	ResNet-101	ResNet-152	ResNeXt-50	ResNet-50	ResNet-50
Leaf mold	**0.9060**	0.8827	0.803	0.8273	0.840	**0.8820**	0.8510
Gray mold	**0.7968**	0.6684	0.449	0.4499	0.620	**0.7960**	0.7620
Canker	**0.8569**	0.7580	0.660	0.7154	0.738	**0.8638**	0.8326
Plague	**0.8762**	0.7588	0.613	0.6809	0.742	**0.8732**	0.8409
Miner	0.8046	0.7884	0.756	0.7793	0.767	0.8812	0.7963
Low temperature	0.7824	0.6733	0.468	0.5221	0.623	0.7545	0.7892
Powdery mildew	0.6556	0.5982	0.413	0.4928	0.505	0.7950	0.8014
Whitefly	0.8301	0.8125	0.637	0.7001	0.720	0.9492	0.8402
Nutritional excess	0.8971	0.7637	0.547	0.8109	0.814	0.9290	0.8553
Background	0.9005	0.8331	0.624	0.7049	0.745	0.8644	0.8841
Total mean AP	**0.8306**	0.7537	0.590	0.6683	0.711	**0.8598**	0.8253

* The bold numbers correspond the more challenging classes and best results among other meta-architectures.
AP: average precision.

The comparative results show that, in our task, plain networks perform better than deeper networks, such as the case of Faster R-CNN with VGG-16 with a mean AP of 83%, compared to the same meta-architecture with ResNet-50 that achieves 75.37% or ResNeXt-50 with 71.1%. In contrast, SSD with ResNet-50 performs at 82.53% and R-FCN with ResNet-50 as feature extractor achieves a mean AP of 85.98%, which is slightly better than Faster R-CNN overall and is comparable in some classes.

Although the mean AP for the whole system shows a performance of more than 80% for the best cases, some diseases, such as leaf mold, gray mold, canker, and plague, show a variable performance. Both Faster R-CNN and R-FCN use the same method of Region Proposal Network (RPN) to extract features from the last layer of the CNN, but using different feature extractors as in our experiment, with VGG-16 and ResNet-50 for Faster R-CNN and R-FCN, respectively, shows comparable results and outstanding performance in the more challenging classes. An early estimation of diseases or pests in the plant could avoid several losses in the whole crop; therefore, we consider leaf mold, gray mold, canker, and pest as the most complex and main classes due to their high intra-class variation (e.g., infection status, location of the infection in the plant, side of leaf, type of fungus, etc.) and some inter-class similarities, especially in the last state of infection when the plant is already dead. Despite the complexity of the scenarios, Faster R-CNN with VGG-16 shows better recognition results especially on the classes above mentioned.

The number of samples is another fact that influences the generation of better results. That could be the case for leaf mold, since our dataset contains a number of samples of this class. The background class is a transversal category that is annotated in most of the images, including healthy parts of the plant as well as parts of the scenario. Furthermore, as we know, the implementation of deep learning systems requires a large number of data that can certainly influence the final performance. In Table 5, we show how the use of a data augmentation technique has allowed our system to improve the Average Precision for each case compared to a previously trained system without using data augmentation.

Table 5. Influence of the Data Augmentation Technique in the Final Results [1].

Class	Without Data Augmentation	With Data Augmentation
Leaf mold	0.6070	0.9060
Gray mold	0.5338	0.7968
Canker	0.5741	0.8569
Plague	0.5870	0.8762
Miner	0.5390	0.8046
Low temperature	0.5242	0.7824
Powdery mildew	0.4392	0.6556
Whitefly	0.5591	0.8301
Nutritional excess	0.6010	0.8971
Background	0.6033	0.9005
Total mean AP	0.5564	0.8306

[1] Experiments using the same meta-architecture and feature extractor (Faster R-CNN with VGG-16).

4.4. Qualitative Results

We evaluate the performance of bounding-box regression and the class score for each class in our Tomato Disease and Pest dataset. As shown in Figure 6, our system is able to effectively detect the class and location of diseases and pests. We compared the estimated results with the ground-truth using an $IoU > 0.5$. Thus, the regions of interest can be estimated while avoiding false positives.

Each class is independent of each other, not only by its origin or cause but also visibly as they show different patterns and characteristics. We find the best results are generated when the main part of the image consists of the target candidate, in contrast with images that include large background regions.

Using meta-architectures and deep feature extractors, the system shows several advantages compared to previous traditional methods when dealing for instance with objects of various sizes (e.g., Gray mold vs. Whitefly), shapes (e.g., Leaf Mold vs. Canker), color (e.g., Plague vs. Leaf mold), etc. Moreover, the proposed approach introduces a fast and effective solution performing at about 160 ms per image.

(a)

(b)

Figure 6. *Cont.*

Figure 6. Detection results of diseases and pests that affect tomato plants with Faster R-CNN and a VGG-16 detector. From left to right: the input image, annotated image, and predicted results. (**a**) Gray mold; (**b**) Canker; (**c**) Leaf mold; (**d**) Plague; (**e**) Leaf miner; (**f**) Whitefly; (**g**) Low temperature; (**h**) Nutritional excess or deficiency; (**i**) Powdery mildew.

4.5. Deep Network Visualization

Understanding a neural network can be interpreted as a deep analysis of how each neuron interacts in the learning process to generate the final results. For that effect, the most popular approach is to use Deconvolutional Neural Networks (DeConv). Using an input image, it aims to highlight which pixels in that image contribute to a neuron firing. This deconvolutional operation can be generated like a convolutional operation but in reverse, such as un-pooling feature maps and convolving un-pooled maps.

As mentioned earlier, diseases and pests in tomato plants can be produced by different causes, such as temperature, humidity, nutrients, lighting conditions, etc. At some point of their infection status, some diseases show similar characteristics or develop visible patterns in the plant that help to distinguish one from another. Therefore, by this experiment, we aim to find a feature map for each class which allows us to understand better their content and representation.

After passing the images by a deconvolutional neural network, which is similar in structure to our main CNN but in a reverse procedure, the final representations are shown in Figure 7. Each feature map illustrates how our neural network system interprets a disease in the context after being classified by a SoftMax function.

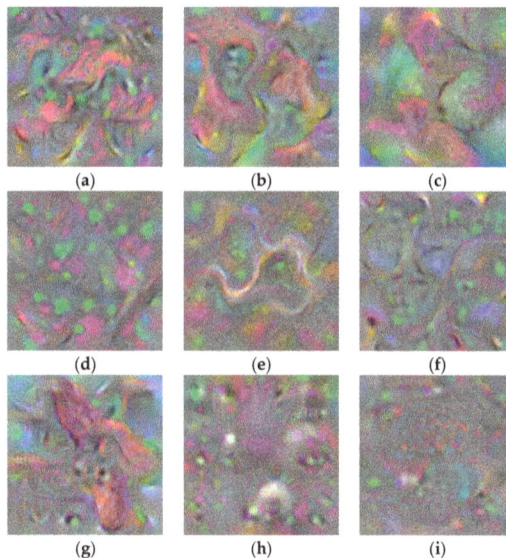

Figure 7. Deep feature maps visualization of diseases and pest (**a**) Canker; (**b**) Gray mold; (**c**) Leaf mold; (**d**) Low temperature; (**e**) Miner; (**f**) Nutritional excess; (**g**) Plague; (**h**) Powdery mildew; (**i**) Whitefly. Each feature map illustrates how our neural network system interprets a disease in the context after being classified by a SoftMax function.

4.6. Diseases Effects in the Plant

The infection symptoms of diseases and pests in the plants start by different ways. It could be started either by a disease originating in the plant itself or infection from other surrounding plants. Therefore, it is useful to identify all the possible causes affecting the plant in order to develop an early detection approach. As shown in Figure 8, diseases and pests can simultaneously affect a plant when it becomes vulnerable due to its condition. For example, Figure 8a shows how the effect of the white fungus, which is a characteristic of powdery mildew, appears to generate spot areas in the leaves where a plague can be developed easily. Furthermore, Figure 8b illustrates the detection results of low temperature, gray mold, and miners in the same plant. Figure 8c,d represent an example of intra-class variations, such as in leaf mold, where the sample leaf corresponds to the same class but with different patterns on its front side and back side. Figure 8e,f show the intra-class variations because of the infection status. Although both images belong to the same class, they visibly show different patterns at an early and the last stage, respectively. Figure 8g,h extend the idea of disease and pest identification to other parts of the plants, such as stem and fruits. Those are also special features that help to identify a pathogen affecting a plant. This experiment gives an idea of how our system is able to efficiently deal with inter- and intra-class variations and its importance as an early detection approach when the symptoms have just appeared.

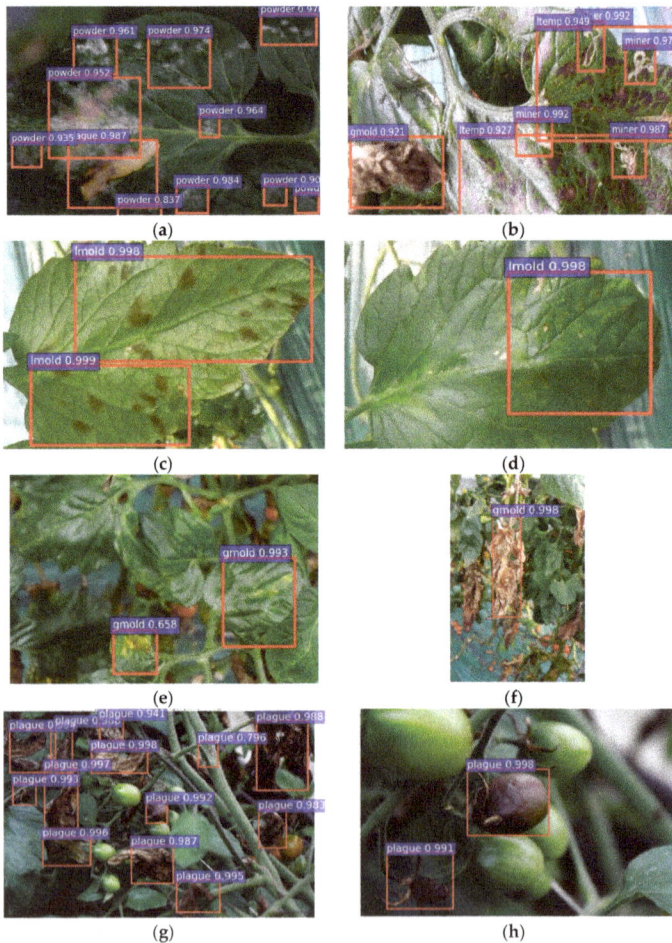

Figure 8. Detection results of inter- and intra-class variation of diseases and pests in the images. (**a**) Two classes affecting the same sample (powdery mildew and pest); (**b**) Three classes in the same sample (Gray mold, low temperature, and miners); (**c**) Leaf mold affecting the back side of the leaf; (**d**) Leaf mold affecting the front side of the leaf; (**e**) Gray mold in the early stage; (**f**) Gray mold in the last stage; (**g**) Plague can be also detected on other parts of the plant, such as fruits or stem; (**h**) Plague affecting the tomato production.

4.7. Confusion Matrix

Due to the complexity of the patterns shown in each class, especially in terms of infection status and background, the system tends to be confused on several classes that results in lower performance. In Figure 9, we present a confusion matrix of the final detection results. Based on the results, we can visually evaluate the performance of the classifier and determine what classes and features are more highlighted by the neurons in the network. Furthermore, it helps us to analysis a further procedure in order to avoid those inter-class confusions. For instance, the canker class shows to be confused in more intensity with gray mold, but also with leaf mold and low temperature. Similarly, the low-temperature class shows confusion with the nutritional excess class.

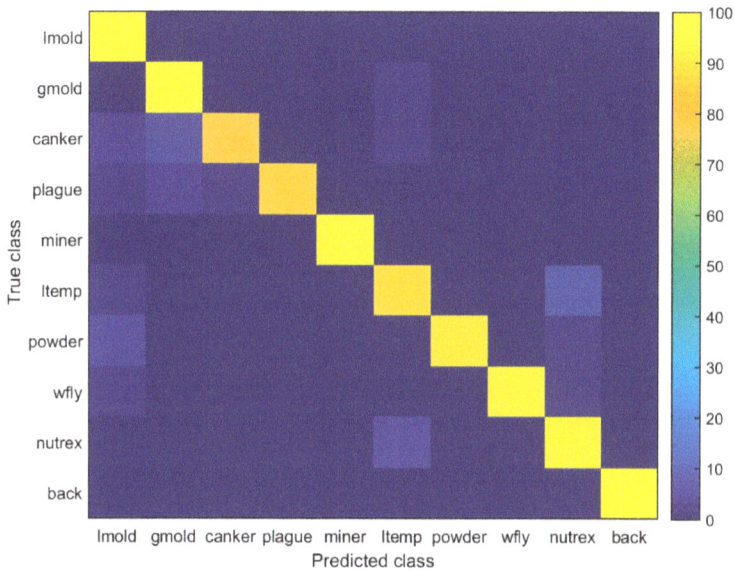

Figure 9. Confusion matrix of the Tomato Diseases and Pests detection results (including background, which is a transversal class containing healthy parts of the plants and surrounding areas, such as part of the greenhouse).

4.8. Failures Analysis and Discussion

Although our system shows an outstanding performance on the evaluated cases, it also presents difficulties in some cases that could be a possible topic for a further study. Due to the lacking number of samples, some classes with high pattern variation tend to be confused with others, resulting in false positives or lower average precision. As shown in Figure 10, for the white fly (e.g., eggs and whiteflies) and leaf mold classes, the presence of targets with different mature status makes their recognition hard when comparing the visible characteristics between them.

Figure 10. A representation of failure cases. (**a**) The intra-class variation makes the recognition harder and results in a low recognition rate. (**b**) Misdetected class due to confusion at earlier infection status (e.g., the real class is leaf mold, but the system recognizes it as canker).

5. Conclusions

In this work, we have proposed a robust deep-learning-based detector for real-time tomato diseases and pests recognition. This system introduces a practical and applicable solution for detecting the class and location of diseases in tomato plants, which in fact represents a main comparable difference with other methods for plant diseases classification. Our detector applied images captured in-place by various camera devices and processed them by a real-time hardware and software system using GPUs, rather than using the process of collecting physical samples (leaves, plants) and analyzing them in the laboratory. Furthermore, our tomato plant diseases and pest dataset contains different task complexities, such as illumination conditions, the size of objects, background variations, etc., included in the surrounding area of the plant. Our goal was to find the more suitable deep-learning architecture for our task. Thus, the experimental results and comparisons between various deep-meta-architectures with feature extractors demonstrated how our deep-learning-based detector is able to successfully recognize nine different categories of diseases and pests, including complex intra- and inter-class variations. In addition, we found that using technique-based data annotation and augmentation results in better performance. We expect that our proposed system will make a significant contribution to the agriculture research area. Future works will be focused on improving the current results, and a promising application will be to extend the idea of diseases and pest recognition to other crops.

Acknowledgments: This work was carried out with the support of the "Cooperative Research Program for Agriculture Science and Technology Development (Project No. PJ0120642016)" Rural Development Administration, Republic of Korea. This research was supported by the "Research Base Construction Fund Support Program" funded by Chonbuk National University in 2016.

Author Contributions: Alvaro Fuentes designed the study, performed the experiments and data analysis, and wrote the paper. Dong Sun Park and Sook Yoon advised in the design of architectures and analyzed to find the best methods for diseases and pest detection. Sang Cheol Kim provided the facilities for data collection and contributed with the information for the data annotation.

Conflicts of Interest: The authors declare no conflict of interest.

References

1.	Mabvakure, B.; Martin, D.P.; Kraberger, S.; Cloete, L.; Van Bruschot, S.; Geering, A.D.W.; Thomas, J.E.; Bananej, K.; Lett, J.; Lefeuvre, P.; et al. Ongoing geographical spread of Tomato yellow leaf curl virus. *Virology* **2016**, *498*, 257–264. [CrossRef] [PubMed]

2.   Canizares, M.C.; Rosas-Diaz, T.; Rodriguez-Negrete, E.; Hogenhout, S.A.; Bedford, I.D.; Bejarano, E.R.; Navas-Castillo, J.; Moriones, E. Arabidopsis thaliana, an experimental host for tomato yellow leaf curl disease-associated begomoviruses by agroinoculation and whitefly transmission. *Plant Pathol.* **2015**, *64*, 265–271. [CrossRef]

3.   The World Bank. Reducing Climate-Sensitive Risks. 2014, Volume 1. Available online: http://documents. worldbank.org/curated/en/486511468167944431/Reducing-climate-sensitive-disease-risks (accessed on 20 June 2017).

4.   Nutter, F.W., Jr.; Esker, P.D.; Coelho, R. Disease assessment concepts and the advancements made in improving the accuracy and precision of plant disease data. *Eur. J. Plant Pathol.* **2006**, *115*, 95–113. [CrossRef]

5.   Munyaneza, J.E.; Crosslin, J.M.; Buchman, J.L.; Sengoda, V.G. Susceptibility of Different Potato Plant Growth Stages of Purple Top Disease. *Am. J. Potato Res.* **2010**, *87*, 60–66. [CrossRef]

6.   Gilbertson, R.L.; Batuman, O. Emerging Viral and Other Diseases of Processing Tomatoes: Biology Diagnosis and Management. *Acta Hortic.* **2013**, *1*, 35–48. [CrossRef]

7.   Diaz-Pendon, J.A.; Canizares, M.C.; Moriones, E.; Bejarano, E.R.; Czosnek, H.; Navas-Castillo, J. Tomato yellow leaf curl viruses: Menage a trois between the virus complex, the plant and whitefly vector. *Mol. Plant Pathol.* **2010**, *11*, 414–450. [CrossRef] [PubMed]

8.   Coakley, S.M.; Scherm, H.; Chakraborty, S. Climate Change and Plant Disease Management. *Annu. Rev. Phytopathol.* **1999**, *37*, 399–426. [CrossRef] [PubMed]

9.   Food and Agriculture Organization of the United Nations. Plant Pests and Diseases. 2017. Available online: http://www.fao.org/emergencies/emergency-types/plant-pests-and-diseases/en/ (accessed on 20 June 2017).

10. Fuentes, A.; Yoon, S.; Youngki, H.; Lee, Y.; Park, D.S. Characteristics of Tomato Plant Diseases—A study for tomato plant disease identification. *Proc. Int. Symp. Inf. Technol. Converg.* **2016**, *1*, 226–231.

11. Food and Agriculture Organization of the United Nations. Value of Agricultural Production-Tomatoes. Food and Agriculture Data. 2015. Available online: http://www.fao.org/faostat/en/#data/QV/visualize (accessed on 9 May 2017).

12. Hanssen, I.; Lapidot, M.; Thomma, B. Emerging Viral Diseases of Tomato Crops. *Mol. Plant Microbe Interact.* **2010**, *23*, 539–548. [CrossRef] [PubMed]

13. Sankaran, S.; Mishra, A.; Ehsani, R. A review of advanced techniques for detecting plant diseases. *Comput. Electron. Agric.* **2010**, *72*, 1–13. [CrossRef]

14. Chaerani, R.; Voorrips, R.E. Tomato early blight (Alternaria solani): The pathogens, genetics, and breeding for resistance. *J. Gen. Plant Pathol.* **2006**, *72*, 335–347. [CrossRef]

15. Alvarez, A.M. Integrated approaches for detection of plant pathogenic bacteria and diagnosis of bacterial diseases. *Annu. Rev. Phytopathol.* **2004**, *42*, 339–366. [CrossRef] [PubMed]

16. Gutierrez-Aguirre, I.; Mehle, N.; Delic, D.; Gruden, K.; Mumford, R.; Ravnikar, M. Real-time quantitative PCR based sensitive detection and genotype discrimination of Pepino mosaic virus. *J. Virol. Methods* **2009**, *162*, 46–55. [CrossRef] [PubMed]

17. Martinelli, F.; Scalenghe, R.; Davino, S.; Panno, S.; Scuderi, G.; Ruisi, P.; Villa, P.; Stropiana, D.; Boschetti, M.; Goudart, L.; et al. Advanced methods of plant disease detection. A review. *Agron. Sust. Dev.* **2015**, *35*, 1–25. [CrossRef]

18. Bock, C.H.; Poole, G.H.; Parker, P.E.; Gottwald, T.R. Plant Disease Sensitivity Estimated Visually, by Digital Photography and Image Analysis, and by Hyperspectral Imaging. *Crit. Rev. Plant Sci.* **2007**, *26*, 59–107.

19. Krizhenvshky, A.; Sutskever, I.; Hinton, G. Imagenet classification with deep convolutional networks. In Proceedings of the Conference Neural Information Processing Systems (NIPS), Lake Tahoe, NV, USA, 3–8 December 2012; pp. 1097–1105.

20. Russakovsky, O.; Deng, J.; Su, H.; Krause, J.; Satheesh, S.; Ma, S.; Huang, Z.; Karpathy, A.; Khosla, A.; Bernstein, M.; et al. ImageNet Large Scale Visual Recognition Challenge. *Int. J. Comput. Vis.* **2015**, *115*, 211–252. [CrossRef]

21. Lin, M.; Chen, Q.; Yan, S. Network in Network. *arXiv* **2013**, arXiv:1312.4400.

22. Simonyan, K.; Zisserman, A. Very deep convolutional networks for large-scale image recognition. *arXiv* **2014**, arXiv:1409.1556.

23. Szegedy, C.; Liu, W.; Jia, Y.; Sermanet, P.; Reed, S.; Anguelov, D.; Erhan, D.; Vanhoucke, V.; Rabinovich, A. Going deeper with convolutions. In Proceedings of the 2015 IEEE Conference on Computer Vision and Pattern Recognition, Boston, MA, USA, 7–12 June 2015; pp. 1–9. [CrossRef]

24. He, K.; Zhang, X.; Ren, S.; Sun, J. Deep residual learning for image recognition. In Proceedings of the 2016 IEEE Conference on Computer, Vision, Pattern Recognition, Las Vegas, NV, USA, 27–30 June 2016; pp. 770–778. [CrossRef]

25. He, K.; Zhang, X.; Ren, S.; Sun, J. Identity Mapping in deep residual networks. *arXiv* **2016**, arXiv:1603.05027.

26. Xie, S.; Girshick, R.; Dollár, P.; Tu, Z.; He, K. Aggregated Residual Transformations for Deep Neural Networks. *arXiv* **2017**, arXiv:1611.05431.

27. Zhang, K.; Sun, M.; Han, T.X.; Yuan, X.; Guo, L.; Liu, T. Residual Networks of Residual Networks: Multilevel Residual Networks. *IEEE Trans. Circ. Syst. Video Technol.* **2017**, *99*. [CrossRef]

28. Zagoruyko, S.; Komodakis, N. Wide Residual Networks. *arXiv* **2016**, arXiv:1605.07146.

29. Huang, J.; Rathod, V.; Sun, C.; Zhu, M.; Korattikara, A.; Fathi, A.; Fischer, I.; Wojna, Z.; Song, Y.; Guadarrama, S.; et al. Speed/accuracy trade-offs for modern convolutional object detectors. In Proceedings of the IEEE Computer Society Conference on Computer Vision and Pattern Recognition, Honolulu, HI, USA, 22–25 July 2017.

30. Ren, S.; He, K.; Girshick, R.; Sun, J. Faster R-CNN: Towards Real-Time Object Detection with Region Proposal Networks. *IEEE Trans. Pattern Anal. Mach. Intell.* **2016**, *39*, 1137–1149. [CrossRef] [PubMed]

31. Liu, W.; Anguelov, D.; Erhan, D.; Szegedy, C.; Reed, S.; Fu, C.; Berg, A.C. SSD: Single Shot MultiBox Detector. In Proceedings of the European Conference on Computer Vision—ECCV, Amsterdam, The Netherlands, 8–16 October 2016; pp. 21–37.

32. Dai, J.; Li, Y.; He, K.; Sun, J. R-FCN: Object Detection via Region-based Fully Convolutional Networks. *arXiv* **2016**, arXiv:1605.06409v2.

33. Irudayaraj, J. Pathogen Sensors. *Sensors* **2009**, *9*, 8610–8612. [CrossRef] [PubMed]

34. Meroni, M.; Rosini, M.; Picchi, V.; Panigada, C.; Cogliati, S.; Nali, C.; Colombo, R. Asse Assessing Steady-state Fluorescence and PRI from Hyperspectral Proximal Sensing as Early Indicators of Plant Stress: The Case of Ozone Exposure. *Sensors* **2008**, *8*, 1740–1754. [CrossRef] [PubMed]

35. Wah Liew, O.; Chong, P.; Li, B.; Asundi, K. Signature Optical Cues: Emerging Technologies for Monitoring Plant Health. *Sensors* **2008**, *8*, 3205–3239. [CrossRef] [PubMed]

36. Mazarei, M.; Teplova, I.; Hajimorad, M.; Stewart, C. Pathogen Phytosensing: Plants to Report Plant Pathogens. *Sensors* **2008**, *8*, 2628–2641. [CrossRef] [PubMed]

37. Ryant, P.; Dolezelova, E.; Fabrik, I.; Baloum, J.; Adam, V.; Babula, P.; Kizek, R. Electrochemical Determination of Low Molecular Mass Thiols Content in Potatoes (*Solanum tuberosum*) Cultivated in the Presence of Various Sulphur Forms and Infected by Late Blight (*Phytophora infestans*). *Sensors* **2008**, *8*, 3165–3182. [CrossRef] [PubMed]

38. Dalal, N.; Trigs, B. Histogram of Oriented Gradients for Human Detection. In Proceedings of the IEEE Computer Society Conference on Computer Vision and Pattern Recognition, San Diego, CA, USA, 20–25 June 2005. [CrossRef]

39. Lowe, D. Distinctive Image Features from Scale-Invariant Keypoints. *Int. J. Comput. Vis.* **2004**, *60*, 91–110. [CrossRef]

40. Cortes, C.; Vapnik, V. Support Vector Networks. *Mach. Learn.* **1995**, *20*, 293–297. [CrossRef]

41. Schapire, R. A Brief Introduction to Boosting. In Proceedings of the Sixteenth International Joint Conference on Artificial Intelligence, Stockholm, Sweden, 31 July 1999; Volume 2, pp. 1401–1406.

42. Pawara, P.; Okafor, E.; Surinta, O.; Schomaker, L.; Wiering, M. Comparing Local Descriptors and Bags of Visual Words to Deep Convolutional Neural Networks for Plant Recognition. In Proceedings of the 6th International Conference on Pattern Recognition Applications and Methods (ICPRAM 2017), Porto, Portugal, 24–26 February 2017; pp. 479–486. [CrossRef]

43. Cugu, I.; Sener, E.; Erciyes, C.; Balci, B.; Akin, E.; Onal, I.; Oguz-Akyuz, A. Treelogy: A Novel Tree Classifier Utilizing Deep and Hand-crafted Representations. *arXiv* **2017**, arXiv:1701.08291v1.

44. Amara, J.; Bouaziz, B.; Algergawy, A. A Deep Learning-based Approach for Banana Leaf Diseases Classification. In *Lecture Notes in Informatics (LNI)*; Gesellschaft für Informatik: Bonn, Germany, 2017.

45. Johannes, A.; Picon, A.; Alvarez-Gila, A.; Echazarra, J.; Rodriguez-Vaamonde, S.; Diez-Navajas, A.; Ortiz-Barredo, A. Automatic plant disease diagnosis using mobile capture devices, applied on a wheat use case. *Comput. Electron. Agric.* **2017**, *138*, 200–209. [CrossRef]

46. Fujita, E.; Kawasaki, Y.; Uga, H.; Kagiwada, S.; Iyatomi, H. Basic investigation on a robust and practical plant diagnostic system. In Proceedings of the 2016 15th IEEE International Conference on Machine Learning and Applications (ICMLA), Anaheim, CA, USA, 18–20 December 2016. [CrossRef]

47. Kawasaki, Y.; Uga, H.; Kagiwada, S.; Iyatomi, H. Basic Study of Automated Diagnosis of Viral Plant Diseases Using Convolutional Neural Networks. In *Advances in Visual Computing, Proceedings of the 11th International Symposium, ISVC 2015, Las Vegas, NV, USA, 14–16 December 2015*; Bebis, G., Ed.; Lecture Notes in Computer Science; Springer: Cham, Switzerland, 2015; Volume 9475, pp. 638–645.

48. Owomugisha, G.; Mwebaze, E. Machine Learning for Plant Disease Incidence and Severity Measurements from Leaf Images. In Proceedings of the 2016 15th IEEE International Conference on Machine Learning and Applications (ICMLA), Anaheim, CA, USA, 18–20 December 2016. [CrossRef]

49. Sladojevic, S.; Arsenovic, M.; Anderla, A.; Culibrk, D.; Stefanovic, D. Deep Neural Networks Based Recognition of Plant Diseases by Leaf Image Classification. *Comput. Intell. Neurosci.* **2016**, *2016*, 3289801. [CrossRef] [PubMed]

50. Mohanty, S.P.; Hughes, D.; Salathe, M. Using Deep Learning for Image-Based Plant Disease Detection. *Front. Plant Sci.* **2016**, *7*, 1419. [CrossRef] [PubMed]

51. Hughes, D.P.; Salathe, M. An open access repository of images on plant health to enable the development of mobile disease diagnostics. *arXiv* **2016**, arXiv:1511.08060v2.

52. Wang, G.; Sun, Y.; Wang, J. Automatic Image-Based Plant Disease Severity Estimation Using Deep Learning. *Comput. Intell. Neurosci.* **2017**, *2017*, 2917536. [CrossRef] [PubMed]

53. Everingham, M.; Van Gool, L.; Williams, C.; Winn, J.; Zisserman, A. The Pascal Visual Object Classes (VOC) Challenge. *Int. Comput. Vis.* **2010**, *88*, 303–338. [CrossRef]

sensors

MDPI

Article

Hyperspectral Imaging Analysis for the Classification of Soil Types and the Determination of Soil Total Nitrogen

Shengyao Jia [1], Hongyang Li [1,2], Yanjie Wang [1], Renyuan Tong [1] and Qing Li [1,*]

[1] College of Mechanical and Electrical Engineering, China Jiliang University, Hangzhou 310018, China; 15a0106134@cjlu.edu.cn (S.J.); 10b0102113@cjlu.edu.cn (H.L.); wangyanjiexx@163.com (Y.W.); tongrenyuan@126.com (R.T.)
[2] College of Computer Science and Technology, Zhejiang University of Technology, Hangzhou 310014, China
* Correspondence: 85a0106008@cjlu.edu.cn; Tel.: +86-571-8687-2385

Received: 4 September 2017; Accepted: 28 September 2017; Published: 30 September 2017

Abstract: Soil is an important environment for crop growth. Quick and accurately access to soil nutrient content information is a prerequisite for scientific fertilization. In this work, hyperspectral imaging (HSI) technology was applied for the classification of soil types and the measurement of soil total nitrogen (TN) content. A total of 183 soil samples collected from Shangyu City (People's Republic of China), were scanned by a near-infrared hyperspectral imaging system with a wavelength range of 874–1734 nm. The soil samples belonged to three major soil types typical of this area, including paddy soil, red soil and seashore saline soil. The successive projections algorithm (SPA) method was utilized to select effective wavelengths from the full spectrum. Pattern texture features (energy, contrast, homogeneity and entropy) were extracted from the gray-scale images at the effective wavelengths. The support vector machines (SVM) and partial least squares regression (PLSR) methods were used to establish classification and prediction models, respectively. The results showed that by using the combined data sets of effective wavelengths and texture features for modelling an optimal correct classification rate of 91.8%. could be achieved. The soil samples were first classified, then the local models were established for soil TN according to soil types, which achieved better prediction results than the general models. The overall results indicated that hyperspectral imaging technology could be used for soil type classification and soil TN determination, and data fusion combining spectral and image texture information showed advantages for the classification of soil types.

Keywords: hyperspectral imaging; soil type classification; total nitrogen; texture features; data fusion

---

1. Introduction

Concerns about the environmental impacts of excessive nitrogen fertilizer application have been growing in recent years. In order to manage nitrogen in an efficient way and fertilize crops according to their demands, it is necessary to obtain detailed information about the total nitrogen (TN) of farmland soil. Traditional chemical analysis methods for TN are complex, time-consuming, costly and poor in real-time. A rapid, nondestructive method should be developed, which is a key step toward the successful implementation of precision farming.

During the last two decades, near-infrared (NIR) spectroscopy has been widely employed as an effective tool for the analysis of soil properties. Numerous studies on the measurement of soil TN have been reported using this technique [1–3]. Nevertheless, the variability of the sample sets (soil texture, moisture content, minerals and organics) greatly complicates the prediction accuracy of the calibration models in soil near-infrared (NIR) spectroscopy analysis [4,5]. The spectral prediction mechanism may vary from one sample set to another. Due to this variability, the soil samples are first classified, then a local spectral model is established for each soil type which can effectively improve the prediction accuracy [6]. On the other hand, soil type classification is an important foundation of soil science, which provides the basis for rational exploitation and scientific management of soil resources. Many researchers have adopted NIR spectroscopy to distinguish soil types [7,8]. The correct classification can reach rates above 80% [9].

Developed from remote sensing, hyperspectral imaging (HSI) has gained extensive attention from different fields such as the food industry [10], agriculture [11], and medical science [12]. Through each measurement by the HSI instrument, both the spectral information and image texture information of the sample can be obtained. Image texture, which is characterized by the relationship of the intensities of neighboring pixels, has been successfully used for the classification of fruit ripeness [13], fish freshness [14], and plant disease degree [15], and the mapping of weed patches [16]. Cai, et al. used image texture features to classify soil samples with different degrees of salinization, and a higher correct classification rate was obtained [17]. They considered that when the soil samples were similar in spectral features, texture features would play a positive role in the sample recognition, and the combined spectral and texture features information can help to improve the classification accuracy. Ma, et al. used analysis of hyperspectral images to distinguish healthy, greening disease infected and zinc-deficient citrus [15]. As the leaf spectra of greening disease infected and zinc-deficient citrus were partially overlapped, and the leaf texture features of greening disease infected and zinc-deficient citrus were similar, the utilization of spectral information or texture features for modelling cannot achieve good classification results in this case, but data fusion combining spectral information and texture features greatly improved the correct classification rate for the three kinds of citrus. To our knowledge, comprehensive utilization of spectral information and image texture features for the classification of soil types has seldom been reported.

Hyperspectral imaging generates an immense amount of data. Some of them may contribute more co-linearity, redundancies, and noise than relevant information to calibration models, which is a huge challenge for the analysis of hyperspectral images [18]. Effective wavelength selection, aiming to select only a few wavelengths which carry most of the useful information with minimum collinearity and redundancy from full spectrum, is believed to reduce amount of data, computational task, and help build a simple and robust model [19,20]. The successive projections algorithm (SPA) is a popular tool for wavelength selection in multivariate calibration and classification [21]. It is able to select a small representative set of spectral wavelengths with a minimum of collinearity. He, et al. used a visible-near infrared HSI technique to detect the tenderness of Atlantic salmon, and SPA was applied to select effective wavelengths [22]. They stated that the number of wavelengths used in the calibration model can be significantly reduced without a decrease in prediction accuracy. In machine visual systems, the most popular method for texture feature analysis is gray level co-occurrence matrix (GLCM) method [23]. GLCM, created through calculating how often a pixel with a particular gray level value occurs at a specified distance and angle from its adjacent pixels, is able to take into account the specific position of a pixel relative to another. In this work, SPA and GLCM were adopted to select effective wavelengths and extract texture features, respectively. The objective of this work was to investigate the feasibility of classifying soil types and determining soil TN content using analysis of hyperspectral images. The specific objectives were to: (1) build classification models for soil types in utilization of spectral information and image texture features; (2) establish robust and accurate calibration models for each soil type to measure soil TN content.

2. Materials and Methods

2.1. Soil Samples and Laboratory Reference Measurement

The study area is located in the city of Shangyu (Zhejiang Province, People's Republic of China, 29°43′38″–30°16′17″ N, 120°36′23″–121°6′9″ E), stretching 60 km from north to south and 40 km from east to west. The climate of this area is subtropical monsoon with an annual average temperature of 16.4 °C and mean annual precipitation of 1400 mm. The southern region of Shangyu is mainly hills and mountains. According to the classification and codes for Chinese soil (National Standard of China, GB/T 17296-2009), the most representative soil type of this region is red soil. The landforms of northern region are river network plains and coastal plains, and the main soil types are paddy soil and coastal saline soil, respectively.

A total of 183 soil samples were sampled from different farmlands of 12 towns in Shangyu, including 84 paddy soil samples, 57 red soil samples and 42 seashore saline soil samples. They were taken from the upper soil layer (0–30 cm) from 2014 to 2016. The samples were collected using a soil-sampling auger. A composite sample was obtained by mixing five soil samples of equal volume, one from the central plot and the remaining four separated by 1 m from each other. To reduce the impact of soil moisture, the soil samples were tiled on a plate and air dried at 80 °C for 60 h. At the 48th h and the 60th h, three samples were randomly selected for weighing. Their weight had barely changed. Then the samples were sieved with a diameter of 1 mm. After that, the samples were air dried again at 60 °C for 48 h to reduce the impact of air moisture during storage. A small portion of each sample was sent to the agricultural testing center of Zhejiang Provincial Academy of Agricultural Sciences (ZPAAS) for soil chemical analyses. The remaining samples were used for HSI measurement. Laboratory reference measurement of soil TN were performed using the Kjeldahl method, as described in Hesse, [24]. Soil TN content was expressed in percentage of its weight to the total weight of dry soil.

2.2. Hyperspectral Image Acquisition

The hyperspectral images of soil samples were captured by a near-infrared HSI system with the wavelength range of 874–1734 nm and 256 bands. The system was composed of an imaging spectrograph (ImSpector N17E; Spectral Imaging Ltd., Oulu, Finland), a CCD camera (Xeva 992; Xenics Infrared Solutions, Leuven, Belgium), two 150W quartz tungsten halogen lamps (Fiber-Lite DC950 Illuminator, Dolan Jenner Industries Inc., Boxborough, MA, USA), and a conveyer belt which was driven by a stepper motor for sample movement (Figure 1). The entire system was fixed in a darkroom. The soil samples were put into Petri dishes with a diameter of 60 mm. The Petri dishes were placed on the conveyer belt for image acquisition. Hyperspectral image provided both spectral and image information simultaneously. Each pixel within the hyperspectral image contained a spectrum at the spectral range of the system, and there was a gray-scale image at each wavelength.

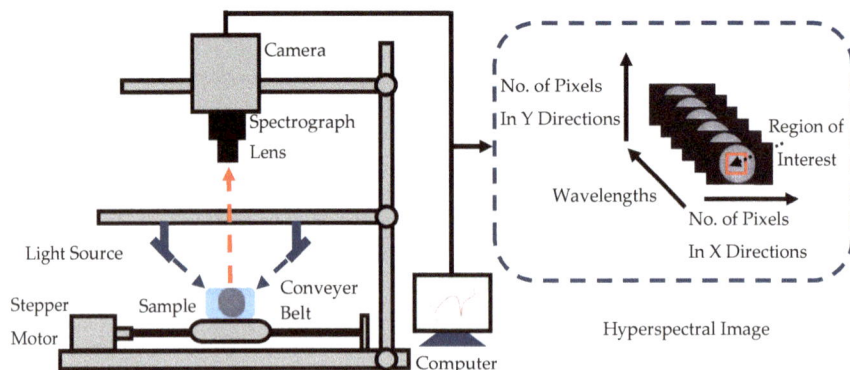

Figure 1. Schematic diagram of the hyperspectral imaging system. This system can obtain images in the spectral region of 874–1734 nm.

To acquire clear and non-deformable hyperspectral images, the moving speed of the conveyer belt, the exposure time of the camera, and the height between the lens of the camera and the sample were set as 24 mm/s, 3 ms, and 30.8 cm, respectively.

The raw hyperspectral image (I0) was corrected by white (W) and dark (D) reference images. The white reference image was obtained using a standard Teflon tile (~99.9% reflectance), and the dark reference image was acquired by turning off the light source and covering the camera lens with its opaque cap. The corrected image (I) was calculated by the following equation:

$$I = \frac{I_0 - D}{W - D} \times 100\% \tag{1}$$

2.3. Spectral Data Extraction, Preprocessing and Effective Wavelength Selection

For each soil sample's hyperspectral image, the region that covered the Petri dish without the edge was selected as the region of interest (ROI). The reflectance values of all pixels in the ROI were averaged to generate only one mean spectrum. Because of the noise in the head and the end of the spectra, only spectra at 975–1645 nm (200 bands) were used for further processing and model establishment. The same procedure was repeated for all ROI images, and a full spectrum matrix 183 samples × 200 bands was constructed. Standard normalized variate (SNV) was used to reduce baseline offset of the spectral matrix, and z-score normalization was used to get all the spectral data to approximately the same scale or to get a more even distribution of the variances and the average values [25].

Effective wavelengths were selected by the SPA method. Generally, SPA comprises two phases. The first phase consists of projections carried out on the spectral matrix, which generate candidate subsets of variables with minimum colinearity. In the second phase, candidate subsets of variables selected in the first phase are used to establish multi-linear regression (MLR) models. The best variable subset was determined on the basis of the root mean square error of leave-one-out cross validation in the calibration set (RMSECV). A detailed description of SPA can be found in literature [26,27].

2.4. Texture Variable Extraction

In creating the GLCM, the direction of 0°, 45°, 90° and 135° and distance of one pixel were applied, and four popular texture variables, such as energy, contrast, homogeneity and entropy were calculated in each direction based on GLCM [28,29]. The mean values of the four directions were used, and four averaged texture variables were obtained from the ROI of one gray-scale image. As the hyperspectral image contained gray-scale images at continuous wavelength bands, a total of 200 gray-scale images have been obtained from a single measurement of one soil sample. Extracting texture features from each gray-scale image would generate a large amount of redundant information which was not useful for modelling. Hence, texture features were only extracted from the gray-scale images at effective wavelengths.

2.5. Establishment of Classification and Regression Models

The main steps of the work were shown in Figure 2. After hyperspectral image acquisition, correction and reflectance preprocessing, the samples of each soil type were randomly spilt into the calibration set and prediction set at a ratio of 2:1 so as to establish classification models: the calibration set was composed of 56 paddy soil samples, 38 red soil samples and 28 seashore saline soil samples, while the prediction set included the remaining 28 paddy soil samples, 19 red soil samples and 14 seashore saline soil samples. Then the SPA method was used to select effective wavelengths based on the calibration set. The reference data y in SPA was category value. The samples of paddy soil, red soil and seashore saline soil were assigned category values of 1, 2 and 3. After effective wavelength selection, texture features were extracted by GLCM. The method of support vector machines (SVM) was used to establish classification models based on the effective wavelengths and texture features. SVM has been proved as a reliable method for classification, dealing with both linear and nonlinear data efficiently [30,31]. In this work, radial basis function kernel was selected as the kernel function, which is the typical general-purpose kernel.

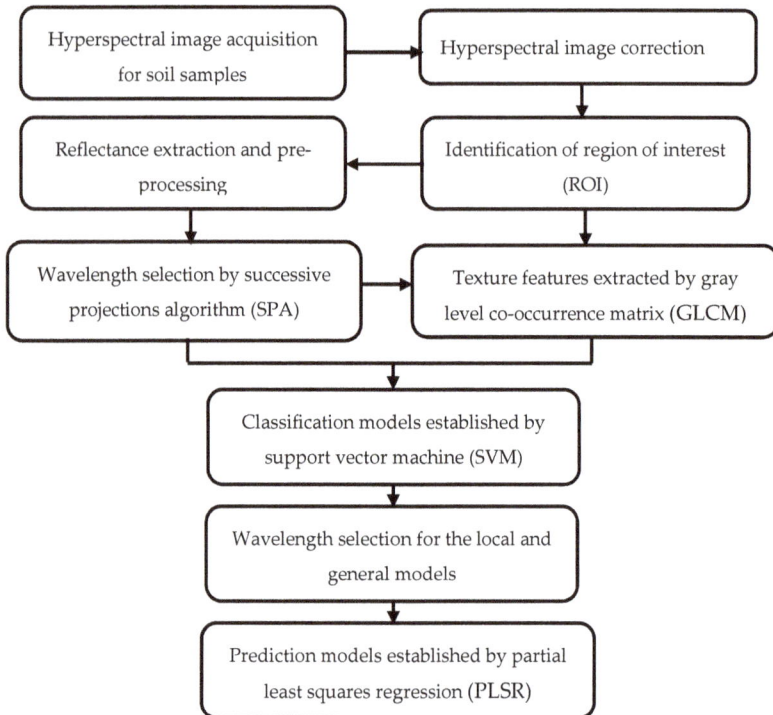

Figure 2. Main steps of this work.

After soil type classification, both general and local models have been established for the prediction of soil TN content. In general models, the total samples were randomly divided into two subsets: the calibration set was composed of 122 samples, while the prediction set included 61 samples. In local models, the samples in each soil type were randomly divided into two subsets at a ratio of 2:1. The sample numbers in the calibration sets were 56, 38 and 28 for paddy soil, red soil and seashore saline soil respectively, while the prediction sets consisted of 28, 19 and 14 samples for paddy soil, red soil and seashore saline soil, respectively. The calibration sets was used to establish calibration models, whereas the prediction sets was used for independent prediction of the established models. The method of SPA was used to select effective wavelengths for each local model and the general models. In this procedure, the reference data y in SPA was soil TN content. The method of partial least squares regression (PLSR) was used to establish prediction models for soil TN based on full spectrum and effective wavelengths, which has been widely applied in many areas [32].

2.6. Performance Assessment and Software

The performance of the established models were evaluated by the root mean squared error of prediction in the prediction set (RMSEP), the residual predictive deviation (RPD) and the coefficient of determination (R^2). Generally, large values of R^2 and RPD, and small value of RMSEP indicate good performances. The hyperspectral image analysis was conducted on ENVI 4.6 (ITT, Visual Information Solutions, Boulder, CO, USA) and Matlab 2010 (The Math Works, Natick, MA, USA). The methods of SVM, SPA were operated in Matlab 2010, and the partial least squares regression (PLSR) models were established in Unscrambler 10.1 (CAMO Inc., Oslo, Norway).

3. Results and Discussion

3.1. Spectral Profiles

Figure 3 shows RGB images of the three soil type samples. It can be noted that the surface of seashore saline soil was rougher than that of paddy soil and red soil. As can be seen in Figure 4a, the average spectrum of each soil type in the range of 975–1645 nm showed similar trend. Significant troughs appeared around 1400 nm in all spectra, which were attributed to the absorption of water in soil. There were some differences in the average spectral baselines. The reflectance value of seashore saline soil was lower than that of paddy soil and red soil, mainly because the light scattering of the surface of seashore saline soil was too intense.

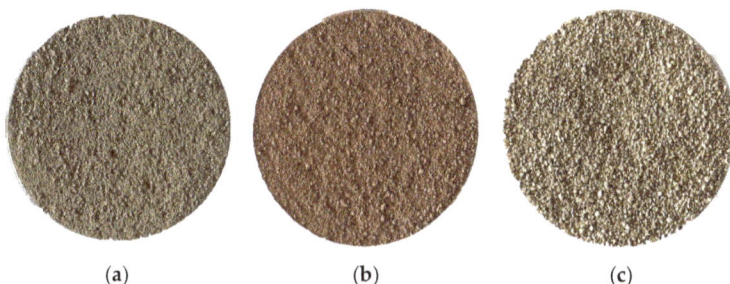

(a)   (b)   (c)

Figure 3. RGB images of paddy soil (**a**), red soil (**b**) and seashore saline soil (**c**) samples.

In order to examine the structure of the spectral data, a principal components analysis was performed on the full spectrum matrix. The principal components analysis scores were submitted to Fisher's linear discriminant analysis (LDA). Because the first four principal components (PCs) of the spectral data can explain nearly 100% of total variance, they were set as input of LDA. Figure 4b shows the samples of paddy soil, red soil and seashore saline soil distinguished by the score plot of Fisher's LDA. The correct classification percentage was 85%. It can be observed that the samples of paddy soil and seashore saline soil were relatively well grouped, while some red soil samples were mixed with the samples of the other two soil types.

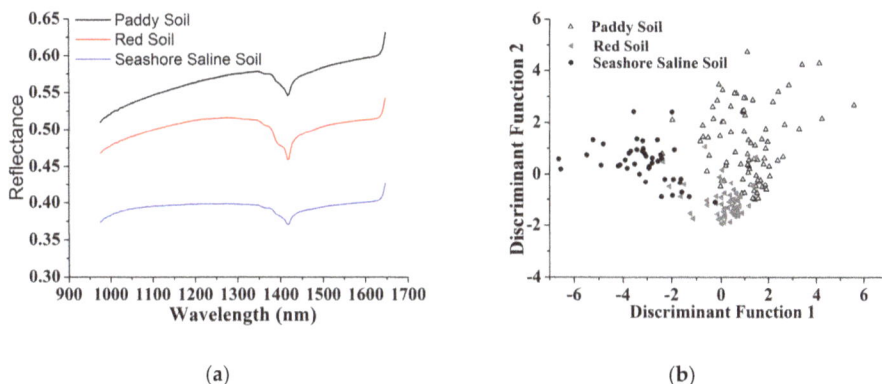

(a)   (b)

Figure 4. (**a**) The average spectrum of each soil type in the wavelength range of 975–1645 nm; (**b**) Grouping of 183 soil samples based on Fisher's LDA using the first four principal components of full spectrum matrix as input.

3.2. Classification for Soil Types

SPA was carried out to select effective variables from the full spectrum. The variation of RMSECV with the number of selected variables for soil type classification is shown in Figure 5a. Let RMSECVmin be the minimum value in the RMSECV sequence. Seven variables were selected through comparison of the RMSECV values which was not significantly larger than RMSEVmin by applying the F-test criterion with a significance level α = 0.25 [32]. Figure 5b presents an overview of the selected variables corresponding to raw spectra. The selected variables around the trough of 1400 nm can be approximately attributed to the absorption of water absorptions in the second overtone region, while the variables selected in the wavelength range of 950–1050 nm were related to overtones of aromatics C-H bond and amine N-H bond in organics [33]. This indicated that considerable differences existed in moisture content and organic ingredients among the samples of the three soil types.

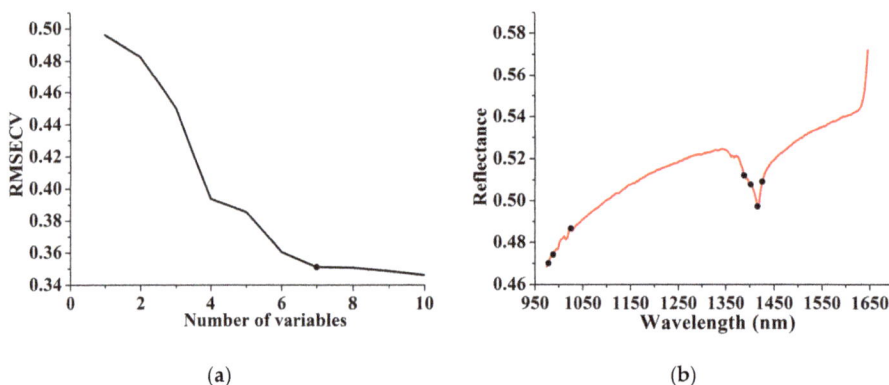

(a)   (b)

Figure 5. RMSECV curves with the number of variables selected by SPA for soil type classification (**a**). The reference data in SPA was category value. The selected variables (shown as dots) corresponding to raw spectra were presented in (**b**).

ROI was defined as a rectangular area in the middle of the sample with 50 × 50 pixels (Figure 1). Four texture features (energy, contrast, homogeneity and entropy) based on GLCM at 7 effective wavelengths were extracted, resulting in a total of 28 texture features (4 texture features × 7 wavelengths) obtained from the ROIs for each soil sample.

Figure 6 shows the mean values of the four texture features of different soil types. It can be seen that energy and homogeneity of seashore saline soil was highest compared with the other two soil types at the effective wavelengths, which indicated that the image texture of seashore saline soil was rougher than that of the other two soil types [13]. A similar conclusion could be also obtained by analyzing the mean values of contrast and entropy. They were the lowest for seashore saline soil, which meant that the image texture of seashore saline soil contained less local variations. In general, the texture features of seashore saline soil were clearly distinguished from those of the other two soil types, and there were no intersections between the texture features of paddy soil and red soil, although they were close at some effective wavelengths. Hence, it was possible for soil type classification based on these statistics.

To build SVM models for soil type classification, the samples of paddy soil, red soil and seashore saline soil were assigned category values of 1, 2 and 3. Table 1 showed the classification results of SVM models using different input variables. First, with full spectrum, the discrimination accuracy was 90.1% for the calibration set and 81.9% for the prediction set. When using spectral effective wavelengths for modelling, similar results were obtained for the calibration set and prediction set, respectively. It can be noted that the samples of paddy soil and seashore saline soil were well classified. Some of them were misclassified with red soil samples, while some red soil samples were misclassified with the samples of the other two soil types. The results were similar to those performed on the full spectrum matrix by LDA. Then, texture features were used for modelling. The discrimination accuracy was 81.9% for the calibration set and 77.0% for the prediction set. The performances were poorer compared with the model established by effective wavelengths. However, the samples of seashore saline soil were well classified from the samples of the other two soil types.

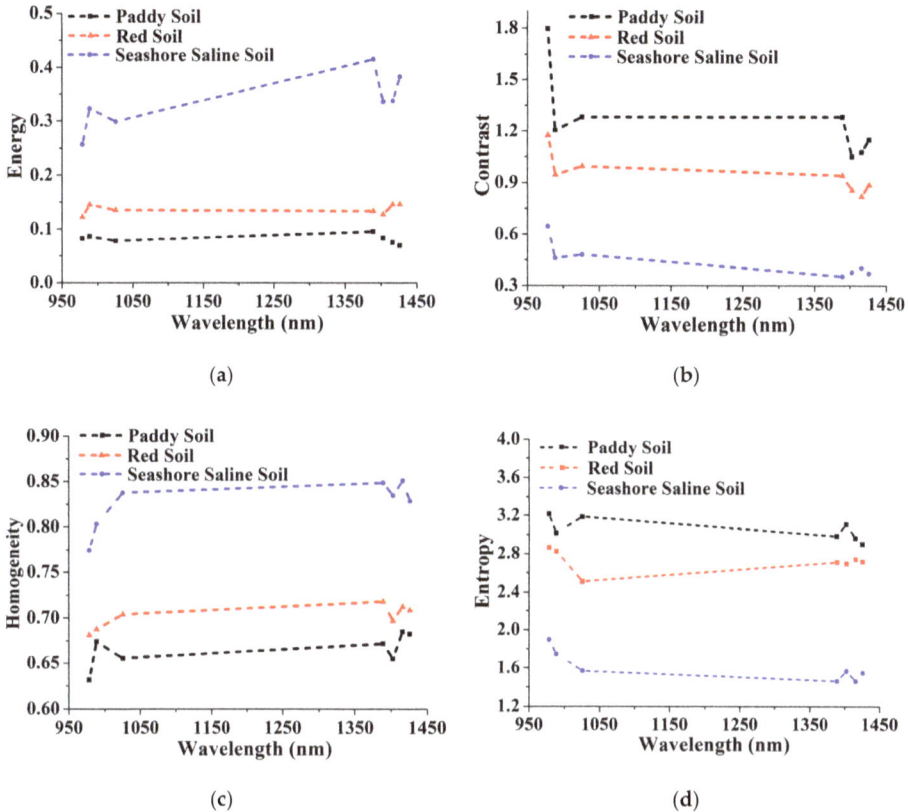

Figure 6. The mean values of (**a**) Energy, (**b**) Contrast, (**c**) Homogeneity and (**d**) Entropy of different soil types at the effective wavelengths.

Finally, both effective wavelengths and texture features were set as input for building SVM models. As can be seen, the discrimination accuracy of the calibration set and prediction set were both improved compared with the models using only spectral effective wavelengths or texture features as input. The samples of paddy soil and seashore saline soil were successfully classified, while some samples of paddy soil and red soil were misclassified, and a few seashore saline soil samples were misclassified as red soil samples. The results indicated that data fusion by combining effective wavelengths and texture features showed advantages for the classification of soil types.

Table 1. Classification results for soil types using SVM models established based on different input variables.

Input variables	(c, g) [a]		Calibration Set				Prediction Set			
			1	2	3	Accuracy	1	2	3	Accuracy
Full spectrum	(30.94, 0.51)		52	3	1	92.8%	22	4	2	78.6%
			2	33	3	86.8%	2	15	2	78.9%
			1	2	25	89.2%	0	1	13	92.8%
		total				90.1%				81.9%
Effective wavelengths	(23.12, 2.42)	1	51	4	1	91.1%	1	24	2	2	85.7%
		2	3	33	2	86.8%	2	2	16	1	84.2%
		3	0	4	24	85.7%	3	1	2	11	78.6%
		total				88.5%					83.6%
Texture features	(90.95, 0.26)	1	47	9	0	83.9%	1	23	5	0	82.1%
		2	10	27	1	71.1%	2	7	12	0	63.1%
		3	1	1	26	92.8%	3	1	1	12	85.7%
		total				81.9%					77.0%
Effective wavelengths and texture features	(190.12, 2.28)	1	54	2	0	96.4%	1	27	1	0	96.4 %
		2	3	34	1	89.4%	2	3	16	0	84.2%
		3	0	1	27	96.4%	3	0	1	13	92.8%
		total				94.2%					91.8%

[a] (c, g) are the parameters of the SVM model, where c is the penalty coefficient, and g is the kernel function parameter.

3.3. Prediction of Soil Total Nitrogen

The statistics values of soil TN content for the calibration sets and prediction sets in each local model and the general models were listed in Table 2. The concentration of soil TN ranged from 0.038% to 0.312%. The range of the prediction sets was covered in the calibration sets.

Table 2. Statistics of reference values of total nitrogen (TN) in the local models and general models.

Property	Calibration Set				Prediction Set			
	NS [a]	Range (%)	Mean (%)	SD [b]	NS	Range (%)	Mean (%)	SD
Paddy soil	56	0.088–0.312	0.170	0.036	28	0.124–0.255	0.174	0.042
Red soil	38	0.056–0.262	0.151	0.041	19	0.102–0.215	0.179	0.030
Seashore saline soil	28	0.038–0.205	0.131	0.031	14	0.055–0.178	0.133	0.030
General models	122	0.038–0.312	0.160	0.041	61	0.042–0.250	0.156	0.037

[a] NS = Number of samples. [b] SD = Standard deviation.

The SPA method was used to select effective variables for the prediction of soil TN from the full spectrum. Figure 7 shows the variation of RMSEV with the number of selected variables for each local model and the general models.

Through comparison of the RMSEV values that were not significantly larger than RMSEVmin by applying the F test criterion with a significance level $\alpha = 0.25$ [32], 15, 18, 17, and 13 variables were selected for paddy soil, red soil, seashore saline soil, and the general models, respectively. The overview of the selected variables corresponding to raw spectra is shown in Figure 8. The selected locations were mainly concentrated in the trough around 1400 nm and the wavelength range of 950 to 1050 nm. The result was supported by Yang et al. [2], who selected similar effective variables for soil TN in the NIR wavelength range. There were some differences among the selected locations for paddy soil, red soil, seashore saline soil and the general models, which indicated that the NIR feature absorptions varied from one soil type to another.

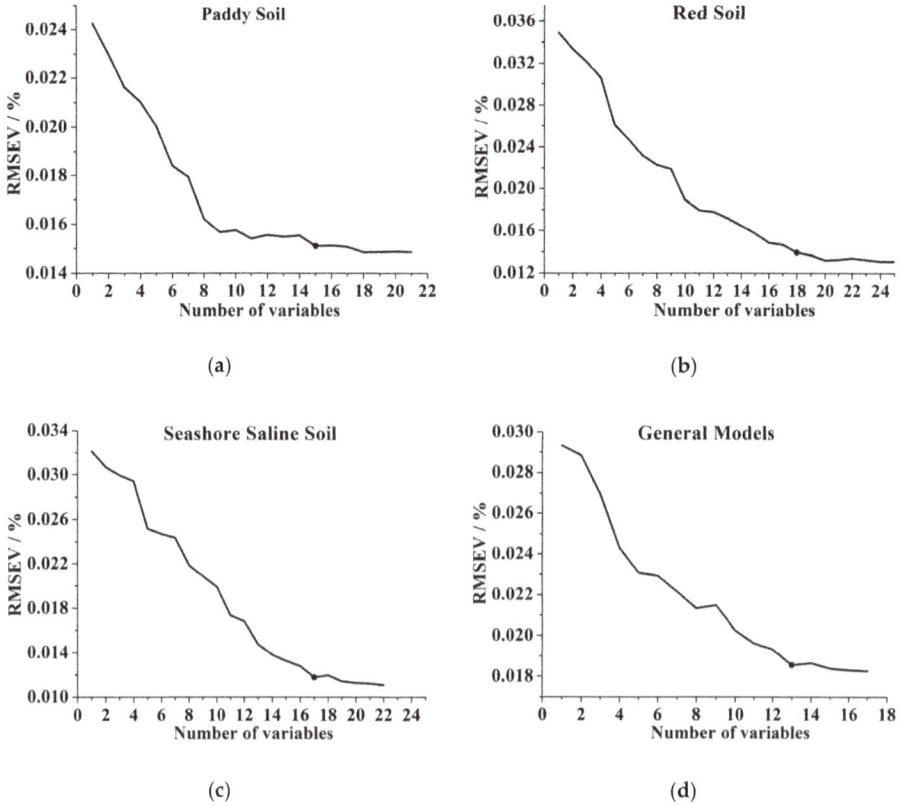

Figure 7. RMSECV curves with the number of variables selected by SPA for (**a**) paddy soil, (**b**) red soil, (**c**) seashore saline soil and (**d**) general models. The reference data in SPA was soil TN content.

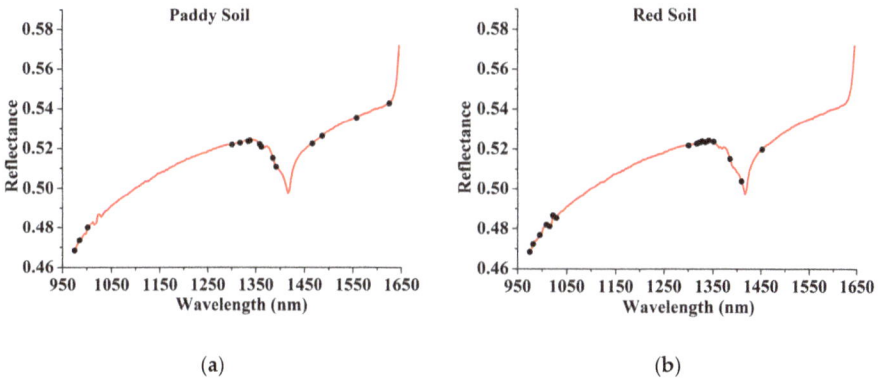

Figure 8. *Cont.*

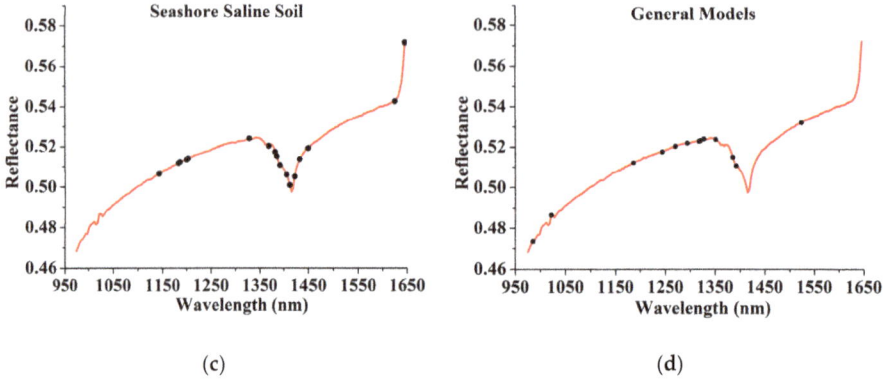

Figure 8. The variables selected by SPA (shown in circle markers) corresponding to raw spectra for paddy soil, red soil, seashore saline soil and general models were presented in (**a–d**), respectively.

The prediction results of soil TN using different sample sets and input variables are listed in Table 3. Regression models were built by PLSR. It can be observed that whether using the full spectrum or effective wavelengths for modelling, the local models which were established individually for each soil type achieved better prediction results than the general models in terms of larger R^2 and RPD values for the prediction sets. The reason can be attributed to the small variability of the samples in the same soil type [5]. Similar results have been reported by Mouazen, et al. [34], who developed local calibration models for the prediction of soil moisture content, so as to increase the accuracy of the NIR measurement. The full spectrum and effective-wavelength models achieved similar predictions for soil TN. However, the number of input variables in the effective-wavelength models was much reduced, and the efficiency has been promoted.

Table 3. Comparison of prediction results for soil nitrogen using different sample sets and input variables.

Property	NS [a]	Input Variables [b]	NV [c]	RMSEP [d]	R^2 [e]	RPD [f]
Paddy soil	28	Full spectrum	200	0.0166	0.83	2.5
		EW	12	0.0155	0.85	2.7
Red soil	19	Full spectrum	200	0.0129	0.80	2.3
		EW	10	0.0136	0.77	2.2
Seashore saline soil	14	Full spectrum	200	0.0118	0.83	2.5
		EW	10	0.0125	0.81	2.4
General models	61	Full spectrum	200	0.0176	0.76	2.1
		EW	14	0.0179	0.74	2.1

[a] NS = Number of samples for the prediction set; [b] EW = effective wavelengths; [c] NV = Number of variables for the established models; [d] RMSEP = Root mean squared error of prediction in the prediction set; [e] R^2 = Coefficient of determination; [f] RDP = Residual predictive deviation.

4. Conclusions

In this work, a HSI system covering the spectral range of 874–1734 nm was used to classify soil types and evaluate soil TN content. The SPA method was applied to select effective wavelengths from the full spectrum, and texture features of energy, contrast, homogeneity and entropy were extracted from the gray-scale images at the effective wavelengths. The classification models for soil types and prediction models for soil TN were established by the methods of SVM and PLSR, respectively. The results showed that:

(1) The classification model established by the combining effective wavelengths and texture features data achieved the optimal results for the classification of red, paddy and seashore saline soil compared with the models established by the effective wavelengths or texture features alone. The correct classification rate was 91.8%.

(2) The soil samples were first classified, then local models were established for soil TN according to soil types, which achieved better prediction results than the general models.

(3) The overall results indicated that it was helpful to use image texture features for soil type classification, and HSI technique could be used for soil type classification and the determination of soil TN.

In future work, more soil samples with a wide range of soil types should be studied to build more robust soil type classification models and more reliable prediction models for soil TN.

Acknowledgments: This work was supported by the National Key Research and Development Program of China (project No. 2017YFC0804604), the Natural Science Foundation of Zhejiang Province, China (project No. LQ16F010006), and the scientific research project of the education department of Zhejiang Province (project No. Y201533855).

Author Contributions: Shengyao Jia proposed the idea and wrote the manuscript. Hongyang Li performed the experiments and processed the data. Yanjie Wang and Renyuan Tong contributed to data processing. Qing Li guided the experiments and writing.

Conflicts of Interest: The authors declare no conflict of interest. The founding sponsors had no role in the design of the study; in the collection, analyses, or interpretation of data; in the writing of the manuscript; or in the decision to publish the results.

References

1. He, Y.; Huang, M.; García, A.; Hernández, A.; Song, H. Prediction of soil macronutrients content using near-infrared spectroscopy. *Comput. Electron. Agric.* **2007**, *58*, 144–153. [CrossRef]
2. Yang, H.; Kuang, B.; Mouazen, A.M. Quantitative analysis of soil nitrogen and carbon at a farm scale using visible and near infrared spectroscopy coupled with wavelength reduction. *Eur. J. Soil Sci.* **2012**, *63*, 410–420. [CrossRef]
3. Morellos, A.; Pantazi, X.-E.; Moshou, D.; Alexandridis, T.; Whetton, R.; Tziotzios, G.; Wiebesohn, J.; Bill, R.; Mouazen, A. Machine learning based prediction of soil total nitrogen, organic carbon and moisture content by using vis-nir spectroscopy. *Biosyst. Eng.* **2016**, *152*, 104–116. [CrossRef]
4. Vohland, M.; Emmerling, C. Determination of total soil organic C and hot water-extractable C from VIS-NIR soil reflectance with partial least squares regression and spectral feature selection techniques. *Eur. J. Soil Sci.* **2011**, *62*, 598–606. [CrossRef]
5. Jia, S.; Li, H.; Wang, Y.; Tong, R.; Li, Q. Recursive variable selection to update near-infrared spectroscopy model for the determination of soil nitrogen and organic carbon. *Geoderma* **2016**, *179*, 211–219. [CrossRef]
6. Mouazen, A.M.; Maleki, M.R.; De Baerdemaeker, J.; Ramon, H. On-line measurement of some selected soil properties using a VIS–NIR sensor. *Soil Tillage Res.* **2007**, *93*, 13–27. [CrossRef]
7. Vasques, G.M.; Demattê, J.A.M.; Viscarra Rossel, R.A.; Ramírez-López, L.; Terra, F.S. Soil classification using visible/near-infrared diffuse reflectance spectra from multiple depths. *Geoderma* **2014**, *223–225*, 73–78. [CrossRef]
8. Lacerda, M.; Demattê, J.; Sato, M.; Fongaro, C.; Gallo, B.; Souza, A. Tropical texture determination by proximal sensing using a regional spectral library and its relationship with soil classification. *Remote Sens.* **2016**, *8*, 701–721. [CrossRef]
9. Viscarra, R.A.; Webster, R. Discrimination of Australian soil horizons and classes from their visible-near infrared spectra. *Eur. J. Soil Sci.* **2011**, *62*, 637–647. [CrossRef]
10. Zhang, C.; Guo, C.; Liu, F.; Kong, W.; He, Y.; Lou, B. Hyperspectral imaging analysis for ripeness evaluation of strawberry with support vector machine. *J. Food Eng.* **2016**, *179*, 11–18. [CrossRef]
11. Gomez, C.; Gholizadeh, A.; Borůvka, L.; Lagacherie, P. Using legacy data for correction of soil surface clay content predicted from VNIR/SWIR hyperspectral airborne images. *Geoderma* **2016**, *276*, 84–92. [CrossRef]

12. Neittaanmaki-Perttu, N.; Gronroos, M.; Tani, T.; Polonen, I.; Ranki, A.; Saksela, O.; Snellman, E. Detecting field cancerization using a hyperspectral imaging system. *Lasers Surg. Med.* **2013**, *45*, 410–417. [CrossRef] [PubMed]

13. Wei, X.; Liu, F.; Qiu, Z.; Shao, Y.; He, Y. Ripeness classification of astringent persimmon using hyperspectral imaging technique. *Food Bioprocess Technol.* **2013**, *7*, 1371–1380. [CrossRef]

14. Zhu, F.; Zhang, D.; He, Y.; Liu, F.; Sun, D.-W. Application of visible and near infrared hyperspectral imaging to differentiate between fresh and frozen–thawed fish fillets. *Food Bioprocess Technol.* **2012**, *6*, 2931–2937. [CrossRef]

15. Ma, H.; Ji, H.-Y.; Won, S.L. Identification of the citrus greening disease using spectral and textural features based on hyperspectral imaging. *Spectrosc. Spectr. Anal.* **2016**, *36*, 2344–2350.

16. Tamouridou, A.A.; Alexandridis, T.K.; Pantazi, X.E.; Lagopodi, A.L.; Kashefi, J.; Moshou, D. Evaluation of UAV imagery for mapping Silybum marianumweed patches. *Int. J. Remote Sens.* **2017**, *38*, 2246–2259.

17. Cai, S.; Zhang, R.; Liu, L.; Zhou, D. A method of salt-affected soil information extraction based on a support vector machine with texture features. *Math. Comput. Model.* **2010**, *51*, 1319–1325. [CrossRef]

18. Dai, Q.; Cheng, J.-H.; Sun, D.-W.; Zhu, Z.; Pu, H. Prediction of total volatile basic nitrogen contents using wavelet features from visible/near-infrared hyperspectral images of prawn (*Metapenaeus ensis*). *Food Chem.* **2016**, *197*, 257–265. [CrossRef] [PubMed]

19. Li, J.; Tian, X.; Huang, W.; Zhang, B.; Fan, S. Application of long-wave near infrared hyperspectral imaging for measurement of soluble solid content (SSC) in pear. *Food Anal. Methods* **2016**, *9*, 3087–3098. [CrossRef]

20. Mollazade, K. Non-destructive identifying level of browning development in button mushroom (agaricus bisporus) using hyperspectral imaging associated with chemometrics. *Food Anal. Methods* **2017**, *10*, 2734–2754. [CrossRef]

21. Galvão, R.K.H.; Araújo, M.C.U.; Fragoso, W.D.; Silva, E.C.; José, G.E.; Soares, S.F.C.; Paiva, H.M. A variable elimination method to improve the parsimony of MLR models using the successive projections algorithm. *Chemom. Intell. Lab. Syst.* **2008**, *92*, 83–91. [CrossRef]

22. He, H.-J.; Wu, D.; Sun, D.-W. Potential of hyperspectral imaging combined with chemometric analysis for assessing and visualising tenderness distribution in raw farmed salmon fillets. *J. Food Eng.* **2013**, *126*, 156–164. [CrossRef]

23. Haralick, R.M.; Shanmugam, K.; Dinstein, I. H. Textural Features for Image Classification. *IEEE Trans. Syst.* **1973**, *3*, 610–621. [CrossRef]

24. Hesse, P.R. *A Textbook of Soil Chemical Analysis*; John Murray: London, UK, 1971.

25. Rahman, M.R.; Shi, Z.H.; Chongfa, C. Soil erosion hazard evaluation—An integrated use of remote sensing, GIS and statistical approaches with biophysical parameters towards management strategies. *Ecol. Model.* **2009**, *220*, 1724–1734. [CrossRef]

26. Ye, S.; Wang, D.; Min, S. Successive projections algorithm combined with uninformative variable elimination for spectral variable selection. *Chemom. Intell. Lab. Syst.* **2008**, *91*, 194–199. [CrossRef]

27. Insausti, M.; Gomes, A.A.; Cruz, F.V.; Pistonesi, M.F.; Araujo, M.C.; Galvao, R.K.; Pereira, C.F.; Band, B.S. Screening analysis of biodiesel feedstock using UV-vis, NIR and synchronous fluorescence spectrometries and the successive projections algorithm. *Talanta* **2012**, *97*, 579–583. [CrossRef] [PubMed]

28. Mendoza, F.; Aguilera, J.M. Application of image analysis for classification of ripening bananas. *J. Food Sci.* **2004**, *69*, E471–E477. [CrossRef]

29. Xie, C.; Shao, Y.; Li, X.; He, Y. Detection of early blight and late blight diseases on tomato leaves using hyperspectral imaging. *Sci. Rep.* **2015**, *5*, 16564. [CrossRef] [PubMed]

30. Li, S.X.; Zhang, Y.J.; Zeng, Q.Y.; Li, L.F.; Guo, Z.Y.; Liu, Z.M.; Xiong, H.L.; Liu, S.H. Potential of cancer screening with serum surface-enhanced Raman spectroscopy and a support vector machine. *Laser Phys. Lett.* **2014**, *11*, 065603. [CrossRef]

31. Langeron, Y.; Doussot, M.; Hewson, D.J.; Duchêne, J. Classifying NIR spectra of textile products with kernel methods. *Eng. Appl. Artif. Intell.* **2007**, *20*, 415–427. [CrossRef]

32. Jia, S.; Yang, X.; Zhang, J.; Li, G. Quantitative analysis of soil nitrogen, organic carbon, available phosphorous, and available potassium using near-infrared spectroscopy combined with variable selection. *Soil Sci.* **2014**, *179*, 211–219. [CrossRef]

33. Rossel, R.A.V.; Behrens, T. Using data mining to model and interpret soil diffuse reflectance spectra. *Geoderma*
     2010, *158*, 46–54. [CrossRef]
34. Mouazen, A.M.; De Baerdemaeker, J.; Ramon, H. Towards development of on-line soil moisture content
     sensor using a fibre-type NIR spectrophotometer. *Soil Tillage Res.* **2005**, *80*, 171–183. [CrossRef]

MDPI

Article

Classification of *Fusarium*-Infected Korean Hulled Barley Using Near-Infrared Reflectance Spectroscopy and Partial Least Squares Discriminant Analysis

Jongguk Lim [1], Giyoung Kim [1], Changyeun Mo [1,*], Kyoungmin Oh [1], Hyeonchae Yoo [1], Hyeonheui Ham [2] and Moon S. Kim [3]

[1] Department of Agricultural Engineering, National Institute of Agricultural Sciences, Rural Development Administration, 310 Nongsaengmyeng-ro, Wansan-gu, Jeonju 54875, Korea; limjg@korea.kr (J.L.); giyoung@korea.kr (G.K.); yoonmine@korea.kr (K.O.); hyeonchae1@naver.com (H.Y.)
[2] Microbial Safety Team, National Institute of Agricultural Sciences, Rural Development Administration, 166 Nongsaengmyeong-ro, Iseo-myeon, Wanju-gun 55365, Korea; hhham@korea.kr
[3] Environmental Microbial and Food Safety Laboratory, Agricultural Research Service, US Department of Agriculture, 10300 Baltimore Avenue, Beltsville, MD 20705, USA; Moon.Kim@ars.usda.gov
* Correspondence: cymoh100@korea.kr; Tel.: +82-63-238-4120; Fax: +82-63-238-4105

Received: 29 August 2017; Accepted: 26 September 2017; Published: 30 September 2017

Abstract: The purpose of this study is to use near-infrared reflectance (NIR) spectroscopy equipment to nondestructively and rapidly discriminate *Fusarium*-infected hulled barley. Both normal hulled barley and *Fusarium*-infected hulled barley were scanned by using a NIR spectrometer with a wavelength range of 1175 to 2170 nm. Multiple mathematical pretreatments were applied to the reflectance spectra obtained for *Fusarium* discrimination and the multivariate analysis method of partial least squares discriminant analysis (PLS-DA) was used for discriminant prediction. The PLS-DA prediction model developed by applying the second-order derivative pretreatment to the reflectance spectra obtained from the side of hulled barley without crease achieved 100% accuracy in discriminating the normal hulled barley and the *Fusarium*-infected hulled barley. These results demonstrated the feasibility of rapid discrimination of the *Fusarium*-infected hulled barley by combining multivariate analysis with the NIR spectroscopic technique, which is utilized as a nondestructive detection method.

Keywords: *Fusarium*; near-infrared; spectroscopy; hulled barely; partial least squares-discriminant analysis

1. Introduction

Fusarium (particularly *F. graminearum*) that occurs in the heads of cereal crops such as barley (*Hordeum vulgare* L.) and wheat (*Triticum aestivum* L.) has been reported to decrease the yield and degrade the quality, resulting in enormous economic losses to farmers [1]. The *Fusarium* pathogen grows rapidly between 10 and 25 °C in a high humidity environment due to heavy rainfall in the heading period of the cereal crops, leading to an increase in the occurrence frequency [2]. The pathogen wintering in seeds, straw, stubbles, and soil after harvest, becomes a primary inoculum, and molds formed from the primary symptoms scatter and rapidly spread [3–5]. Damaged ears develop a brown discoloration at the early stage, and the sheaths of ears become gradually covered with red conidiospores. In Korea, the incidence of *Fusarium* infection in barley increased greatly between 1963 and 1998. In 1998, the incidence of *Fusarium* infection damaged 39,202 hectares of fields, corresponding to 47.8% of the total cultivation area. Particularly, if mycotoxins such as deoxynivalenol (DON), nivalenol (NIV), and zearalenone (ZEA) produced by the *Fusarium* are mixed with the normal cereals and supplied in the finished product, this could cause intoxication of livestock as well as diseases such as vomiting, diarrhea, immunosuppression, and cancer in humans [6–8]. The most severe concern relating to *Fusarium* infection is that mycotoxins occurring due to *Fusarium* infection and remaining as a carcinogen in the feed could cause serious damage to livestock and humans [9]. Therefore, a preliminary inspection of barley, wheat, maize, and other cereal crops is essential to prevent *Fusarium* and mycotoxins from being introduced in the finished products.

Typical contamination inspections of *Fusarium* and mycotoxin use high performance liquid chromatography (HPLC), gas chromatography (GC), and enzyme-linked immunosorbent assay (ELISA). These destructive chemical analysis methods are demanding in terms of cost, time, and effort, and do not facilitate prompt detection due to small sample size and complicated pretreatment processes [10]. Nondestructive and rapid spectroscopic analysis techniques have emerged as alternatives to overcome these disadvantages.

Spectroscopic analysis techniques are nondestructive measurement methods that can be used for rapid quantitative analysis. These include NIR spectroscopy, Raman spectroscopy, and other spectroscopic methods according to light characteristics and analyte type, and can be used for pathogen detection as well as internal and external quality analysis of agricultural products [11]. NIR spectroscopic analysis uses the wavelength range from 700 nm to 2500 nm. Although the near-infrared region of the spectrum was discovered 160 years ago, its practical analytical applications were first performed by a U.S. chemist, Norris. In the 1960s, Norris proved the potential of NIR spectroscopy through quantitative analysis of moisture and protein content in wheat [12]. The later development of chemometric techniques expanded the applications of NIR spectroscopy. The chemometric technique analyzes wavelength-based data from a spectrometer to accurately predict the components in a short time [13]. Analysis methods such as principal component analysis (PCA), partial least squares (PLS), partial least squares-discriminant analysis (PLS-DA), and linear discriminant analysis (LDA) have been used for model development [14].

The NIR spectroscopic technique has recently been applied to detect mold and mycotoxin in cereal crops such as wheat, barley, and maize [15–17]. Delwiche used the NIR technique to discriminate between wheat *Fusarium* and other molds with accuracies of 95% and 98%, respectively [18]. Polder showed PLS results using a transmission image in the visible range to detect *Fusarium* head blight (FHB, also known as 'scab' or ear blight) [19]. Liu used a 1064-nm NIR laser to reduce the fluorescence that occurs in barley and wheat, in order to detect a low concentration of mycotoxin [20]. Gaspardo used the Fourier transform (FT)-NIR technique to predict the concentrations of the mycotoxins fumonisin (B_1) and fumonisin (B_2) in maize kernels [21].

This study aimed to develop a technique for discriminating between *Fusarium*-infected hulled barley and normal hulled barley by using a NIR spectrometer. For this purpose, various spectral pretreatments were applied to the reflectance spectra from 1175 nm to 2170 nm. The study was further aimed to nondestructively detect *Fusarium*-infected hulled barley by developing a PLS-DA model.

2. Materials and Methods

2.1. Materials

The samples used in this experiment were 515 kernels of hulled barley samples that were collected from five Korean provinces in 2014. The samples were divided into a control group and experimental groups, as shown in Table 1. The control group samples that were not infected with *Fusarium* in the preliminary investigation were 127 kernels of one group (JN151) collected from Jeonnam Province. The experimental group samples containing *Fusarium*-infected hulled barley include 80 kernels of one group (GN121) from Gyeongnam Province as well as 95 kernels, 108 kernels, and 105 kernels of three groups (JB021, JB061, and JB094) from Jeonbuk Province.

As shown in Table 2, the hulled barley samples used in the NIR spectroscopic experiments were stored as individual kernels in a storage box with a 10×10 grid structure, and refrigerated at 4 °C.

Table 1. Collection province and quantity of hulled barley samples used in experiment.

Province	Groups	Sample Group Based on Region	Number of Kernels
Jeonnam	Control group	JN151	127
Gyeongnam		GN121	80
Jeonbuk	Experimental group	JB021	95
		JB061	108
		JB094	105
Total		5	515

Table 2. Hulled barley samples separated by kernel into grid boxes according to groups.

Control Group	Experimental Groups			
JN151 (127)	GN121 (80)	JB021 (95)	JB061 (108)	JB094 (105)

2.2. Near-Infrared Measurement System

The NIR measurement system, as shown in Figure 1, consists of an indium gallium arsenide (InGaAs) linear array spectrometer (AvaSpec-NIR256-2.2TC, Avantes BV Inc., Apeldoorn, The Netherlands) based on the symmetrical Czerny-Tunner optical design with a 50-mm focal length, a 10 W tungsten-halogen lamp light source (AvaLight-HAL, Avantes BV Inc.), a bifurcated fiber optic probe (FCR-7IR400-2-ME, Avantes BV Inc.) covering 350 nm to 2000 nm, a specimen stage, and a computer. As shown in Table 3, the NIR spectrometer is thermoelectrically cooled by a double-stage Peltier effect device, and has 256 light-sensitive pixels (50×500 μm pixel size, 3.4 nm average pixel interval) covering the spectral range from 1175 nm to 2170 nm. The light source provides illumination between 360 nm and 2500 nm, and uses a subminiature A (SMA) connector to maximize light coupling into light-conducting optic fibers with a diameter of up to 600 μm. The probe has a Y-style structure with a total length of 2 m, with 400-μm-diameter fibers.

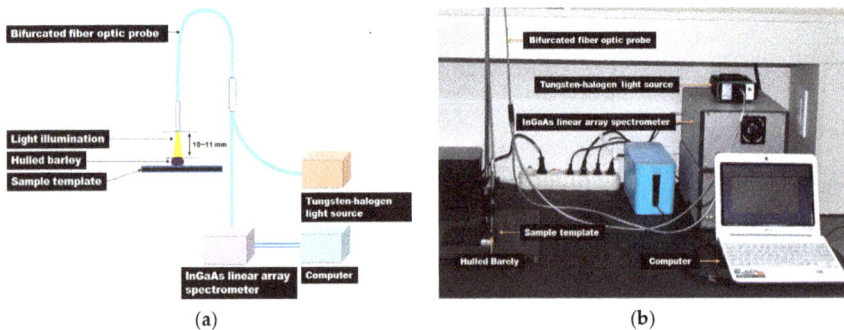

(a) (b)

Figure 1. (a) Schematic diagram and (b) photo of the NIR measurement system for the discrimination of hulled barley infected with *Fusarium*.

Seven optic fibers are arranged on the stainless-steel tip (length 50 mm, diameter 6.35 mm) at the end of the probe. Among the seven-fiber optics, six optical fibers, which are arranged in a circle on the outer side, transmit the light from the light source to the hulled barley samples, and the optical fiber at the center transmits the light reflected from the hulled barley samples to the InGaAs spectrometer. The distance between the hulled barley samples and the probe end was maintained at 10 mm to 11 mm, vertically.

Table 3. Specifications of NIR spectrometer used for classification of *Fusarium*-infected hulled barley.

Model (Manufacturer)	Ava Spec-NIR256-2.2TEC (Avantes BV, Apeldoorn, The Netherlands)
Appearance	
Spectral range	1175–2170 nm
Detection sensor	InGaAs linear array
Pixel pitch	3.4 nm
Pixel size	50×500 μm
Total pixel count	256
Minimum exposure time	1 ms
Signal to noise ratio	4100:1
PC interface	USB 2.0
Dimensions	$315 \times 235 \times 135$ mm
Weight	5.1 kg

2.3. Medium Preparation for Fusarium Culture of Hulled Barley

After obtaining the NIR reflectance spectra, the hulled barley samples further underwent a culture experiment to visually determine whether a *Fusarium* infection was present. The medium used in this experiment was a mixture of materials from Difco^{TM} Potato Dextrose Agar (PDA, Difco Labs, Detroit, MI, USA) and Bacto^{TM} Agar (BD Biosciences, San Jose, CA, USA) to smoothly culture and easily solidify *Fusarium*. The medium was prepared by mixing 26 g/1000 mL of Difco^{TM} Potato Dextrose Agar and 5.6 g/1000 mL Bacto^{TM} Agar in distilled water, and then sterilized at high pressure. The solution was then cooled, and antibiotics that had been stored in a frozen state were injected into the medium solution to inhibit the growth of other molds. The antibiotics used in this experiment were neomycin sulfate (MP Biomedicals Inc., Solon, OH, USA) and streptomycin sulfate (Biosesang, Seongnam, Korea). Neomycin (0.5 g/50 mL) was dissolved into distilled water, further filtered through a 0.2 μm filter, and injected at a ratio of 12 mL per 1000 mL and 2.5 g/50 mL of streptomycin was dissolved in distilled water, further filtered through a 0.2 μm filter, and injected at a ratio of 20 mL per 1000 mL.

The medium solution was solidified in a Petri dish for 2 to 3 h. Subsequently, seven kernels of hulled barley samples were placed on each Petri dish, as shown in Figure 2. The Petri dishes were sealed with tape to prevent infection by foreign germs, and then incubated at 25 °C for 2 to 4 days in a darkened incubator. Any *Fusarium* growth was then classified visually.

(a)   (b)

Figure 2. (**a**) Location and (**b**) configuration of kernels on Petri dish to culture *Fusarium* of hulled barley.

2.4. Acquisition and Pretreatment of NIR Reflectance Spectra

A hulled barley has an embryo in the lower part of the pericarp of the grain as well as a valley extending horizontally from the embryo, which is called the crease. In this experiment the side with the crease region was denoted as the front side and the opposite side as the back side, as shown in Figure 3. The measurement of one hulled barley kernel sample was repeated three times for the front side and for the back side, leading to a total of six spectra for each sample.

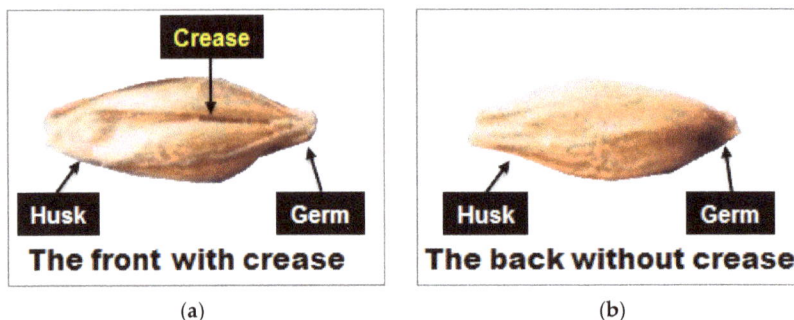

Figure 3. Denotation of (**a**) front side and (**b**) back side of hulled barley by the presence of the crease.

As shown in Figure 4, the hulled barley kernels were transferred to the optimal samples template for measurements. The integration time was 200 ms, and the spectra were cumulatively averaged 3 times. The smoothing pretreatment was applied to the basic reflectance spectra.

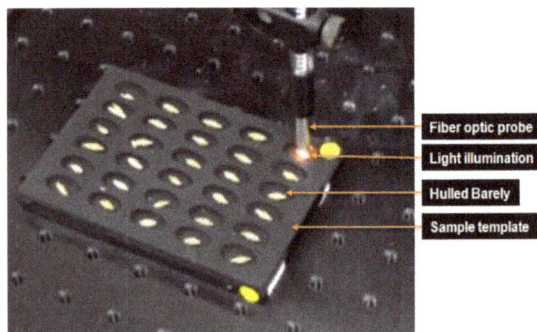

Figure 4. Probe and light illumination in NIR reflectance spectrum acquisition for hulled barley.

Reflectance spectral data obtained from hulled barley samples require calibration due to the irregular surface shape, and the data are pretreated using various mathematical methods to develop the best discriminant prediction model [22,23]. Several preprocessing techniques were applied to the obtained reflectance spectra to reduce the systematic noise and variation caused by the light source. The mathematical pretreatments used in this experiment include first-order derivative, second-order derivative, third-order derivative, mean normalization, maximum normalization, range normalization, mean scattering correction (MSC), baseline, and standard normal variate (SNV).

2.5. Fusarium Discrimination Prediction Model Development

In the PLS model, actually measured concentration information were used as dependent variables to develop a linear regression model with spectrum data, which are independent variables, and to predict actual concentration values [24–27]. PLS-DA is based on the classical partial least squares regression to construct prediction models [28]. In this regard, PLS-DA develops a regression model by designating the groups to be discriminated as dummy constants instead of concentration values as dependent variables. For a development of a PLS-DA model, this study designated as dependent variables with constant value '0' the reflectance spectra obtained from hulled barley samples not infected with *Fusarium*, and as dependent variables with constant value '1' the reflectance spectra obtained from hulled barley samples infected with *Fusarium* after the cultivation. A cross-validation method was applied to validate the PLS-DA models, and the accuracy of the developed models was determined by its coefficient of determination for calibration (R_C^2), the standard error of calibration (SEC), the coefficient of determination for validation (R_V^2), the standard error of prediction (SEP), and the optimal factor (F). Data pretreatment, model development, and validation using the obtained spectrum were performed by using multivariate data analysis software (Unscrambler v9.2, Camo Co., Oslo, Norway).

3. Results

3.1. Culture Results of Hulled Barley

3.1.1. Culture Results of Hulled Barley Classified as Control Group

After seven hulled barley samples cultured in each Petri dish were incubated for between 2 and 4 days in an incubator, the occurrence of *Fusarium* was visually observed to determine whether a sample was infected. Figure 5 shows the selected culture results (1 to 7, 50 to 56, 99 to 105, and 120 to 126) for 127 hulled barley samples that were used in the control group, indicating that *Fusarium* spores were not observed.

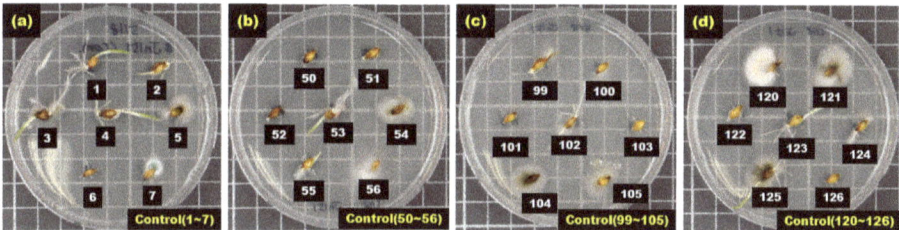

Figure 5. Culture results for hulled barley (JN151) in the control group, showing that *Fusarium* was not observed.

3.1.2. Culture Results of Hulled Barley Classified as Experimental Group

Figure 6 shows the selected culture results for hulled barley samples in the *Fusarium*-infected sample groups (GN121, JB021, JB061, JB094), in which *Fusarium* spores were actively generated and dispersed within the Petri dish, indicating that these hulled barley samples were infected with *Fusarium*.

Table 4 summarizes the results of the culture tests on the control and experimental sample groups, and shows that some hulled barley in the infected sample groups did not develop *Fusarium*. In the GN121 sample group, 46 kernels were infected among the total 80 kernels. In the JB021 sample group, 82 kernels were infected among the total 95 kernels. Furthermore, in the JB061 and JB094 sample groups, 82 kernels of hulled barley samples among the 108 kernels, and 88 kernels among the 105 kernels were infected, respectively. To discriminate *Fusarium*-infected kernels by using the NIR spectroscopic technique, the kernels were divided into 127 kernels in the control group and 298 (46 + 82 + 82 + 88) *Fusarium*-infected kernels in the contaminated sample group.

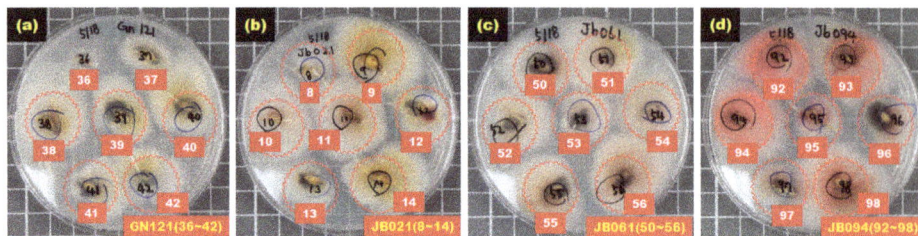

Figure 6. Culture results for hulled barley (GN121, JB021, JB061, JB094) in the experimental group in which *Fusarium* was observed.

Table 4. Culture results for control group and experimental group of hulled barley samples infected with *Fusarium*.

Experimental Group	Number of Grains (Sample Group)	Culture Results	
		[a] NIF	[b] IF
Hulled barley group classified as not infected with *Fusarium*	127 (JN151)	**127**	0
Hulled barley group classified as infected with *Fusarium*	80 (GN121)	34	46
	95 (JB021)	13	82
	108 (JB061)	26	82
	105 (JB094)	17	88

Note: [a] NIF, hulled barley not infected with *Fusarium*; [b] IF, hulled barley infected with *Fusarium*.

3.2. Spectral Characteristics of Hulled Barley

The experiment measured the NIR reflectance spectra of the total 515 kernels including 388 kernels in the four experimental groups classified as containing *Fusarium*-infected samples, and 127 kernels in the control group. A discriminant prediction model was developed using the NIR reflectance spectra. Figure 7 shows the 2550 reflectance spectra obtained by measuring both the front and back of each kernel three times. The maximum reflectance of the hulled barley was observed around 1575 nm, a rising peak was observed around 1315 nm, and the reflectance intensity then decreased around 1935 nm.

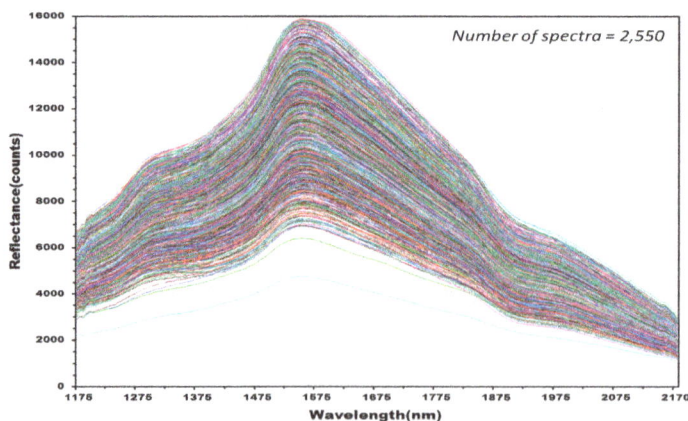

Figure 7. NIR reflectance spectra obtained from hulled barley samples (515 kernels) in both the control and experimental groups.

Figure 8 shows 762 reflectance spectra obtained from 127 kernels of normal hulled barley samples in the control group. Figure 9 shows 1788 reflectance spectra obtained from the 298 kernels infected with *Fusarium*. The raw reflectance spectra showed no difference in the reflectance spectrum intensity between the uninfected group and the group infected with *Fusarium*. Thus, it was difficult to discriminate the infection with the reflectance spectrum alone.

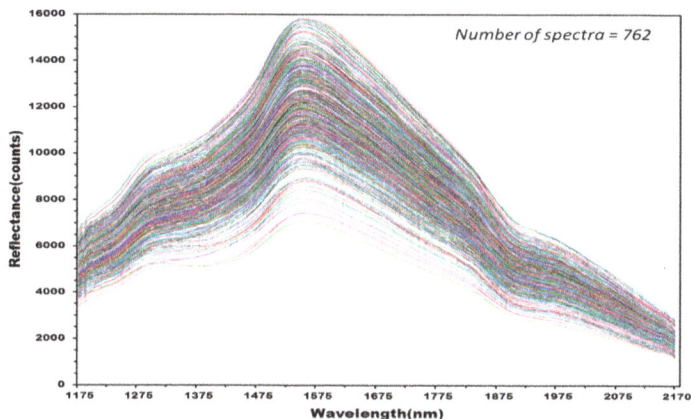

Figure 8. Total NIR reflectance spectra obtained from normal hulled barley samples (127 kernels) in the control group.

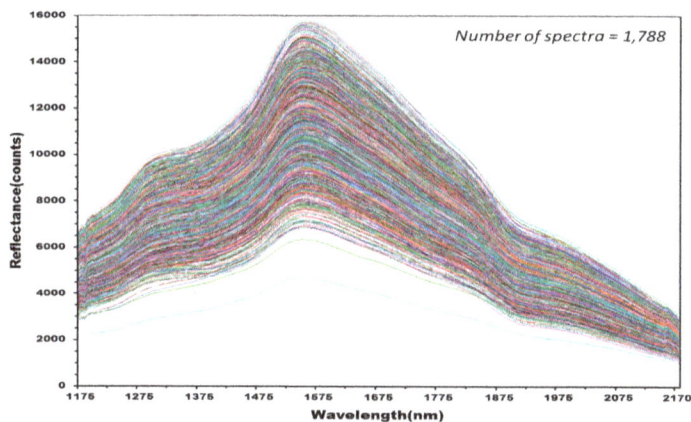

Figure 9. Total NIR reflectance spectra obtained from *Fusarium*-infected hulled barley samples (298 kernels) in the experimental group.

Figure 10 shows the average reflectance spectra for 2550 reflectance spectra obtained from 510 kernels of hulled barley samples used to develop the PLS-DA model, comprised of (a) 1275 reflectance spectra obtained from the front side of the kernels (with crease), and (b) 1275 reflectance spectra obtained from the back-side samples (without crease). As shown in each average reflectance spectra, the reflectance intensity was overall higher on the back side (b) than on the front side (a). The reflectance difference could be due to a decrease in the quantity of reflectance because the valley formed by the crease scatters relatively more light. The average (c) of the total NIR reflectance spectra obtained from both the front side and the back side showed the median reflectance value between (a) and (b).

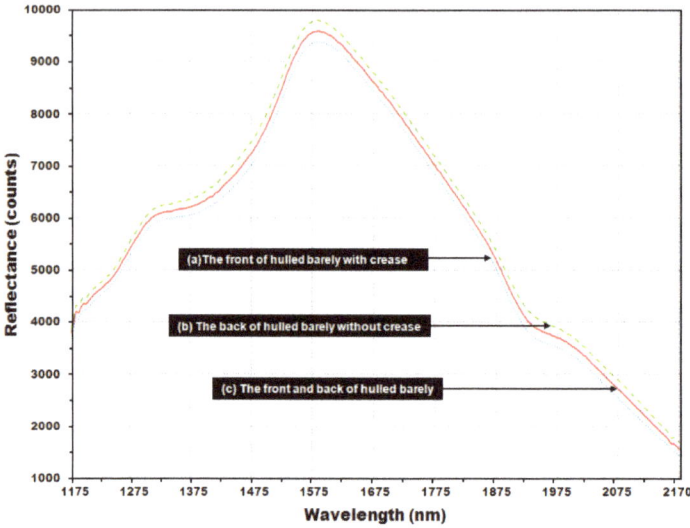

Figure 10. Average NIR reflectance spectra obtained from (**a**) the front side of the hulled barley with crease; (**b**) the back side of the hulled barley without crease; and (**c**) both front and back sides of the hulled barley.

3.3. Prediction Results of PLS-DA Model for Fusarium-Infected Huske Barley

This study applied the PLS-DA method to discriminate *Fusarium*-infected hulled barley by using the reflectance spectra and the culture results of the hulled barley samples. Independent variables used for PLS-DA model development are the reflectance spectra obtained from hulled barley samples. As the dependent variables, the random dummy variables were designated as '0' for the reflectance spectra obtained from the normal samples that were not infected with *Fusarium*, and the random dummy variables were designated as '1' for the *Fusarium*-infected samples to develop a discriminant model.

Table 5 shows the results of the PLS-DA models developed by applying various pretreatments to the reflectance spectra of normal hulled barley samples from the control group as well as the reflectance spectra of *Fusarium*-infected hulled barley samples from the experimental group.

Table 5. Performance of the PLS-DA calibration and validation models for *Fusarium*-infected hulled barley samples as well as classification accuracy for normal and infected hulled barleys.

Pretreatment	[a] F	[b] PC		[c] PV			
		R_C^2	SEC	R_V^2	SEP	[d] CCR	
						NIF	IF
The front and back measurement results of hulled barley							
Non-pretreatment	17	0.870	0.165	0.865	0.168	99.61	99.50
1st order Derivative	14	0.924	0.126	0.919	0.131	99.87	**100.00**
2nd order Derivative	10	**0.950**	**0.103**	**0.948**	**0.105**	**100.00**	**100.00**
3rd order Derivative	9	**0.942**	**0.110**	**0.941**	**0.112**	**100.00**	**100.00**
Mean Normalization	17	0.868	0.166	0.863	0.170	99.87	99.55
Maximum Normalization	17	0.865	0.168	0.860	0.172	99.74	99.38
Range Normalization	17	0.860	0.171	0.855	0.175	99.87	99.44
MSC	16	0.852	0.176	0.846	0.180	99.87	99.22
Baseline	14	0.838	0.184	0.835	0.186	99.48	98.99
SNV	17	0.853	0.176	0.847	0.179	99.87	99.16

Table 5. *Cont.*

Pretreatment	[a] F	[b] PC		[c] PV			
		R_C^2	SEC	R_V^2	SEP	[d] CCR	
						NIF	IF
The front measurement results with crease of hulled barley							
Non-pretreatment	17	0.872	0.164	0.862	0.170	99.21	99.66
1st order Derivative	15	0.938	0.114	0.929	0.122	99.74	100.00
2nd order Derivative	10	0.948	0.104	0.944	0.108	99.74	100.00
3rd order Derivative	9	0.943	0.109	0.939	0.113	100.00	100.00
Mean Normalization	16	0.858	0.172	0.847	0.179	99.48	99.78
Maximum Normalization	16	0.854	0.175	0.843	0.182	99.48	99.66
Range Normalization	17	0.858	0.173	0.847	0.179	99.48	99.33
MSC	16	0.845	0.180	0.833	0.187	99.21	99.22
Baseline	17	0.863	0.169	0.853	0.176	99.48	99.66
SNV	17	0.846	0.180	0.834	0.187	99.21	99.22
The back measurement results without crease of hulled barley							
Non-pretreatment	17	0.922	0.128	0.915	0.133	100.00	99.89
1st order Derivative	12	0.941	0.111	0.933	0.119	100.00	99.89
2nd order Derivative	10	0.962	0.089	0.959	0.093	100.00	100.00
3rd order Derivative	9	0.954	0.098	0.951	0.102	100.00	100.00
Mean Normalization	13	0.881	0.158	0.874	0.163	100.00	99.66
Maximum Normalization	13	0.878	0.160	0.871	0.165	100.00	99.55
Range Normalization	13	0.877	0.160	0.871	0.165	99.74	99.55
MSC	14	0.887	0.154	0.879	0.159	100.00	99.66
Baseline	14	0.897	0.147	0.890	0.152	100.00	99.78
SNV	14	0.878	0.160	0.871	0.165	100.00	99.55

Notes: [a] F, number of factors; [b] PC, performance of calibration; [c] PV, performance of validation; [d] CCR, correct classification rate.

3.3.1. Prediction Results of PLS-DA Model Using Reflectance Spectra Obtained from Front Side and Back Side of Hulled Barley

Figure 11 shows the calibration and validation results of a PLS-DA model developed without applying any pretreatment to the reflectance spectra obtained from the front side and back side of the hulled barley used in the experiment. In the validation model, three reflectance spectra among the 762 reflectance spectra obtained from the hulled barley not infected with *Fusarium*, were predicted (false positive) with a reference value of 0.5 or more, indicating an accuracy of 99.61%. On the other hand, two reflectance spectra among the 1788 reflectance spectra obtained from the hulled barley infected with *Fusarium*, were predicted (false negative) with a reference value of 1.5 or more, and 7 reflectance spectra were predicted (false negative) with a reference value of 0.5 or less, indicating an accuracy of 99.50%.

This study applied various spectrum pretreatments to improve the accuracy of the prediction model. Figure 12 shows the prediction results of the PLS-DA model, which was developed by applying the second-order derivative pretreatment to the raw reflectance spectra for the front and back sides of the hulled barley obtained from the control group and the experimental group. The results showed that the calibration and validation models had an excellent discrimination accuracy of 100%. In the validation model, to which the second-order derivative pretreatment was applied, R^2 was improved to 0.948, which was better than that of the raw spectral data. Moreover, the SEP decreased to 0.105, all indicating that the spectrum pretreatment was effective.

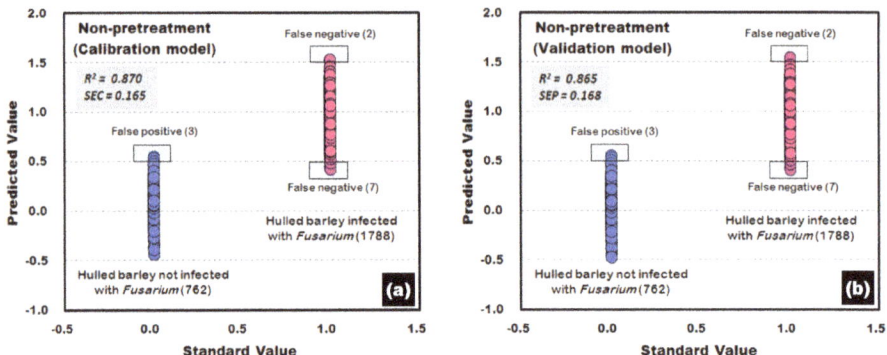

Figure 11. *Fusarium* discrimination calibration (**a**) and validation (**b**) results of the PLS-DA model developed using raw reflectance spectra obtained from the front side and back side of the hulled barley.

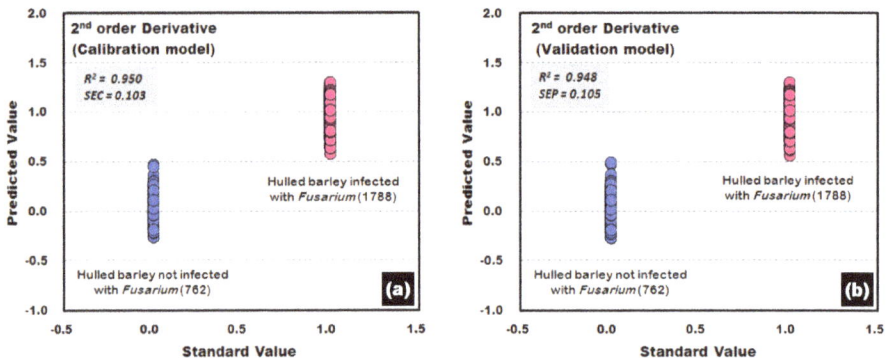

Figure 12. *Fusarium* discrimination calibration (**a**) and validation (**b**) results of PLS-DA model developed by applying the second-order derivative pretreatment to the reflectance spectra obtained from the front side and back side of hulled barley.

3.3.2. Prediction Results of PLS-DA Model Using Reflectance Spectra Obtained from Front Side of Hulled Barley

Figure 13 shows the results of PLS-DA model, which was developed by using the raw reflectance spectra that were obtained from the front side with crease on the hulled barley samples in both the control group and the experimental group. In the developed validation model, three reflectance spectra among the 381 reflectance spectra obtained from the front side of the hulled barley that were not infected with *Fusarium* were predicted with a reference value of 0.5 or more, indicating an accuracy of 99.21%. Among the 894 reflectance spectra obtained from the back side of the *Fusarium*-infected hulled barley, one reflectance spectrum was predicted as false negative with a reference value of 1.5 or more, and two reflectance spectra were predicted as false negative with a reference value of 0.5 or less, indicating an accuracy of 99.66%.

Figure 14 shows the calibration and validation results of the PLS-DA model that was developed by applying the third-order derivative pretreatment to the raw reflectance spectra obtained from the front side of the hulled barley, indicating 100% discrimination accuracy. The results of the developed validation model show that R^2 was improved to 0.939 and that SEP decreased to 0.113.

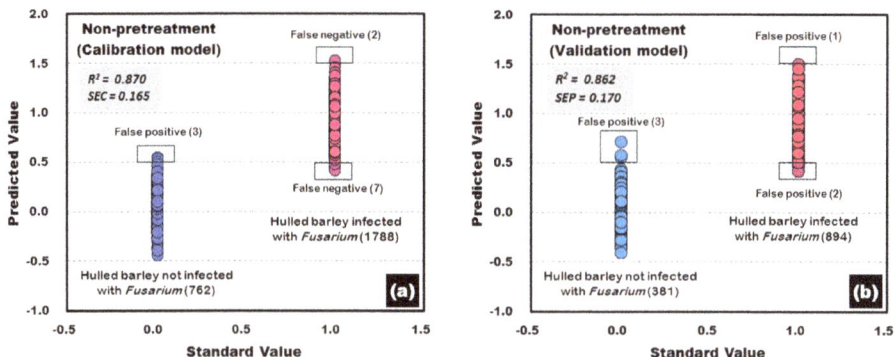

Figure 13. *Fusarium* discrimination calibration (**a**) and validation (**b**) results of PLS-DA model developed using the raw reflectance spectra obtained from the front side of hulled barley.

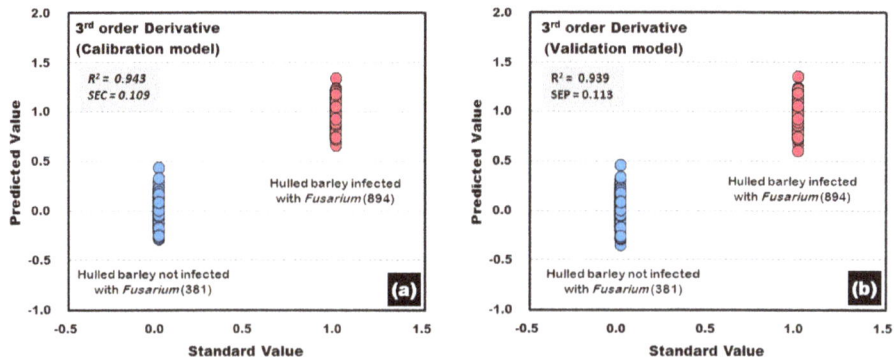

Figure 14. *Fusarium* discrimination calibration (**a**) and validation (**b**) results of PLS-DA model developed by applying the third-order derivative pretreatment to the reflectance spectra obtained from the front side of hulled barley.

3.3.3. Prediction Results of PLS-DA Model Using Reflectance Spectra Obtained from the Back Side of the Hulled Barley

Figure 15 shows the calibration and validation results of the PLS-DA model. To obtain the results, the raw reflectance spectra acquired from the back side of the hulled barley (without crease) was used to develop the PLS-DA model. The results showed an accuracy of 99.89% because one reflectance spectrum from each of the calibration and validation models deviated from the reference value (false positive) out of the 894 reflectance spectra for *Fusarium*-infected hulled barley samples.

The PLS-DA model was developed by applying the second-order derivative pretreatment acquired from the side of the hulled barley without the crease. The calibration and validation results of the PLS-DA model are shown in Figure 16. Of the 894 reflectance spectra for *Fusarium*-infected hulled barley samples, one reflectance spectrum from each of the calibration and validation models deviated from the reference value as a false positive. The results show an accuracy of 99.89%.

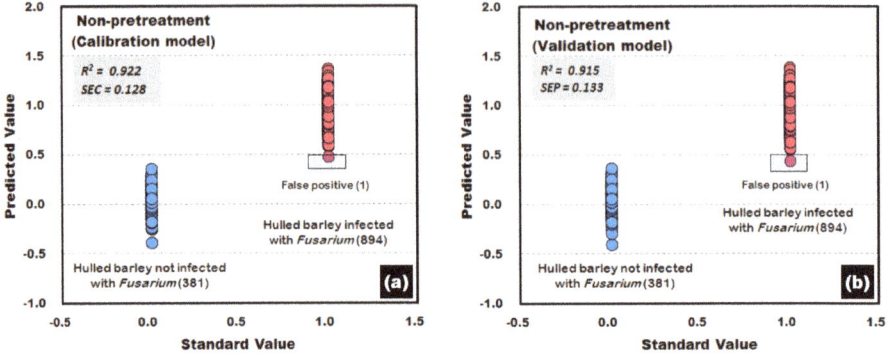

Figure 15. *Fusarium* discrimination calibration (**a**) and validation (**b**) results of PLS-DA model developed by using the raw reflectance spectra obtained from the back side of hulled barley.

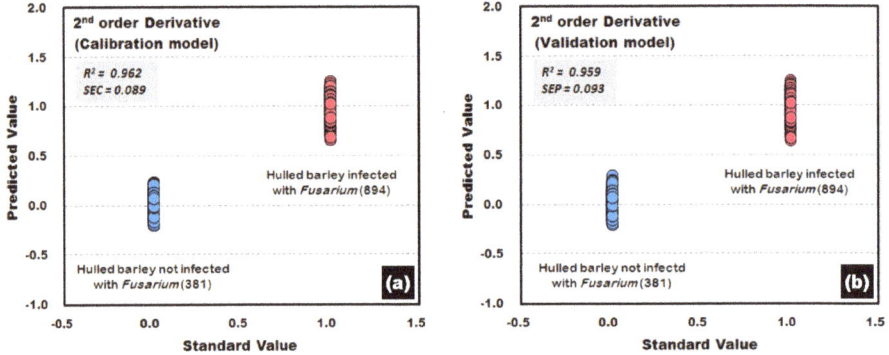

Figure 16. *Fusarium* discrimination calibration (**a**) and validation (**b**) results of the PLS-DA model developed by using the raw reflectance spectra obtained from the back side of the hulled barley.

4. Conclusions and Outlook

This paper indicates potential advantages of NIR spectroscopy for the discrimination of *Fusarium*-infected Korean hulled barley. This study nondestructively discriminated *Fusarium*-infected hulled barley and normal hulled barley by using NIR spectroscopy from 1175 nm to 2170 nm to measure the reflectance spectrum intensity of the hulled barley. The study further applied various spectrum pretreatments to develop the discriminant prediction models and validate the performance and discrimination accuracy. 127 kernels of hulled barley samples not infected with *Fusarium* and 298 kernels confirmed with *Fusarium* spores by a culture test were used as experiment samples. Each measurement was repeated three times for both sides of the hulled barley kernels to obtain 2550 reflectance spectra. Various mathematical pretreatments were applied to the obtained reflectance spectra to minimize the instability of the light source as well as several changes that occur in the size and shape of the samples, and the discrimination performance and accuracy for each pretreatment were examined by using the PLS-DA method to develop discriminant prediction models. To verify the accuracy of the developed PLS-DA model, additional specimens should be obtained to verify the sensitivity and specificity. These studies using NIR reflectance spectroscopy indicate the potential of this spectroscopic technique for agricultural application in the near future.

In the *Fusarium* discrimination performance of the developed PLS-DA validation model by using the 2550 raw reflectance spectra, the discrimination accuracy for the normal hulled barley samples was 99.61% and the accuracy for *Fusarium*-infected hulled barley samples was 99.50%. Validation was performed to develop a technique for nondestructive prediction. After the application of secondary derivative pretreatment to improve discrimination accuracy, the results showed 100% discrimination accuracy for discriminating normal and *Fusarium*-infected hulled barley samples.

The PLS-DA validation prediction model developed by applying 1275 raw reflectance spectra obtained from the front side of the hulled barley, showed accuracies of 99.21% and 99.66% for discriminating the normal and *Fusarium*-infected hulled barleys, respectively. The PLS-DA model developed by applying the third-order derivative pretreatment showed 100% accuracy for classification. The discrimination results of the PLS-DA validation model developed by using the raw reflectance spectra obtained from the back side of the hulled barley (without crease) with the best discrimination prediction results, showed an accuracy of 100% for the normal samples and 99.89% for the *Fusarium*-infected samples. Furthermore, the results with the application of the second and the third-order derivative pretreatments showed 100% classification accuracy.

As shown in the results from this study, the PLS-DA prediction model, developed by applying the reflectance spectra obtained from the NIR spectroscopic equipment with the optimal mathematical pretreatment, is anticipated to achieve nondestructive and rapid detection of *Fusarium*-infected hulled barley.

Acknowledgments: This study was carried out with the support of "Research Program for Agricultural Science & Technology Development (Project No. PJ01096901)", National Institute of Agricultural Sciences, Rural Development Administration, Republic of Korea.

Author Contributions: Jongguk Lim and Changyeun Mo contributed to the conception of the study; Giyoung Kim and Moon S. Kim contributed significantly to theoretical analysis and manuscript preparation; Kyoungmin Oh and Hyeonheui Ham performed the test; Hyeonche Yoo analyzed the data; All authors approved the final version.

Conflicts of Interest: The authors declare no conflict of interest.

References

1. Jansen, C.; Wettsteint, D.V.; Schafer, W.; Kogel, K.H.; Felk, A.; Maier, F.J. Infection patterns in barley and wheat spikes inoculated with wild-type and trichodiene synthase gene disrupted *Fusarium* graminearum. *Proc. Natl. Acad. Sci. USA* **2005**, *102*, 16892–16897. [CrossRef] [PubMed]
2. Doohan, F.M.; Brennan, J.; Cooke, B.M. Influence of climatic factors on *Fusarium* species pathogenic to cereals. *Eur. J. Plant Pathol.* **2003**, *109*, 755–768. [CrossRef]
3. Jouany, J.P. Methods for preventing, decontaminating and minimizing the toxicity of mycotoxins in feeds. *Anim. Feed Sci. Technol.* **2007**, *137*, 342–362. [CrossRef]
4. Wagacha, J.M.; Muthomi, J.W. *Fusarium* culmorum: Infection process, mechanisms of mycotoxin production and their role in pathogenesis in wheat. *Crop. Prot.* **2007**, *26*, 877–885. [CrossRef]
5. Del Ponte, E.M.D.; Fernandes, J.M.C.; Bergstrom, G.C. Influence of Growth Stage on *Fusarium* Head Blight Deoxynivalenol Production in Wheat. *J. Phytopathol.* **2007**, *155*, 577–581. [CrossRef]
6. Miedaner, T. Review Breeding wheat and rye for resistance to *Fusarium* diseases. *Plant Breed.* **1997**, *116*, 210–220. [CrossRef]
7. Richard, J.L. Some major mycotoxins and their mycotoxicoses—An overview. *Int. J. Food Microbiol.* **2007**, *119*, 3–10. [CrossRef] [PubMed]
8. Girolamo, A.D.; Lippolis, V.; Nordkvist, E.; Visconti, A. Rapid and non-invasive analysis of deoxynivalenol in durum and common wheat by Fourier-Transform Near Infrared (FT-NIR) spectroscopy. *Food Addit. Contam. Part A* **2009**, *26*, 907–917. [CrossRef] [PubMed]
9. Bryden, W.L. Mycotoxin contamination of the feed supply chain: Implications for animal productivity and feed security. *Anim. Feed Sci. Technol.* **2012**, *173*, 134–158. [CrossRef]
10. Koppen, R.; Koch, M.; Siegel, D.; Merkel, S.; Maul, R.; Nehls, I. Determination of mycotoxins in foods: Current state of analytical methods and limitations. *Appl. Microbiol. Biotechnol.* **2010**, *86*, 1595–1612. [CrossRef] [PubMed]
11. Herrero, A.M. Raman spectroscopy a promising technique for quality assessment of meat and fish: A review. *Food Chem.* **2008**, *107*, 1642–1651. [CrossRef]

12. Norris, K.H. Reports on the design and development of a new moisture meter. *Agric. Eng.* **1964**, *45*, 370–372.
13. Adams, M.J. *Chemometrics in Analytical Spectroscopy*, 2nd ed.; Royal Society of Chemistry: Cambridge, UK, 2004. [CrossRef]
14. Lorente, D.; Aleixos, N.; Gómez-Sanchis, J.; Cubero, S.; García-Navarrete, O.L.; Blasco, J. Recent Advances and Applications of Hyperspectral Imaging for Fruit and Vegetable Quality Assessment. *Food Bioprocess. Technol.* **2011**, *5*, 1121–1142. [CrossRef]
15. Cen, H.; Yong, H. Theory and application of near infrared reflectance spectroscopy in determination of food quality. *Trends Food Sci. Technol.* **2007**, *18*, 72–83. [CrossRef]
16. Giacomo, D.R.; Stefania, D.Z.A. A multivariate regression model for detection of fumonisins content in maize from near infrared spectra. *Food Chem.* **2013**, *141*, 4289–4294. [CrossRef] [PubMed]
17. Sirisomboon, C.D.; Putthang, R.; Sirisomboon, P. Application of near infrared spectroscopy to detect aflatoxigenic fungal contamination in rice. *Food Control* **2013**, *33*, 207–217. [CrossRef]
18. Delwiche, S.R. Classification of scab- and other mold-damaged wheat kernels by near-infrared reflectance spectroscopy. *Trans. ASAE* **2003**, *46*, 731–738. [CrossRef]
19. Polder, G.; Van Der Heijden, G.W.A.M.; Waalwijk, C.; Young, A.I.T. Detection of *Fusarium* in single wheat kernels using spectral imaging. *Seed Sci. Technol.* **2005**, *33*, 655–668. [CrossRef]
20. Liu, Y.; Delwiche, S.R.; Dong, Y. Feasibility of FT–Raman spectroscopy for rapid screening for DON toxin in ground wheat and barley. *Food Addit. Contam. Part A* **2009**, *26*, 1396–1401. [CrossRef]
21. Gaspardo, B.; Del Zotto, S.; Torelli, E.; Cividino, S.R.; Firrao, G.; Della Riccia, G.; Stefanon, B. A rapid method for detection of fumonisins B_1 and B_2 in corn meal using Fourier transform near infrared (FT-NIR) spectroscopy implemented with integrating sphere. *Food Chem.* **2012**, *135*, 1608–1612. [CrossRef] [PubMed]
22. Mo, C.Y.; Kim, G.Y.; Lee, K.J.; Kim, M.S.; Cho, B.K.; Lim, J.G.; Kang, S.W. Non-Destructive Quality Evaluation of Pepper (*Capsicum annuum L.*) Seeds Using LED-Induced Hyperspectral Reflectance Imaging. *Sensors* **2014**, *14*, 7489–7504. [CrossRef] [PubMed]
23. Lim, J.G.; Kim, G.Y.; Mo, C.Y.; Kim, M.S. Design and Fabrication of a Real-Time Measurement System for the Capsaicinoid Content of Korean Red Pepper (*Capsicum annuum L.*) Powder by Visible and Near-Infrared Spectroscopy. *Sensors* **2015**, *15*, 27420–27435. [CrossRef] [PubMed]
24. Lee, K.J.; Hruschka, W.R.; Abbott, J.A.; Noh, S.H.; Park, B.S. Predicting the soluble solids of apples by near infrared spectroscopy(II)—PLS and ANN models. *J. Biosyst. Eng.* **1998**, *23*, 571–582.
25. Son, J.R.; Lee, K.J.; Kang, S.W.; Yang, G.M.; Seo, Y.M. Development of prediction model for sugar content of strawberry using NIR spectroscopy. *Food Eng. Prog.* **2009**, *13*, 297–301.
26. Alexandrakis, D.; Downey, G.; Scannell, A.G.M. Detection and identification of bacteria in an isolated system with near-infrared spectroscopy and multivariate analysis. *J. Agric. Food Chem.* **2008**, *56*, 3431–3437. [CrossRef] [PubMed]
27. Lim, J.G.; Mo, C.Y.; Kim, G.Y.; Kang, S.W.; Lee, K.J.; Kim, M.S.; Moon, J.H. Non-destructive and Rapid Prediction of Moisture Content in Red Pepper (*Capsicum annuum L.*) Powder Using Near-infrared Spectroscopy and a Partial Least Squares Regression Model. *J. Biosyst. Eng.* **2014**, *39*, 184–193. [CrossRef]
28. Wold, S.; Sjostrom, M.; Eriksson, L. PLS-regression: A basic tool of chemometrics. *Chemom. Intell. Lab. Syst.* **2001**, *58*, 109–130. [CrossRef]

sensors

MDPI

Article

Application of Multilayer Perceptron with Automatic Relevance Determination on Weed Mapping Using UAV Multispectral Imagery

Afroditi A. Tamouridou [1,2], Thomas K. Alexandridis [2], Xanthoula E. Pantazi [1], Anastasia L. Lagopodi [3], Javid Kashefi [4], Dimitris Kasampalis [2], Georgios Kontouris [2] and Dimitrios Moshou [1,*]

[1] Agricultural Engineering Laboratory, Faculty of Agriculture, Aristotle University of Thessaloniki, 54124 Thessaloniki, Greece; tamouridoualex@gmail.com (A.A.T.); renepantazi@gmail.com (X.E.P.)
[2] Laboratory of Remote Sensing and GIS, Faculty of Agriculture, Aristotle University of Thessaloniki, 54124 Thessaloniki, Greece; thalex@agro.auth.gr (T.K.A.); dkasampa@agro.auth.gr (D.K.); giorgoskontouris@gmail.com (G.K.)
[3] Plant Pathology Laboratory, Faculty of Agriculture, Aristotle University of Thessaloniki, 54124 Thessaloniki, Greece; lagopodi@agro.auth.gr
[4] USDA-ARS-European Biological Control Laboratory, 54623 Thessaloniki, Greece; jkashefi@ars-ebcl.org
* Correspondence: dmoshou@agro.auth.gr; Tel.: +30-231-0-998264; Fax: +30-231-0-998729

Received: 16 June 2017; Accepted: 6 October 2017; Published: 11 October 2017

Abstract: Remote sensing techniques are routinely used in plant species discrimination and of weed mapping. In the presented work, successful *Silybum marianum* detection and mapping using multilayer neural networks is demonstrated. A multispectral camera (green-red-near infrared) attached on a fixed wing unmanned aerial vehicle (UAV) was utilized for the acquisition of high-resolution images (0.1 m resolution). The Multilayer Perceptron with Automatic Relevance Determination (MLP-ARD) was used to identify the *S. marianum* among other vegetation, mostly *Avena sterilis* L. The three spectral bands of Red, Green, Near Infrared (NIR) and the texture layer resulting from local variance were used as input. The *S. marianum* identification rates using MLP-ARD reached an accuracy of 99.54%. The study had an one year duration, meaning that the results are specific, although the accuracy shows the interesting potential of *S. marianum* mapping with MLP-ARD on multispectral UAV imagery.

Keywords: remote sensing; precision agriculture; crop monitoring; data fusion

1. Introduction

Weed detection has conventionally been carried out through either ground-based platforms or remotely sensed images acquired from aircrafts, unmanned aerial vehicles (UAV) or satellites [1,2] through examination of remote sensing pixel data unable to provide precise crop detection due to pixel variation and spectral behavior resemblance [3]. The main aim of the current study is to provide a robust solution in utilizing remote sensing data for the purpose of plant species identification to contribute to the larger field of precision agriculture practices.

Silybum marianum is a weed accountable for major loss of crop yield and is tough to eradicate. Herbicides are costly and pollute the rural and natural ecosystems. Due to its height and thorny leaves, it may also inhibit the movement of livestock across grasslands. Khan et al. (2011) [4] confirmed the allelopathic effects of cold water extracts of *Silybum marianum* triggering harmful effects on germination percentage, germination stage, germination index and seed vigor index on *Phaseolus vulgaris, Vigna radiata, Cicer arietinum*, and *Glycine max*. For these reasons, it is significant to map the levels and distribution of the weed's presence so as to define a suitable management practice.

Previous successful applications of UAV in weed mapping include Tamouridou et al. (2017) [5] who evaluated the optimum scale for mapping weed patches; Pena 2013 [6] who described weed mapping in early-season maize fields using object-based analysis of UAV images; Torres-Sanchez (2013) [7] who provided configuration and specifications of an UAV for early site specific weed management.

Lamb and Brown (2001) [8] have observed significant variance among plant species rendering separation between weeds and crop plants problematic. Slaughter et al. (2004) [9] studied the spectral reflectance of field- grown tomato and nightshade leaves, which are difficult to discriminate since they belong to the same taxonomic family (Solanaceae). They proved that a broadband Red-Green-Blue (RGB) color classifier reached an accuracy of 76% for species identification in leave-one-out cross- validation tests. Moreover, accurate species recognition (98% or higher) in leave-one-out cross-validation tests was achieved by narrowband classifiers using near-infrared reflectance.

In order to attain real-time recognition of goal plants and to spread over specific application of various management practices in a more effective and specific way, sensors and machine architectures were utilized [10]. Weed species recognition with sensors [11] is critical for the application of required chemical herbicide and spraying dosage. Discrimination between crops and weeds based on spectral leaf reflectance, has been tried in previous studies. Borregaard et al. (2000) [12] showed that it is possible to successfully discriminate among crop and plants as well as between weed species. Additional approaches targeting at weed recognition have been presented, taking into account the leaf size and shape analysis. A method for automatic machine vision based on weed species grouping via active shape modelling (ASM) has reached a correctness between 65% and 90%. Moreover, it has been revealed that spectral reflectance characteristics are adequate to discriminate between different weed species as in Moshou et al. (2001) [13] who realised a successful discrimination between maize crop and weeds with an accurate classification rate of 96% for maize and 90% for weeds. Regarding the discrimination of sugar-beet from weeds, the precision rate was 98% and 97% respectively. Nonetheless, these results pointed solely at crop- weed perception, not weed classes recognition.

Apart from ASM approaches, various classification algorithms are applied in remote sensing mapping applications. In several studies the effectiveness of Artificial Neural Networks (ANN) when applied in remote sensing classification is indicated [14]. The MLP classifiers are networks where the nodes receive inputs coming from previous layers. As a result, the information flows in a single direction to the output layer [14]. In the intermediate layer(s) the number of nodes is related to the complexity and the performance of a neural network model to express the upcoming relationships and structures built-in a training data set which describe the generalization capability of the network [15]. The number of nodes in hidden layers is relevant to achieve classification of more complex, and low quality satellite images.

The current work proposes a data fusion approach directed on weed recognition. By using MLP-ARD *S. marianum* was identified among other vegetation. The data fusion approach consisted of combining the spectral and textural features derived from a UAV multispectral camera.

2. Materials and Methods

In the present paper, the proposed MLP-ARD classifier was evaluated for performing fusion of features from UAV images for the accurate recognition of *S. marianum*. The imaging of *S. marianum* was achieved by using a fixed wing UAV and a multispectral camera. The fusion structures that were used were constructed from the mixture of spectral and textural information. The MLP-ARD classifier was used to categorise the data into *S. marianum* and other plants.

To assess the capability of weed recognition, spectral signatures were gathered during a field operation, both from *S. marianum* plants and from other vegetation on 29 May 2015 using a UniSpec-DC portable spectroradiometer (PP Systems, Inc., Amesbury, MA, USA), with a resolution of 10 nm in the range of 310 nm and 1100 nm. From the various plant species, four spectral signatures were taken in the field from the upper surface area of the leaves.

2.1. Study Area

This study was performed at a 10.1 ha field located in the area of Thessaloniki, Greece (40°34′14.3″ N 22°59′42.6″ E) (Figure 1). The ground topography is about level and the elevation is 75 m. This field belongs to the American Farming School of Thessaloniki and for the past decade it has never been cultivated. In this left aside field, graminaceous weeds are growing, which comprise the field's main vegetation with huge patches of *S. marianum*. Weeds include: *Avena sterilis* and *Conium maculatum* L.

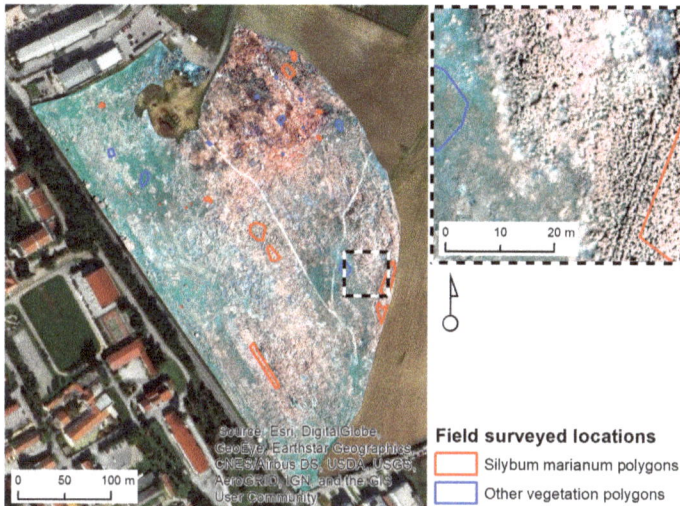

Figure 1. Orientation map and field surveyed locations in the study area.

2.2. Preparation of Datasets

Since the study focuses on the efficient combination of spectral and textural information, the reference data gathering process took place in the study areas and is described below. An eBee fixed wing UAV (http://www.geosense.gr/en/ebee/) with a Canon S110 NIR camera (12 Mpixels) acquired the remote sensing images on 19 May 2015. The spectral bands included are green (560 ± 25 nm), red (625 ± 45 nm) and near-infrared (850 ± 50 nm), the original resolution was 0.1 m, and was rescaled to 0.5 m, as demonstrated by Tamouridou et al. (2017) [5]. Beyond the analysis of the spectral information, a texture layer was created based on the NIR layer using the local variance algorithm (moving window size 7×7 pixels) depicting the structural patterns of vegetation.

Data was required in order to identify and pinpoint different vegetation categories represented in the UAV image. During a field operation using a handheld Trimble GPS, the location of the largest *S. marianum* patches (which are visible and easily identified) and other predominant vegetation categories were identified. Their locations were logged in the GPS and were used during the construction of the training sets that were fed to the MLP-ARD for image classification. The GPS model that was used in the process, is the GPS Trimble GeoXH 2008 (Sunnyvale, CA, USA). It features EGNOS error correction, which is a combination of field computer Trimble GPS with Microsoft Windows Mobile version 6 software. GeoXH utilizes EVEREST and H-Star™ technology in order to achieve an >30 cm accuracy. Each of the pixels of the UAV image contains specific spectral values corresponding to the afore mentioned spectral regions (Green, Red, NIR). The MLP-ARD algorithm is looking for similarities in those spectral values in order to successfully classify each pixel. The location of the sampling of the two vegetation categories is depicted in the polygons of Figure 1. The polygons were carefully positioned in order to represent areas of the whole region. The pixels from those polygons were used to construct the datasets. The first group of polygons, representing *S. marianum*, consisted of 17 polygons, with a total area of 0.12 ha. The second group, representing the other vegetation types, consisted of 13 polygons with a total area of 0.04 ha.

The afore mentioned polygons were used for the construction of data sets. The constructed datasets comprised of 4745 pixels of the *S. marianum* category and of 1434 pixels of the category of other types of vegetation respectively (Figure 1). The two calibration datasets comprised of 2868 pixels for acquiring equal datasets of *S. marianum* and other vegetation categories. From the calibration set a training dataset was formed containing a random selection of 70% (2008 pixels) for model calibration and setting the rest 30% (860 pixels) for validation.

In Figure 2, the spectral signatures of three vegetation species (*Avena sterilis, Conium maculatum*), are depicted. *A. sterilis* shows analogous spectral reflectance features with *S. marianum* rendering the separation a challenge. The three species, though, were easier to discriminate between in the NIR spectrum. This suggests that the NIR region, and perhaps other features (e.g., texture) can be utilized to make the class separation possible.

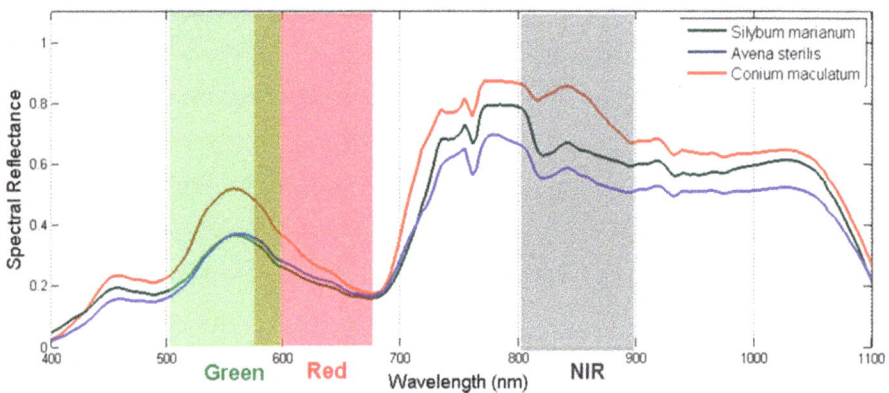

Figure 2. Vegetation species demonstrated similar spectral reflectance with *Silybum marianum*.

2.3. Methods of Data Fusion

The intention of this data processing is to develop a model between spectral and textural information of each MS camera pixel and the label of it as belonging to a *S. marianum* or other plant. MLP is a neural network cosisting of multiple layers of neurons. ARD automatically finds relevant neurons by modulating their connections strength so as to maximise relevant neurons in terms of evidence presented from the data, while a weighted evidence of multifeature sets amounts to fusion since adaptive weights are determined to combine effectively input features.

The main goal is to determine a mapping between spectral and textural input data points of each pixel to the plant identity. IT learns without external supervision. The obstacle is overfitting which is avoided by the bayesian regularization applied during learning which restricts ovefitting.

A feed-forward multilayer perceptron neural network (MLP) with one hidden layer and output unit [16] was used by using a given training set:

$$D = \{x_i, t_i\}_{i=1}^{N} \tag{1}$$

where x_i represents the observation data $x_i = \left(x_i^{(1)}, \cdots, x_i^{(n)}\right) \in R^n$ while t_i is the target category labels $t_i \in \{0, 1\}$. As result, the MLP classification system forms the following non-linear mapping:

$$y(x) = f_2(v f_1(u x)) \tag{2}$$

where $y(x) \in R$, and υ and u are the respective weight vectors of the hidden and output layer, that establish the weight vector ω. These are estimated during the training process, while hyperbolic transfer functions f_1 and f_2 are used for complex non-linear mapping and inserted to the hidden layer. Then, a logistic transfer function is employed in the output layer which facilitates the output of the MLP which can be defined as $y(x)$ for estimating the form $p(t = 1|x)$ in which ω is randomly adapted and incrementally adjusted to minimize a cross-entropy objective function G [16]:

$$G = -\sum_{i=1}^{N}\{t_i \ln(y_i) + (1 - t_i) \ln(1 - y_i)\} \tag{3}$$

For avoiding over-fitting of the objective function regularization weight is realized by the following function:

$$F(\omega) = G + \sum_k \alpha_k Ew_{(k)} \tag{4}$$

in which $Ew_{(k)} = \frac{1}{2}\sum_j \omega_j^2$, weights $\omega_j \in R$ and j represents an index estimating all weights of weight class W(k).

Automatic relevance determination (ARD) attains weight regularization by considering $k = 1, \ldots, n + 3$ weight classes inside weight vector ω, each associated with a weight decay hyperparameter (α_k). Each hyperparameter (α_k) is related with each predictor variable $x(k)$ ($k = 1, \ldots, n$) and more explicit, with the weights connected to the hidden neurons. Three extra weight decay hyperparameters including weight classes, are demonstrated: The first, associated with the synaptic connection bias, the second is related to the connections from hidden neurons to the output neuron while the third one is associated with the connection that is linking the hidden layer bias neuron and the output neuron [16].

The above described MLP-ARD was employed in dealing with to the UAV image pixels for identifying *S. marianum* weeds and other vegetation according to the calibration dataset. Accuracy assessment was accomplished by assessing the confusion tables on the validation dataset of 860 pixels (441 were from *S. marianum* and 419 from other vegetation).

3. Results

The results from the processing of the pixel data with MLP ARD for *S. marianum* identification vs. other vegetation are accessible in Table 1. It is obvious that the MLP-ARD classifier has managed to classify correctly >99% of the pixels of both categories, and only few pixels (<1%) were misclassified.

Table 1. Confusion matrix of the MLP-ARD for *S. marianum* and other plants. Percentages are estimated on actual observation sums.

Categories	Network Prediction	
	S. marianum (%) 441 pixels	Other plants (%) 419 pixels
S. marianum	99.55	0.45
Other plants	0.48	99.52

For the visualisation of the MLP-ARD function, the Hinton diagram weight component maps were formed (Figure 3). The Hinton diagram depicts the values of the weights as a combination of spectral and textural features and hidden neurons. Positive weight is depicted in white colour while negative weight is depicted in black. It can be assumed that the weights of less relevant features have been restricted by ARD, while the weights of more active features have been amplified with the condition that the relative magnitudes of the weights are illustrated by the size of the squares. Thus, the status of individual features can be inferred through the Hinton diagrams.

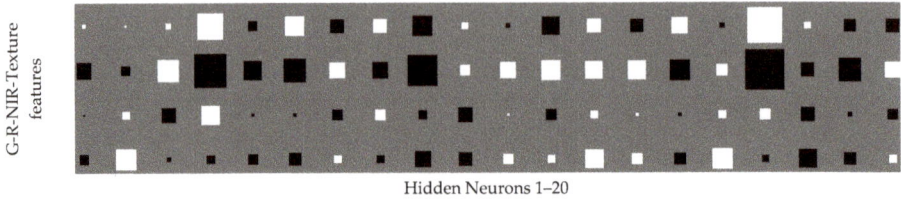

Figure 3. Hinton diagram of the trained MLP-ARD. The vertical axis corresponds to feature while the horizontal shows hidden neurons.

Figure 3 illustrates how the ARD capability has influenced the weight distribution. It shows larger weights that correspond to NIR and texture features. Moreover, the weights are for most hidden neurons for the NIR and texture of opposite sign which means that the overlap of the two features is minimal and that their activity is synergistic. This result in terms of classification shows strong synergy of the texture and NIR components which means that the classifier performs fusion of the components.

A quantification of the influence of individual features can be assessed by observation of the L2 norms of the weights and the associated values of the hyperparameters (a) that are calculated from Equation (4). At the final period of the training large norms are associated to smaller hyperparameter values which in turn are linked to larger variances for the weights. In Figure 4 one can assess the values of the hyperparameters for each weight group that link a specific input feature to the units of the hidden layer.

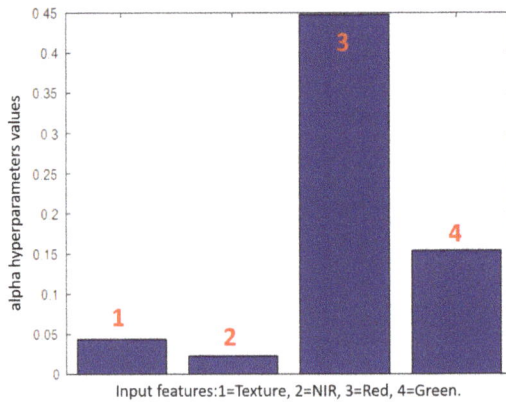

Figure 4. The alpha hyperparameters are shown for the four input features (1 = Texture, 2 = NIR, 3 = Red, 4 = Green).

The inverse behaviour of hyperparameters with respect to weights in the trained network is evident. This means that minimal hyperparameters (a) correspond to large weight values. To the contrary, large hyperparameters (a) are associated to weights that are inert so they are not frequently used during training to construct the mapping. In this way, ARD process achieves a soft selection of weights that are more closely related to the classification result. However, the weight groups that link the input layer to the hidden layer following this soft selection can visually depict the influence of each feature has on the classification result. Features that have low importance will remain with large hyperparameters (a) and will have small weights which is a result that is related to their minimal contribution. The opposite occurs when the hyperparameters (a) converge to small values near to zero, indicating large weight values connected to important features.

The calibration dataset was used to obtain the model. In the case of the whole field the trained model was deployed operationally by using feature vectors from the whole field. The deployment was performed by using new test feature vectors as input to the trained fixed model and obtaining the response of the MLP in the form of binary vectors [1 0] for *S. marianum* and [0 1] for other vegetation. Then, the output vectors were color coded into green and yellow pixels forming the final classification map covering the whole field. The trained MLP-ARD model was fed with feature vectors from the whole field with 782,838 feature vectors formed from UAV images from the whole field and comprising four components (green, red, NIR and texture). The map demonstrates the location of the patches of *S. marianum* and other vegetation in the field so the type of vegetation for each location was not obtained. Each input vector produced a classification corresponding to 1 for *Silybum* and 2 for other vegetation. In Figure 5 the classified UAV image of the study area is shown. The map shown in Figure 5 was produced by the MLP-ARD. Afterwards, in order to validate its success, the aforementioned map was compared to the points of the validation dataset which was sampled to be representative of the whole area. The map's accuracy was found to be 99.55%. Large patches of *S. marianum* is evident on the central and eastern parts of the study area, which are interleaved by large patches of other vegetation types on the south east. Only small sporadic patches of *S. marianum* appear in the western part of the study area.

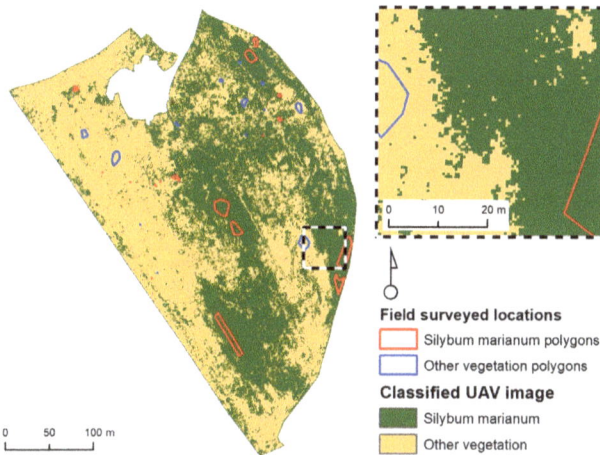

Figure 5. *S. marianum* weed mapping based on MLP-ARD prediction. Green is *S. marianum* and yellow is other vegetation.

4. Discussion

The regularisation procedure has the potential to constrain the model complexity. For example, a cost function could accommodate an additional function by adding a weight constraining term. Individual constraining terms associated to the weights offer an advantage according to the evidence framework of MacKay [17].

By employing the evidence framework there is no actual need to foresee a validation set. In the current work different weight groups were equipped with individual constraining parameters. Such a regularisation process was employed by implementing the ARD approach through which a soft feature selection was achieved. The term "soft" selection is founded on the idea of maintaining the extra features in the network whereas constraining their influence to minimal values.

The field in question has not been cultivated in the past decade, so the present vegetation consists of endemic weed species growing in random patterns. The vegetation nearby the *S. marianum* patches consists mainly of graminaceous weeds that generate a flat surface with low texture (as observed in

Figure 1). On the other hand, the *S. marianum* patches consist of large discrete plants with high texture. This effect explains the positive relation of texture to *S. marianum* appearance.

During the specific season when the field data was obtained, the bulk of the area's vegetation consisting of *A. sterilis* was already drying up compared to the *S. marianum* weeds that were still vigorous. Therefore, the *S. marianum* patches were the only areas in the field where high NIR reflection was observed. This effect substantiates the positive relation of NIR to *S. marianum* presence. These effects and observations are relevant to the general conditions of the present study (e.g., season, other vegetation) and are not necessarily representative of every situation.

By assimilating all the above it can be decided that both the texture and NIR features have a synergistic fusion effect on the classification accuracy while on the other hand MLP-ARD exploits this synergy by fusing the most active synaptic contributions to accurately classify *S. marianum* presence vs. other vegetation. As shown in Figure 3, the synergistic effect of the two features (NIR and texture) lays on the complementary of their weights, meaning that the classifier performs fusion of the components, thus affirming the main aim of the study which was discrimination of vegetation species by sensor fusion.

5. Conclusions

In the current paper it was established that it is possible to map *S. marianum* by using the MLP-ARD classifier. A NIR multispectral camera on a UAV was employed for the acquisition of high-resolution images. As functional input to the classifier, the three spectral bands (green-red-NIR) and the texture feature based on the local variance of the NIR layer were used. The classification rates obtained using MLP-ARD network reached high overall accuracy and reliability by achieving a 99.55% correct *S. marianum* detection. The results indicate that the on-line classification of *S. marianum* with MLP-ARD can be used operationally for performing UAV based weed mapping for various applications, including coverage assessment, eradication programs and assessment of treatment effectiveness by using change detection.

Acknowledgments: Authors are grateful to Geosense (http://www.geosense.gr) for providing the eBee (SenseFly S.A., Cheseaux-Lausanne, Switzerland) and Pix4Dmapper Pro (Pix4D, Lausanne, Switzerland) processing. We are thankful to the American Farm School of Thessaloniki for providing access to the study area, and to Vasilis Kaprinis and Kostantinos Georgiadis for their assistance in the field surveys.

Author Contributions: T.K.A., A.L.L., A.A.T. and J.K. conceived and designed the experiments; A.A.T., T.K.A. and D.K. developed the algorithms; X.E.P., A.A.T. and D.M. coded and debugged the algorithms; A.A.T. and G.K. performed the experiments; A.A.T., T.K.A., D.K. and G.K. analysed the data; J.K. provided field data; A.A.T., T.K.A., D.K. and G.K. contributed to the field survey; all authors wrote the paper.

Conflicts of Interest: The authors declare no conflict of interest.

References

1. Rasmussen, J.; Nielsen, J.; Garcia-Ruiz, F.; Christensen, S.; Streibig, J.C. Potential uses of small unmanned aircraft systems (UAS) in weed research. *Weed Res.* **2013**, *53*, 242–248. [CrossRef]
2. Thorp, K.R.; Tian, L.F. A review on remote sensing of weeds in agriculture. *Precis. Agric.* **2004**, *5*, 477–508. [CrossRef]
3. Peña-Barragán, J.M.; Ngugi, M.K.; Plant, R.E.; Six, J. Object-based crop identification using multiple vegetation indices, textural features and crop phenology. *Remote Sens. Environ.* **2011**, *115*, 1301–1316. [CrossRef]
4. Khan, M.; Farrukh, H.; Shahana, M. Allelopathic potential of Rhazya stricta Decne on germination of Pennisetum typhoides. *Int. J. Biosci.* **2011**, *1*, 80–85.
5. Tamouridou, A.A.; Alexandridis, T.K.; Pantazi, X.E.; Lagopodi, A.L.; Kashefi, J.; Moshou, D. Evaluation of UAV imagery for mapping Silybum marianum weed patches. *Int. J. Remote Sens.* **2017**, *38*, 2246–2259. [CrossRef]
6. Peña, J.M.; Torres-Sánchez, J.; de Castro, A.I.; Kelly, M.; López-Granados, F. Weed mapping in early-season maize fields using object-based analysis of unmanned aerial vehicle (UAV) images. *PLoS ONE* **2013**, *8*, e77151. [CrossRef] [PubMed]

7. Torres-Sánchez, J.; López-Granados, F.; De Castro, A.I.; Peña-Barragán, J.M. Configuration and specifications of an unmanned aerial vehicle (UAV) for early site specific weed management. *PLoS ONE* **2013**, *8*, e58210. [CrossRef] [PubMed]

8. Lamb, D.W.; Brown, R.B. PA—Precision agriculture: Remote-sensing and mapping of weeds in crops. *J. Agric. Eng. Res.* **2001**, *78*, 117–125. [CrossRef]

9. Slaughter, D.C.; Lanini, W.T.; Giles, D.K. Discriminating weeds from processing tomato plants using visible and near-infrared spectroscopy. *Am. Soc. Agric. Biol. Eng.* **2004**, *47*, 1907–1911. [CrossRef]

10. Tellaeche, A.; Pajares, G.; Burgos-Artizzu, X.P.; Riberio, A. A computer vision approach for weeds identification through support vector machines. *Appl. Soft Comput.* **2011**, *11*, 908–915. [CrossRef]

11. Moshou, D.; Ramon, H.; De Baerdemaeker, J. A weed species spectral detector based on neural networks. *Precis. Agric.* **2002**, *3*, 209–223. [CrossRef]

12. Borregaard, T.; Nielsen, H.; Nørgaard, L.; Have, H. Crop-weed discrimination by line imaging spectroscopy. *J. Agric. Eng. Res.* **2000**, *75*, 389–400. [CrossRef]

13. Moshou, D.; Vrindts, E.; De Ketelaere, B.; De Baerdemaeker, J.; Ramon, H. A Neural Network based plant classifier. *Comput. Electron. Agric.* **2001**, *31*, 5–16. [CrossRef]

14. Córcoles, J.I.; Ortega, J.F.; Hernández, D.; Moreno, M.A. Estimation of leaf area index in onion (*Allium cepa* L.) using an unmanned aerial vehicle. *Biosyst. Eng.* **2013**, *115*, 31–42. [CrossRef]

15. Kavzoglu, T. Increasing the accuracy of neural network classification using refined training data. *Environ. Model. Softw.* **2009**, *24*, 850–858. [CrossRef]

16. Bishop, C.M. *Neural Networks for Pattern Recognition*, 1st ed.; Oxford University Press: New York, NY, USA, 1995; ISBN 0 19 8538 64 2 (Pbk).

17. MacKay, D.J.C. A practical Bayesian framework for back-propagation networks. *Neural Comput.* **1992**, *4*, 448–472. [CrossRef]

![sensors logo] *sensors*

MDPI

Article

Assessing White Wine Viscosity Variation Using Polarized Laser Speckle: A Promising Alternative to Wine Sensory Analysis

Christelle Abou Nader [1,*], Hadi Loutfi [1], Fabrice Pellen [2], Bernard Le Jeune [2], Guy Le Brun [2,*], Roger Lteif [3] and Marie Abboud [1,*]

[1] Physics Department, UR TVA, Faculty of Science, Saint Joseph University, B.P. 11-514-Riad El Solh, Beirut 1107 2050, Lebanon; hadi.loutfi@net.usj.edu.lb
[2] Laboratoire OPTIMAG (EA 938), Université de Bretagne Occidentale, 29238 Brest CEDEX 3, France; fabrice.pellen@univ-brest.fr (F.P.); bernard.lejeune@univ-brest.fr (B.L.J.)
[3] Chemistry Department, UR TVA, Faculty of Science, Saint Joseph University, B.P. 11-514-Riad El Solh, Beirut 1107 2050, Lebanon; roger.lteif@usj.edu.lb
* Correspondence: abounaderchristelle@gmail.com (C.A.N.); guy.lebrun@univ-brest.fr (G.L.B.); marie.abboud@usj.edu.lb (M.A.); Tel.: +961-1-421-375 (M.A.)

Academic Editor: Dimitrios Moshou
Received: 11 September 2017; Accepted: 10 October 2017; Published: 13 October 2017

Abstract: In this paper, we report measurements of wine viscosity, correlated to polarized laser speckle results. Experiments were performed on white wine samples produced with a single grape variety. Effects of the wine making cellar, the grape variety, and the vintage on wine Brix degree, alcohol content, viscosity, and speckle parameters are considered. We show that speckle parameters, namely, spatial contrast and speckle decorrelation time, as well as the inertia moment extracted from the temporal history speckle pattern, are mainly affected by the alcohol and sugar content and hence the wine viscosity. Principal component analysis revealed a high correlation between laser speckle results on the one hand and viscosity and Brix degree values on the other. As speckle analysis proved to be an efficient method of measuring the variation of the viscosity of white mono-variety wine, one can therefore consider it as an alternative method to wine sensory analysis.

Keywords: speckle; diffusion; scattering; biological sensing

1. Introduction

The wine industry is a major global economic hub. All stakeholders, including producers, need to improve controls of wine composition as well as its botanical and geographical origins [1,2], to improve the transparency of transactions and ultimately to prevent fraud. Even though novel methods for the assessment of wine quality and attributes are currently being considered [3–6], wine tasting through sensory analysis is still the most commonly used practice for the description of wine [7–10]. Two types of tasting are usually performed: (i) a horizontal one, where different bottles of wine produced within the same year (i.e., same vintage) are compared; and (ii) a vertical one, where a given wine is monitored while it undergoes aging.

The composition of the finished wine is a very complicated mixture of grape original components and those that were produced during the vinification of the wine, starting with fermentation as the first step, continuing along through the whole cellar operations, and ending up when the wine is aged in the bottle. Wine is composed of water (~84%), ethanol (~15%), and other minor components (~1%). Recently, it was shown that mouth feel, alcohol content, and other properties of wine can be determined using viscosity measurements [11–14]. In addition to ethanol, other components can have a significant effect on viscosity, mainly sugar and glycerol [15]. Mixtures of water and ethanol are more viscous than either liquid by itself, with the most viscous mixture occurring at 40% alcohol. At the typical 12% alcohol content of commercial wine and at 25 °C, the viscosity would be about 1.4 mPa.s vs. 0.89 mPa.s for pure water [15,16]. Sugar also contributes to the wine viscosity [11,12]. Glycerol is present at about one-tenth of the concentration of alcohol in most wines, this being higher for *botrytis*-affected grapes. While glycerol is very viscous by itself, it does not contribute significantly to the actual viscosity at the concentrations found in most wines [17]. Wine viscosity can thus be considered as a relevant criterion for the classification of wine, since viscosity values vary with parameters including grape variety, vintage, and temperature [11–14]. Information related to wine astringency, fluidity, dewatering, thickness, alcoholic, and sugar content can thus be obtained. Oenologists usually estimate the alcoholic and sugar content of wine during fermentation by measuring the density of the wine must, before the alcoholic fermentation and at the end of the fermentation process. Unfortunately, the predicted values of alcohol content often overestimate the real values in commercial wines [18]. One can also qualitatively assess alcohol and sugar content by twirling a glass of wine [19]. In addition to the release of the wine's aromatic compounds in the air, twirling causes a portion of the wine to adhere to the walls of the glass. Evaporation-driven Marangoni flows near the meniscus of water–alcohol mixtures drive liquid upward, forming a thin liquid film, and a rim forms near the moving contact line. Eventually, the rim undergoes an instability, forming drops that roll back into bulk reservoir, forming the so-called "wine tears" or "wine legs." When the wine is more viscous, the number of droplets increases, and the "tears" take longer time to form.

As the viscosity of the wine is of the order of only a few mPa.s, and low viscosity variations are measured when the terroir, grape variety, vintage, or wine temperature are considered [11–14], a sensitive and yet efficient method to determine this relative variation is thus needed. Since a polarized laser speckle method showed a high sensitivity for the measurement of small variations in viscosity [20], we employed it in our study to measure the viscosity of mono-variety white wines, and hence to characterize commercial wines. Speckle is produced when coherent light is scattered off an illuminated diffusing medium. Speckle pattern results from the interference of light scattered by the assembly of many optical inhomogeneities distributed randomly in space, such as particles in suspension [21]. In our work, we evaluate and compare various approaches of the polarized speckle field. Speckle patterns are analyzed by evaluating the spatial contrast [20,22], the temporal correlation coefficient [23], and the temporal history of the speckle pattern (THSP) [23–25].

In Section 2, we describe the wine samples and the speckle experimental setup. We also present the optical parameters extracted from the speckle patterns and other standard experiments. Experimental results are displayed in Section 3, along with the principal component analysis (PCA) that highlights the correlation between laser speckle results on one hand, and viscosity and Brix degree values on the other. Conclusions are drawn and future perspectives are considered in the last section.

2. Materials and Methods

2.1. White Wine Samples

A dozen bottles of commercial mono-variety white wine, i.e. each wine is produced using one grape variety, were chosen for this study. Several grape varieties (Chardonnay, Merwah, Pinot Gris and Sauvignon Blanc), vintages (2004, 2007, 2009, and 2012), and wine-making cellars (Château Bybline, Château Clos Saint Thomas, Château Florentine, Château Khoury, and Domaine Wardy) were considered. After the bottles of wine provided by the Lebanese wine making cellars were received, bottles were stored at room temperature (20 °C). For each wine bottle, and hence for each wine sample, we performed speckle, viscosity, alcohol content, and Brix degree measurements under 20 °C, within 4 to 5 h after the wine bottle was opened.

2.2. Speckle Experimental Setup and Parameters

The speckle experimental setup is illustrated in Figure 1. A volume of 0.3 mL of polystyrene microspheres (Polysciences Inc., Warrington, PA, USA), with an average diameter $\Phi = 0.22$ μm and a refractive index of 1.59 suspended in water with an initial concentration of 0.00453 mg/mL, was added to a volume of 1.7 mL of white wine in a quartz cell. In fact, total attenuation coefficients (scattering and absorption coefficients), for all considered white wine samples, ranged from 10^{-3} to 2.10^{-2} mm^{-1} as measured using a Beer-Lambert experimental setup for collimated transmission [21]. Such scattering characteristics generated a backscattered signal that is too weak. In order to increase the scattered signal, and in order to allow the measurement of the speckle field with a sufficient signal-to-noise ratio, extrinsic scattering particles were added. The addition of microspheres to white wine made it possible to increase scattering and to obtain a total attenuation coefficient between 3 and 4 mm^{-1}. This addition yielded an enhancement of the light diffused signal in low scattering media and hence allowed speckle measurements. This operation did not affect the viscosity of the wine sample. In fact, the volumic fraction of added microspheres in the total volume of the sample was approximately 10^{-6}. Hence, one can reasonably consider that the viscosity of wine samples with added microspheres is equal to that of the wine samples [26]. A 15 mW He–Ne laser emitting at 632.8 nm, delivering a 1 mm wide linear polarized beam at I_0/e^2, where I_0 is the maximum laser intensity, with a coherence length of about 20 cm, illuminated the quartz cell. Backscattered light was captured by a high-speed recording Complementary Metal Oxide Semiconductor (CMOS) camera (MotionBLITZ EoSens mini, pixel size of 14 μm × 14 μm with a global electronic shutter) with an exposure time of 0.3 ms, appropriately chosen given the samples' viscosity of only a few mPa.s, and a rate of 2130 fps. A polarizer and an analyzer were used in order to select the linear parallel or perpendicular scattered light with respect to the incident beam. Two quarter-wave plates were used to generate a circular polarization of incident light and to ensure a collection of circular backscattered light. The angle θ between the laser-sample axis and the sample-camera axis was equal to 20° in order to avoid specular reflection on the quartz cell. The speckle images were acquired using linear parallel (LP) and circular crossed (CC) light polarization configurations. The choice of these polarizations was made given the samples' diffusing nature, in order to allow enough backscattered light intensity to reach the camera, which was set at a relatively short exposure time and in order to take into account both surface and volume diffusion, respectively [21,27].

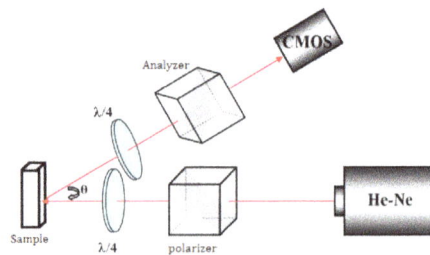

Figure 1. Schematic view of the speckle experiment setup. λ/4 is a quarter-wave plate.

Analysis of the speckle patterns, produced by each sample, was performed while considering a spatial and a temporal approach.

We undertake the spatial analysis by calculating the contrast C of a single speckle image [20,22]. The contrast is defined as the ratio between the intensity standard deviation σ and the mean intensity of the speckle pattern image, as follows:

$$C = \frac{\sigma_I}{\langle I \rangle} = \frac{\sqrt{\langle I \rangle^2 - \langle I \rangle^2}}{\langle I \rangle}. \tag{1}$$

For the temporal approach, speckle images were analyzed by computing the speckle intensity correlation on one hand [23], and extracting the temporal history of the speckle pattern (THSP) matrix on the other [24,25].

For each sample, we acquired a temporal series of speckle images with a frame rate of 2130 fps. We computed the speckle intensity correlation $C(\tau, \Delta x, \Delta y)$ with the mean of a cross correlation analysis between the first speckle image intensity $I(t_0, x_0, y_0)$ and the k-th speckle image intensity $I(t, x, y)$ using

$$C(\tau, \Delta x, \Delta y) = \frac{\langle I(t_0, x_0, y_0) I(t, x, y) \rangle - \langle I(t_0, x_0, y_0) I(t, x, y) \rangle}{\left\{ \left[\langle I^2(t_0, x_0, y_0) \rangle - \langle I(t_0, x_0, y_0) \rangle^2 \right] \left[\langle I^2(t, x, y) \rangle - \langle I(t, x, y) \rangle^2 \right] \right\}^{\frac{1}{2}}} \tag{2}$$

where $\tau = t - t_0$ is the time, $\Delta x = x - x_0$, $\Delta y = y - y_0$, $(\Delta x, \Delta y) = (m\Delta r, n\Delta r)$, m and n are the pixel positions in the image, and Δr is the pixel size [23]. An example of the temporal correlation curve is plotted in Figure 2a as a function of time, giving an idea of the analyzed sample activity evolution. Taking into account the analyzed light polarizations (linear parallel and cross circular) in a backscattered geometry [27], as well as the low scattering nature of our samples [20], we regard the collected speckle fields to be mainly generated by photons that have undergone a simple diffusion. Therefore, an exponential fit of the correlation curve $ae^{b\tau} + d$, where $b = -1/\tau_c$, and $d = 1 - a$, allows for an estimation of the speckle decorrelation time constant τ_c. As shown in previous studies [20,22], the speckle decorrelation time constant τ_c is linearly correlated to the samples' viscosity.

Furthermore, as part of the temporal analysis, the THSP matrix was extracted from a series of speckle images. After recording N continuous speckle images, the middle column is chosen from each one. These columns are then positioned side by side, according to their chronological order, to create a new image called the THSP. The objective of the THSP is to study intensity variations in a horizontal direction. As each row of the matrix represented the intensity variation of a speckle column with time, the THSP could monitor horizontal time fluctuations [24]. If a slow activity is present in the diffusing sample, the speckle temporal variations associated with the sample are slow and the THSP shows stretching shapes. If the phenomenon is very dynamic and the grain size is almost equal to camera pixel size, the THSP is similar to an ordinary speckle image. An example of a THSP image for a white wine sample is illustrated in Figure 2b. The THSP allows us to calculate the inertia moment *IM* defined by

$$IM = \sum M_{i,j}(i - j)^2 \tag{3}$$

where $M_{i,j}$ is the (i, j) pixel gray value of the normalized co-occurrence matrix (MCOM) [24,28]. This value indicates the number of occurrences of a certain intensity value i, followed instantly by an intensity value j. *IM* is useful for estimating a sample's total activity. When the activity of a sample decreases, its variations become slower, and *IM* thus decreases [23,28,29].

As samples scattering and absorption properties might influence the parameters calculated via speckle images [20,30], scattering and absorption coefficients were measured for all wine samples considered in our study. The results show almost the same scattering and absorption coefficients for all samples, allowing us to attribute changes in optical parameters exclusively to changes in the samples' viscosity.

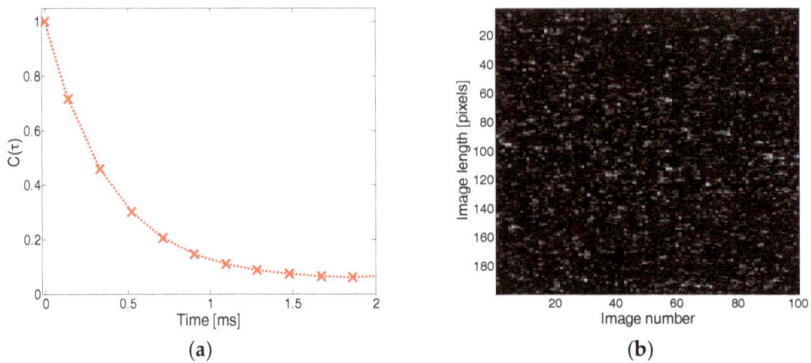

Figure 2. (**a**) Temporal correlation curve $C(\tau)$ and (**b**) the temporal history of the speckle pattern (THSP) image for a 1.7 mL white wine sample with 0.3 mL of added microspheres with a diameter of 0.22 μm.

2.3. Viscosity Measurement

The viscosity of the wine samples was measured through a calibrated glass capillary Ostwald viscometer. Liquid, with no-added microspheres, was drawn into the upper bulb via suction, and then allowed to flow down through the capillary into the lower bulb. The time δt taken for the level of the liquid to pass between two marks, one above and one below the upper bulb (see Figure 3), and measured with a stopwatch, is proportional to the viscosity η. Consequently, the viscosity η is deduced using the following formula:

$$\eta = K\,\delta t\,\rho \tag{4}$$

where δt is the measured time, K is an instrumental constant, and ρ is the liquid density. This measurement was repeated four times for each of the considered samples, and the average value is given in the results section.

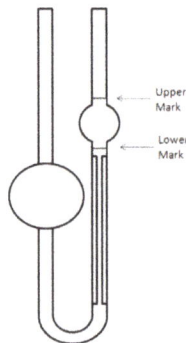

Figure 3. Sketch of an Oswald viscometer showing the upper and the lower marks.

2.4. Alcohol Content and Brix Degree Measurements

Alcohol content was determined using a density meter (Anton-Paar BMA 4500M) after the wine was distilled. Values correspond to the dissolved alcohol in wine.

The Brix degree was measured using a portable refractometer (FG-103) that permits measurements in the range 0–32%.

To account for repeatability, all measurements were performed four times, and mean values are given in the results section.

3. Results and Discussion

3.1. Viscosity Measurement Using Speckle

We adopted various approaches of the polarized speckle field for the optical measurements in the wine samples with different grape varieties, vintages, and wine making cellars. Both speckle parameters, the spatial contrast C, and the decorrelation time τ_c in LP and CC polarization configurations, were considered for the viscosity variations of the different samples of white wine.

Figure 4 shows the evolution of the contrast C as a function of the measured viscosity. For both of the considered light polarization configurations (LP and CC), we found that the contrast values increases when the viscosity of the wine increases. As shown in other studies, when the sample viscosity increases, the Brownian motion of the suspended particles in the sample shows down, yielding a decrease in the speckle intensity fluctuations. As a result, the contrast C increases. This behavior has been previously observed, and the contrast of the speckle images has been correlated to samples viscosities [20]. Furthermore, we found that contrast values in the LP light configuration are slightly larger than those obtained in the CC light polarization configuration. This is consistent with the polarization memory effect of backscattered light by a turbid medium, which depends on the size of the scatterers [27,31]. In the case of small particles with respect to an optical wavelength, as in our situation, scattering is equally likely in all directions. Linear polarization is hence maintained for longer than circular one [27]. Backscattering does not affect the linear character of linearly polarized light, but it reverses the helicity of circularly polarized light and randomizes it more rapidly. Contrarily, when the scattering medium contains large particles with respect to optical wavelength, it has been shown that for pathlengths of the order of a transport mean free path, linearly polarized light is randomized, whereas circularly polarized light preserves its original polarization [32], yielding better speckle image contrast when circularly polarized light is used, as reported in a previous study [33].

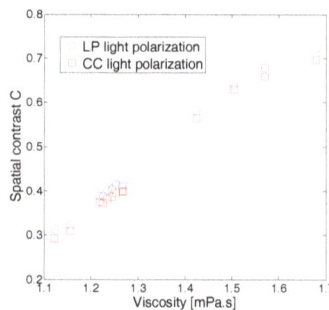

Figure 4. Variation of the speckle spatial contrast C as a function of white wine viscosity. Standard deviations are approximately ±0.001 mPa.s for the viscosity, and ±0.002 for the contrast C. Error bars are not displayed on the figures for the sake of clarity.

For the temporal approach, the decorrelation time τ_c is extracted using an exponential fit of the correlation coefficient curve as already mentioned. In fact, τ_c is directly proportional to the stirring motion of the particles in a solution [20,31,34]. In addition, it was proven that the speckle decorrelation time increases linearly with the viscosity of the solution [20]. As presented in Figure 5, results similar to those obtained using the spatial approach were also found for different wine samples having different viscosities: as the wine viscosity increases, speckle decorrelation time also increases. Furthermore, τ_c values in the LP light configuration are slightly larger than the results obtained in CC light polarization configuration. Here, and similarly to the spatial contrast approach, since linear polarization is preserved for longer paths than the circular one because of the small dimension of optical scatterers, larger decorrelation times are measured when linearly polarized light is used [27,32].

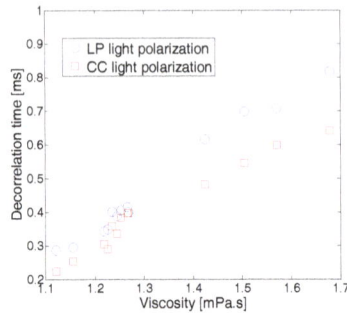

Figure 5. Variation of the speckle decorrelation time τ_c as a function of white wine viscosity. Symbols correspond to experimental values and lines correspond to linear fits. Standard deviations are approximately ± 0.001 mPa.s for the viscosity, and ± 0.006 ms for the decorrelation time τ. Error bars are not displayed on the figures for the sake of clarity.

We discuss in the following the efficiencies of both procedures, particularly in terms of sensitivity and the duration of the measurement. When wine viscosity covers the range [1.1, 1.7] mPa.s, and therefore over a relative variation of about 55%, the contrast shows an increase by a factor 2.3 in both the LP and the CC polarization configurations. Meanwhile, the decorrelation time varies by a factor of 3, ranging from 0.22 to 0.65 ms and from 0.28 to 0.82 ms in CC and LP polarization configurations, respectively. This shows the efficiency and sensitivity of the temporal approach in detecting viscosity variations, as small as they may be. From another perspective, if one compares the duration of each type of measurement, one can clearly see that the spatial approach presents a substantial advantage with respect to the speckle temporal approach and viscosity measurement using Ostwald viscometer. In fact, the latter methods are time-consuming, whereas the spatial contrast approach can be performed in less than 1 s since it relies on the acquisition of a single speckle image, allowing for the evaluation of the viscosity value at a glance [20].

3.2. The Effect of the Wine Making Cellar on Wine Viscosity

In order to evaluate the effect of the winery domain on white wine characteristics, we considered samples obtained from different wine making cellars, Château Florentine, Domaine Wardy, and Château Clos Saint Thomas, with identical vintage (2012) and grape variety (Chardonnay).

Results indicate that the spatial contrast C and the decorrelation time τ_c values for LP and CC polarization configurations are substantially identical with a standard deviation in the range of ± 0.002 for C and ± 0.006 ms for τ_c. In addition, the three samples have exactly the same Brix degree value (6.90), while viscosity and alcohol content are identical with standard deviations in the range of ± 0.001 mPa.s and 0.3%, respectively. We can deduce that the viscosity, the alcohol content, and the Brix degree are not affected by the wine making cellar as long as the same grape variety and vintage are considered. This may be justified by the fact that similar production processes are applied while producing commercial wines.

3.3. The Effect of the Grape Variety on Wine Viscosity

As the wine making cellar has little effect on wine viscosity, after setting 2012 as the year of production, we considered four different grape varieties, Sauvignon Blanc, Merwah, Chardonnay, and Pinot Gris, to determine their effect on wine viscosity.

As shown in Table 1a, the viscosity changes from one grape variety to another. This variation is consistent with analysis carried out in a previous study on French mono-variety wines [14]. In fact, each grape variety has its own features during fermentation. This behavior is demonstrated by the fluctuation in Brix degree values and alcohol content [11–13].

Table 1. Results of the effect of the grape variety on white wine (**a**) viscosity, Brix degree, and alcohol content and (**b**) speckle results in linear parallel (LP) and circular crossed (CC) light polarization configurations. Standard deviations are approximately ±0.001 mPa.s for the viscosity, ±0.01 for the Brix degree, ±0.02% for the alcohol content, ±0.002 for the contrast C, ±0.006 ms for the decorrelation time τ_C, and ±170 for the inertia moment *IM*.

Winery	Clos Saint Thomas	Bybline	Wardy	Florentine	Clos Saint Thomas	Khoury
			(a)			
grape variety	Sauvignon Blanc	Merwah	Chardonnay	Chardonnay	Chardonnay	Pinot Gris
viscosity (mPa.s)	1.225	1.253	1.266	1.268	1.262	1.505
Brix (°Brix)	6.95	6.50	6.90	6.90	6.90	9.50
alcohol content (%)	13.75	12.06	12.68	13.10	12.62	13.87
			(b)			
grape variety	Sauvignon Blanc	Merwah	Chardonnay	Chardonnay	Chardonnay	Pinot Gris
C_{LP}	0.3874	0.4154	0.4108	0.4170	0.4127	0.6408
C_{CC}	0.3715	0.3954	0.3987	0.3999	0.3976	0.6289
τ_{CLP} (ms)	0.3470	0.4050	0.4148	0.4000	0.4113	0.6973
τ_{CCC} (ms)	0.2895	0.3833	0.3967	0.3968	0.3973	0.5454
IM_{LP}	4333	3970	3876	3877	3860	1842
IM_{CC}	3549	3499	3281	3229	3269	1371

In addition to the measurements of the alcohol content and Brix degree, the speckle analysis results also show a variation of the contrast C, the speckle decorrelation time τ_c, and the inertia moment *IM* values (see results displayed in Table 1b). Since wine samples present low scattering and absorption coefficients, we can attribute changes in optical parameters to changes in the samples' viscosity. In our results, while wine viscosity increases from 1.2254 to 1.5046 mPa.s., the contrast of the speckle images rises significantly and the decorrelation time increases, as microspheres added to the wine tend to move slower [20,30]. Moreover, the results show a decrease in the values of *IM*. In fact, this speckle parameter can be linked to the sample's total speckle activity. When the viscosity of the sample becomes higher, the Brownian motion of microbeads added to the wine is slowed down, resulting in lower *IM* values [23,28,29].

Plotting the variation of each parameter as a function of viscosity would lead to a high number of similar graphs that would be too cumbersome if shown at once. Thus, for the sake of clarity, and in order to establish all possible correlations between experimental data, we used principal component analysis (PCA) as a statistical method. PCA helps reduce the number of variables and spots a relationship between them [35]. The calculus was made using the XLSTAT program. The speckle parameters (the contrast C, the inertia moment *IM*, and the decorrelation time τ_c) in both LP and CC light polarization configurations, as well as viscosity, Brix degree, and alcohol content were taken into account.

We present in Table 2 the correlation between the variables. A correlation is found between the viscosity and the Brix degree, with a Pearson correlation coefficient equal to 0.942. This correlation shows a connection between the viscosity and the sugar concentration in the wine. The correlations between inertia moment values and all other parameters, when taken one by one, are negative. This simply indicates that these parameters vary in opposite directions, while being highly correlated. Finally, high correlation values (with values of Pearson correlation coefficient ranging between 0.946 and 0.997) are obtained between viscosity and all speckle parameters (contrast and decorrelation time in LP and CC light polarization configurations).

The Pearson circle contains the different variables used in this study (Figure 6). The statistical calculation shows a first principal component F1 at 89.16% and a second principal component F2 at 6.90%. The F1 and F2 interrelation was set at 96.06%. In addition, the decorrelation time τ_c and the spatial contrast C in the two light polarization configurations (LP and CC) are highly correlated to the first component axis F1. However, the inertia moment IM is negatively correlated to τ_c and C, and hence F1. Indeed, when τ_c and C values increase because of slower dynamics in the sample, IM values decrease as previously stated. Finally, viscosity and Brix degree values are strongly correlated with F1, whereas alcohol content is distant from both F1 and F2. In fact, the Brix degree reflects the amount of residual sugars in the wine. When the latter increases, the viscosity also increases. However, the dissolved alcohol affects the wine viscosity to a lesser extent. This is in perfect agreement with previously reported results in white wine, where, across a large range of sugar concentrations, it was revealed that sugar drastically influences the viscosity, whereas ethanol has only a moderate effect [11,12].

Table 2. Pearson correlation coefficient matrix between values of speckle parameters, viscosity, Brix degree, and alcohol content measurements for all wine samples.

Variable	Viscosity	Brix Degree	Alcohol Content	C_{LP}	C_{CC}	τ_{CLP}	τ_{CCC}	IM_{LP}	IM_{CC}
Viscosity	1	–	–	–	–	–	–	–	–
Brix degree	0.942	1	–	–	–	–	–	–	–
Alcohol content	0.618	0.718	1	–	–	–	–	–	–
C_{LP}	0.996	0.958	0.625	1	–	–	–	–	–
C_{CC}	0.997	0.959	0.618	1	1	–	–	–	–
τ_{CLP}	0.994	0.942	0.623	0.994	0.992	1	–	–	–
τ_{CCC}	0.946	0.821	0.52	0.93	0.928	0.955	1	–	–
IM_{LP}	−0.935	−0.946	−0.685	−0.947	−0.941	−0.962	−0.897	1	–
IM_{CC}	−0.865	−0.917	−0.713	−0.88	−0.874	−0.9	−0.813	0.982	1

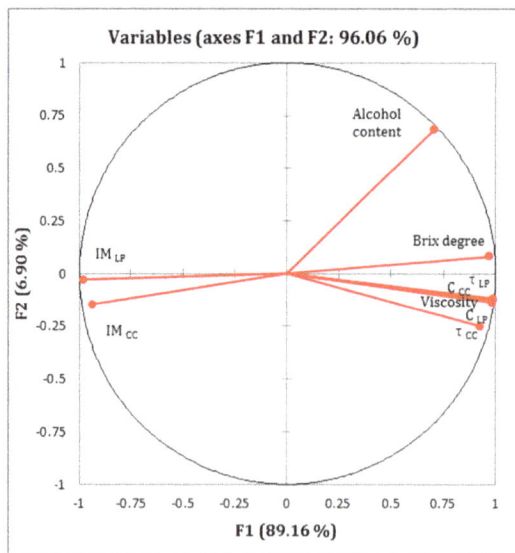

Figure 6. Pearson correlation circle representing the variables projected on the first principal component F1 and the second principal component F2 for all wine samples.

4. Conclusions

In this paper, we show that it is possible to estimate wine viscosity by using the polarized laser speckle method and by calculating either the spatial contrast of a speckle pattern, either the speckle decorrelation time between a series of speckle patterns. This study allowed us to conclude that wines

Sensors **2017**, *17*, 2340

produced from different wine-making cellars have the same viscosity and alcoholic content. This may be justified by the fact that similar production processes are applied while producing commercial wines. However, the variation of the grape variety has an important influence on the white wine viscosity. This variation is in agreement with previous studies on French white, red, and rosé wines [14]. Moreover, speckle spatial and temporal parameters proved to be sensitive to wine viscosity variations, as we showed that the spatial contrast and the speckle decorrelation time increase with the viscosity of the wine samples. Optical parameters (spatial contrast C, speckle decorrelation time τ_c, and inertia moment IM) were correlated to other parameters issued from standard measurements (viscosity, alcoholic content, and Brix degree) by means of a statistical approach. PCA showed high correlation levels between the different speckle parameters on one hand and the wine viscosity and Brix degree that can be linked to the sugar content on the other.

In the near future, we consider applying this study to red and rosé wines. Moreover, we envisage an extension of our study to the case of wine samples issued from a mix of different grape varieties. In the long term, we are confident that the polarized laser speckle method can be used to characterize grape varieties, to date wine, and even to detect wine alteration and, eventually, fraud. The polarized laser speckle method could also constitute an alternative to commonly used methods of testing wines, such as sensorial analysis, even though not all tasting-related aspects would not be considered.

Since causal links exist between the viscosity and the microstructure of biological fluids, one can similarly tackle other food-related applications within inspection activities [36]. For instance, in the quality control sector, the measurement of viscous properties can serve as an indirect tool for defining product consistency and the quality of various foodstuffs such as milk [37], ketchup [38], oil [39], and honey [40].

Acknowledgments: The authors acknowledge the Council for Research of Saint Joseph University for funding the project, and the Lebanese wine making cellars (Château Bybline, Château Clos Saint Thomas, Château Florentine, Château Khoury, and Domaine Wardy) for providing bottles of wine.

Author Contributions: All authors have made significant contributions to this article and participated actively to the conception and design of the experiments. C. Abou Nader, H. Loutfi, and M. Abboud conducted experiments and performed data analysis. All authors contributed to discussing the results and writing the manuscript.

Conflicts of Interest: The authors declare no conflict of interest.

References

1. Van Leeuwen, C.; Friant, P.; Choné, X.; Tregoat, O.; Koundouras, S.; Dubourdieu, D. Influence of climate, soil, and cultivar on terroir. *Am. J. Enol. Vitic.* **2004**, *55*, 207–217.
2. Bejjani, J.; Balaban, M.; Rizk, T. A sharper characterization of the geographical origin of Lebanese wines by a new interpretation of the hydrogen isotope ratios of ethanol. *Food Chem.* **2014**, *165*, 134–139. [CrossRef] [PubMed]
3. Chung, S.; San Park, T.; Hyun Park, S.; Yong Kim, J.; Park, S.; Son, D.; Min Bae, Y.; In Cho, S. Colorimetric sensor array for white wine tasting. *Sensors* **2015**, *15*, 18197–18208. [CrossRef] [PubMed]
4. Di Gennaro, S.F.; Matese, A.; Mancin, M.; Primicerio, J.; Palliotti, A. An open-source and low-cost monitoring system for precision enology. *Sensors* **2014**, *14*, 23388–23397. [CrossRef] [PubMed]
5. Aguilera, T.; Lozano, J.; Paredes, J.A.; Alvarez, F.J.; Suarez, J.I. Electronic nose based on independent component analysis combines with partial least squares and artificial neural networks for wine prediction. *Sensors* **2012**, *12*, 8055–8072. [CrossRef] [PubMed]
6. Gutierrez, M.; Llobera, A.; Ipatov, A.; Vila-Planas, J.; Minguez, A.; Demming, A.; Buttgenbach, S.; Capdevila, F.; Domingo, C.; Jimenez-Jorquera, C. Application of an E-tongue to the analysis of monovarietal and blends of white wines. *Sensors* **2011**, *11*, 4840–4857. [CrossRef] [PubMed]
7. Murray, J.M.; Delahunty, C.M.; Baxter, I.A. Descriptive sensory analysis: Past, present and future. *Food Res. Int.* **2001**, *34*, 461–471. [CrossRef]
8. Legin, A.; Rudnitskaya, A.; Lvova, L.; Vlasov, Yu.; Di Natale, C.; D'Amico, A. Evaluation of Italian wine by the electronic tongue: Recognition, quantitative analysis and correlation with human sensory perception. *Anal. Chim. Acta* **2003**, *484*, 33–44. [CrossRef]

9.   Cliff, M.A.; King, M.C.; Schlosser, J. Anthocyanin, phenolic composition, colour measurement and sensory analysis of BC commercial red wines. *Food Res. Int.* **2007**, *40*, 92–100. [CrossRef]

10. Lima Ferreira, M.; Amaral, B.; Salagoïty, M.H.; Lagrèze, C.; de Revel, G.; Médina, B. *Document sur L'analyse Sensorielle du vin. Partie I: Conditions Générales Pour la Réalisation de Tests D'analyse Sensorielle*; FV 1356; O.I.V.: Paris, France, 2016; 3–9.

11. Burns, D.J.W.; Noble, A.C. Evaluation of the separate contributions of viscosity and sweetness of sucrose to perceived viscosity, sweetness and bitterness of vermouth. *J. Texture Stud.* **1985**, *16*, 365–381. [CrossRef]

12. Nurgel, C.; Pickering, G. Contribution of glycerol, ethanol and sugar to the perception of viscosity and density elicited by model white wines. *J. Texture Stud.* **2005**, *36*, 303–323. [CrossRef]

13. Yanniotis, S.; Kotseridis, G.; Orfanidou, A.; Petraki, A. Effect of ethanol, dry extract and glycerol on the viscosity of wines. *J. Food Eng.* **2006**, *81*, 399–403. [CrossRef]

14. Siret, R.; Madieta, E.; Symonaux, R.; Jourjon, F. Mesures rhéologiques de la texture et de la viscosité des vins. Corrélations avec l'analyse sensorielle. In Proceedings of the 31st World Congress of Vine and Wine, 6th General Assembly of the O.I.V., Verona, Italy, 15–20 June 2008.

15. Margalit, Y. *Concepts in Wine Chemistry*, 3rd ed.; Board and Bench Publishing: San Francisco, CA, USA, 2016; pp. 152, 210–212.

16. Pickering, G.J.; Heatherbell, D.A.; Vanhanen, L.P.; Barnes, M.F. The effect of ethanol concentration on the temporal perception of viscosity and density in white wine. *Am. J. Enol. Vitic.* **1998**, *49*, 306–318.

17. Noble, A.C.; Bursick, G.F. The contribution of glycerol to perceived viscosity and sweetness in white wine. *Am. J. Enol. Vitic.* **1984**, *35*, 110–112.

18. Ribéreau-Gayon, P.; Dubourdieu, D.; Donèche, B.; Lonvaud, A. *Handbook of Enology Volume 1 The Microbiology of Wine and Vinifications*, 2nd ed.; Wiley & Sons: Hoboken, NJ, USA, 2006; Chapter 3; pp. 80–81.

19. Venerus, D.C.; Simavilla, D.N. Tears of wine: New insights on an old phenomenon. *Sci. Rep.* **2015**, *5*, 161–162. [CrossRef] [PubMed]

20. Abou Nader, C.; Pellen, F.; Roquefort, P.; Aubry, T.; Le Jeune, B.; Le Brun, G.; Abboud, M. Evaluation of low viscosity variations in fluids using temporal and spatial analysis of the speckle pattern. *Opt. Lett.* **2016**, *41*, 2521–2524. [CrossRef] [PubMed]

21. Abou Nader, C.; Nassif, R.; Pellen, F.; Le Jeune, B.; Le Brun, G.; Abboud, M. Influence of size, proportion, and absorption coefficient of spherical scatterers on the degree of light polarization and the grain size of speckle pattern. *Appl. Opt.* **2015**, *54*, 10369–10375. [CrossRef] [PubMed]

22. Fercher, A.F.; Briers, J.D. Flow visualisation by means of single-exposure speckle photography. *Opt. Commun.* **1981**, *37*, 326–330. [CrossRef]

23. Nassif, R.; Abou Nader, C.; Pellen, F.; Le Brun, G.; Abboud, M.; Le Jeune, B. Retrieving controlled motion parameters using two speckle pattern analysis techniques: Spatiotemporal correlation and the temporal history speckle pattern. *Appl. Opt.* **2013**, *52*, 7564–7569. [CrossRef] [PubMed]

24. Haralick, R.M.; Shanmugam, K.; Dinstein, I. Textural features for image classification. *IEEE Trans. Syst. Man Cybern.* **1973**, *3*, 610–621. [CrossRef]

25. Oulamara, A.; Tribillon, G.; Duvernoy, J. Biological activity measurements on botanical specimen surfaces using a temporal decorrelation effect of laser speckle. *J. Mod. Opt.* **1989**, *36*, 165–179. [CrossRef]

26. Einstein, A. Eine neue Bestimmung der Moleküldimensionen. *Ann. Phys.* **1906**, *324*, 289–306. [CrossRef]

27. Morgan, S.P.; Ridgway, M.E. Polarization properties of light backscattered from two layer scattering medium. *Opt. Exp.* **2000**, *7*, 395–402. [CrossRef]

28. Arizaga, R.; Trivi, M.; Rabal, H. Speckle time evolution characterization by the co-occurrence matrix analysis. *Opt. Laser Technol.* **1999**, *31*, 163–169. [CrossRef]

29. Nassif, R.; Abou Nader, C.; Afif, C.; Pellen, F.; Le Brun, G.; Le Jeune, B.; Abboud, M. Detection of golden apples' climacteric peak by laser biospeckle measurements. *Appl. Opt.* **2014**, *53*, 8276–8282. [CrossRef] [PubMed]

30. Hajjarian, Z.; Nadkarni, S.K. Correction of optical absorption and scattering variations in laser speckle rheology measurements. *Opt. Exp.* **2014**, *22*, 6349–6361. [CrossRef] [PubMed]

31. Piederriere, Y.; Boulevert, F.; Cariou, J.; Guern, Y.; Le Brun, G. Backscattered speckles size as a function of polarization: Influence of particle-size and concentration. *Opt. Exp.* **2005**, *13*, 5030–5039. [CrossRef]

32. MacKintosh, F.C.; Zhu, J.X.; Pine, D.J.; Weitz, D.A. Polarization memory of multiply scattered light. *Phys. Rev. B* **1989**, *40*, 9342–9345. [CrossRef]

33. Ni, X.; Alfano, R.R. Time-resolved backscattering of circularly and linearly polarized light in a turbid medium. *Opt. Lett.* **2004**, *29*, 2773–2775. [CrossRef] [PubMed]

34. Potanin, A.A. On the self-consistent calculations of the viscosity of colloidal dispersions. *J. Colloid Interface Sci.* **1993**, *156*, 143–152. [CrossRef]

35. Abdi, H.; Hervé, L. Principal component analysis. *WIREs Comput. Stat.* **2010**, *2*, 433–459. [CrossRef]

36. Brown, A. *Understanding Food: Principles and Preparation*, 5th ed.; Cengage Learning: Boston, MA, USA, 2015; Chapter 2; p. 26.

37. Fernández-Martín, F. Influence of temperature and composition on some physical properties of milk and milk concentrates. II. Viscosity. *J. Dairy Res.* **1972**, *39*, 75–82. [CrossRef]

38. McCarthy, K.L.; McCarthy, M.J. Relationship between in-line viscosity and Bostwick measurement during ketchup production. *J. Food Sci.* **2009**, *74*, E291. [CrossRef] [PubMed]

39. Bonnet, J.P.; Devesvre, L.; Artaud, J.; Moulin, P. Dynamic viscosity of olive oil as a function of composition and temperature: A first approach. *Eur. J. Lipid Sci. Technol.* **2011**, *113*, 1019–1025. [CrossRef]

40. El Sohaimya, S.A.; Masryb, S.H.D.; Shehataa, M.G. Physicochemical characteristics of honey from different origins. *Ann. Agric. Sci.* **2015**, *60*, 279–287. [CrossRef]

sensors

MDPI

Article

Evaluating Oilseed Biofuel Production Feasibility in California's San Joaquin Valley Using Geophysical and Remote Sensing Techniques

Dennis L. Corwin [1,*], Kevin Yemoto [2], Wes Clary [3], Gary Banuelos [4], Todd H. Skaggs [1], Scott M. Lesch [5] and Elia Scudiero [1]

[1]   USDA-ARS, U.S. Salinity Laboratory, 450 West Big Springs Road, Riverside, CA 92507, USA;
      Todd.Skaggs@ars.usda.gov (T.H.S.); Elia.Scudiero@ars.usda.gov (E.S.)
[2]   USDA-ARS, Water Management Systems Research, 2150 Center Ave., NRRC Building D,
      Fort Collins, CO 80526, USA; Kevin.Yemoto@ars.usda.gov
[3]   Department of Earth and Planetary Sciences, Northrop Hall, 221 Yale Blvd. NE, University of New Mexico,
      Albuquerque, NM 87131, USA; wesclary@unm.edu
[4]   USDA-ARS, Water Management Research Unit, San Joaquin Valley Agricultural Sciences Center, 9611 South
      Riverbend Ave., Parlier, CA 93648, USA; Gary.Banuelos@ars.usda.gov
[5]   Riverside Public Utilities—Resources Division, 3435 14th St., Riverside, CA 92501, USA;
      SLesch@riversideca.gov
[*]   Correspondence: Dennis.Corwin@ars.usda.gov; Tel.: +1-951-369-4819

Received: 2 September 2017; Accepted: 5 October 2017; Published: 14 October 2017

Abstract: Though more costly than petroleum-based fuels and a minor component of overall military fuel sources, biofuels are nonetheless strategically valuable to the military because of intentional reliance on multiple, reliable, secure fuel sources. Significant reduction in oilseed biofuel cost occurs when grown on marginally productive saline-sodic soils plentiful in California's San Joaquin Valley (SJV). The objective is to evaluate the feasibility of oilseed production on marginal soils in the SJV to support a 115 ML yr^{-1} biofuel conversion facility. The feasibility evaluation involves: (1) development of an Ida Gold mustard oilseed yield model for marginal soils; (2) identification of marginally productive soils; (3) development of a spatial database of edaphic factors influencing oilseed yield and (4) performance of Monte Carlo simulations showing potential biofuel production on marginally productive SJV soils. The model indicates oilseed yield is related to boron, salinity, leaching fraction, and water content at field capacity. Monte Carlo simulations for the entire SJV fit a shifted gamma probability density function: $Q = 68.986 + \text{gamma}(6.134, 5.285)$, where Q is biofuel production in ML yr^{-1}. The shifted gamma cumulative density function indicates a 0.15–0.17 probability of meeting the target biofuel-production level of 115 ML yr^{-1}, making adequate biofuel production unlikely.

Keywords: apparent soil electrical conductivity; EC$_a$-directed soil sampling; electromagnetic induction; proximal sensor; response surface sampling; salt tolerance; boron tolerance; soil mapping; soil salinity; spatial variability

1. Introduction

The United States is the world's largest consumer of crude oil, resulting in two major problems: (1) low energy security and (2) high greenhouse gas (GHG) emissions. To increase national energy independence and decrease GHG emissions, the U.S. Congress enacted the Energy Independence and Security Act (EISA) in 2007 (110. P.L. 140). The EISA aims to increase the production of clean renewable fuels within the USA. Biofuels, such as biodiesel and renewable jet fuel from oilseed crops, are an alternative to petroleum-based fuels [1]. As part of EISA, the Renewable Fuel Standard (RFS) mandates the production of 137 billion liters of biofuel annually by 2022.

1.1. U.S. Military's Need for Alternative Fuels

The Department of Defense is the leading consumer of fuel within the USA. The U.S. military consumes 23 billion liters of aviation fuel a year. The United States military desires a secure fuel source that is not threatened or controlled by world events. Biofuels developed from crops grown within the borders of the USA are a secure source of fuel uninfluenced by the same world events that affect petroleum-based fuels. To diversify their fuel sources the U.S. military set a goal of 5% of their yearly aviation fuel needs (1.15 billion liters per year) from biofuel.

Even though alternative fuels may not be price competitive, there is a long-term commitment by the military to use diversification in fuel sources as a means of reducing risk [2]. The military's reliance on alternative fuels is strategic to help to ensure their operational readiness by increasing the ability to use multiple reliable fuel sources, thereby reducing dependence on any single fuel source that would make military decisions vulnerable to foreign manipulation. Interest in biofuels stems (1) from their potential to improve U.S. energy security because they are from renewable domestic sources that are theoretically unlimited over time and (2) from their potential to reduce GHG emissions, which is largely dependent on how the biofuels are produced and what land-use or land-cover changes occur [3].

1.2. Need for a Feasibility Study of Biofuel Production on Marginal Land in the San Joaquin Valley

Technology is available to produce biofuels, such as biodiesel and renewable jet fuel, from oilseed crops. The common oilseed crops for temperate regions include canola (rapeseed), sunflower, soybean, flax, safflower, and mustard. Some oilseed crops, such as mustard, show considerable salt tolerance.

Because of its climate, which enables year-round crop growth, and its reasonably secure source of irrigation water from surface, ground, and/or degraded water sources, California's San Joaquin Valley (SJV) is an ideal agricultural area for a secure source of biofuel. In the SJV, biofuel feedstock that can grow on marginally productive soil of poor quality, particularly saline soils, is advantageous for cost reduction since marginal soils are usually fallow or produce yields too low to be profitable. The U.S. Department of Agriculture and the U.S. Navy's Office of Naval Research identified the SJV as a potentially strategic location for biofuel production to meet 10% of the yearly biofuel production of aviation fuel (115 ML yr^{-1}).

Marginally productive lands in California are an ample potential resource that can be used to great advantage to reduce cost. In the early 1980s, Backlund and Hoppes [4] estimated that the entire SJV had approximately 8.9×10^5 ha of marginally productive saline-sodic soil, much of which resided on the west side of the SJV (WSJV). Current estimates of salt-affected soil for the WSJV using National Resource Conservation Service's (NRCS) Soil Survey Geographic Database (SSURGO; https://websoilsurvey.nrcs.usda.gov/) are 3.6×10^5 ha. Recent estimates using satellite imagery place this even higher at 5.5×10^5 ha [5].

Regardless of the estimate that is accepted, there is extensive salt-affected soil within the SJV to grow salt-tolerant oilseed crops for conversion to biofuel that would not compete with food crops for land use. Salt and drought tolerant biofuel feedstock, such as mustard oilseed, has tremendous potential when grown on marginally productive salt-affected soils. Specially bred mustard varieties, such as Ida Gold mustard (*Sinapis alba* L.), are salt tolerant and produce reasonably high oil yields. Moreover, after the oil is pressed out from its seeds, the residual Ida Gold mustard seed meal can act as an effective biodegradable bioherbicide and can provide Se as a nutritional supplement in livestock feed. Se-enriched seed meal is a unique and extra cash-value product that can only be produced in the Se-laden soils of the WSJV.

Legislative mandates and incentives, volatility in oil prices, and advances in research and technology are driving the expectations of major increases in biofuel production as a viable alternative fuel source. USDOE and USDA [6] recommend the need for greater research to identify viable biofuel feedstock production and management systems to support bio-refineries at commercially viable capacities, i.e., 115 ML yr^{-1} or more. However, it is unknown if sufficient oilseed production could support a biofuel conversion facility of sufficient capacity to be economically viable, i.e., cost less than \$1 L^{-1}. Before considering an in-depth analysis of the economic viability and commercialization of biofuel, answers to basic questions are needed. The most fundamental question is, can sufficient biofuel feedstock be grown to support a biofuel conversion facility within agricultural regions of the USA, whether the Southwest, Midwest, or Southeast? All are potential agricultural areas proposed for biofuel production. More specifically, can Ida Gold mustard oilseed grow with sufficient yields on marginally productive salt-affected soils in the SJV to support a 115 ML yr^{-1} conversion facility?

1.3. Objective

It is the objective of this study (1) to formulate a crop yield model relating Ida Gold mustard oilseed yield to edaphic properties for the WSJV and (2) to use the crop yield model to predict the yield of Ida Gold mustard oilseed on salt-affected soils (i.e., soils with an EC$_e$ greater than 4 dS m^{-1}) for evaluating the feasibility of oilseed production on marginal soils to support a 115 ML yr^{-1} biofuel conversion facility in the SJV.

2. Materials and Methods

The feasibility evaluation involves four steps: (1) development of an Ida Gold mustard oilseed yield model for marginal SJV soils using apparent soil electrical conductivity (EC$_a$) directed soil sampling; (2) identification of marginally productive salt-affected soils for oilseed production in the SJV; (3) development of a spatial database of edaphic factors influencing mustard yield for the SJV derived from satellite imagery and the SSURGO database and (4) applying the Ida Gold mustard oilseed yield model on marginally productive soil for the SJV and performing Monte Carlo simulations to show the range and probability of potential biofuel production in the region. Figure 1 provides a flow chart showing the four steps and the flow of information.

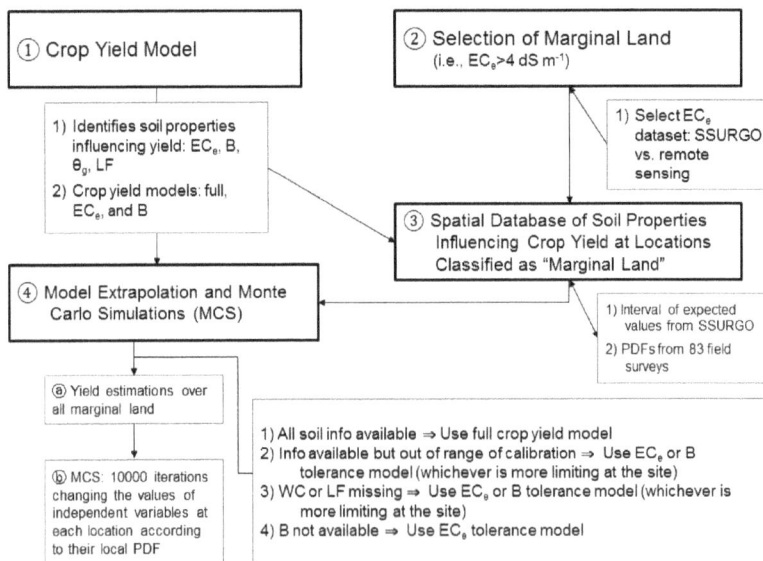

Figure 1. Flow chart of the feasibility evaluation of oilseed production on marginal soils in the San Joaquin Valley showing the four steps and flow of information.

2.1. Development of a Field-Based Ida Gold Mustard Oilseed Yield Model

A field experiment was conducted to identify the edaphic properties that influence oilseed yield of Ida Gold mustard. The approach of Corwin et al. [7] was followed. The approach uses geospatial electromagnetic induction measurements of EC_a to direct soil sampling for determining the soil properties influencing crop yield. The approach is based on the concept that if a correlation exists between crop yield and EC_a, then EC_a is measuring, either directly or indirectly, one or more soil properties that are influencing the crop yield. By conducting an EC_a survey to direct soil sampling, crop yield/soil sampling sites can be identified that provide a range of edaphic properties and their influence on yield.

The development of a field-based Ida Gold mustard oilseed yield model involves the following steps: (1) selection of an appropriate field that has a full range of edaphic properties that are suspected of influencing Ida Gold mustard oilseed yield; (2) conducting an intensive EC_a survey of the field; (3) identification of sites within the field where crop yield and soil core samples are taken that reflect the range and variability of edaphic influences on oilseed yield; (4) analysis of chemical and physical properties of the soil cores thought to influence yield and (5) statistical analysis and crop yield model formulation.

2.1.1. Study site description

The study site was a 16.2-ha field (latitude-longitude coordinates: 37°02′02.97″N, 120°47′31.56″) located west of Los Banos in Merced County, California on the WSJV (Figure 2). The site provided a range of soil properties thought to influence the yield of Ida Gold mustard oilseed. In particular, the field was characterized by a broad range of salinity and boron values. Salinity and boron are properties common to marginally productive soils in the SJV that are known to significantly influence Ida Gold mustard oilseed yield. The soil at the study site is a Britto clay loam. The soil taxonomic class is fine, smectitic, thermic Typic Natraqualfs. The Britto series consists of deep, very poorly drained soils with high concentrations of salt and alkali in the lower horizons. The soil ranges from moderately saline to strongly saline (8–16 dS m^{-1}). The texture in the top 55 cm is a clay loam and 0.55–1.55 m is a sandy clay loam. The parent material is alluvium derived from sedimentary rock. The mean annual precipitation is 25 cm.

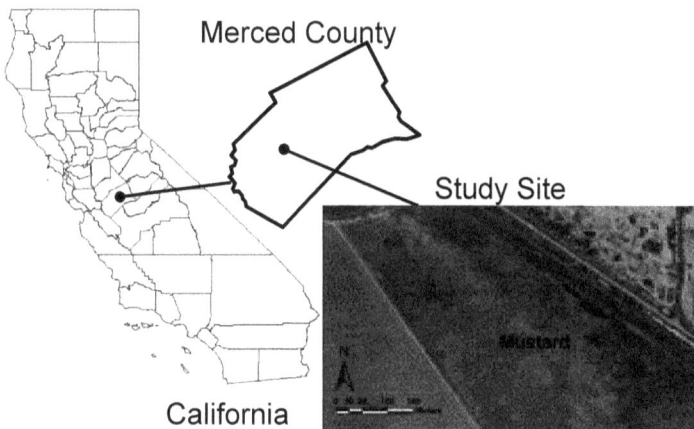

Figure 2. Map showing the location of the 16.2-ha field in California's Merced County.

2.1.2. Preliminary and Intensive Apparent Soil Electrical Conductivity Surveys

Preliminary and intensive EC_a surveys were conducted on 14 January and 28 January 2014, following a pre-plant irrigation to bring the water content in the root zone to field capacity. The methods and materials used in the EC_a surveys followed the protocols and guidelines outlined in Corwin and Lesch [8–10]. An EM38 Dual Dipole electrical conductivity meter (Geonics Ltd., Mississaugua, Ontario, Canada. Product identification is provided for the benefit of the reader and does not imply endorsement by USDA.) connected to a GPS and mounted on a non-metallic sled was used in the EC_a surveys. Geospatial EC_a measurements in the vertical (EM_v) and horizontal (EM_h) coil configurations were taken simultaneously every 3–5 m. Each EC_a measurement was geo-referenced using GPS. The GPS receiver accuracy had sub-meter accuracy. The preliminary EC_a survey determined whether the study site provided a sufficiently wide range in soil properties influencing Ida Gold mustard oilseed yield to meet the objective of formulating a crop yield model. The preliminary survey consisted of making six east-west traverses. From the geospatial EC_a data set, six locations were selected to take core samples (0–1.5 m depth increment), which were immediately analyzed for pH_e, saturation percentage (SP), B, and EC_e.

Following the preliminary EC_a survey, two separate intensive EC_a surveys were conducted. One EC_a survey for the entire 16.2 ha and another survey confined to the southeastern corner of the field. Figure 3 shows maps of a composite of the EC_a survey data for EM_v and EM_h. The reason for taking the two intensive EC_a surveys was because the preliminary cursory EC_a survey indicated the greatest variability in salinity over the range that would affect Ida Gold mustard oilseed yield was in the southeast corner of the field; consequently, a separate survey and crop yield/soil sampling were performed for the southeast corner. The intention was to provide a range of soil properties, particularly with regard to salinity and boron, which would influence Ida Gold mustard oilseed yield to varying degrees thereby providing the data to formulate a more robust statistical model of crop yield.

2.1.3. Soil and Ida Gold Mustard Oilseed Yield Sampling Design

Once the two intensive EC_a surveys were completed, ESAP software version 2.10R [11–13] was used to identify sites where crop yield and soil core samples were taken based on the spatial variation in the EC_a survey data. The ESAP software package uses a model-based sampling strategy (i.e., response surface sampling design) to identify sample site locations. The ESAP software identifies sites that characterize the range and variation in the geospatial EC_a measurements, reflecting the observed spatial variability in EC_a, while minimizing any clustering of the sample sites by maximizing the spatial uniformity of the sampling design across the study area. A detailed discussion of the application of the response surface sampling design using EC_a survey data is in Lesch et al. [12].

The sample design for the 16.2-ha field consisted of 20 sample site locations for the entire field and 20 sample site locations for the southeast corner as selected by ESAP (Figure 3). Soil cores were taken at the 40 sites with a Giddings rig at six depth increments: 0–0.15, 0.15–0.30, 0.30–0.60, 0.60–0.90, 0.90–1.20, and 1.20–1.50 m. Duplicate soil samples were taken at eight sample site locations within 1 m of the original core to establish local-scale variability as explained in Corwin and Scudiero [14]. All soil samples were bagged in zip-lock bags and stored in an ice chest until refrigerated. A total of 288 soil samples were taken (6 depths at each site, 40 sites, and 8 duplicate sites). The depth to the water table was recorded as <1.5 m or >1.5 m.

At each of the 40 sample-site locations biomass and Ida Gold mustard oilseed yield were determined by hand within a 1 m^2 area where each soil sample location was the centroid of the 1 m^2 plant sample area. The biomass and oilseed yield were collected on 28–29 May 2014. Six of the 40 sample site locations had no Ida Gold mustard oilseed yield. All subsequent referral to yield is with respect to oilseed yield.

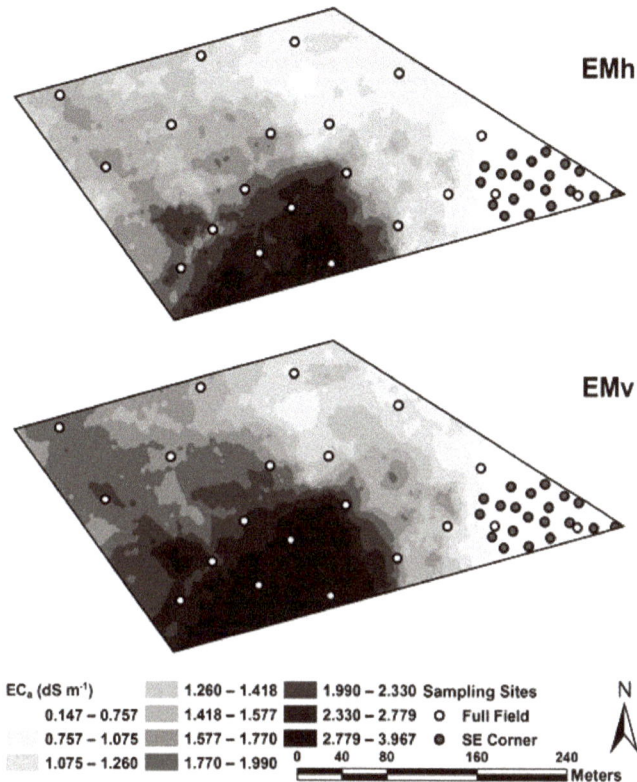

Santos Farms, Los Banos, CA
EC$_a$, Combined Surveys, January 28, 2014

Figure 3. Maps of the apparent soil electrical conductivity (EC$_a$) surveys taken with electromagnetic induction in the horizontal (EM$_h$) and vertical (EM$_v$) coil configurations, combining entire (full) 16.2-ha field and southeast (SE) corner surveys. Soil core sample sites are indicated by the clear and filled-in circles. The clear circles are soil sample sites selected from the full field EC$_a$ survey and the filled-in circles are from the SE corner EC$_a$ survey.

2.1.4. Soil Chemical and Physical Analyses

In the field, a subsample (100–300 g) of each soil core sample was taken for soil moisture determination. The subsamples were weighed in the field to minimize error due to moisture evaporation. The subsamples were subsequently oven-dried at 110°C for 24 h and weighed again to determine θ_g. Saturation pastes were prepared for all 288 soil samples and saturation paste extracts were obtained following the procedure of Rhoades [15]. The saturation extracts were analyzed for the following properties: EC$_e$, SP, pH$_e$, 5 major anions (Cl$^-$, HCO$_3^-$, PO$_4^{3-}$, NO$_3^-$, SO$_4^{2-}$), 4 major cations (Na$^+$, K$^+$, Ca^{2+}, Mg^{2+}), B, and sodium adsorption ratio (SAR). The chemical analysis procedures followed were those found in Sparks [16]. Leaching fraction (LF), defined as the ratio of the quantity of water draining past the root zone to that infiltrated into the soil's surface, was estimated using two techniques: (1) the ratio of the EM$_h$ EC$_a$ divided by EM$_v$ EC$_a$ and (2) the ratio of Cl concentration in the irrigation water and Cl concentration at 1.2–1.5 m. The LF reflects the excess water applied to translocate salts from the root zone. Each property selected for analysis had the potential to influence Ida Gold mustard oilseed yield.

2.1.5. Statistical Analysis and Ida Gold Oilseed Yield Model Formulations

Simple correlations were determined between yield and the edaphic properties of θ_g, EC_e, SP, pH_e, Cl^-, HCO_3^-, PO_4^{3-}, NO_3^-, SO_4^{2-}, Na^+, K^+, Ca^{2+}, Mg^{2+}, B, SAR, and LF. Correlations between the edaphic properties and EC_a and between oilseed yield and EC_a were determined. Scatter plots of yield vs. individual soil-related properties were also obtained.

The correlation analyses and scatter plots served as a basis for the development of the yield model for Ida Gold mustard oilseed. The correlations were useful to determine what properties were likely significant in influencing oilseed yield, while the scatter plots helped to determine the general form of the oilseed yield model. The oilseed yield model was developed using (spatial) multiple regression techniques [17]. The edaphic properties were the regressor or independent variables, and yield was the response or dependent variable. Backward variable selection was used to screen out the clearly nonsignificant edaphic properties with t-score values below 1.8. This predictor screening helped to filter out any multicollinearity in the regressor variables. Statistical data analyses were performed on the individual depth increment data (i.e., 0–0.15, 0.15–0.3, 0.3–0.6, 0.6–0.9, 0.9–1.2, and 1.2–1.5 m) and composite depth increment data (i.e., 0–0.15, 0–0.3, 0–0.6, 0–0.9, 0–1.2, and 0–1.5 m). Multiple regression modeling was performed on the composite depth increment data, and the increment characterized by the best goodness-of-fit was retained for further analyses.

In some instances, sufficient input data for the Ida Gold mustard oilseed yield model did not exist or fell outside the range of data that were used to develop the oilseed yield model. In those instances, the two-piece linear salt tolerance model of Maas and Hoffman [18] was used. For soil salinities exceeding the threshold salinity level, relative yield is estimated by the following equation:

$$Y_r = 100 - b\left(\overline{EC_e} - a\right) \tag{1}$$

where Y_r is the relative crop yield, a is the salinity threshold (dS m^{-1}), b is the slope expressed in yield decrement percentage per dS m^{-1}, and $\overline{EC_e}$ is the mean electrical conductivity of the saturation extract for the root zone (dS m^{-1}). Maas and Hoffman [18] proposed that crop salt tolerance was represented by two linear lines, one a tolerance plateau with a slope of zero and the other shown in Equation (1) as a salinity dependent line whose slope was the yield reduction per unit increase in salinity. The point where both lines intersect is the salinity threshold, which represents the maximum soil salinity that does not reduce yield. The parameters a and b were determined from a compilation of salt tolerance data available for Ida Gold mustard oilseed for the SJV, including the data collected within this study and the work of Maas [19] and Grieve et al. [20]. If salinity was not limiting yield, then the B level determined the oilseed yield when B data were available. The three-piece trace element tolerance model presented in Page et al. [21] and first suggested by Burton et al. [22] was used. The salt and B tolerance data to develop these models were obtained from the 40 soil core and oilseed yield sample sites identified from EC_a-directed soil sampling and from 10 supplemental sites. The locations of the supplemental sites were from a transect covering a range of yields. Salt tolerance data were from those sites varying in oilseed yield where all soil properties were optimal except salinity, which varied over a wide range, while B tolerance data were from those sites where all soil properties were optimal except B, which varied over a wide range. If no salinity and B data were available, then all properties were considered optimal for yield.

2.2. Identification of Salt-Affected Soils for Oilseed Production in the SJV

To assess the potential production of Ida Gold oilseed in the SJV, the yield model was applied over all salt-affected soils in the valley. Bohn et al. [23] defines salt-affected soils as soils with a root-zone salinity (i.e., EC_e) above 4 dS m^{-1}. Above 4 dS m^{-1} very sensitive, sensitive, and moderately salt-sensitive crops will show yield decrements. From 4–8 dS m^{-1} the yields of many crops are restricted, from 8–16 dS m^{-1} only salt tolerant crops yield satisfactorily, and above 16 dS m^{-1} only a few very tolerant crops produce satisfactory yields [24].

One means of identifying salt-affected soils in the SJV is the use of SSURGO. However, the accuracy and reliability of soil salinity in SSURGO is dubious because salinity is a spatially and temporally variable soil property influenced by crop and irrigation management strategies. Recent NRCS reports (e.g., [25]), provided salinity estimations only for non-irrigated farmland because the influence of irrigation on soil salinity cannot be accounted for using traditional soil survey protocols. Therefore, an evaluation of SSURGO's accuracy with respect to soil salinity is needed. To evaluate the accuracy of SSURGO, salinity assessment surveys were performed on 22 agricultural fields (total area: 542 ha) scattered throughout the WSJV. In 2013, intensive EC_a surveys conducted at the 22 fields, collected 41,779 EC_a readings at an average density of 175 readings ha^{-1}. Simultaneous EM_h and EM_v measurements of EC_a were taken. Across the 22 fields, 267 soil-sampling locations were identified using ESAP. Soil cores were taken to a depth of 1.2 m, representing the root-zone depth. Details of the EC_a survey and soil sampling are in Scudiero et al. [26].

Soil samples were analyzed for salinity (EC_e; dS m^{-1}), gravimetric water content (θ_g; g g^{-1}), pH$_e$, and saturation percentage (SP) using procedures presented in *Methods of Soil Analysis Part 3* [16] and *Part 4* [27]. The SSURGO salinity data for the 22 fields were compared to both root-zone EC_e hard and EC_a soft data. The hard data (i.e., laboratory measurements of EC_e of the 0–1.2 m soil samples) were averaged for each field. The soft data (i.e., geospatial measurements of EC_a) were calibrated to the EC_e data using spatial linear regression models [17] with an overall R^2 = 0.93 by Scudiero et al. [26].

A map of EC_e using the soft data and EC_e-EC_a calibration prepared for each field established spatial patterns for comparison to SSURGO map units. If the SSURGO database proved unreliable, then the regional-scale salinity assessment approach developed by Lobell et al. [28], which combines EC_a-directed soil sampling with satellite imagery, would be used. This work has already been completed and published by Scudiero et al. [29] for the WSJV.

2.3. Development of a Spatial Database of Edaphic Properties Influencing Ida Gold Mustard Yield for the SJV

The most extensive spatial database of edaphic properties in the USA is the SSURGO Database. Collection of the information in SSURGO occurred by walking over the land and observing variation in the soil and vegetation to delineate map units. Characterization of soil properties within map units occurred by the collection of numerous soil samples and their analysis in the laboratory. Occasionally observation trenches are dug to characterize the horizonation. Soil maps outline areas called map units, which describe soils and other components that have unique properties, interpretations, and productivity. Collection of the information occurred at scales ranging from 1:12,000 to 1:63,360. The soils maps are intended for natural resource planning and management. Examples of information available from the database include available water capacity, texture, pH, electrical conductivity, and frequency of flooding; yields for cropland, woodland, rangeland, and pastureland; and limitations affecting recreational development, building site development, and other engineering uses. SSURGO map data can be viewed in the Web Soil Survey or downloaded in ESRI® Shapefile format. Attribute data can be downloaded in text format. For the marginal soils of the SJV, information for water capacity, texture, pH, and frequency of flooding; yields for cropland, woodland, rangeland, and pastureland; and limitations affecting recreational development, building development, and other engineering uses was obtained through SSURGO.

To supplement the SSURGO data an extensive spatial database of quantitative soils information (i.e., EC_e, pH$_e$, saturation percentage, B, available water content, LF) for the SJV exists. The supplemental data set, collected over a period of 25 years by Corwin and colleagues, is a compilation of data that appeared in publications by Bourgault et al. [30], Corwin [31], Corwin and Lesch [8–10,32], Corwin et al. [7,33–36], Lesch and Corwin [17], Lesch et al. [37–39], Loague et al. [40], Rhoades et al. [41], Sanden et al. [42], and Scudiero et al. [26,29,43]. The supplemental data set consisted of edaphic property data from 83 fields within the SJV ranging in spatial extent from 0.4 to 65 ha with from 6 to 72 sample sites within a field. Soil samples were collected at 0.3-m increments to a minimum depth of 1.2 m and occasionally to 1.5 and 1.8 m. The supplemental data were used to determine frequency distributions (i.e., histograms), averages, ranges, and standard deviations for those properties found to influence Ida Gold mustard oilseed yield in the SJV. From this statistical information probability density functions (PDFs) were developed for LF, θ_g, and B for the composite depth increment of 0–1.2 m, which was found to be the root zone depth for Ida Gold mustard oilseed. In general, SSURGO provides water content and B ranges associated with soil type. The PDFs, defined within the ranges of water content and B provided by SSURGO, were used as input for Monte Carlo simulations with the oilseed yield model. In instances where ranges of water content or B were not given in SSURGO, then B was assumed optimal and water content was estimated from a pedotransfer function using SSURGO texture data.

The use of degraded soil is crucial to driving down the cost of biofuel production in the SJV. Subsequently, the reliability of spatial salinity data for the SJV was of paramount importance for identifying salt-affected soils and for model input data. There were concerns regarding the reliability of the salinity ranges provided in SSURGO for the root zone due to anthropogenic influences (e.g., leaching of salts due to irrigation); consequently, as discussed in detail in Section 2.2 an evaluation of the reliability of SSURGO root-zone salinity was conducted by comparison to salinity ground-truth measurements of 22 fields in the WSJV presented by Scudiero et al. [26]. If SSURGO salinity in the root zone proved unreliable, then salinity predictions from the regional-scale soil salinity model of Scudiero et al. [29] were used. Scudiero et al. [29] found that as salinity increased, the prediction error increased, and quantified the error within salinity categories (i.e., 0–2, 2–4, 4–8, 8–16, >16 dS m^{-1}). To incorporate this uncertainty into the Monte Carlo simulations, PDFs were established for each salinity category. The PDFs were defined by the average residuals and standard deviation of the residuals between observed salinities from the ground-truth salinity measurements of Scudiero et al. [26] and predicted salinities from the regional-scale soil salinity model of Scudiero et al. [29].

Leaching fraction is a difficult edaphic property to obtain and is not found in SSURGO. The supplemental data set of Corwin and colleagues was used to determine the frequency distribution, average, range, and standard deviation of LF for the SJV. Leaching fraction was determined from a ratio of EM_h EC_a to EM_v EC_a and from a ratio of Cl in irrigation water to Cl concentration below the root zone). Only those LFs where EM and Cl ratios agreed to within 5% were used.

2.4. Feasibility of Biofuel Production for the SJV: Application of the Yield Model and Monte Carlo Simulations

Monte Carlo simulations with the Ida Gold mustard oilseed yield model were performed with 10,000 iterations to provide a range of potential yields (kg ha^{-1}) and probability of those yields, which are easily converted to L ha^{-1} of biofuel. This was done first for the WSJV and then for the entire SJV. The mean, median, standard deviation, skewness, and kurtosis of the Monte Carlo simulation distribution were calculated to characterize quantitatively the PDF and subsequently derive the cumulative density function (CDF) of biofuel production. Once the CDF is known, the probability (and thereby feasibility) of oilseed production in the SJV to support sufficiently a conversion facility becomes evident.

Sufficient input data was not always available for each field in the WSJV or SJV or sometimes the input data was outside the range of data used to develop the crop yield model for Ida Gold mustard oilseed. For these instances, an alternative model was needed (see Figure 1). When a complete set of input data was available at a field location, then the full crop yield model was used. For instances where input data was available but was outside the range of data from which the crop yield model was formulated, then either the EC_e (i.e., Equation (1)) or B tolerance model was used, whichever was more limiting at the site. For instances where insufficient input data existed and only EC_e and B input data were available, then either the EC_e (i.e., Equation (1)) or B tolerance model was used, whichever was more limiting at the site. If insufficient input data existed and only EC_e input data was available, then the EC_e tolerance model (Equation (1)) was used.

3. Results and Discussion

The preliminary EC_a survey revealed that the greatest variation in salinity over the range that would result in oilseed yield decrements was in the southeast corner of the study site; consequently, two intensive EC_a surveys were conducted with separate soil sampling designs for each survey. One EC_a survey and associated soil sampling covered the full field and the other focused on the SE corner. Figure 3 shows maps of the horizontal coil configuration (EM_h) and vertical coil configuration (EM_v) EC_a surveys. The combined soil sampling designs (i.e., full field and southeast corner) provided a full range of soil properties and oilseed yields to build a robust Ida Gold mustard oilseed yield model. The only property potentially influencing Ida Gold mustard oilseed yield that did not vary significantly at the study site was texture. In general, the fine-textured soils (mainly loams and clay loams), which predominate the WSJV, do not vary to a significant extent because the soil is a consequence of lacustrine deposits and of fine-grained alluvium material originating from the Coastal Ranges [44,45].

Geospatial EC_a measurements, both EM_h and EM_v, were higher on the west side of the 16.2-ha field than on the east side (see Figure 3). The lowest EC_a measurements were in the southeast corner, where EC_a ranged from 0.15 to 0.76 dS m^{-1}. Over this range of EC_a the yield was found to vary the greatest (see Figure 4) and therefore provided the most useful data for oilseed yield model formulation. The range of EC_a over the entire 16.2 ha was 0.15 to 3.97 dS m^{-1}. Oilseed yield over the EC_a range of 0.76 to 3.97 dS m^{-1} tended to be low. Yield significantly correlated to both EM_h and EM_v EC_a, with correlation coefficients of 0.68 and 0.51, respectively. The higher correlation of EM_h EC_a to oilseed yield suggests that the root zone for Ida Gold mustard was around 1 m since the EM_h measurement penetrates to a depth of approximately 1 m, while EM_v penetrates to approximately 1.5 m. The fact that EC_a and oilseed yield correlated indicates that EC_a must be measuring a soil property or properties that influence oilseed yield; therefore, the response surface sampling design will successfully map the property or properties [9].

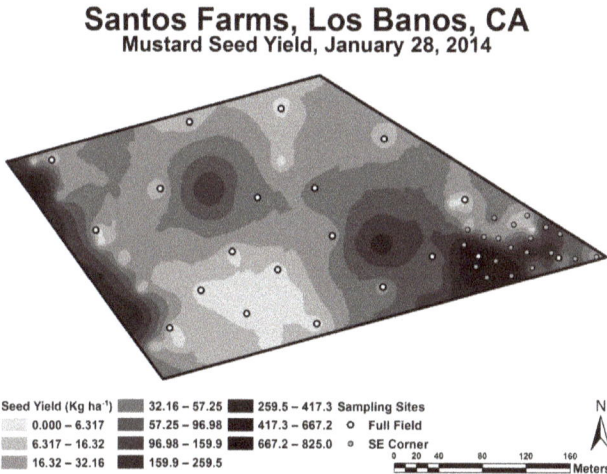

Figure 4. Map of Ida Gold mustard oilseed yield. Oilseed yield sample sites (1 m^2 sample area) are indicated by the clear and filled-in circles. The clear circles are oilseed yield sample sites selected from the full field EC_a survey and the filled-in circles are from the SE corner EC_a survey.

The measured edaphic properties that were felt to potentially influence Ida Gold mustard oilseed yield included: EC_e, θ_g, SP, pH_e, trace elements (B, Se, As, Mo), major cations (Na$^+$, K$^+$, Ca^{2+}, Mg^{2+}), major anions (Cl$^-$, HCO$_3^-$, PO$_4^{3-}$, NO$_3^-$, SO$_4^{2-}$), SAR, micro-elevation, depth to groundwater, and groundwater EC. Table 1 is a summary by depth of these edaphic properties, except for micro-elevation, depth to groundwater, and groundwater EC.

Table 1. Mean and range statistics of soil edaphic properties for the depths within the root zone of 0–0.15, 0.15–0.3, 0.3–0.6, 0.6–0.9, 0.9–1.2, and 1.2–1.5 m of the combined full-field (20 soil sample locations) and southeast-corner (20 soil sample locations) surveys.

Soil Property [†]	No. of Sample Sites	Mean	Minimum	Maximum	Range	Standard Deviation	Standard Error	Coefficient of Variation	Skewness	Kurtosis
Depth 0–0.15 m										
θ_g, kg kg⁻¹	40	0.10	0.01	0.27	0.26	0.05	0.01	47.9	1.11	1.86
SP	40	53.41	44.26	65.45	21.19	5.49	0.87	10.3	0.40 ‡	−0.81 ‡
EC_e, dS m⁻¹	40	10.29	2.54	33.00	30.46	7.74	1.22	75.2	0.92	0.08 ‡
pH_e	40	7.29	6.64	8.45	1.80	0.33	0.05	4.5	1.29	3.29
Cl^-, meq L⁻¹	40	64.07	12.67	204.95	192.28	57.52	9.09	89.8	0.98	−0.36 ‡
HCO_3^-, meq L⁻¹	40	2.91	1.28	7.59	6.31	1.54	0.24	52.8	1.32	1.41 ‡
PO_4^{3-}, meq L⁻¹	40	0.23	0.04	0.70	0.66	0.14	0.02	60.3	1.52	3.18
NO_3^-, meq L⁻¹	40	4.19	0.69	17.99	17.31	4.35	0.69	103.7	1.32	1.08 ‡
SO_4^{2-}, meq L⁻¹	40	51.76	7.20	343.94	336.74	61.99	9.80	119.8	3.03	12.12
SAR	40	6.53	2.26	40.33	38.07	7.73	1.22	118.4	2.90	9.39
Na^+, meq L⁻¹	40	10.97	5.21	48.07	42.86	8.20	1.30	74.7	2.80	10.06
K^+, meq L⁻¹	40	62.14	11.58	360.84	349.26	66.69	10.54	107.3	2.57	9.28
Ca^{2+}, meq L⁻¹	40	1.85	1.00	4.11	3.11	0.70	0.11	37.9	1.91	3.98
Mg^{2+}, meq L⁻¹	40	19.89	4.42	57.50	53.08	13.96	2.21	70.2	0.80	−0.25 ‡
B, mg L⁻¹	40	33.93	5.24	105.38	100.14	29.27	4.63	86.3	0.93	−0.35 ‡
Depth 0.15–0.3 m										
θ_g, kg kg⁻¹	40	0.13	0.03	0.24	0.21	0.06	0.01	43.1	0.24 ‡	−1.29 ‡
SP	40	53.86	39.02	63.83	24.81	6.47	1.02	12.0	−0.21 ‡	−0.67 ‡
EC_e, dS m⁻¹	40	9.50	2.01	33.70	31.69	7.73	1.22	81.4	1.14	0.93 ‡
pH_e	40	7.34	6.70	8.62	1.92	0.32	0.05	4.4	1.54	5.24
Cl^-, meq L⁻¹	40	42.42	8.04	132.50	124.46	39.31	6.22	92.7	1.03	−0.44 ‡
HCO_3^-, meq L⁻¹	40	2.33	0.78	5.01	4.24	1.12	0.18	48.2	0.84	−0.26 ‡
PO_4^{3-}, meq L⁻¹	40	0.20	0.00	0.93	0.93	0.15	0.02	78.2	2.93	12.69
NO_3^-, meq L⁻¹	40	2.24	0.58	6.26	5.68	1.65	0.26	73.7	0.86	−0.51 ‡
SO_4^{2-}, meq L⁻¹	40	71.28	6.93	370.34	363.41	78.73	12.45	110.5	1.89	4.25
SAR	40	10.71	1.99	58.32	56.34	15.06	2.38	140.6	2.33	4.60
Na^+, meq L⁻¹	40	12.58	5.15	48.78	43.63	9.73	1.54	77.3	2.20	5.03
K^+, meq L⁻¹	40	66.41	10.03	360.01	349.98	72.63	11.48	109.4	2.12	5.73
Ca^{2+}, meq L⁻¹	40	1.20	0.43	3.80	3.37	0.56	0.09	46.8	2.80	11.32
Mg^{2+}, meq L⁻¹	40	15.63	3.02	36.93	33.91	10.64	1.68	68.1	0.41 ‡	−1.23 ‡
B, mg L⁻¹	40	30.09	3.69	89.71	86.01	27.12	4.29	90.1	0.84	−0.70 ‡
Depth 0.3–0.6 m										
θ_g, kg kg⁻¹	40	0.18	0.12	0.25	0.13	0.04	0.01	23.4	0.04 ‡	−1.34 ‡
SP	40	54.85	44.90	63.64	18.74	4.99	0.79	9.1	−0.20 ‡	−0.77 ‡
EC_e, dS m⁻¹	40	12.64	1.28	47.00	45.72	11.49	1.82	90.9	1.45	1.69
pH_e	40	7.53	6.32	8.76	2.44	0.42	0.07	5.6	−0.10 ‡	1.97
Cl^-, meq L⁻¹	40	52.37	4.44	186.89	182.46	55.67	8.80	106.3	1.20	0.11 ‡
HCO_3^-, meq L⁻¹	40	1.71	0.67	3.83	3.16	0.58	0.09	33.8	1.66	4.61
PO_4^{3-}, meq L⁻¹	40	0.18	0.00	0.74	0.74	0.21	0.03	117.7	1.35	1.08 ‡
NO_3^-, meq L⁻¹	40	1.33	0.52	4.60	4.08	0.83	0.13	62.2	1.93	5.06
SO_4^{2-}, meq L⁻¹	40	121.15	4.58	643.29	638.71	140.95	22.29	116.3	2.10	4.59
SAR	40	19.04	1.83	103.37	101.54	25.51	4.03	133.9	2.13	3.85
Na^+, meq L⁻¹	40	17.59	5.69	69.43	63.74	15.58	2.46	88.6	2.26	4.66
K^+, meq L⁻¹	40	108.07	7.28	598.83	591.55	133.51	21.11	123.5	2.16	4.59
Ca^{2+}, meq L⁻¹	40	0.91	0.18	3.50	3.32	0.77	0.12	84.7	1.70	2.64
Mg^{2+}, meq L⁻¹	40	17.35	1.47	44.45	42.98	11.72	1.85	67.6	0.15 ‡	−0.93 ‡
B, mg L⁻¹	40	41.29	1.80	141.13	139.33	38.29	6.05	92.7	1.08	0.64 ‡

Table 1. *Cont.*

Soil Property †	No. of Sample Sites	Mean	Minimum	Maximum	Range	Standard Deviation	Standard Error	Coefficient of Variation	Skewness	Kurtosis
Depth 0.6–0.9 m										
θ_g, kg kg⁻¹	40	0.21	0.17	0.26	0.09	0.03	0.004	12.9	0.14 ‡	−1.38 ‡
SP	40	54.14	46.03	70.77	24.74	5.80	0.92	10.7	0.72 ‡	0.37 ‡
EC_e, dS m⁻¹	40	12.49	1.57	46.00	44.43	11.21	1.77	89.8	1.36	1.27 ‡
pH_e	40	7.74	6.80	8.93	2.13	0.44	0.07	5.6	−0.13 ‡	0.84 ‡
Cl⁻, meq L⁻¹	39	49.10	1.67	205.33	203.67	51.18	8.09	104.2	1.35	1.05 ‡
HCO₃⁻, meq L⁻¹	40	1.85	0.80	3.90	3.10	0.60	0.10	32.7	1.34	2.22
PO₄³⁻, meq L⁻¹	40	0.15	0.00	0.56	0.56	0.17	0.03	116.3	1.09	0.12 ‡
NO₃⁻, meq L⁻¹	40	0.66	0.20	1.74	1.54	0.34	0.05	52.0	1.53	2.38
SO₄²⁻, meq L⁻¹	40	117.17	7.11	504.59	497.48	135.06	21.35	115.3	1.75	2.32
SAR	40	18.94	1.99	80.32	78.33	22.26	3.52	117.5	1.76	2.10
Na⁺, meq L⁻¹	40	18.98	7.23	61.72	54.48	14.74	2.33	77.6	1.91	2.74
K⁺, meq L⁻¹	40	108.10	10.43	450.35	439.92	125.52	19.85	116.1	1.57	2.16
Ca²⁺, meq L⁻¹	40	0.81	0.14	3.15	3.01	0.75	0.12	92.0	1.57	2.05
Mg²⁺, meq L⁻¹	40	15.36	1.80	35.49	33.68	10.85	1.72	70.6	0.09 ‡	−1.64
B, mg L⁻¹	40	36.89	2.36	152.45	150.09	38.00	6.01	103.0	1.38	1.59
Depth 0.9–1.2 m										
θ_g, kg kg⁻¹	40	0.22	0.18	0.26	0.08	0.02	0.003	9.1	0.43 ‡	−0.74 ‡
SP	40	51.21	39.47	75.44	35.96	8.80	1.39	17.2	1.03	0.41 ‡
EC_e, dS m⁻¹	40	9.78	2.24	41.50	39.26	9.07	1.43	92.8	1.66	2.79
pH_e	40	7.90	7.10	8.30	1.21	0.31	0.05	3.9	−0.72 ‡	−0.31 ‡
Cl⁻, meq L⁻¹	40	33.06	4.44	163.47	159.03	33.83	5.35	102.3	2.07	5.24
HCO₃⁻, meq L⁻¹	40	2.15	1.10	3.88	2.77	0.78	0.12	36.3	0.48 ‡	−1.15 ‡
PO₄³⁻, meq L⁻¹	40	0.10	0.00	0.39	0.39	0.12	0.02	113.3	1.03	0.00 ‡
NO₃⁻, meq L⁻¹	39	0.46	0.15	1.30	1.16	0.26	0.04	55.5	1.82	3.72
SO₄²⁻, meq L⁻¹	40	90.62	10.68	418.43	407.76	102.12	16.15	112.7	1.68	2.37
SAR	40	14.18	2.20	66.34	64.14	17.39	2.75	122.6	1.68	1.75
Na⁺, meq L⁻¹	40	16.75	5.00	54.08	49.08	12.61	1.99	75.3	1.82	2.27
K⁺, meq L⁻¹	40	80.93	14.28	389.51	375.22	94.43	14.93	116.7	1.85	2.82
Ca²⁺, meq L⁻¹	40	0.60	0.12	2.12	2.00	0.51	0.08	85.4	1.39	1.31 ‡
Mg²⁺, meq L⁻¹	40	13.70	2.02	28.78	26.76	10.53	1.66	76.9	0.10 ‡	−1.89
B, mg L⁻¹	40	24.95	3.28	118.77	115.49	26.92	4.26	107.9	1.99	4.60
Depth 1.2–1.5 m										
θ_g, kg kg⁻¹	40	0.24	0.21	0.30	0.09	0.02	0.004	9.8	0.61 ‡	0.03 ‡
SP	40	48.13	36.59	67.57	30.98	8.61	1.36	17.9	0.85	−0.35 ‡
EC_e, dS m⁻¹	40	7.91	2.09	33.80	31.71	7.24	1.15	91.6	1.75	3.17
pH_e	40	7.98	7.39	8.34	0.95	0.27	0.04	3.4	−0.68 ‡	−0.67 ‡
Cl⁻, meq L⁻¹	40	22.92	3.39	117.63	114.23	24.12	3.81	105.2	2.55	7.36
HCO₃⁻, meq L⁻¹	40	2.18	0.97	3.86	2.88	0.75	0.12	34.6	0.47 ‡	−0.80 ‡
PO₄³⁻, meq L⁻¹	40	0.09	0.00	0.39	0.39	0.11	0.02	128.0	1.31	0.96 ‡
NO₃⁻, meq L⁻¹	40	0.36	0.14	1.04	0.90	0.21	0.03	56.7	1.65	2.78
SO₄²⁻, meq L⁻¹	40	72.13	9.72	336.59	326.88	78.87	12.47	109.3	1.56	2.11
SAR	40	10.72	1.72	50.15	48.43	13.78	2.18	128.5	1.69	1.69
Na⁺, meq L⁻¹	40	14.55	4.28	45.79	41.51	10.28	1.63	70.7	1.78	2.16
K⁺, meq L⁻¹	40	62.22	14.00	328.97	314.97	72.11	11.40	115.9	2.01	4.04
Ca²⁺, meq L⁻¹	40	0.49	0.10	1.71	1.61	0.43	0.07	87.1	1.29	0.83 ‡
Mg²⁺, meq L⁻¹	40	11.76	1.49	25.69	24.20	9.67	1.53	82.3	0.29 ‡	−1.88
B, mg L⁻¹	40	18.26	2.93	81.23	78.31	18.69	2.95	102.3	1.95	4.33

† Definitions: θ_g = gravimetric water content, SP = saturation percentage, EC_e = electrical conductivity of the saturation extract, SAR = sodium adsorption ratio. ‡ Significant. Skewness is significant if skewness divided by standard error of skewness > 2. Kurtosis is significant if kurtosis divided by stand error of kurtosis > 2.

Table 2. Soil edaphic property mean and range statistics of the composite depths of (a) 0–1.5 m and (b) 0–1.2 m for the combined full-field (20 soil sample locations) and southeast-corner (20 soil sample locations) surveys.

Soil Property [†]	No. of Sample Sites	Mean	Min.	Max.	Range	Standard Deviation	Standard Error	Coefficient of Variation	Skewness	Kurtosis
(a) Depth 0–1.5 m										
θ_g, kg kg⁻¹	40	0.18	0.14	0.24	0.10	0.03	0.005	16.1	0.26[‡]	−1.17[‡]
SP	40	52.97	43.46	64.56	21.10	4.93	0.78	9.3	0.10[‡]	−0.57[‡]
EC_e, dS m⁻¹	40	10.70	2.05	36.22	34.17	9.03	1.43	84.4	1.30	1.13[‡]
pH_e	40	7.63	7.18	8.53	1.35	0.27	0.04	3.6	0.68[‡]	1.80
Cl^-, meq L⁻¹	40	44.09	8.16	152.11	143.95	40.08	6.34	90.9	1.08	0.08[‡]
HCO_3^-, meq L⁻¹	40	2.12	1.18	3.40	2.22	0.60	0.09	28.4	0.30[‡]	−0.72[‡]
PO_4^{3-}, meq L⁻¹	40	0.16	0.01	0.58	0.57	0.14	0.02	91.5	1.27	1.03[‡]
NO_3^-, meq L⁻¹	40	1.40	0.48	4.52	4.04	0.96	0.15	68.8	1.22	1.18[‡]
SO_4^{2-}, meq L⁻¹	40	92.43	7.95	420.42	412.47	101.32	16.02	109.6	1.75	2.63
SAR	40	31.86	3.48	112.13	108.66	28.73	4.54	90.2	1.10	0.74[‡]
Na^+, meq L⁻¹	40	14.30	2.17	65.31	63.14	17.76	2.81	124.2	1.91	2.59
K^+, meq L⁻¹	40	15.73	6.86	54.73	47.86	12.08	1.91	76.8	2.04	3.37
Ca^{2+}, meq L⁻¹	40	85.14	11.56	398.95	387.39	95.56	15.11	112.2	1.85	2.88
Mg^{2+}, meq L⁻¹	40	0.94	0.32	3.00	2.68	0.58	0.09	61.8	1.83	3.66
B, mg L⁻¹	40	15.57	2.65	35.06	32.41	10.45	1.65	67.1	0.15[‡]	−1.63
(b) Depth 0–1.2 m										
θ_g, kg kg⁻¹	34[£]	0.19	0.12	0.25	0.13	0.05	0.01	24.9	−0.12[‡]	−1.60[‡]
SP	34[£]	53.23	45.83	61.61	15.77	4.01	0.69	7.5	0.29[‡]	−0.54[‡]
EC_e, dS m⁻¹	34[£]	9.97	1.84	29.97	28.13	6.29	1.08	63.0	0.99	1.57[‡]
pH_e	34[£]	7.74	7.06	8.74	1.68	0.33	0.06	4.2	0.46[‡]	1.50[‡]
Cl^-, mEq L⁻¹	34[£]	51.83	7.87	279.17	271.30	54.37	9.32	104.9	2.53	8.49
PO_4^{3-}, mEq L⁻¹	34[£]	0.19	0.00	1.11	1.11	0.24	0.04	126.4	2.01	4.89
SO_4^{2-}, mEq L⁻¹	34[£]	67.16	6.40	165.86	159.46	41.82	7.17	62.3	0.41[‡]	−0.36[‡]
SAR	34[£]	12.06	6.08	22.47	16.40	4.02	0.69	33.4	0.71[‡]	0.12[‡]
Na^+, mEq L⁻¹	34[£]	63.68	9.91	166.12	156.21	41.48	7.11	65.1	0.71[‡]	0.00[‡]
K^+, mEq L⁻¹	34[£]	0.92	0.39	3.91	3.51	0.60	0.10	65.1	4.05	19.62
Ca^{2+}, mEq L⁻¹	34[£]	17.65	2.68	53.90	51.22	11.34	1.95	64.3	0.87	1.64[‡]
Mg^{2+}, mEq L⁻¹	34[£]	33.94	3.23	112.97	109.75	25.44	4.36	75.0	0.95	1.27[‡]
B, mg L⁻¹	34[£]	10.03	2.14	24.24	22.10	6.06	1.04	60.4	0.60[‡]	−0.59[‡]
LF [§]	34[£]	0.27	0.08	0.61	0.53	0.15	0.02	53.1	0.61[‡]	−0.30

[†] Definitions: θ_g = gravimetric water content, SP = saturation percentage, EC_e = electrical conductivity of the saturation extract, SAR = sodium adsorption ratio, LF = leaching fraction. [‡] Significant. Skewness is significant if skewness divided by standard error of skewness > 2. Kurtosis is significant if kurtosis divided by stand error of kurtosis > 2. [§] Leaching fraction was determined by dividing the Cl^- concentration of the irrigation water by the Cl^- concentration of the saturation extract at the 1.2–1.5 m depth increment at each site where the oilseed yield was greater than zero. [£] Sites where oilseed yield was greater than zero. These sites were used for Ida Gold mustard oilseed yield model development.

Table 1 reveals patterns in the field-wide soil profile. Field-wide average soil salinity (EC_e) increases with depth up to the 0.3–0.6, levels off at the 0.3–0.6 and 0.6–0.9 m depth increments, and then decreases with depth. Saturation percentage (SP) is reasonably constant over depth ranging from means of 48.13% at 1.2–1.5 m to 54.85% at 0.3–0.6 m, indicating uniform texture through the soil profile. Gravimetric water content (θ_g) at field capacity increases with depth from 0.10 kg kg^{-1} at 0–0.15 m to 0.24 kg kg^{-1} at 1.2–1.5 m. Boron levels tend to be lower below 0.3–0.6 m. pH increases with depth from 7.29 at 0–0.15 m to 7.98 at 1.2–1.5 m. SAR increases with depth from 6.53 at 0–0.15 m to 19.04 at 0.3–0.6 m, then decreases to 10.72 at 1.2–1.5 m.

Table 2a presents mean and range statistics, standard deviation, standard error, coefficient of variation, skewness, and kurtosis for the composite 0–1.5 m depth. The highest coefficients of variation (CVs) are for EC_e, various cations and anions (e.g., Na^+, Ca^{2+}, SO_4^{2-}, PO_4^-, and Cl^-), and SAR, while the lowest CVs are for pH_e, SP, and θ_g. All edaphic properties in Table 2a are positively skewed. Most properties show a positive kurtosis except θ_g, SP, HCO_3^-, and B. The range, minimum, and maximum of the edaphic properties in Table 2a are of particular interest because they confirm that the study site is well suited for developing a crop yield model based on edaphic properties since a wide range of edaphic conditions influencing oilseed yield are present. For instance, EC_e, pH_e, B, and SAR cover broad ranges from low to very high. For the composite depth of 0–1.5 m, EC_e ranged from 2.05 to 36.22 dS m^{-1}, pH_e ranged from 7.18 to 8.53, B ranged from 2.65 to 35.06 mg L^{-1}, and SAR ranged from 3.48 to 112.13. The SP is the only soil property that is narrow in range, nonetheless it reflects a texture that is typical of the WSJV.

3.1. Ida Gold Mustard Oilseed Yield Models for Marginal SJV Soils

Exploratory statistical analyses revealed that Ida Gold mustard oilseed yield was most significantly correlated to individual edaphic properties for the top 1.2 m. Table 2b presents mean and range statistics, standard deviation, standard error, coefficient of variation, skewness, and kurtosis for the composite 0–1.2 m depth. Table 3 presents simple correlations between edaphic properties and both EC_a and oilseed yield for the 0–1.2 m depth increment. The edaphic properties most significantly correlated to EC_a include θ_g, EC_e, B, SAR, LF (determined by the ratio EM_h EC_a /EM_v EC_a), and SP. The edaphic properties most significantly correlated to oilseed yield include θ_g, EC_e, B, LF, and SP.

Table 3. Correlation coefficients between edaphic properties and both EC_a and oilseed yield that are significantly correlated. [†]

Edaphic Property [‡]	EC_a	Oilseed Yield
θ_g	0.73 **	0.46 **
EC_e	0.98 **	−0.41 **
Boron	0.88 **	−0.32 *
pH_e	−0.09	0.28
SAR	0.87 **	−0.30
LF	0.80 **	0.55 **
SP	0.45 **	−0.45 **

[†] Using 34 locations where yield > 0. [‡] Averaged over 0–1.2 m. Definitions: θ_g = gravimetric water content, SP = saturation percentage, EC_e = electrical conductivity of the saturation extract, SAR = sodium adsorption ratio. * Significant at $p < 0.05$ level. ** Significant at $p < 0.01$ level.

Corwin and Lesch [9] indicate that the depth increment or composite depth increment associated with the best-fitting yield model (i.e., highest R^2) reflects the root zone of the crop. The top 1.2 m resulted in the most statistically significant and best-fit Ida Gold mustard oilseed yield model (see Equation (3)). Consequently, the 0–1.2 m soil interval was taken to represent the root zone of Ida Gold mustard at the study site. Subsequently, all data presented and discussed are with respect to the 0–1.2 m composite depth or to individual depth increments that lie within the composite depth of 0–1.2 m.

Exploratory statistical analyses revealed that θ_g, EC_e, B, and LF were the most influential edaphic properties on oilseed yield. Scatter plots of these properties vs oilseed yield indicated quadratic relationships of EC_e and B to oilseed yield and linear relationships of LF and θ_g to yield. Based on the initial exploratory correlation and multiple linear regression analysis, the following regression model structure was proposed to describe edaphic property effects on oilseed yield:

$$Y = \beta_0 + \beta_1(B) + \beta_2(B)^2 + \beta_3(EC_e) + \beta_4(EC_e)^2 + \beta_5(LF) + \beta_6(\theta_g) + \varepsilon \tag{2}$$

where Y is the Ida Gold mustard oilseed yield (kg ha^{-1}); B is boron concentration (mg L^{-1}); EC_e is electrical conductivity of the saturation extract (dS m^{-1}); LF is the leaching fraction; θ_g is the gravimetric water content (kg kg^{-1}); β_0, β_1, β_2, . . . , β_6 are the regression model parameters, and ε is the random error component, initially assumed to be normally distributed and spatially independent. Ordinary least squares (OLS) regression techniques resulted in a fitted regression equation with a $R^2 = 0.89$ and adjusted $R^2 = 0.78$. Adjusting for spatial autocorrelation using the maximum-likelihood approach resulted in the following Ida Gold mustard oilseed yield model (Equation (3)):

$$Y = 146.4(B) - 18.3(B)^2 + 83.0(EC_e) - 6.1(EC_e)^2 + 1301.0(LF) + 319.8(\theta_g) + 30.1 \tag{3}$$

Equation (3) represents the most parsimonious and robust model for marginally productive salt-affected soils of the WSJV. Any locations where no oilseed yield was obtained were not used in the model development. The LF and θ_g parameters are highly significant at or near the 0.01 level, and the EC_e (linear and quadratic) and B (linear and quadratic) parameter estimates are significant at or near the 0.05 level. The LF and θ_g parameters are both positive, implying that the yield increased as either LF or θ_g increased, which is physically sound since increased leaching reduces osmotic stress and increased water content increases the plant-available water reducing matric stress. The positive linear and negative quadratic EC_e terms imply that the yield increased at low EC_e up to a point of maximum yield with respect to EC_e and decreased beyond the maximum. The point of maximum yield with respect to EC_e was calculated by setting the first partial derivative of the fitted regression to zero with respect to EC_e, which resulted in a value of 6.8 dS m^{-1}. Similarly, the positive linear and negative quadratic B terms imply that the yield increased under low B up to a point of maximum yield with respect to B and decreased beyond the maximum. The point of maximum yield with respect to B was 4 mg L^{-1}.

In those instances where sufficient input data for Equation (3) did not exist or fell outside the range of data that was used to develop the model, the two-piece linear salt tolerance model (i.e., Equation (1)) of Maas and Hoffman [18] was used to predict oilseed yield. Salt tolerance data at the study site established a salinity threshold of 8.3 dS m^{-1}, which is the term a in Equation (1), and a yield decrement slope of 17%, which is the term b in Equation (1) (Figure 5a). The salinity threshold of 8.3 dS m^{-1} corresponds reasonably well with the salinity of maximum oilseed yield of 6.8 dS m^{-1} in Equation (3). Equation (1) established the upper limit of the salinity range of salt-affected soils that would grow Ida Gold mustard oilseed. A 17% yield decrement for each 1 dS m^{-1} increase in root-zone soil salinity beyond 8.3 dS m^{-1} resulted in no oilseed yield above 14.3 dS m^{-1}. It is important to note that the salinity range of 4–14.3 dS m^{-1} was not necessarily economically viable. Once the feasibility of reaching the 115 ML yr^{-1} goal is established, then the maximum yield decrement that is economically viable could be determined. Viability would take into account other economically relevant factors such as (1) selling the seed meal remaining, after the oilseed has been pressed to extract the oil, as Se-enriched meal for livestock or as a herbicide used in organic agriculture and (2) incorporation of gasification to generate power to run the oil press.

The two-piece linear salt tolerance model (i.e., Equation (1)) of Maas and Hoffman [18] was not the best model to fit the data as seen in Figure 5a. A quadratic model (Figure 5b) actually fits the data best:

$$Y = 74.0 + 254.6\ EC_e - 18.8\ EC_e^2 \quad (R^2 = 0.87) \tag{4}$$

By taking the derivative of Equation (4) with respect to EC_e and setting it equal to 0, EC_e corresponds to the peak mustard oilseed yield, which is 6.8 dS m^{-1}. This is equal to the EC_e producing the maximum yield in Equation (3). Similarly, a quadratic relationship between yield and EC_e was also found by Corwin et al. [7] for cotton seed yield. The explanation for a quadratic relationship between EC_e and yield of cotton seed and mustard oilseed is that when the plant is osmotically stressed it puts energy into the development of the reproductive part of the plant rather than vegetative tissue; consequently, as EC_e increases up to the salinity threshold biomass yield of both crops remains constant or may decrease, while mustard oilseed and cotton seed yield steadily increase. After the salinity threshold is reached then oilseed and cotton seed yield and the biomass of both crops decrease as salinity increases. This quadratic relationship indicates that more breeding research is needed to obtain new cultivars of mustard that have their peaks of potential yield at different ranges of salinity to maximize production in the SJV.

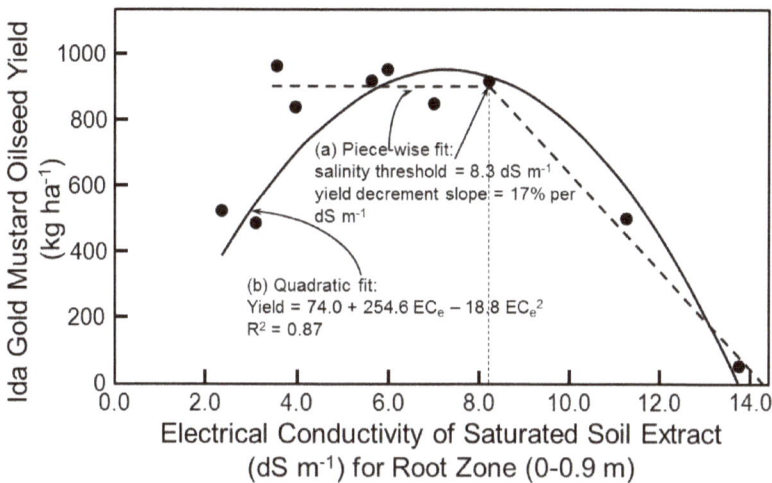

Figure 5. Salt tolerance models for Ida Gold mustard oilseed: (**a**) two-piece linear salt tolerance model (thick dashed line) of Maas and Hoffman [18] and (**b**) quadratic salt tolerance model (solid line). Thin dashed line indicates the salinity threshold of the two-piece linear salt tolerance model.

Boron tolerance studies at the field site showed an excellent fit of the three-piece linear model with an optimum range for B of 4.2–8 mg L^{-1} (Figure 6a). Below 4.2 mg L^{-1} of B and above 8 mg L^{-1} the yield of Ida Gold mustard oilseed drops, where B < 4.2 mg L^{-1} is a B deficiency and B > 8 mg L^{-1} is a B toxicity. Above 8 mg L^{-1} the drop was 28% for every 1 mg L^{-1} increase in B with no yield occurring above 11.6 mg L^{-1} of B. A quadratic model (Equation (5)) produced a comparable fit to the B tolerance data (Figure 6b):

$$Y = 555.2B - 42.4B^2 - 418.0 \quad (R^2 = 0.90) \tag{5}$$

By taking the derivative of Equation (5) with respect to B and setting it equal to 0, B corresponds to the peak oilseed yield, which is 6.5 mg L^{-1}.

The piece-wise linear models of Figures 5a and 6a for salinity and B, respectively, are traditional models found throughout the literature. The piece-wise linear models were used in the Monte Carlo simulations. However, the quadratic models provided slightly better fits to the data; consequently, a second set of Monte Carlo simulations was performed with the salinity and B quadratic models, Equations (4) and (5), respectively.

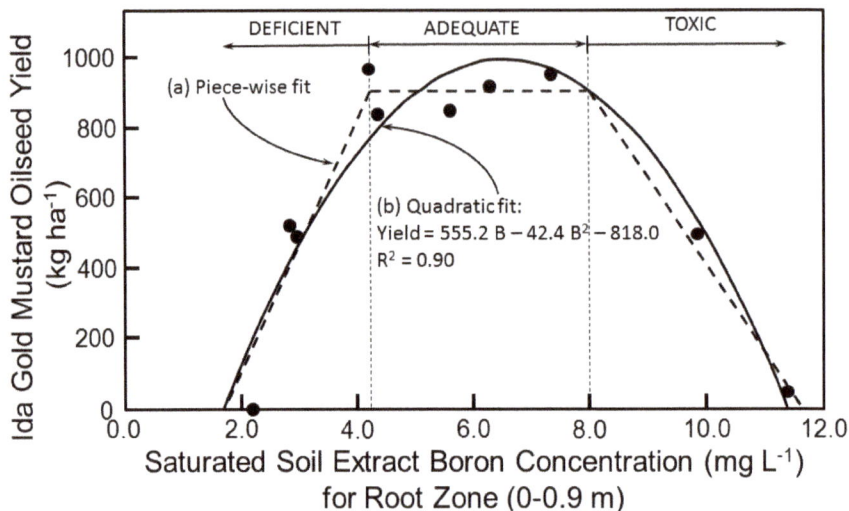

Figure 6. Boron tolerance models for Ida Gold mustard oilseed: (**a**) three-piece linear B tolerance model (thick dashed line) and (**b**) quadratic B tolerance model (solid line). Thin dashed lines indicate the B deficiency and toxicity thresholds.

3.2. Evaluation of SSURGO Soil Salinity Accuracy and Identification of Salt-Affected Soils

An evaluation of the accuracy of SSURGO spatial data for root-zone soil salinity revealed that only 5 out of 22 fields assessed the mean salinity or range in salinity accurately, suggesting that the more transient salinity levels and patterns in the root zone are not captured in the one-time measurements of NRCS soil surveys. However, SSURGO was able to assess 15 out of 22 fields accurately for salinity below the root zone, indicating that the salt levels below the root zone remained relatively unchanged and unaffected by anthropogenic influences. The failure of SSURGO to provide accurate root-zone soil salinity spatial data necessitated a reliance on the regional-scale salinity model developed by Scudiero et al. [29] as input data for the Ida Gold mustard oilseed yield model.

Because of the inaccuracy of salinity in SSURGO the Scudiero et al. [29] regional-scale salinity model was used to identify salt-affected soils (EC > 4 dS m^{-1}) for the SJV. Figure 7 shows the extent of salt-affected soils for the WSJV as estimated by the Scudiero et al. [29] regional-scale salinity model. A comparison of total land cover greater than 4 dS m^{-1} for SSURGO and the Scudiero et al. [29] regional-scale salinity model indicates that SSURGO estimated 33% less salt-affected land for the WSJV. Furthermore, the distribution of salt-affected soils was concentrated in contiguous patterns along the eastern half of the WSJV for SSURGO, whereas the Scudiero et al. [29] salinity patterns were more diffusely spread as shown in Figure 7. Only fields identified as salt-affected were subsequently used in the Monte Carlo simulations. Any field where the field average root zone EC$_e$ was estimated to be <4 dS m^{-1} was disregarded and not included in the Monte Carlo simulations.

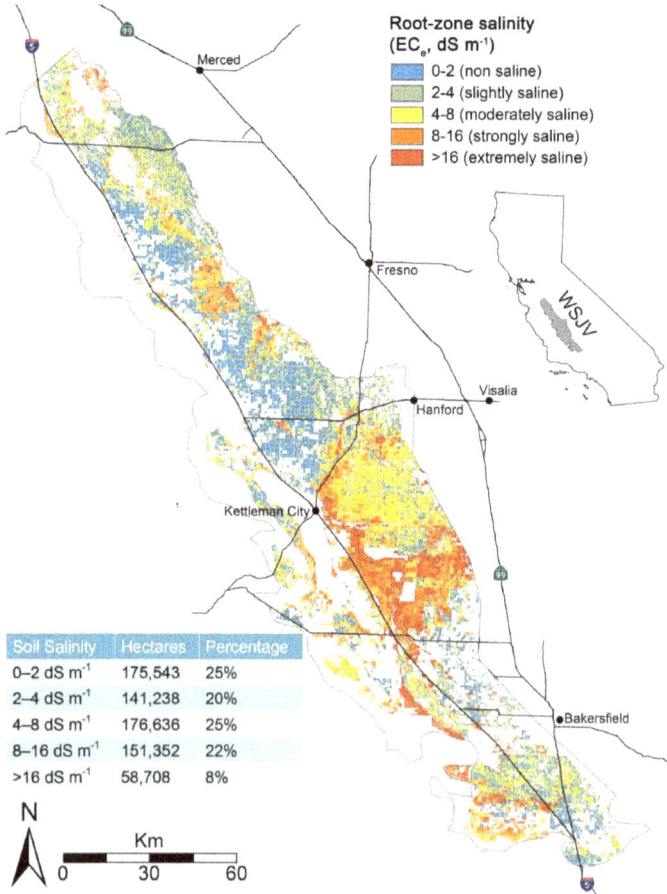

Figure 7. Map of salt-affected soils for the west side of California's San Joaquin Valley (WSJV) estimated from the regional-scale soil salinity model of Scudiero et al. [5,29].

3.3. Input Data for Ida Gold Mustard Oilseed Yield Model

The reliability and accuracy of the spatial data that serves as input into any model are just as critical as the model itself, exemplified by the old adage, garbage in garbage out. Sensitivity analysis of the crop yield model provides an indication of the input variables that need the greatest level of accuracy and therefore need particular scrutiny when building a spatial database of inputs for the crop yield model. Sensitivity analysis (Table 4) established the degree of influence that each edaphic property had on oilseed yield in Equation (3). The influence was determined by calculating how much the predicted oilseed yield changed when the value for each independent variable in Equation (3) was individually shifted by 1 standard deviation from its mean level or point of maximum yield with respect to the edaphic property. The means and standard deviations were obtained for the 0–1.2 m depth increment excluding any points where no yield occurred (Table 2b). A baseline yield was used as the point of reference to establish the percentage of change. A baseline value of 6.8 dS m^{-1} was used for salinity, rather than the mean EC_e level of 10.0 dS m^{-1} from Table 2b because the value 6.8 dS m^{-1} represents the point of maximum yield with respect to the quadratic salinity response pattern. For the same reason a B baseline of 4.0 mg L^{-1} was used in the sensitivity analysis. The calculated percentage yield change shown in Table 4 indicates that B is the most significant factor influencing Ida Gold mustard oilseed yield, followed by EC_e, then LF, and finally θ_g.

Table 4. Degree of predicted oilseed yield sensitivity to 1 standard deviation (SD) change in each edaphic property of Equation (3). [†]

Parameter Sensitivity [‡]	Calculated Yield (kg ha^{-1})	Conversion to Biofuel (L ha^{-1})	Percentage Change (%)	Boron (mg L^{-1})	EC_e (dS m^{-1})	LF	θ_g (kg kg^{-1})
Baseline	1017.3	178.3		4.0	6.8	0.27	0.19
B + 1 SD	345.2	60.5	−66.1	10.1	6.8	0.27	0.19
EC_e + 1 SD	776.2	136.0	−23.7	4.0	13.1	0.27	0.19
LF + 1 SD	1212.3	212.5	19.2	4.0	6.8	0.42	0.19
θ_g + 1 SD	1033.3	181.1	1.6	4.0	6.8	0.27	0.24

[†] For the 0–1.2 m composite depth increment. [‡] Definitions: B = boron, EC_e = electrical conductivity of the saturation extract, LF = leaching fraction, θ_g = gravimetric water content.

In the case of EC_e, oilseed yield model input values were obtained from the 30 x 30 m predictions from Scudiero et al. [29] and used to determine the average EC_e for each field for the root zone (0–1.2 m) within the SJV. For the Monte Carlo simulations, PDFs were defined by the average residual and standard deviation of the residuals for each category of salinity (i.e., 0–1, 1–2, 2–3, 3–4, 4–5, 5–6, 6–7, 7–8, 8–9, 9–10, 10–11, 11–12, 12–13, 13–14, 14–15, 15–16, and >16 dS m^{-1}) using the predicted and ground-truth EC_e data from Scudiero et al. [5,29]. The input EC_e of each pixel in the Monte Carlo simulations was determined from the predicted EC_e and residual PDF. Table 5 is a summary of the average residuals, standard deviation of the residuals, and data count for each salinity category and for the entire data set. All PDFs were normally distributed. The PDFs for categories 0–1, 1–2, 2–3, and 3–4 d S m^{-1} were not used since only those fields with an average root zone EC_e of greater than 4 dS m were used in the Monte Carlo simulations. There was not a substantial difference in the Monte Carlo simulation findings using the PDF for the entire data set in Table 5 as compared to the individual salinity categories; consequently, all subsequent Monte Carlo simulation discussions will be for the use of the PDF for the entire EC_e residual data set.

Table 5. Summary of the average residual, standard deviation of the residual, and data count for each category of predicted salinity (i.e., EC_e) from Scudiero et al. [29], indicating the uncertainty of the EC_e prediction and used in defining the EC_e residual PDFs.

Lower Limit of EC_e Interval (dS m^{-1})	Average EC_e Residual (dS m^{-1})	Standard Deviation of the EC_e Residuals (dS m^{-1})	Data Count
0	−2.29	3.38	131
1	−0.38	2.13	577
2	−0.68	2.39	623
3	−0.32	2.21	584
4	−0.86	2.62	351
5	−0.97	2.45	245
6	−0.95	2.49	298
7	−0.75	2.69	267
8	−0.86	3.01	215
9	−0.52	2.67	143
10	−0.50	3.03	82
11	1.27	2.38	83
12	2.18	1.51	121
13	3.00	1.69	193
14	3.64	1.65	127
15	4.91	2.07	77
16	6.57	2.60	194
Entire data set	0.14	3.11	4311

Definition: EC_e = electrical conductivity of the saturation extract.

The edaphic property data collected and presented by Corwin and colleagues (Bourgault et al., [30]; Corwin [31]; Corwin and Lesch [8–10,32,46]; Corwin et al. [7,33–36]; Lesch and Corwin [17]; Lesch et al. [37–39]; Loague et al. [40]; Rhoades et al. [41]; Sanden et al. [42]; and Scudiero et al. [26,29,43]) over 2.5 decades of salinity assessment field studies were used to develop the PDFs for B, LF, and θ_g for the SJV, which were subsequently used as input data in the Monte Carlo simulations. The PDFs for B, LF, and θ_g were log normally distributed.

Oilseed yield must be converted to biofuel yield. The oil content of Ida Gold mustard oilseed was 26% and the extraction efficiency was 64%. Subsequently, for every 1000 kg ha^{-1} of oilseed produced then 166.4 kg ha^{-1} of 100% biofuel resulted, which represented 175.3 L ha^{-1} of biofuel. It is of importance to note that transesterified biofuel is generally blended with diesel (1:5 ratio of biofuel to diesel) and sold as 20% biofuel in California.

3.4. Monte Carlo Simulations Feasibility of Biofuel Production for the SJV

Monte Carlo simulations for the WSJV never produced a simulation that resulted in more than 71.9 ML yr^{-1} of biofuel produced, which is well below the goal of 115 ML yr^{-1}. Subsequently, Monte Carlo simulations were performed for the entire SJV. To do this, the regional-scale salinity model of Scudiero et al. [29] was applied to the entire SJV to identify salt-affected soils, which totaled 9.7×10^5 ha. Monte Carlo simulations resulted in the PDF and associated CDF shown in Figure 8a,b, respectively. The PDF and CDF are based on Monte Carlo simulations that incorporate the piece-wise models of salinity and B tolerance. The histogram of Monte Carlo simulations is best fit with the shifted gamma PDF in Equation (6) (Figure 8a):

$$Q = 68.986 + \text{gamma}(6.134, 5.285) \tag{6}$$

where Q is the biofuel production in ML yr^{-1}. A comparison of the means, medians, standard deviations, skewness, and kurtosis for the measured Monte Carlo simulation data and for the estimates from Equation (6) shows excellent agreement, with means of 101.4 and 101.4 ML yr^{-1}, medians of 97.8 and 99.4 ML yr^{-1}, standard deviations of 14.1 and 14.1 ML yr^{-1}, skewness of 0.87 and 0.87, and kurtosis of 4.17 and 4.14, respectively (Figure 8a). From the CDF (Figure 8b) there is a 17% probability of meeting the minimum production level when all salt-affected soils in the SJV are utilized for oilseed production. When the Monte Carlo simulations include quadratic salt and B tolerance models (i.e., Equations (5) and (6), respectively), there is little difference, with a 15% probability.

Several potential weaknesses in the approach need discussion as well as how the impacts of these weaknesses were mitigated in the Monte Carlo simulations. First, the oilseed yield model was developed from a single field with a fine-texture soil that did not vary significantly. Even though the study site may have been representative of many of the fine-textured soils of the WSJV, it was not representative of coarse-textured soils found in the east side of the SJV; consequently, the use of the oilseed yield model for the entire SJV is dubious. Second, the regional-scale salinity model of Scudiero et al. [29] was developed from a database that did not include tree crops or vineyards, which makes the input salinities for the oilseed yield model dubious for areas containing vineyards or tree crops. Scudiero et al. [5] showed that the regional-scale salinity model of Scudiero et al. [29] over estimates salinity levels for orchards and vineyards, which would render reduced oilseed yield estimates of Ida Gold mustard when planted between tree rows or would identify these lands as salt-affected when they are not. Third, applying the oilseed yield model outside the range of data that was used to develop it is problematic.

To rectify these problems, several precautions were taken. No areas where orchards or vineyards occurred were included in the Monte Carlo simulations. In fields where the texture was coarse, the oilseed yield model of Equation (3) was not applied. Rather, the two-piece linear salt tolerance model of Maas and Hoffman [18] presented in Equation (1) or the three-piece linear B tolerance model presented in Figure 6 were used, depending on which was most limiting to oilseed yield. Similarly, in cases where the input data for the oilseed yield model (Equation (3)) fell outside the range of data used to develop the model, then the two-piece linear salt tolerance model of Maas and Hoffman [18] or the three-piece linear B tolerance model were used instead.

San Joaquin Valley Biofuel Production PDF and CDF

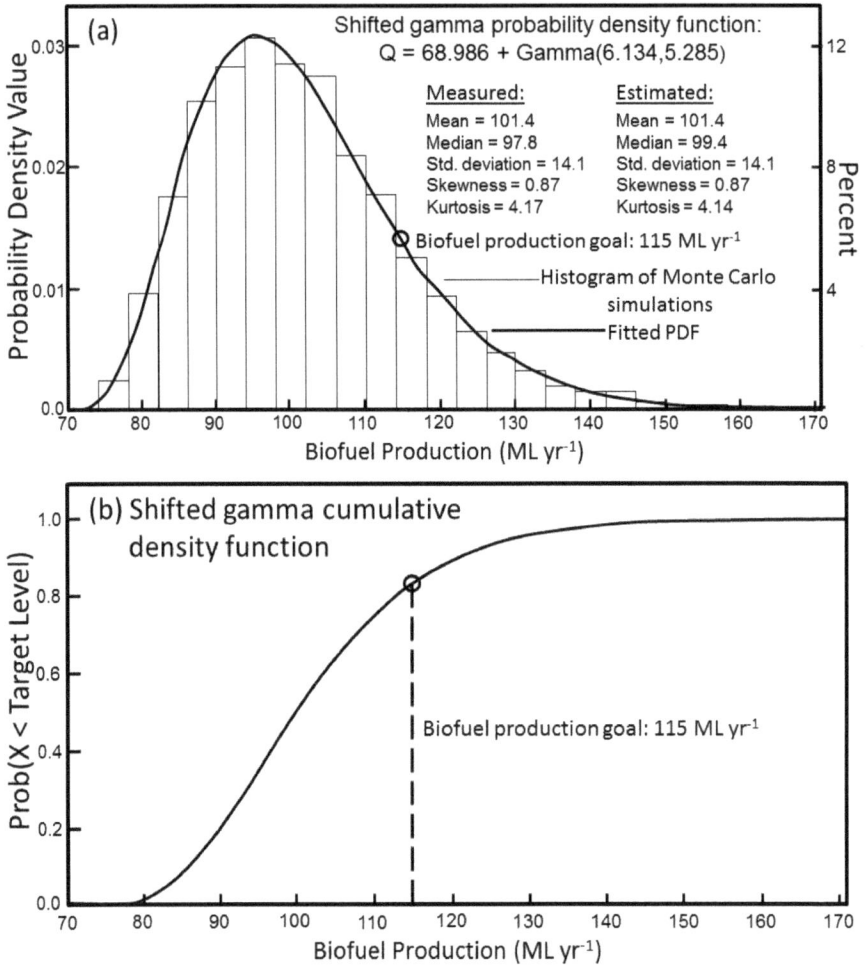

Figure 8. Biofuel production (**a**) histogram of Monte Carlo simulations and associated probability density function (PDF) and (**b**) cumulative density function (CDF) for Ida Gold mustard oilseed from San Joaquin Valley salt-affected soils (i.e., soils with salinity of >4 dS m^{-1}) based on 10,000 Monte Carlo simulations.

Even though the feasibility of meeting desired oilseed production levels is dubious, there are circumstances that could improve the likelihood of meeting the 115 ML lower limit of production. Water in terms of water content and water available for leaching are two impactful properties pertaining to oilseed yield in the model. Maintaining the root zone water content at high levels with adequate leaching could drive up yields sufficiently high to make oilseed a viable biofuel in the SJV. In particular, high frequency irrigation to maintain root zone water contents near field capacity would have significant impact. However, this is only feasible when the SJV has sufficient irrigation water supplies, which may be the exception rather than the rule as shown by the recent 6-year California drought (2011–2016).

Orchards, whether in salt-affected ($EC_e > 4$ dS m^{-1}) or non-salt-affected ($EC_e \leq 4$ dS m^{-1}) soils, could play a major role in enhancing the feasibility of biofuel production in the SJV. Planting mustard oilseed between rows in orchards, sometimes referred to as intercropping or alley cropping, provides tremendous acreage for crop growth on lands that are otherwise left fallow in many instances. Intercropping has several additional advantages, including weed control, wind and water erosion control, dust management, improved percolation, more effective use of land and resources, and additional crop revenue. There are roughly 130,000 ha of orchards in the SJV. By planting an oilseed crop between rows as a secondary crop, biofuel production could increase roughly 15–30%, which would increase the probability of meeting the 115 ML yr^{-1} target from 15–17% to 60–80% for the entire SJV. Even though intercropping with oilseed may make biofuel production feasible, it certainly does not necessarily make it economically viable. Furthermore, secondary crops from intercropping are a management challenge for producers filled with additional management problems.

Aside from meeting the 115 ML yr^{-1} target, economic viability is a concern. Growing oilseed on marginally productive soil is not the only means of lowering cost. Another resource that can lower biofuel production cost is the reuse of degraded water for irrigation. Corwin [31] showed in a long-term (12 years) study that the reuse of 3–5 dS m^{-1} drainage water was a viable means of reclaiming saline-sodic soil on the WSJV, supporting a salt-tolerant crop during the reclamation process, and providing financial return to the producer on otherwise non-productive farmland. In addition, drainage water is a particularly valuable alternative water resource during drought years in the SJV, provided drainage water is in sufficient supply.

Even though biofuel feasibility evaluation is a valuable application of sensor technology, the unique aspect of this study is that it presents an innovative regional-scale approach for modeling the interaction of edaphic properties on crop yield by characterizing the spatial variability of soil properties influencing yield using proximal and satellite sensors. In essence, each sampling location identified from an EC_a survey serves as an independent crop yield study looking at the interaction of edaphic properties on yield, thereby rendering a more robust model with lower long-term labor requirements than plant salt (or B) tolerance studies provided in the past. Furthermore, past plant salt tolerance studies did not evaluate the interaction of edaphic properties on crop yield.

The same approach can be used to predict drought impacts on crop productivity within an agricultural region, to identify reclamation needs to optimize crop productivity, and to manage resources (e.g., irrigation water, land, crop selection) and edaphic properties (e.g., salinity, B, etc.) to optimize crop yield. Because of its regional-scale application and quantitative probability assessment, the approach provides land and water resource specialists with a tool to make regional- and landscape-scale resource decisions with a clear understanding of the likelihood of impact on agricultural productivity.

4. Conclusions

This study exemplifies the combined use of proximal and satellite sensors to answer a practical agricultural question of national significance with strategic military implications concerning the feasibility of biofuel production in California's San Joaquin Valley to help meet the aviation fuel needs of the U.S. military. A quantitative assessment of the probability of attaining a minimum of 115 ML yr^{-1} of biofuel from oilseed indicates that there is a 15–17% probability of meeting the minimum production level when all salt-affected soils in the SJV are utilized for oilseed production (Figure 8). The low probability of meeting the minimum production level and the fact that all salt-affected soils throughout the SJV, not just WSJV salt-affected soils, would need to be cropped with Ida Gold mustard to produce sufficient oilseed biofuel to support a 115 ML conversion plants makes the idea untenable. Furthermore, a significant portion of the 115 ML of biofuel would be produced on soils of moderate salinity (i.e., 4–6 dS m^{-1}), moderate B levels (i.e., 4–8 mg L^{-1}), high LF, and high available water content, which are soils that can either be easily reclaimed to produce higher cash crops or used to grow salt and B tolerant crops of higher cash return than oilseed, such as pistachio.

Acknowledgments: The authors wish to acknowledge the Office of Naval Research for providing the funds to support this work (FMMI number: 3200001344). These crucial funds ultimately led to the development of landscape- and regional-scale salinity assessment technology using EC_a-directed soil sampling. The authors thank Michael Bagtang for his diligent work in the field and laboratory. The authors also wish to thank David Santos of Santos Farming, Los Banos, CA, for the use of his land to perform the field-scale salt tolerance study, which was crucial to the success of this work.

Author Contributions: Dennis L. Corwin, Elia Scudiero, and Todd H. Skaggs conceived and designed the experiments; Kevin Yemoto, Wes Clary, and Gary Banuelos performed the experiments; Dennis L. Corwin and Scott M. Lesch analyzed the data; Dennis L. Corwin wrote the paper.

Conflicts of Interest: The authors declare no conflict of interest. The founding sponsors had no role in the design of the study; in the collection, analyses, or interpretation of data; in the writing of the manuscript, and in the decision to publish the results.

Abbreviations

The following abbreviations are used in this manuscript:

EC_e	Electrical Conductivity of the Saturation Extract (dS m^{-1})
EM_h, EC_a	Measured with Electromagnetic Induction in the Horizontal Coil Configuration (dS m^{-1})
EM_v, EC_a	Measured with Electromagnetic Induction in the Vertical Coil Configuration (dS m^{-1})
EMI	Electromagnetic Induction
SJV	San Joaquin Valley
WSJV	West Side of the San Joaquin Valley

References

1. Naik, S.N.; Goud, V.V.; Rout, P.K.; Dalai, A.K. Production of first and second generation biofuels: A comprehensive review. *Renew. Sustain. Energy Rev.* **2010**, *14*, 578–597. [CrossRef]
2. Bruno, M.; Washington, G.W. Energy equation: Could abundant natural gas force the Pentagon to slow its drive toward sustainable fuels? *Aviation Week Space Technol.* **2013**, *2013*, 40–41.
3. National Academy of Sciences (NAS). Renewable Fuel Standard: Potential economic and environmental effects of U.S. biofuel policy. Committee on Economic and Environmental Impacts of Increasing Biofuels Production. 2011. Available online: http://deis.nas.edu/banr (accessed on 20 February 2017).
4. Backlund, V.L.; Hoppes, R.R. Status of soil salinity in California. *Calif. Agric.* **1984**, *38*, 8–9.
5. Scudiero, E.; Corwin, D.L.; Anderson, R.G.; Yemoto, K.K.; Clary, W.A.; Wang, Z.; Skaggs, T.H. Remote sensing is a viable tool for mapping soil salinity in agricultural lands. *Calif. Agric.* **2017**, *2017*. [CrossRef]
6. USDOE and USDA. Sustainability of biofuels: Future research opportunities. Report from the October 2008 Workshop, DOE/SC-0114; U.S. Department of Energy Office of Science and U.S. Department of Agriculture, 2009. Available online: http://genomicsgtl.energy.gov/biofuels/sustainability/ (accessed on 20 February 2017).
7. Corwin, D.L.; Lesch, S.M.; Shouse, P.J.; Soppe, R.; Ayars, J.E. Identifying soil properties that influence cotton yield using soil sampling directed by apparent soil electrical conductivity. *Agron. J.* **2003**, *95*, 352–364. [CrossRef]
8. Corwin, D.L.; Lesch, S.M. Application of soil electrical conductivity to precision agriculture: Theory, principles, and guidelines. *Agron. J.* **2003**, *95*, 455–471. [CrossRef]
9. Corwin, D.L.; Lesch, S.M. Characterizing soil spatial variability with apparent soil electrical conductivity: I. Survey protocols. *Comput. Electron. Agric.* **2005**, *46*, 103–133. [CrossRef]
10. Corwin, D.L.; Lesch, S.M. Protocols and guidelines for field-scale measurement of soil salinity distribution with EC_a-directed soil sampling. *J. Environ. Eng. Geophys.* **2013**, *18*, 1–25. [CrossRef]
11. Lesch, S.M.; Strauss, D.J.; Rhoades, J.D. Spatial prediction of soil salinity using electromagnetic induction techniques: 1. Statistical prediction models: A comparison of multiple linear regression and cokriging. *Water Resour. Res.* **1995**, *31*, 373–386. [CrossRef]
12. Lesch, S.M.; Strauss, D.J.; Rhoades, J.D. Spatial prediction of soil salinity using electromagnetic induction techniques: 2. An efficient spatial sampling algorithm suitable for multiple linear regression model identification and estimation. *Water Resour. Res.* **1995**, *31*, 387–398. [CrossRef]

13. Lesch, S.M.; Rhoades, J.D.; Corwin, D.L. ESAP−95 version 2.10R: User manual and tutorial guide. In *Research Report No. 146*; U.S. Salinity Laboratory: Riverside, CA, USA, 2000; p. 161.

14. Corwin, D.L.; Scudiero, E. Field-scale apparent soil electrical conductivity. In *Methods of Soil Analysis*; Logsdon, S., Ed.; SSSA: Madison, WI, USA, 2016; Volume 1, pp. 1–29. [CrossRef]

15. Rhoades, J.D. Salinity: Electrical conductivity and total dissolved solids. In *Methods of Soil Analysis: Part 3 – Chemical Methods*; SSSA Book Series No. 5; Sparks, D.L., Ed.; SSSA: Madison, WI, USA, 1996; pp. 417–435.

16. *Methods of Soil Analysis Part 3 – Chemical Methods*; SSSA Book Series No. 5; Sparks, D.L. (Ed.) SSSA: Madison, WI, USA, 1996.

17. Lesch, S.M.; Corwin, D.L. Prediction of spatial soil property information from ancillary sensor data using ordinary linear regression: Model derivations, residual assumptions and model validation tests. *Geoderma* **2008**, *148*, 130–140. [CrossRef]

18. Maas, E.V.; Hoffman, G. Crop salt tolerance – current assessment. *J. Irrig. Drain. Div. Am. Soc. Civ. Eng.* **1977**, *103*, 115–134.

19. Maas, E.V. Crop salt tolerance. In *Agricultural Salinity Assessment and Management*, 1st ed.; Tanji, K.K., Ed.; ASCE: New York, NY, USA, 1996; pp. 262–304.

20. Grieve, C.M.; Grattan, S.R.; Maas, E.V. Plant salt tolerance. In *Agricultural Salinity Assessment and Management*, 2nd ed.; Wallender, W.W., Tanji, K.K., Eds.; ASCE: Reston, VA, USA, 2012; pp. 405–459.

21. Page, A.L.; Chang, A.C.; Adriano, D.C. Deficiencies and toxicities of trace elements. In *Agricultural Salinity Assessment and Management*, 1st ed.; Tanji, K.K., Ed.; ASCE: New York, NY, USA, 1996; pp. 138–160.

22. Burton, K.W.; Morgan, E.; Roig, A. The influence of heavy metals upon growth of sitka-spruce in South Wales forests. I. Upper critical and foliar critical concentration. *Plant Soil* **1983**, *73*, 327–336. [CrossRef]

23. Bohn, H.L.; McNeal, B.L.; O'Connor, G.A. *Soil Chemistry*; Wiley-Interscience: New York, NY, USA, 1979; p. 329.

24. U.S. Salinity Laboratory Staff. *Diagnosis and Improvement of Saline and Alkali Soils*; Agriculture Handbook No. 60; Richards, L.A., Ed.; U.S. Government Printing Office: Washington, DC, USA, 1954; p. 160.

25. Arroues, K.D. *Soil Survey of Fresno County, California, Western Part*; U.S. Department of Agriculture, Natural Resources Conservation Service: Washington, DC, USA, 2006.

26. Scudiero, E.; Skaggs, T.H.; Corwin, D.L. Regional scale soil salinity evaluation using Landsat 7, Western San Joaquin Valley, California, USA. *Geoderma Reg.* **2014**, *2–3*, 82–90. [CrossRef]

27. *Methods of Soil Analysis Part 4 – Physical Methods*; SSSA Book Series: 5; Dane, J.H.; Topp, G.C. (Eds.) SSSA: Madison, WI, USA, 2002.

28. Lobell, D.B.; Lesch, S.M.; Corwin, D.L.; Ulmer, M.; Anderson, K.; Potts, D.; Doolittle, J.; Matos, M.; Baltes, M. Regional-scale assessment of soil salinity in the Red River Valley using multi-year MODIS EVI. *J. Environ. Qual.* **2010**, *39*, 35–41. [CrossRef] [PubMed]

29. Scudiero, E.; Skaggs, T.H.; Corwin, D.L. Regional-scale soil salinity assessment using Landsat ETM+ canopy reflectance. *Remote Sens. Environ.* **2015**, *169*, 335–343. [CrossRef]

30. Bourgault, G.; Journel, A.G.; Lesch, S.M.; Rhoades, J.D.; Corwin, D.L. Geostatistical analysis of a soil salinity data set. *Adv. Agron.* **1997**, *58*, 241–292.

31. Corwin, D.L. Field-scale monitoring of the long-term impact and sustainability of drainage water reuse on the west side of California's San Joaquin Valley. *J. Environ. Monit.* **2012**, *14*, 1576–1596. [CrossRef] [PubMed]

32. Corwin, D.L.; Lesch, S.M. Delineating site-specific management units with proximal sensors. In *Geostatistical Applications for Precision Agriculture*; Oliver, M.A., Ed.; Springer: London, UK, 2010; pp. 139–165.

33. Corwin, D.L.; Vaughan, P.J.; Wang, H.; Rhoades, J.D.; Cone, D.G. Predicting areal distributions of salt-loading to the groundwater. In Proceedings of the ASAE Winter Meeting, Chicago, IL, USA, 12–17 December 1993.

34. Corwin, D.L.; Carrillo, M.L.K.; Vaughan, P.J.; Rhoades, J.D.; Cone, D.G. Evaluation of GIS-linked model of salt loading to groundwater. *J. Environ. Qual.* **1999**, *28*, 471–480. [CrossRef]

35. Corwin, D.L.; Kaffka, S.R.; Hopmans, J.W.; Mori, Y.; Lesch, S.M.; Oster, J.D. Assessment and field-scale mapping of soil quality properties of a saline-sodic soil. *Geoderma* **2003**, *114*, 231–259. [CrossRef]

36. Corwin, D.L.; Lesch, S.M.; Oster, J.D.; Kaffka, S.R. Monitoring management-induced spatio-temporal changes in soil quality with soil sampling directed by apparent soil electrical conductivity. *Geoderma* **2006**, *131*, 369–387. [CrossRef]

37. Lesch, S.M.; Rhoades, J.D.; Corwin, D.L. Mapping soil salinity using calibrated electromagnetic measurements. *Soil Sci. Soc. Am. J.* **1992**, *56*, 540–548. [CrossRef]

38. Lesch, S.M.; Rhoades, J.D.; Corwin, D.L. Statistical modeling and prediction methodologies for large scale spatial soil salinity characterization: A case study using calibrated electromagnetic measurements within the Broadview Water District. In *Research Report No. 131*; U.S. Salinity Laboratory: Riverside, CA, USA, 1994; p. 44.

39. Lesch, S.M.; Corwin, D.L.; Robinson, D.A. Apparent soil electrical conductivity mapping as an agricultural management tool in arid zone soils. *Comp. Electron. Agric.* **2005**, *46*, 351–378. [CrossRef]

40. Loague, K.; Blanke, J.S.; Mills, M.B.; Diaz-Diaz, R.; Corwin, D.L. Data related uncertainty in near-surface vulnerability assessments for agrochemicals in the San Joaquin Valley. *J. Environ. Qual.* **2012**, *41*, 1427–1436. [CrossRef] [PubMed]

41. Rhoades, J.D.; Corwin, D.L.; Lesch, S.M. Effect of soil EC_a - depth profile pattern on electromagnetic induction measurements. In *Research Report No. 125*; U.S. Salinity Laboratory: Riverside, CA, USA, 1991; p. 108.

42. Sanden, B.L.; Ferguson, L.; Kallsen, C.; Corwin, D.L. Large-scale utilization of saline groundwater for development and irrigation of pistachios *(P. integerrima)* interplanted with cotton *(G. barbadense)*. *Acta Hortic.* **2007**, *ISHS 792*, 551–558. [CrossRef]

43. Scudiero, E.; Skaggs, T.H.; Corwin, D.L. Simplifying field-scale assessment of spatiotemporal changes of soil salinity. *Sci. Total Environ.* **2017**, *587–588*, 273–281. [CrossRef] [PubMed]

44. Nelson, J.W.; Guernsey, J.E.; Holmes, L.C.; Eckmann, E.C. *Reconnaissance Soil Survey of the Lower San Joaquin Valley, California*; U.S. Department of Agriculture–Bureau of Soils: Washington, DC, USA, 1915; p. 157.

45. Letey, J. Soil salinity poses challenges for sustainable agriculture and wildlife. *Calif. Agric.* **2000**, *54*, 43–48. [CrossRef]

46. Corwin, D.L.; Lesch, S.M. Characterizing soil spatial variability with apparent soil electrical conductivity: II. Case study. *Comput. Electron. Agric.* **2005**, *46*, 135–152. [CrossRef]

![sensors logo] *sensors*

MDPI

Article

Evaluation of Sensible Heat Flux and Evapotranspiration Estimates Using a Surface Layer Scintillometer and a Large Weighing Lysimeter

Jerry E. Moorhead [1,*], Gary W. Marek [1], Paul D. Colaizzi [1], Prasanna H. Gowda [2], Steven R. Evett [1], David K. Brauer [1], Thomas H. Marek [3] and Dana O. Porter [4]

1   USDA-ARS Conservation and Production Research Laboratory, PO Drawer 10, Bushland, TX 79012, USA; gary.marek@ars.usda.gov (G.W.M.); paul.colaizzi@ars.usda.gov (P.D.C.); steve.evett@ars.usda.gov (S.R.E.); david.brauer@ars.usda.gov (D.K.B.)
2   USDA-ARS Grazinglands Research Laboratory, 7207 West Cheyenne St., El Reno, OK 73036, USA; prasanna.gowda@ars.usda.gov
3   Texas A&M AgriLife Research, 6500 Amarillo Blvd W, Amarillo, TX 79106, USA; tmarek@ag.tamu.edu
4   Texas A&M AgriLife Extension Service, 1102 E FM 1294, Lubbock, TX 79403, USA; dporter@ag.tamu.edu
*   Correspondence: jed.moorhead@ars.usda.gov; Tel.: +1-806-356-5704

Received: 8 September 2017; Accepted: 12 October 2017; Published: 14 October 2017

Abstract: Accurate estimates of actual crop evapotranspiration (ET) are important for optimal irrigation water management, especially in arid and semi-arid regions. Common ET sensing methods include Bowen Ratio, Eddy Covariance (EC), and scintillometers. Large weighing lysimeters are considered the ultimate standard for measurement of ET, however, they are expensive to install and maintain. Although EC and scintillometers are less costly and relatively portable, EC has known energy balance closure discrepancies. Previous scintillometer studies used EC for ground-truthing, but no studies considered weighing lysimeters. In this study, a Surface Layer Scintillometer (SLS) was evaluated for accuracy in determining ET as well as sensible and latent heat fluxes, as compared to a large weighing lysimeter in Bushland, TX. The SLS was installed over irrigated grain sorghum (*Sorghum bicolor* (L.) Moench) for the period 29 July–17 August 2015 and over grain corn (*Zea mays* L.) for the period 23 June–2 October 2016. Results showed poor correlation for sensible heat flux, but much better correlation with ET, with r^2 values of 0.83 and 0.87 for hourly and daily ET, respectively. The accuracy of the SLS was comparable to other ET sensing instruments with an RMSE of 0.13 mm·h^{-1} (31%) for hourly ET; however, summing hourly values to a daily time step reduced the ET error to 14% (0.75 mm·d^{-1}). This level of accuracy indicates that potential exists for the SLS to be used in some water management applications. As few studies have been conducted to evaluate the SLS for ET estimation, or in combination with lysimetric data, further evaluations would be beneficial to investigate the applicability of the SLS in water resources management.

Keywords: irrigation; energy balance; water management; semi-arid regions

---

1. Introduction

In arid and semi-arid regions of the world such as the Texas High Plains, finite groundwater resources are being mined, often with little to no surface water available as an alternate irrigation source. In the Texas High Plains, irrigation pumping for agricultural crop production accounts for the overwhelming majority of total groundwater withdrawals [1]. Effective irrigation management is essential for extending the longevity of limited water resources in these intensively irrigated production areas. Most effective irrigation management (scheduling) strategies rely on accurate estimates of evapotranspiration (ET) to account for crop water use and evaporative water losses. ET is a term that represents water lost through evaporation from the soil or plant surface and water used through plant transpiration. In the Texas High Plains, ET is the largest water loss component in the soil water budget [1,2]; hence, accurate ET estimates are vital for determining crop water demand and managing irrigation. Groundwater recharge in the Texas High Plains, and the surrounding Southern Ogallala Aquifer region can be as low as ~11 mm year^{-1} [3]. With such small recharge rates, the Ogallala Aquifer is essentially a finite water resource. Assuming an 11 mm recharge rate across all 5.4 million ha (13.4 million ac) in the northern half of the Texas panhandle (Texas Water Development Board Region A) would supply 0.60 km^3 (484,000 ac-ft) of recharge to the aquifer. In contrast, 2010 agricultural water withdrawals were estimated to be 1.81 km^3 (1.47 million ac-ft) for the same area [1]. Water conservation, in part through ET-based irrigation scheduling, is paramount to extending the longevity of this limited water resource for future generations.

Accurate measurement or estimation of ET can be difficult; however, numerous instruments and methods are available. One widely used method is the reference ET (ET$_{ref}$) and crop coefficient product. Meteorological data are used to estimate ET$_{ref}$, which corresponds to the water demand of a reference crop, either a short, clipped grass or alfalfa [4]. To obtain ET for a specific crop using reference ET, a crop coefficient (K$_c$) is used to adjust reference ET to crop specific ET, or ET$_c$; thus, ET$_c$ = K$_c$ × ET$_{ref}$ [5]. For irrigation scheduling, ET$_c$ is typically calculated at a daily time step. ET$_c$ estimates the amount of water that would be used by that crop if there were no water limitation for crop growth. Actual (field-based) crop ET may be less than ET$_c$ due to stresses from insufficient available water, nutrients, pests, etc. As such, actual ET would potentially be more useful than ET$_{ref}$ or ET$_c$ for irrigation management. An issue with using actual ET in irrigation scheduling is that ET can be very difficult to determine accurately. Where ET$_{ref}$ can be calculated from weather parameters using a weather station with a reference surface, efforts to determine actual ET not only require more advanced (and expensive) instrumentation but still only result in an estimate of ET. Current technologies for determining ET estimates include lysimeters, Bowen Ratio, Eddy Covariance (EC), scintillometry, field water balance using soil water measurements, remote sensing models, and others.

The aforementioned methods can all be used to estimate ET; however there are disadvantages of each. With the soil water balance approach, drainage and runoff components can be difficult to accurately determine. Although they are commonly considered relatively small in arid and semiarid regions, they need to be accounted for to obtain the best accuracy. In addition, soil water measurements are valid for a small point in a field creating spatial representation concerns. Weighing lysimeters are the most accurate method of assessing ET [6], but they are very expensive to install, maintain, and operate. In addition, they require a high level of operational knowledge and data processing experience to obtain accurate and representative measurements. Large weighing lysimeters are typically considered research tools and are not practical for generalized irrigation scheduling. The Bowen Ratio method has been used to determine ET from the energy balance, but it is an indirect measurement and can have issues of instrument bias and data discontinuity when the Bowen Ratio approaches −1 [7]. EC is a method of estimating turbulent fluxes and ET, but is known to have significant energy balance closure errors [8–10]. Remote sensing models have been extensively used for ET mapping; however, they have been commonly evaluated using EC [11–13]. Some studies have evaluated remote sensing ET using lysimeters [14–16], though lysimeter data are not available in many regions. Other, simpler methods to achieve highly accurate ET data would greatly benefit future research. Scintillometers are another type of indirect measurement instrument that has been extensively used for surface atmospheric dynamics research, but they can also have large errors. EC and scintillometry are two of the more common turbulent flux and ET methods typically used. They are relatively inexpensive and easy to deploy and maintain; and provide spatially averaged data. Scintillometers measure contributions to fluxes over a fixed path length and EC can measure contributions to fluxes over a variable area influenced by wind movement. The spatial average nature of EC and scintillometers account for variation within the area of measurement and provide a degree of representativeness. The main issue with the EC method is the lack of accuracy, as reported throughout the literature. Although scintillometers are being used in atmospheric research [17], applications in irrigation management research are relatively new and few in number. A thorough evaluation of scintillometers is needed to determine if they are suitable to measure ET for the purposes of irrigation scheduling and developing crop coefficients. Scintillometers are commonly used with the energy balance equation to determine ET from scintillometer measurements.

1.1. Energy Balance

In many studies, scintillometers are used with the energy balance. The energy balance equation is:

$$LE = R_n - H - G, \qquad (1)$$

where R_n is the net radiation, LE is the latent heat flux, H is the sensible heat flux, and G is the soil heat flux (all with units of W·m^{-2}) [18,19]. The energy balance explains the dynamics for the dispersion of radiant energy from the sun to the land surface. Energy from the sun is either reflected or absorbed. The portion absorbed is the R_n, where it is positive when energy moves toward the surface (plant canopy) and negative when it moves from the surface. The absorbed R_n can then be distributed to the soil as G, to the air as H, or provide energy to evaporate water as LE. Sign conventions for Equation (1) vary, but in this study, G is positive when flux moves toward the soil surface, LE is positive when flux moves toward the plant canopy, and H is positive as flux moves from the canopy to the air.

1.2. Scintillometry

A scintillometer consists of a transmitter and receiver, separated by a specified path length. Scintillometry, as applied to agricultural and other landscapes, uses a beam of electromagnetic radiation of known wavelength transmitted across a relatively large distance (100 m–4.5 km). The beam intensity fluctuates due to absorption and diffraction as it encounters eddies in the air. These fluctuations, or scintillations, can be used to determine the structural parameter of the refractive index of air, which can be used to calculate the structural parameters for temperature and humidity, and H. The calculations to obtain H from scintillometers are based on the Monin–Obukhov Similarity Theory (MOST). Details of MOST can be found in Arya [20], Foken [21], Hartogensis [22], McAneney et al. [23] and Monin et al. [24].

The main output of a scintillometer is the natural logarithm of the received beam intensity. Since the SLS uses two parallel beams, the calculation methodology is different. By using two beams, the covariance between the two can be calculated as:

$$B_{12} = 4\pi^2 K^2 \int_0^L \int_0^\infty k\varnothing_n(k) J_0(kd) \sin^2\left[\frac{k^2(L-x)}{2KL}\right] \frac{4J_1^2\left(\frac{KDx}{2L}\right)}{\left(\frac{kDx}{2L}\right)^2} dk dx, \qquad (2)$$

where B_{12} is the covariance between the beams, K is the wavenumber ($K = 2\pi/\lambda$ rad m^{-1}), D is the detector diameter (mm), d is the beam separation distance (mm), L is the path length (m), x is the coordinate along the path, k is the von Karman constant, φ_n is the three dimensional spectrum of the refractive index inhomogeneities, and J_0 and J_1 are Bessel functions of the first kind [25]. The term φ_n is calculated as:

$$\varnothing_n(k) = 0.33 C_n^2 k^{-\frac{11}{3}} f_\varnothing(kl_0), \qquad (3)$$

where C_n^2 is the structural parameter of the refractive index of air (m$^{-2/3}$) and $f_\varphi(kl_0)$ is the refractive index decay function. Equation (3) can be inserted into Equation (2) to define the covariance and the variance if C_n^2, the inner scale of turbulence (l_0), and the instrument physical dimensions are known, thus allowing the use of B_{12} and B_1 or B_2 to derive C_n^2 and l_0 [25]. Since the SLS measures variances in intensity, σ_i^2, rather than variances of log amplitude, B_i, σ_i^2 must be converted to B_i:

$$B_i = \frac{1}{4} \log\left(1 + \frac{\sigma_i^2}{\langle I_i \rangle^2}\right), \qquad (4)$$

where I_i is the intensity [25]. From the SLS, C_T^2 can be calculated from C_n^2 as:

$$C_T^2 = C_n^2 T^4 (ap)^{-2}, \qquad (5)$$

where T is the temperature in Kelvin, p is the atmospheric pressure in mbar, and a is a constant of 7.89×10^{-5} K/mbar at the 670 nm wavelength [25].

One advantage SLS offers over other point source measurements, is that the fluxes can be determined over shorter lengths and at heights closer to the surface [26]. Also, the SLS determines the l_0, which is proportional to the dissipation rate of the turbulent kinetic energy, ε, which are used to determine the friction velocity (u_*) and temperature scale (T_*). Sensible heat flux is calculated from u_* and T_* as:

$$H = -\rho C_p u_* T_*, \qquad (6)$$

where C_p is the specific heat of air (kJ·kg^{-1}) and ρ is the air density (kg·m^{-3}). The complete process to determine H from the SLS is provided in Savage [27] and Scintec [25].

Many different models of scintillometers are available, which differ based on their wavelength and aperture diameter. The large-aperture scintillometer (LAS) has a wavelength around 880 nm and an aperture of 10–30 cm. Microwave scintillometers (MWS) have an aperture around 30 cm and a wavelength of 1–3 mm. Several studies have evaluated scintillometry using the LAS and/or the MWS due to their long operational range. The MWS is noted to have the ability to measure LE directly [28–30] as the wavelength is sensitive to humidity fluctuations. Samain et al. [31] found RMSE values of approximately 14 W·m^{-2} for both LE and H between ETLook, TOPMODEL-based Land-Atmosphere Transfer Scheme (TOPLATS), and a LAS. Samain and Pauwels [32] found an RMSE of 0.1 mm·d^{-1} between an LAS and ET calculated from the Penman–Monteith equation. Meijninger et al. [33] found LE from a single LAS and a dual LAS–MWS setup were within 25% of the LE obtained from EC. Guyot et al. [34] found good agreement from a LAS as compared to EC and the water balance approach. Yee et al. [35] found a standalone LAS to be more suitable than the two-wavelength approach or a standalone MWS for a semi-arid environment. Studies evaluating a LAS or validating using instruments other than a lysimeter are abundant in the literature [28,36–43].

The SLS has been available for many years; however, SLS agricultural and crop-based studies are not abundant in the literature. Research related to the LAS or MWS vastly outweighs research using the SLS. Most SLS literature is related to atmospheric research and the number of publications is still small compared to other instruments. A few studies illustrate the benefits of the SLS over other flux measurement instruments, such as EC. Odhiambo and Savage [44] provided benefits and disadvantages of SLS, as well as an overview of scintillometry and various scintillometers. They note that the operation of a scintillometer using a path average approach provides the benefit of allowing for shorter time intervals. In addition, scintillometry does not require corrections (such as frequency response corrections for EC), that are generally subjective, to be applied. Lastly, Odhiambo and Savage [44] mentioned that since the variance of the logarithm of beam amplitude is the measurement used, any calibration multipliers and constants cancel or are removed by band-pass filtering. The disadvantages given by Odhiambo and Savage [44] include using MOST, which requires the effective beam height and zero-plane displacement height. In addition, the scintillometer itself cannot determine the direction of H, so the temperature difference from additional temperature sensors or other methods must be used to determine the flux direction.

Savage et al. [45] investigated H using a SLS, in addition to EC and Bowen Ratio, over natural grassland over a longer time period. They noted that most SLS studies were rather short, commonly with a time frame of a few days to a few weeks. Therefore, they performed an analysis using more than 30 months of data. They used averaging periods of two minutes for energy balance components, ET_{ref} estimates, and SLS measurements, 20 min for Bowen Ratio, and 30 min for EC. Their regression analysis of daily ET between Bowen Ratio to SLS (BR-SLS) showed an r^2 of 0.87, while a similar analysis of EC to SLS (EC-SLS) showed improved correlation with an r^2 of 0.96. The ET estimate comparison showed an RMSE of 0.58 mm·d^{-1} for the SLS-BR and 0.31 mm·d^{-1} for SLS-EC. Comparisons to ET_{ref} were not made.

De Bruin et al. [46] also found good correlation between H from an EC system and SLS near Uppsala, Sweden over a mixed landscape encompassing mostly agricultural land. Their regression analysis showed the flux from the EC was 1% greater and the flux comparison between EC and SLS had an r^2 of 0.95. They also compared u_* from the SLS and EC and found an r^2 of 0.87.

Nakaya et al. [47] compared an SLS with an EC system over a deciduous forest in Japan. They found the SLS to overestimate H and the LE, as compared to their EC system. Since EC systems often underestimate LE, the SLS values could have been closer to the real LE. They also observed that the SLS underestimated u_*, especially when the values of u_* were large. They concluded that the cause in the underestimation of u_* is due to systematic errors in the determination of l_0.

Hartogensis et al. [48] analyzed data from a SLS using data from the Co-operative Atmosphere Surface Exchange Study (CASES-99) experiment conducted in Kansas. They noted the benefits associated with using an SLS over EC, including a shorter averaging interval, shorter measurement height, and shorter path length, while also noting some disadvantages to the SLS. They noted that conversion from l_0 and C_n^2 to H is done empirically using MOST. This can result in a bias in l_0, causing an overestimation of u_* for small u_* values and underestimation of u_* for large u_* values [46]. The CASES-99 experiment deployed both a SLS and an EC system. Hartogensis, De Bruin and Van de Wiel [48] found that from the shorter averaging interval from the SLS (6 s for CASES-99), intermittent turbulence could be detected, whereas the longer averaging interval for the EC (30 min) would mask some of the intermittent turbulence. Comparing the ε, they showed the bias in the l_0 caused a bias in the SLS ε where an overestimation of ε occurred for small values of ε and an underestimation occurred for large values of ε. Comparing ε obtained from SLS and EC showed good agreement between the two instruments. Comparing C_T^2 obtained from SLS and EC showed less agreement where the SLS C_T^2 was consistently larger [48]. In their analysis of H, they found the SLS to overestimate the EC H for fluxes greater than -50 W·m^{-2}. They concluded that most of the error in the SLS data, as compared to EC, can be improved upon by taking the aperture diameter as 2.6 mm instead of the manufacturer reported 2.7 mm. This was tested in response to a finding in De Bruin et al. [46] where empirical functions were proposed to account for noise and inactive turbulence (where turbulence does not create mixing).

The previously described studies reiterate that evaluation of the SLS has used EC as validation. SLS studies involving lysimeter comparisons were not found in the literature and leave an opportunity for evaluation against highly accurate data. In addition, the SLS has not been widely evaluated for determining ET. In this study, an SLS was evaluated against data from a large weighing lysimeter for determination of hourly and daily H and ET for agricultural crops.

2. Materials and Methods

2.1. Site Description

This study was conducted at the USDA-ARS Conservation and Production Research Laboratory, located 17 km west of Amarillo, TX. The region is classified as semi-arid with approximately 450 mm (17.7 in.) average annual precipitation. The study site is located inside a 19-ha square-shaped field, which is split into four, 4.7 ha quadrants (see Figure 1). Each quadrant is roughly 200 m by 220 m and contains a large weighing lysimeter located in the center of the quadrant. The east quadrants are irrigated with a subsurface drip irrigation system, and the west quadrants are irrigated using a low pressure lateral move sprinkler irrigation system. The NE quadrant was used for this study as it provides the greatest fetch with respect to the predominant wind direction (south southwest). In the 2015 summer growing season, the field was planted with grain sorghum (*Sorghum bicolor* (L.) Moench) at a rate of 210,000 seeds per ha (85,000 seeds per acre), which was fully irrigated using subsurface drip lines buried at 30 cm (12 in.) depth and 152 cm (60 in.) lateral spacing, centered in alternate interrows. In the 2016 growing season, the field was planted with grain corn (*Zea mays* L.) at a rate of 104,000 seeds per ha (42,100 seeds per acre), which was also fully irrigated using the previously described subsurface drip irrigation system.

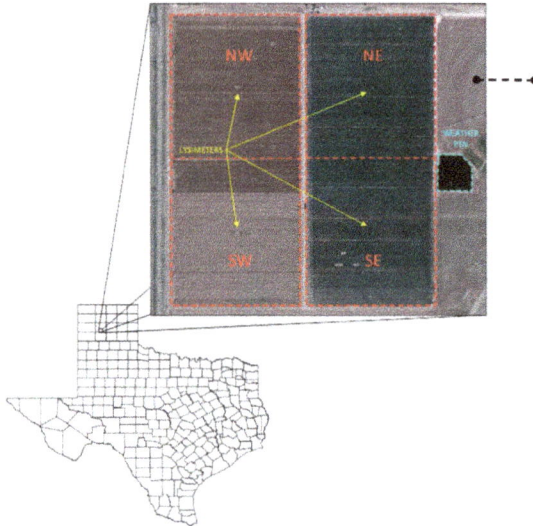

Figure 1. The study location with placement of large weighing lysimeters at the centers of four square fields at the USDA-ARS Conservation and Production Research Laboratory, Bushland, TX. The research weather pen is shown adjacent to the east side of the east lysimeter fields. The dashed black line in the NE field illustrates the path of the scintillometer radiation.

2.2. Lysimeter Description

Hourly H and ET derived from SLS data (SLS-20, Scintec AG, Rottenburg, Germany) were evaluated by comparing against lysimeter data. Estimates from the SLS are denoted "sls" subscript and measurements by the lysimeter are denoted by "lys" subscript. The large weighing lysimeter measures 3 m by 3 m on the surface by 2.3 m deep over a fine sand drainage base. It contains an undisturbed monolith of Pullman silty clay loam soil. The soil container rests on a large balance scale equipped with a counterbalance and load cell system. Initial design and installation details of the lysimeter were provided by Marek et al. [49] and Schneider et al. [50]. The lysimeter was later equipped with drainage effluent tanks suspended from the lysimeter by load cells for separate measurement of drainage mass without changing total lysimeter mass. Load cell output is measured and recorded by a precision datalogger (CR7 in 2016, upgraded to CR6 for 2016, Campbell Scientific, Logan, UT, USA). Load cell voltage outputs are converted to mass using calibration equations, and five-minute means are used to develop a base dataset for subsequent processing [51]. Lysimeter mass in kg is converted to a mass-equivalent relative lysimeter storage value (mm of water) by dividing the mass by the relevant surface area of the lysimeter (~9 m^2) and the density of water (1000 kg·m^{-3}). See Evett et al. [6] for a thorough description of the calculations, and of the lysimeter operation, sensors, and ancillary equipment. Equivalent mass values allow for changes in lysimeter mass to be expressed in terms of water flux, defined as mm of water lost or gained per unit time. The lysimeter datalogger mass resolution is better than 0.001 mm when converted to equivalent depth of water. Lysimeter accuracy is, however, determined by the RMSE of calibration, which ranged from 0.01 mm to 0.05 mm [6,51]. Lysimetric data quality assurance and quality control (QA/QC) and data processing techniques were discussed by Marek et al. [52].

In a few instances, field operations prohibited accurate measurements with the lysimeter. For example, maintenance on the lysimeter, or instruments mounted near the lysimeter, required personnel to step onto the lysimeter container, temporarily increasing the mass. Other operations include draining the percolation storage tanks, which causes an overall decrease in the mass, irrigation applications, and foot traffic associated with taking neutron probe soil water measurements. The amount of water drained is measured, but data recorded during the process of draining the tanks is not valid and usable. For sub-daily data, periods of precipitation or irrigation reduce data availability since the precipitation/irrigation cannot be accounted for in the ET_{lys} for short periods, such as hourly or 30 min intervals. In the instance where operations, precipitation, or irrigation limited accuracy of collected data, the data from those time intervals were omitted. Precipitation and irrigation events were the dominant reason for data exclusion in this study. In the instances where lysimeter data were omitted, the corresponding SLS data also were excluded. For daily and longer time steps, the data can be evaluated and/or corrected to account for periods when hourly data are not available. This involves subtracting the lysimeter mass value at midnight from the mass at midnight of the previous day. With this process, temporary changes in mass, such as taking neutron probe readings, can be disregarded. Permanent changes in mass such as irrigation, precipitation, and emptying drainage tanks can be accounted for by adding/subtracting the mass of water added or removed to the midnight-midnight mass difference. Details on the evaluation and correction of daily lysimeter data can be found in Marek et al. [52].

2.3. Scintillometer Installation

The SLS was installed in the NE lysimeter field with an east-west path length of 100 m passing approximately 25 m north of the lysimeter (Figure 1). The measurement height was 1.73 m in 2015 and 2.84 m in 2016. The maximum crop heights were 1.3 m and 2.3 m for grain sorghum in 2015 and corn in 2016, respectively. The instrument height of 1.73 m was the highest the SLS could be raised with the tripod supplied with the instrument. To compensate for the taller corn crop, a taller tripod was fabricated. The 2.84 m height in 2016 was the tallest the fabricated tripod could achieve. The SLS was connected to a custom built PC running SLS configuration and processing software (SRun version 1.28, Scintec AG, Rottenburg, Germany). The SLS was oriented with the transmitter facing east so the path was as close as possible to perpendicular to the predominant wind direction. The SLS measurements for the period of 29 July–22 August 2015 and 23 June–2 October 2016 were used for this study. Sensible heat flux was determined using SRun software, and H_{sls} data were provided for assumptions according to stable (H_{stable}) and unstable ($H_{unstable}$) atmospheric conditions. Using scintillometry, the value of H can be determined, but not the direction of the flux [44]. The flux direction must be determined during data post-processing using supplemental instrument data. In stable atmospheric conditions, the H will be negative while H will be positive in unstable conditions. Several methods are reported in the literature for determining flux direction [36,44,53,54]. One is based on the view that the atmosphere is typically stable at night and becomes unstable during the day [41]. To account for this pattern, assuming H_{stable} for nighttime hours and $H_{unstable}$ for daytime can be used to assign H direction (H_{sun}). Another method of determining flux direction (H_{dT}) is to use the temperature gradient as determined from two co-located air temperature sensors at different heights [43]. The temperature gradient provides an indication of the stability of the atmosphere near the surface, as well as an indication as to the direction of sensible heat flux since sensible heat will flow across the gradient from high to low. In stable atmospheric conditions, the gradient indicates air temperature increases with height. This will cause heat energy to flow down into the crop canopy and indicate negative H. In an unstable atmosphere, temperature decreases with height, so heat will move vertically and away from the canopy, resulting in positive H. In 2015, air temperature was measured at 1.2 and 2.8 m, while in 2016 air temperature was measured at 2.8 and 6.4 m. The air temperature measurement heights provided a measurement at or below and above the SLS beam for both years. Both of the described methods were used to determine ET_{sls} and compared to the lysimeter to determine the effects of H_{sls} direction methodology. For H_{sun}, the daytime and nighttime hours were determined using sunrise and sunset times for the study period (7:00 a.m. and 9:00 p.m. CDT, on average, respectively).

Additional instruments included air temperature and relative humidity sensors (HMP155, Vaisala, Helsinki, Finland), four soil heat flux plates (HFT-3, Radiation Energy Balance Systems, Bellevue, Washington, DC, USA), four soil water sensors (TDR315, Acclima, Meridian, ID, USA), an infrared thermometer (IRT), and a net radiometer (Q*7, Radiation Energy Balance Systems). The layout of the soil sensors within the lysimeter is presented in Figure 2. The instruments were connected to a datalogger (CR6, Campbell Scientific) with a measurement frequency of six seconds and averaged to an hourly interval.

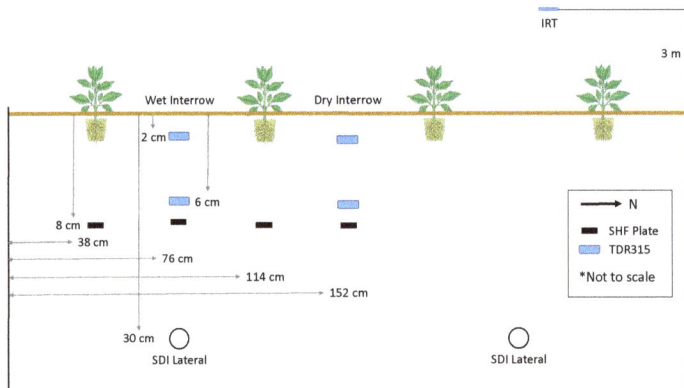

Figure 2. Soil sensor placement within the lysimeter. Shown are relative placement of soil heat flux (SHF) plates, model TDR315 soil water content sensors, and subsurface drip irrigation lines. The TDR315 sensors give accurate soil temperature readings, eliminating the need for thermocouples where they are used.

The soil water and temperature data were used to calculate G at the soil surface by the calorimetric method, and surface G was used in all energy balance calculations. The calorimetric method used in this study was described by Colaizzi et al. [55]; briefly, it used the soil water and temperature measurements to calculate the change in soil heat storage between the surface and the depth of the soil heat flux plates in 1 h time steps. Using measured R_n and surface G provided the available energy to accompany the H measurement and LE calculation. Data from the R_n and surface G instruments on the lysimeter were used with the SLS data.

H was back-calculated from the lysimeter by converting the ET_{lys} to *LE*, summing *LE* with surface soil heat flux and measured net radiation, and treating the residual of this energy balance to H. LE was calculated from ET by:

$$LE = \frac{ET(mm)\lambda}{time(sec)},\tag{7}$$

where λ is the latent heat of vaporization ($J \cdot kg^{-1}$). The latent heat of vaporization was calculated from the surface temperature (T_s °C) by:

$$\lambda = (2.501 - 0.0236T_s) \times 10^6.\tag{8}$$

A south-facing, nadir IRT installed 3 m above the lysimeter was used to determine the surface temperature, which was then used to calculate the latent heat of vaporization for each hourly period. Using the ET_{lys} data, the LE was calculated by multiplying the ET_{lys} by the latent heat of vaporization and dividing by 3600 s to convert to the hourly time period.

During nighttime hours, *LE*, and subsequently ET, become very small [56]. The much smaller values can exhibit much more relative variation; even though the magnitude of the differences may be small, relative percentage differences can be large. To determine the effects of including nighttime ET_{sls}, evaluations were also conducted using only daytime ET_{sls} data. This analysis provided an indication as to how much of the overall variation was influenced by the much smaller nighttime values.

As most water management practices and modeling efforts use a daily time step, the 24 hourly periods were summed to provide an evaluation for daily ET_{sls} data. The 24-h summation was performed for both ET_{lys} and ET_{sls}. Since some hourly intervals were omitted due to operations on the lysimeter, daily lysimeter data (midnight–midnight mass change) as determined by the processes outlined in Marek et al. [52] were also used to evaluate the 24-h summations from the SLS. The statistics of root mean square error (RMSE), percent RMSE (%RMSE), and regression analysis (H_{sls} regressed to H_{lys}, ET_{sls} regressed to ET_{lys}) were used as the basis of evaluation. The %RMSE provides a relational value that allows for comparison of results of different magnitudes. The %RMSE was calculated by dividing the RMSE by the average of the measured lysimeter data.

3. Results

Daily ET_{lys} values for 2015 and 2016, as well as crop height and leaf area index (LAI) are presented in Figure 3. An exceptionally wet year occurred in 2015 for the study location, which may have caused greater evaporation and thus higher ET_{lys} values as compared to 2016 for the same date range. Since the lysimeter field is irrigated using subsurface drip, the soil surface is typically dry and evaporation is minimized. When a precipitation event occurs, the soil and plant surfaces become wet and experience evaporation that does not occur with irrigation as the irrigation water typically does not reach the soil surface. In addition, during the data period for 2015, the sorghum crop was still in a vegetative growth stage with higher relative ET. The data for 2016 encompasses vegetative growth of the corn crop, but also, late season reproductive growth and grain filling. During the corn grain fill stage, ET_{lys} is lower and possibly reduced the overall hourly average values. Both years exhibit the pattern of lower ET_{lys} during the night hours, increasing from sunrise to a maximum around 14:00 and then decreasing. Although ET_{lys} is drastically reduced during the night, the measured ET_{lys} values illustrate that a small amount of ET does occur at night, which is not negligible [56].

Energy balance closure was 87% (based on the slope of the regression equation) when the available energy (AE: R_n-G) was regressed against H_{dT} + LE_{lys}. The energy balance closure error was similar to the %RMSE for H_{dt} (87% and 83%, respectively). This similarity leads to the inference that the energy balance closure error results from error in H_{dt} as compared to H_{lys}. The results of the error analyses for the turbulent fluxes and ET_{sls} are presented in Table 1. Even though the H_{sls} had weak correlation with the H_{lys}, and a large error, the resulting ET_{sls} had a much smaller error, especially for daily ET_{sls}. This indicates that errors in determining H do not have a drastic impact on ET_{sls}. In irrigated agriculture, H is commonly small, especially compared to LE. Large errors in a small component may not result in large error in the final product. Results for the AE and incoming solar irradiance regressed against ET_{lys} are also included in Table 1. This analysis showed the extent of error associated with estimating ET without measuring H. The slopes for the regression equations from AE and irradiance are provided in Table 1; however, the slope values are much larger than for ET and H since AE and irradiance values are considerably larger than ET. Errors for ET were larger using AE and irradiance as predictors of ET, which indicates that although large errors exist for H_{sls}, ET_{sls} errors are still lower when accounting for the H component. The error rates and small error suggest that potential exists for additional research and potential improvements in SLS measurements for H, and possibly the resulting ET_{sls}.

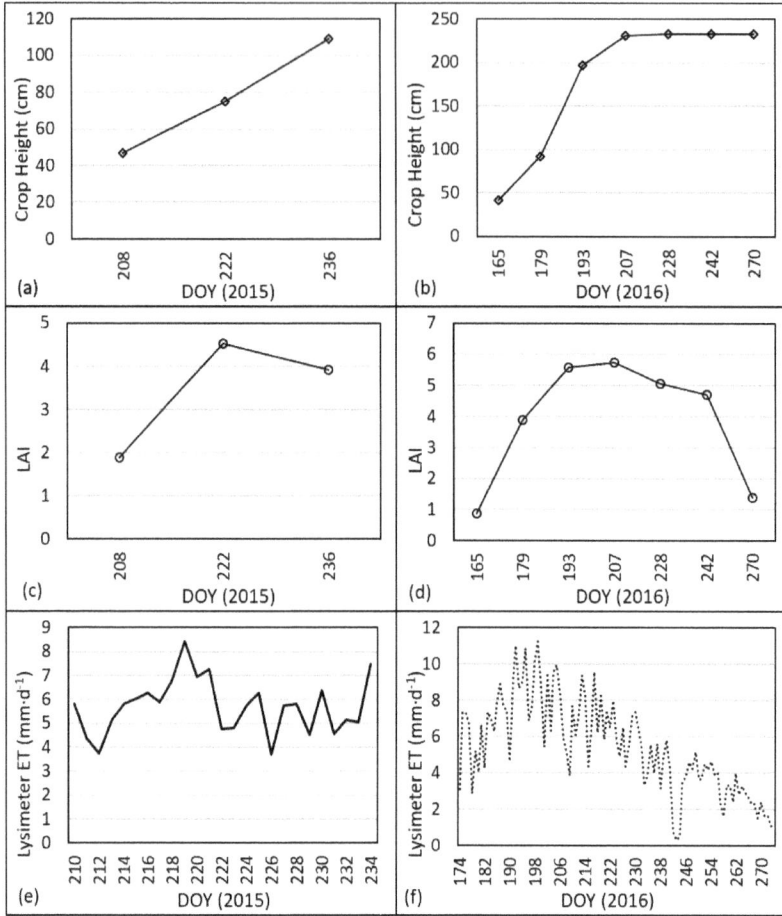

Figure 3. Crop height for 2015 (**a**) and 2016 (**b**), LAI for 2015 (**c**) and 2016 (**d**), and daily lysimeter ET for the 2015 (**e**) and 2016 (**f**) study periods.

Table 1. Regression statistics from comparison of hourly and daily H and ET data between SLS and lysimeter.

Data	RMSE	%RMSE	R^2	Slope
H_{dT}	83.8 W·m^{-2}	83	0.19	0.31
ET (hourly)	0.10 mm	40	0.89	0.97
ET (daily)	0.75 mm	13	0.87	0.91
AE	0.16 mm	63	0.72	697
Irradiance	0.15 mm	62	0.72	920

The average daily energy balance components for the lysimeter are presented in Figure 4. From the figure, R_n and LE had the largest values, and H and G had much lower values. Since R_n is the main energy input, when one component (LE) increases, another component (H) should decrease. This is not always the case as the environment in the Texas High Plains commonly generates advection where heat energy is transferred from a warmer, typically drier adjacent field to cooler irrigated fields. Evidence of advection is also seen in cases where *LE* exceeds R_n [7,57]. The influx of heat energy from nearby (warmer) fields increases the energy available for evaporation of water during transpiration, thus increasing the latent heat flux.

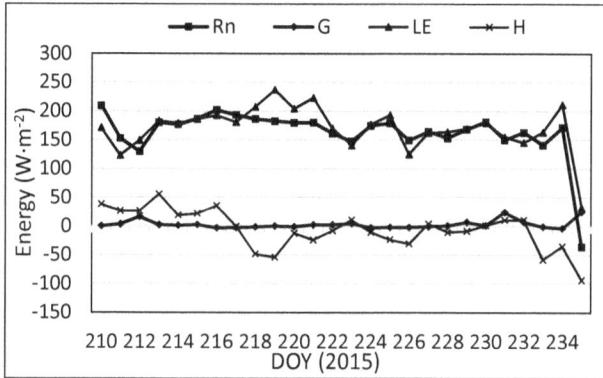

(a)

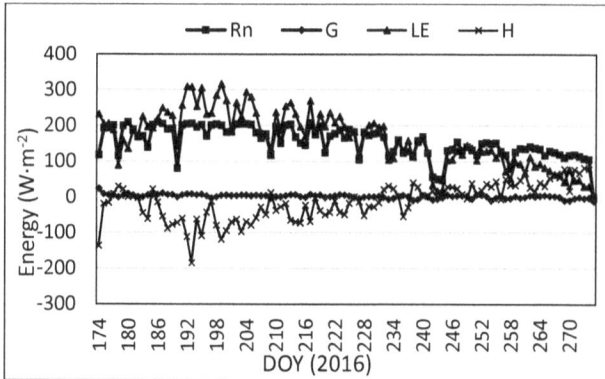

(b)

Figure 4. Average daily energy balance components on the lysimeter for (**a**) 2015; (**b**) 2016.

Figure 5 presents average daily data for all four components of the energy balance equation as determined by the SLS. For the SLS, the temperature difference was used to determine flux direction for H shown in Figure 5. In some instances, *LE* is greater than R_n during the daytime, which in this case, corresponds to instances where LE was greater than R_n with the lysimeter. This usually indicates the occurrence of advection rather than an *LE* error since the same higher LE was measured on the lysimeter for the same dates.

(a)

(b)

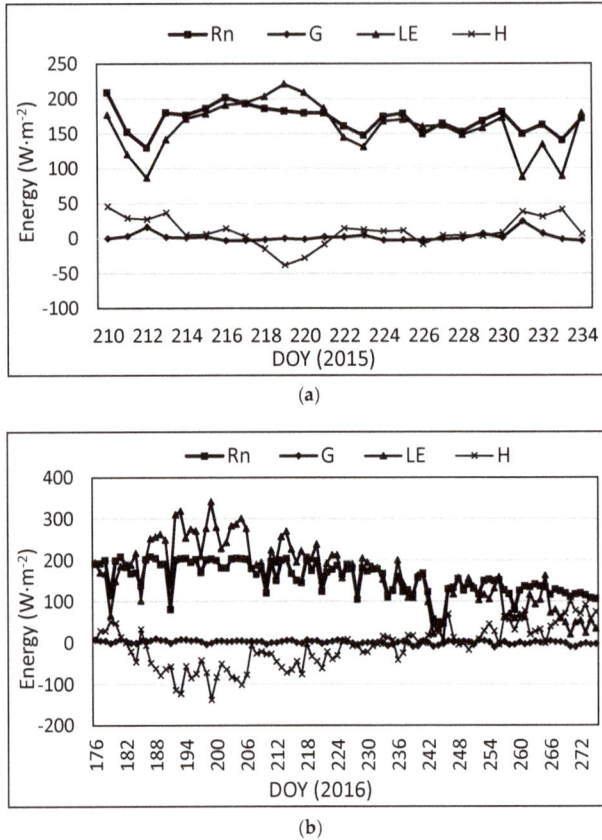

Figure 5. Average daily energy balance components for the SLS for (**a**) 2015; (**b**) 2016.

The results from the statistical analyses of H showed that using the temperature gradient produced a lower error than using H_{sun}. The assumption that the atmosphere was stable at night and unstable during the day proved to not always be true. Comparisons of H_{dT} and H_{sun} are presented in Figure 6. The SLS overestimated H as compared to the lysimeter, which is illustrated by Figure 6 as well as the slope of the regression equation (0.31 and 0.06, respectively). H_{dt} from the SLS ranged from -347.7 to 475.8 $W·m^{-2}$ with a mean of -5.1 $W·m^{-2}$ and a standard deviation of 92.8 $W·m^{-2}$. H_{sun} from the SLS ranged from -226.6 to 475.8 $W·m^{-2}$ with a mean of 33.6 $W·m^{-2}$ and a standard deviation of 93.7 $W·m^{-2}$. H determined from the lysimeter ranged from -761.7 to 426.4 $W·m^{-2}$ with a mean of -11.4 $W·m^{-2}$ and a standard deviation of 130.1 $W·m^{-2}$. The lysimeter showed a broader range of H values compared to the SLS, although mostly with regard to negative flux values. Although weak correlation is shown in Figure 6, the statistical analysis showed significant correlation ($p < 0.05$) for all data periods (2015 only, 2016 only, and 2015–2016) for H. The regression analyses showed all slopes were significant and all intercepts were significant ($p < 0.05$) with the exception of H_{dT} for the combined 2015–2016 dataset ($p = 0.27$). The measurement footprints of the lysimeter and SLS are not exactly the same. The lysimeter measures a 3 m by 3 m square whereas the SLS measures an average across the 100 m path, which is 25 m north of the lysimeter. Although care is taken to ensure the lysimeter is representative of the surrounding field, the difference in measurement footprint may contribute to the error in H.

(a)

(b)

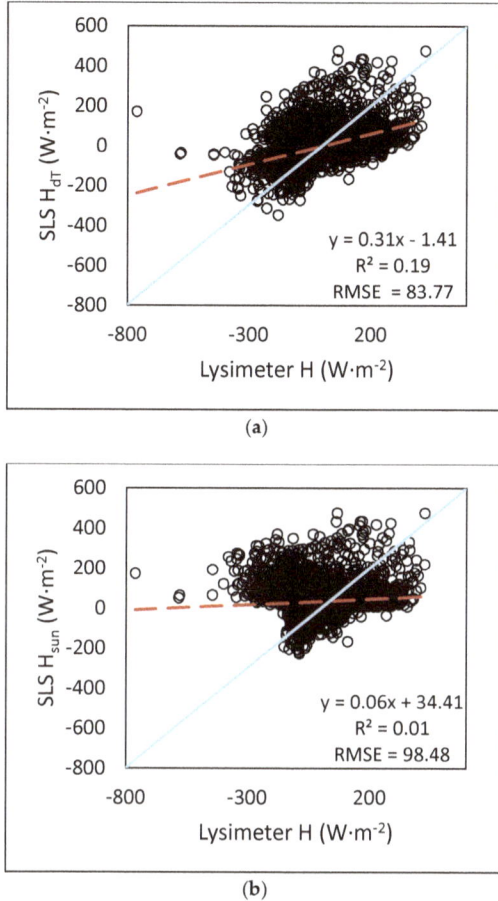

Figure 6. Comparison of hourly H_{dT} and H_{sun} from the SLS to lysimeter H for the combined 2015–2016 data based on lysimeter data regressed to (**a**) H_{dT}; (**b**) H_{sun}.

Average hourly H directions, as indicated by the temperature gradient, for the study period are presented in Figure 7. A clear diurnal pattern is found with the flux direction, as it is common to find a stable atmosphere during nighttime hours and an unstable atmosphere during the daytime. Overall, stable conditions occurred more than unstable atmospheric conditions, with 31% and 42% of the hourly periods under unstable conditions for 2015 and 2016, respectively. Having a greater occurrence of stable atmospheric conditions was not expected as the atmosphere is typically unstable during the day, and the day length was longer than the length of nighttime period. Seeing more stable conditions than unstable indicates that stable atmospheric conditions must have occurred during some of the daytime hours, as also indicated by Figure 7. With 2015 being one of the wettest years on record for the study area, the relative humidity was greater for 2015 than 2016. The average relative humidity for the 2015 study period was 70% whereas the average humidity for 2016 was 59%. In addition, the average wind speed was lower for 2015 compared to 2016 (3.33 m·s^{-1} and 3.81 m·s^{-1}, respectively). The lower wind speeds and greater humidity in 2015 may have allowed the atmosphere to stay more stable than in 2016. The greater humidity could have allowed the atmosphere to hold more heat and the lower wind speed could have reduced turbulence and atmospheric mixing.

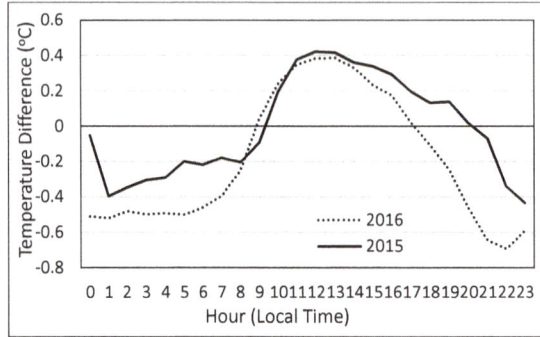

Figure 7. Temperature difference (calculated as the difference between values from two air temperature sensors at different heights—the lower sensor minus the higher sensor) and used as the indicator for flux direction.

In further analysis, hourly ET_{sls} was then calculated using the LE from the SLS and evaluated against lysimeter data. Hourly ET_{sls} correlated well with hourly ET_{lys} (see Figure 8). Coefficients of determination ranged from 0.88 to 0.95 with regression slopes ranging from 0.75 to 0.96. Although the slopes are close to 1, hourly ET_{sls} is slightly overestimated as compared to the lysimeter. RMSE values ranged from 0.08 to 0.10 mm·h^{-1}, although the relative errors indicated by the %RMSE were much larger, ranging from 25 to 41%. The 2015 dataset resulted in smaller error than the 2016 dataset with %RMSE of 25% and 40%, respectively. This is possibly due to the timing of the 2015 data. The same period of 29 July–22 August for 2016 had a similar error to 2015 (0.09 mm, 30%). The period before 29 July, from 23 June to 28 July 2016 had an error of 0.11 (33%) and the period after 22 August from 23 August to 2 October 2016 had an error of 0.10 (72%). Although the RMSE values for the three periods in 2016 were similar, the %RMSE values are different. The sorghum crop in 2015 reached maximum plant height around 21 August, so the 2015 data were collected over at least a partially growing crop. The corn crop in 2016 reached maximum height around 15 July, so both later periods included the same crop height; however, the latter period was during grain fill when the corn ET values were less. For the period of 29 July–22 August 2016, average daily ET_{lys} was 6.73 mm; for the period of 23 August–2 October 2016, the average hourly ET_{lys} was 3.16 mm, roughly half of that for the middle period (see Figure 3a).

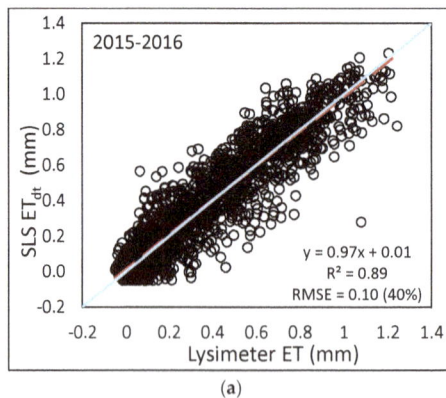

(a)

Figure 8. *Cont.*

(b)

(c)

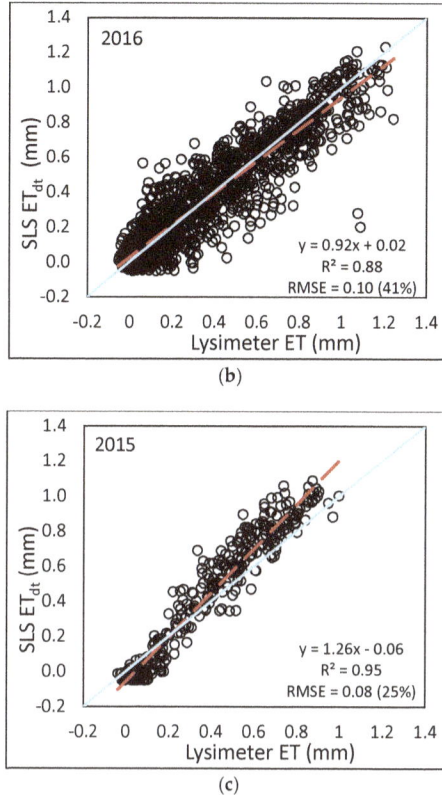

Figure 8. Comparison of hourly ET from SLS and lysimeter using (**a**) combined 2015–2016 data; (**b**) 2016 only; (**c**) 2015 only.

In addition to the crop growth, the weather may have contributed the differences in accuracy of ET_{sls} for 2015 and 2016. Comparing ET_{ref} (which is calculated based on weather measurements) for the same date range of 29 July–22 August in 2015 and 2016, 2016 had larger values than 2015 (see Figure 9). Differences in ET_{ref} between the years were significant with a *p*-value of 0.04. Comparing maximum and minimum air temperatures as well as incoming solar radiation, 2016 showed much more variation in daily total incoming radiation. Incoming radiation data for 2015 were consistently higher and less variable than 2016, which may indicate more occurrences of clear skies in 2015. In addition to differences in incoming solar radiation, the difference between maximum and minimum air temperatures was greater in 2016 that in 2015. Although differences in temperatures were evident, statistical analysis showed no significant difference in maximum air temperature, but significant differences were present with minimum air temperature (*p*-value = 0.027). Wind speed measured at 2 m also showed significant differences (*p*-value = 0.030), with 2016 having higher wind speeds than 2015. The broader range of temperatures and higher wind speeds in 2016 may have had an adverse effect on SLS measurements.

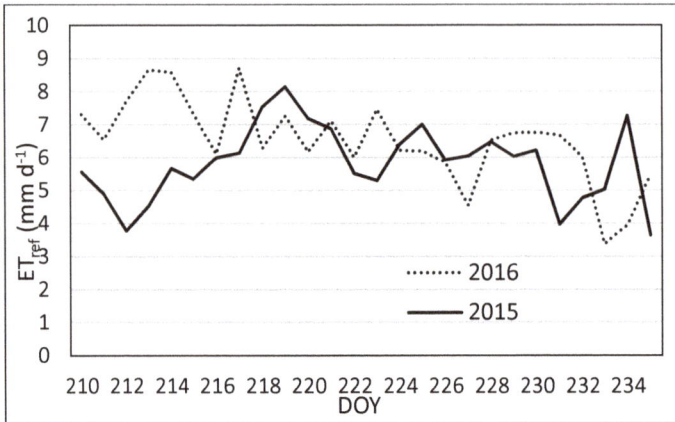

Figure 9. Daily ET_{ref} for the same DOY in 2015 and 2016, used as an indicator for differences in weather parameters.

In Figure 8, a cluster of values was found near zero for both the SLS and lysimeter. These small values likely occurred at night when photosynthesis dramatically slowed and less transpiration occurred. To investigate potential bias from the inclusion of these small values, the daytime ET_{sls} data were separated from the nighttime ET_{sls}. The daytime ET_{sls} data were evaluated against daytime lysimeter measurements. Using only the daytime ET_{sls} resulted in a slight change in the regression equations, as well as increased variation, indicated by the reduced r^2 values (see Figure 10). Although the variation increased, the RMSE decreased from 25 to 18%, 41 to 29%, and 40 to 31% for the combined 2015 only, 2016 only, and 2015–2016 data, respectively. The RMSE was larger using only daytime data; however the magnitude of ET_{sls} values were larger for the daytime, resulting in a smaller %RMSE.

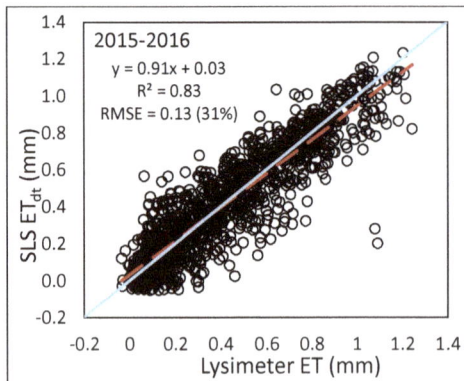

(a)

Figure 10. *Cont.*

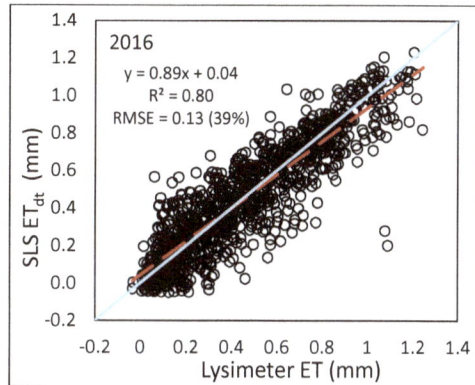

(b)

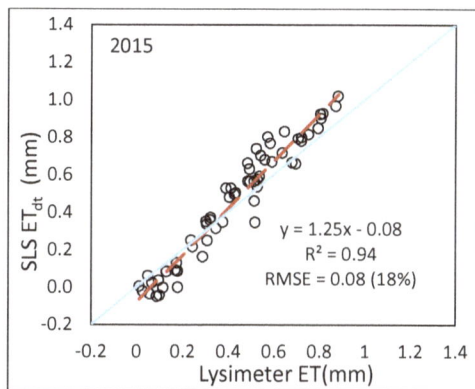

(c)

Figure 10. Comparison of daytime hourly ET from the SLS to lysimeter data using (**a**) 2015–2016 combined data; (**b**) 2016; (**c**) 2015.

Regressions using the summed daily data (n = 24, 100, 124 for 2015 only, 2016 only, and 2015–2016, respectively) resulted in a slightly less correlation than the hourly regressions using hourly data. Coefficients of variation for the 2015 only, 2016 only, and combined 2015–2016 daily data were 0.85, 0.88, and 0.87, respectively, where the hourly data had r^2 values of 0.95, 0.88, and 0.89. However, the slopes of the regressions were much closer to one for the daily sums, indicating that the regression equations were able to explain more of the variation (see Figure 11). All correlations were statistically significant with p-values less than 0.05. In addition, all slopes and intercepts (values presented in figures) were significantly different from one and zero, respectively. Regression on daily ET_{sls} had an RMSE of 0.68 mm·d^{-1} for the combined 2015–2016 data, which corresponded to less than half of the percent error for hourly ET_{sls} (14% compared to 31%). The reduction in error is likely due to the SLS having values greater and less than the lysimeter. At the hourly time step, these values contributed to the error; however, at the daily time step, the greater and lesser values canceled out after summation. It is important to note that the daily sums do not include all 24 data hours for all days. In the daily sums, the hourly data corresponding to missing data were omitted and only hours for which there were acceptable data were summed. Even though usable data for some days were not complete, the sums for both the SLS and the lysimeter still contained the same number of hourly values.

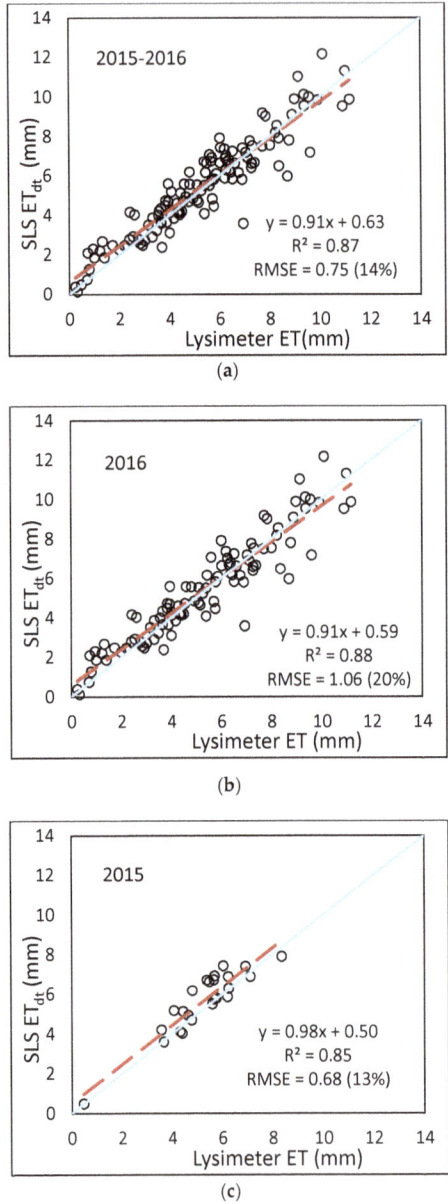

Figure 11. Comparison of summed daily ET from the SLS to summed hourly lysimeter data using (a) combined 2015–2016 data; (b) 2016 only; (c) 2015 only.

The results of the daily lysimeter (mass change from midnight to midnight) analysis are presented in Figure 12. For 2015, the results were similar to the summed hourly data; however there were some differences in the error for 2016 and the combined 2015–2016 data. In the hourly sum analysis, periods of precipitation or irrigation (the main reason lysimeter data were omitted) were not included in the analysis, therefore, the effects of the soil (or subsoil) wetting and drying were not considered. In the daily analysis, these periods were included, and the scintillometer may not have captured the change in soil and atmosphere dynamics caused by the soil wetting. With drip irrigation, the soil surface is typically dry and evaporation is minimized. With a rainfall event, the soil surface is wetted and evaporation increases thus reducing H and increasing *LE*. These transition periods, where energy shifts from H to *LE*, may not be adequately captured by the SLS.

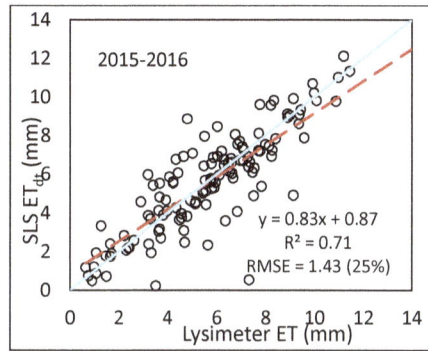

(a)

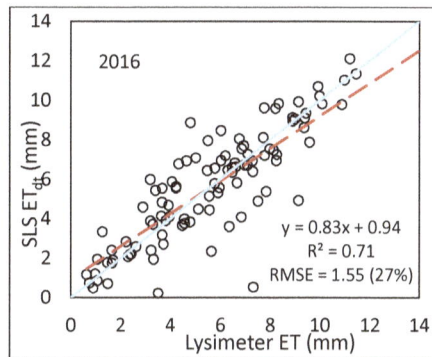

(b)

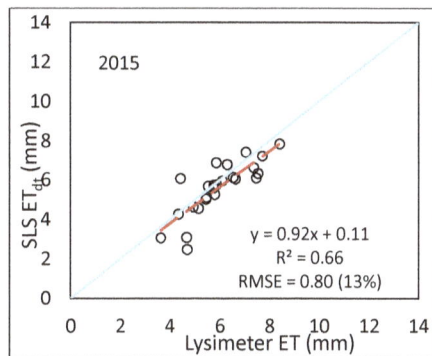

(c)

Figure 12. Comparison of summed daily ET from the SLS to daily (mass change from midnight to midnight) lysimeter data using (**a**) combined 2015–2016 data; (**b**) 2016 only; (**c**) 2015 only.

4. Conclusions

A surface layer scintillometer (SLS) was installed in the subsurface drip-irrigated NE lysimeter field at the USDA-ARS-CPRL in Bushland, TX to evaluate accuracy of the SLS to determine sensible and latent heat fluxes. Summed daily ET from the SLS (ET_{sls}) had as RMSE of 0.75 mm·d^{-1} as compared to the lysimeter with the regression slope (0.91) significantly different from one (p-value < 0.0001) and the intercept not significantly different from zero. Hourly ET regressed between the SLS and the lysimeter showed a slope of 0.91 and intercept of 0.03, both being significantly different from one and zero, respectively (p-values < 0.0001). Since the SLS cannot determine the direction of H, different post-processing techniques were evaluated. Using the difference between two air temperature sensors provides the temperature gradient and subsequent direction of H. The temperature gradient provided the best indication of the direction of H. Further analysis indicated that during the study period, stable atmospheric conditions were consistently present, which could have negatively affected the results from the SLS. Although the statistical performance of H_{sls} was not good, ET_{sls} correlated well, with r^2 values greater than 0.80, and RMSE values of 0.13 to 0.75 mm for hourly and daily ET_{sls}, respectively. For summed, daily ET_{sls}, the error was reduced to roughly half the hourly error at 14%. The reduction in error was likely due to overestimations and underestimations canceling out upon summation. As mentioned in the introduction section, a need exists for portable instruments that can provide highly accurate ET data for validation data and other research purposes. The SLS presents the potential to fill this need. With error rates as low as 14%, the SLS exhibits potential for use in water management activities such as developing crop coefficients or irrigation scheduling when daily data can be used. However, for shorter time steps, error rates are larger and comparable to error rates from other instruments such as EC, and more research is needed to identify the causes of the errors in hourly data and potential improvements in accuracy of the method. Evaluating data with larger H values, such as dryland conditions, may provide more information regarding H discrepancies.

Acknowledgments: We gratefully acknowledge support from the USDA-ARS Ogallala Aquifer Program, a consortium between USDA-Agricultural Research Service, Kansas State University, Texas A&M AgriLife Research, Texas A&M AgriLife Extension Service, Texas Tech University, and West Texas A&M University.

Author Contributions: J.E.M., P.D.C., P.H.G. and G.W.M. conceived and designed the experiments; J.E.M., G.W.M. and S.R.E. performed the experiments; J.E.M., P.D.C. and S.R.E. analyzed the data; T.H.M., D.O.P. and D.K.B. contributed materials; J.E.M. wrote the paper.

Conflicts of Interest: The authors declare no conflict of interest.

References

1. Marek, T.; Amosson, S.; Bretz, F.; Guerrero, B.; Kotara, R. *2011 Panhandle Regional Water Plan Task 2 Report: Agricultural Water Demand Projections*; Texas A&M AgriLife Research and Extension Center: Amarillo, TX, USA, 2009.
2. Moorhead, J.E.; Gowda, P.H.; Singh, V.P.; Porter, D.O.; Marek, T.H.; Howell, T.A.; Stewart, B. Identifying and evaluating a suitable index for agricultural drought monitoring in the Texas high plains. *J. Am. Water Resour. Assoc.* **2015**, *51*, 807–820. [CrossRef]
3. Scanlon, B.; Reedy, R.; Gates, J.; Gowda, P. Impact of agroecosystems on groundwater resources in the central high plains, USA. *Agric. Ecosyst. Environ.* **2010**, *139*, 700–713. [CrossRef]
4. Allen, R.G.; Walter, I.A.; Elliot, R.L.; Howell, T.A.; Itenfisu, D.; Jenson, M.E.; Snyder, R.L. *The Asce Standardized Reference Evapotranspiration Equation*; American Society of Civil Engineers: Reston, VA, USA, 2005.
5. Allen, R.G.; Pereira, L.S.; Raes, D.; Smith, M. *Crop Evapotranspiration—Guidelines for Computing Crop Water Requirements—FAO Irrigation and Drainage Paper 56*; FAO: Rome, Italy, 1998; p. 300.
6. Evett, S.R.; Schwartz, R.C.; Howell, T.A.; Baumhardt, R.L.; Copeland, K.S. Can weighing lysimeter ET represent surrounding field et well enough to test flux station measurements of daily and sub-daily ET? *Adv. Water Resour.* **2012**, *50*, 79–90. [CrossRef]

7. Todd, R.W.; Evett, S.R.; Howell, T.A. The bowen ratio-energy balance method for estimating latent heat flux of irrigated alfalfa evaluated in a semi-arid, advective environment. *Agric. For. Meteorol.* **2000**, *103*, 335–348. [CrossRef]

8. Foken, T. The energy balance closure problem: An overview. *Ecol. Appl.* **2008**, *18*, 1351–1367. [CrossRef] [PubMed]

9. Foken, T.; Wimmer, F.; Mauder, M.; Thomas, C.; Liebethal, C. Some aspects of the energy balance closure problem. *Atmos. Chem. Phys.* **2006**, *6*, 4395–4402. [CrossRef]

10. Oncley, S.P.; Foken, T.; Vogt, R.; Kohsiek, W.; DeBruin, H.A.R.; Bernhofer, C.; Christen, A.; van Gorsel, E.; Grantz, D.; Feigenwinter, C.; et al. The energy balance experiment ebex-2000. Part I: Overview and energy balance. *Bound.-Layer Meteorol.* **2007**, *123*, 1–28. [CrossRef]

11. Wagle, P.; Bhattarai, N.; Gowda, P.H.; Kakani, V.G. Performance of five surface energy balance models for estimating daily evapotranspiration in high biomass sorghum. *ISPRS J. Photogramm. Remote Sens.* **2017**, *128*, 192–203. [CrossRef]

12. Chávez, J.; Neale, C.M.; Hipps, L.E.; Prueger, J.H.; Kustas, W.P. Comparing aircraft-based remotely sensed energy balance fluxes with eddy covariance tower data using heat flux source area functions. *J. Hydrometeorol.* **2005**, *6*, 923–940. [CrossRef]

13. Gonzalez-Dugo, M.; Neale, C.; Mateos, L.; Kustas, W.; Prueger, J.; Anderson, M.; Li, F. A comparison of operational remote sensing-based models for estimating crop evapotranspiration. *Agric. For. Meteorol.* **2009**, *149*, 1843–1853. [CrossRef]

14. Gowda, P.; Senay, G.; Howell, T.; Marek, T. Lysimetric evaluation of simplified surface energy balance approach in the Texas high plains. *Appl. Eng. Agric.* **2009**, *25*, 665–669. [CrossRef]

15. Gowda, P.H.; Howell, T.A.; Chávez, J.L.; Paul, G.; Moorhead, J.E.; Holman, D.; Marek, T.H.; Porter, D.O.; Marek, G.H.; Colaizzi, P.D. A decade of remote sensing and evapotranspiration research at usda-ars conservation and production research laboratory. In Proceedings of the 2015 ASABE/IA Irrigation Symposium: Emerging Technologies for Sustainable Irrigation-A Tribute to the Career of Terry Howell, Sr., Long Beach, CA, USA, 10–12 November 2015; pp. 1–14.

16. Allen, R.G.; Tasumi, M.; Morse, A.; Trezza, R.; Wright, J.L.; Bastiaanssen, W.; Kramber, W.; Lorite, I.; Robison, C.W. Satellite-based energy balance for mapping evapotranspiration with internalized calibration (metric)—Applications. *J. Irrig. Drain. Eng.* **2007**, *133*, 395–406. [CrossRef]

17. Moene, A.; Meijninger, W.; Hartogensis, O.; Kohsiek, W.; De Bruin, H. *A Review of the Relationships Describing the Signal of a Large Aperture Scintillometer*; Internal Report 2004/2; Meteorology and Air Quality Group, Wageningen University: Wageningen, The Netherlands, 2004.

18. Bastiaanssen, W.G.; Menenti, M.; Feddes, R.; Holtslag, A. A remote sensing surface energy balance algorithm for land (sebal). 1. Formulation. *J. Hydrol.* **1998**, *212*, 198–212. [CrossRef]

19. Allen, R.G.; Tasumi, M.; Trezza, R. Satellite-based energy balance for mapping evapotranspiration with internalized calibration (metric)—Model. *J. Irrig. Drain. Eng.* **2007**, *133*, 380–394. [CrossRef]

20. Arya, S.P. Micrometeorology and atmospheric boundary layer. *Pure Appl. Geophys.* **2005**, *162*, 1721–1745. [CrossRef]

21. Foken, T. 50 years of the monin-obukhov similarity theory. *Bound.-Layer Meteorol.* **2006**, *119*, 431–447. [CrossRef]

22. Hartogensis, O. Exploring Scintillometry in the Stable Atmospheric Surface Layer. Ph.D. Thesis, Wageningen University, Wageningen, The Netherlands, 2006.

23. McAneney, K.; Green, A.; Astill, M. Large-aperture scintillometry: The homogeneous case. *Agric. For. Meteorol.* **1995**, *76*, 149–162. [CrossRef]

24. Monin, A.S.; Lumley, J.L.; Yaglom, A.M. *Statistical Fluid Mechanics: Mechanics of Turbulence*; MIT Press: Cambridge, MA, USA, 1971; Volume 1.

25. Scintec. *SLS Hardware Manual*; Scintec AG: Rottenburg, Germany, 2011; Volume 1.04.

26. De Bruin, H. Introduction: Renaissance of scintillometry. *Bound.-Layer Meteorol.* **2002**, *105*, 1–4. [CrossRef]

27. Savage, M.J. Estimation of evaporation using a dual-beam surface layer scintillometer and component energy balance measurements. *Agric. For. Meteorol.* **2009**, *149*, 501–517. [CrossRef]

28. Meijninger, W.; Hartogensis, O.; Kohsiek, W.; Hoedjes, J.; Zuurbier, R.; De Bruin, H. Determination of area-averaged sensible heat fluxes with a large aperture scintillometer over a heterogeneous surface—Flevoland field experiment. *Bound.-Layer Meteorol.* **2002**, *105*, 37–62. [CrossRef]

29. Green, A.E.; Green, S.R.; Astill, M.S.; Caspari, H.W. Estimation latent heat flux form a vineyard using scintillometry. *Terr. Atmos. Ocean. Sci.* **2000**, *11*, 525–542. [CrossRef]

30. Kohsiek, W.; Herben, M. Evaporation derived from optical and radio-wave scintillation. *Appl. Opt.* **1983**, *22*, 2566–2570. [CrossRef] [PubMed]

31. Samain, B.; Simons, G.W.; Voogt, M.P.; Defloor, W.; Bink, N.-J.; Pauwels, V. Consistency between hydrological model, large aperture scintillometer and remote sensing based evapotranspiration estimates for a heterogeneous catchment. *Hydrol. Earth Syst. Sci.* **2012**, *16*, 2095–2107. [CrossRef]

32. Samain, B.; Pauwels, V. Impact of potential and (scintillometer-based) actual evapotranspiration estimates on the performance of a lumped rainfall-runoff model. *Hydrol. Earth Syst. Sci.* **2013**, *17*, 4525–4540. [CrossRef]

33. Meijninger, W.; Beyrich, F.; Lüdi, A.; Kohsiek, W.; Bruin, H.D. Scintillometer-based turbulent fluxes of sensible and latent heat over a heterogeneous land surface—A contribution to litfass-2003. *Bound.-Layer Meteorol.* **2006**, *121*, 89–110. [CrossRef]

34. Guyot, A.; Cohard, J.-M.; Anquetin, S.; Galle, S.; Lloyd, C.R. Combined analysis of energy and water balances to estimate latent heat flux of a sudanian small catchment. *J. Hydrol.* **2009**, *375*, 227–240. [CrossRef]

35. Yee, M.S.; Pauwels, V.R.; Daly, E.; Beringer, J.; Rüdiger, C.; McCabe, M.F.; Walker, J.P. A comparison of optical and microwave scintillometers with eddy covariance derived surface heat fluxes. *Agric. For. Meteorol.* **2015**, *213*, 226–239. [CrossRef]

36. Beyrich, F.; De Bruin, H.; Meijninger, W.; Schipper, J.; Lohse, H. Results from one-year continuous operation of a large aperture scintillometer over a heterogeneous land surface. *Bound. -Layer Meteorol.* **2002**, *105*, 85–97. [CrossRef]

37. Beyrich, F.; Kouznetsov, R.D.; Leps, J.-P.; Lüdi, A.; Meijninger, W.M.; Weisensee, U. Structure parameters for temperature and humidity from simultaneous eddy-covariance and scintillometer measurements. *Meteorol. Z.* **2006**, *14*, 641–649. [CrossRef]

38. Cain, J.; Rosier, P.; Meijninger, W.; De Bruin, H. Spatially averaged sensible heat fluxes measured over barley. *Agric. For. Meteorol.* **2001**, *107*, 307–322. [CrossRef]

39. De Bruin, H.; Meijninger, W.; Kohsiek, W.; Beyrich, F.; Moene, A.; Hartogensis, O. Turbulent surface fluxes on kilometre scale obtained with scintillometry: A review. In Proceedings of the Computational Methods in Water Resources XVI International Conference, Copenhagen, Denmark, 19–20 June 2006.

40. Green, A.; Astill, M.; McAneney, K.; Nieveen, J. Path-averaged surface fluxes determined from infrared and microwave scintillometers. *Agric. For. Meteorol.* **2001**, *109*, 233–247. [CrossRef]

41. Kleissl, J.; Gomez, J.; Hong, S.-H.; Hendrickx, J.; Rahn, T.; Defoor, W. Large aperture scintillometer intercomparison study. *Bound. -Layer Meteorol.* **2008**, *128*, 133–150. [CrossRef]

42. Kleissl, J.; Hartogensis, O.; Gomez, J. Test of scintillometer saturation correction methods using field experimental data. *Bound.-Layer Meteorol.* **2010**, *137*, 493–507. [CrossRef]

43. Liu, S.; Xu, Z.; Wang, W.; Jia, Z.; Zhu, M.; Bai, J.; Wang, J. A comparison of eddy-covariance and large aperture scintillometer measurements with respect to the energy balance closure problem. *Hydrol. Earth Syst. Sci.* **2011**, *15*, 1291–1306. [CrossRef]

44. Odhiambo, G.; Savage, M. Surface layer scintillometry for estimating the sensible heat flux component of the surface energy balance. *S. Afr. J. Sci.* **2009**, *105*, 208–216. [CrossRef]

45. Savage, M.; Odhiambo, G.; Mengistu, M.; Everson, C.; Jarmain, C. Measurement of grassland evaporation using a surface-layer scintillometer. *Water SA* **2010**, *36*, 1–8. [CrossRef]

46. De Bruin, H.; Meijninger, W.; Smedman, A.-S.; Magnusson, M. Displaced-beam small aperture scintillometer test. Part i: The wintex data-set. *Bound. -Layer Meteorol.* **2002**, *105*, 129–148. [CrossRef]

47. Nakaya, K.; Suzuki, C.; Kobayashi, T.; Ikeda, H.; Yasuike, S. Application of a displaced-beam small aperture scintillometer to a deciduous forest under unstable atmospheric conditions. *Agric. For. Meteorol.* **2006**, *136*, 45–55. [CrossRef]

48. Hartogensis, O.; De Bruin, H.; Van de Wiel, B. Displaced-beam small aperture scintillometer test. Part II: Cases-99 stable boundary-layer experiment. *Bound. -Layer Meteorol.* **2002**, *105*, 149–176. [CrossRef]

49. Marek, T.H.; Schneider, A.D.; Howell, T.A.; Ebeling, L.L. Design and construction of large weighing monolithic lysimeters. *Trans. ASAE* **1988**, *31*, 477–484. [CrossRef]

50. Schneider, A.; Marek, T.; Ebeling, L.; Howell, T.; Steiner, J. Hydraulic pulldown procedure for collecting large soil monoliths. *Trans. ASAE* **1988**, *31*, 1092–1097. [CrossRef]

51. Howell, T.A.; Schneider, A.D.; Dusek, D.A.; Marek, T.H.; Steiner, J.L. Calibration and scale performance of bushland weighing lysimeters. *Trans. ASAE* **1995**, *38*, 1019–1024. [CrossRef]

52. Marek, G.W.; Evett, S.R.; Gowda, P.H.; Howell, T.H.; Copeland, K.S.; Baumhardt, R.L. Post-processing techniques for reducing errors in weighing lysimeter evapotranspiration (ET) datasets. *Trans. ASAE* **2014**, *57*, 499–515.

53. Nakaya, K.; Suzuki, C.; Kobayashi, T.; Ikeda, H.; Yasuike, S. Spatial averaging effect on local flux measurement using a displaced-beam small aperture scintillometer above the forest canopy. *Agric. For. Meteorol.* **2007**, *145*, 97–109. [CrossRef]

54. Samain, B.; Defloor, W.; Pauwels, V.R. Continuous time series of catchment-averaged sensible heat flux from a large aperture scintillometer: Efficient estimation of stability conditions and importance of fluxes under stable conditions. *J. Hydrometeorol.* **2012**, *13*, 423–442. [CrossRef]

55. Colaizzi, P.D.; Evett, S.R.; Agam, N.; Schwartz, R.C.; Kustas, W.P. Soil heat flux calculation for sunlit and shaded surfaces under row crops: 1. Model development and sensitivity analysis. *Agric. For. Meteorol.* **2016**, *216*, 115–128. [CrossRef]

56. Tolk, J.A.; Howell, T.A.; Evett, S.R. Nighttime evapotranspiration from alfalfa and cotton in a semiarid climate. *Agron. J.* **2006**, *98*, 730–736. [CrossRef]

57. Gavilán, P.; Berengena, J. Accuracy of the Bowen ratio-energy balance method for measuring latent heat flux in a semiarid advective environment. *Irrig. Sci.* **2007**, *25*, 127–140. [CrossRef]

MDPI

Article

Comparison between Random Forests, Artificial Neural Networks and Gradient Boosted Machines Methods of On-Line Vis-NIR Spectroscopy Measurements of Soil Total Nitrogen and Total Carbon

Said Nawar [1,2,3,*] **and Abdul M. Mouazen** [1]

[1] Department of Soil Management, Ghent University, Coupure 653, 9000 Gent, Belgium; abdul.mouazen@ugent.be
[2] Cranfield Soil and AgriFood Institute, School of Water, Energy and Environment, Cranfield University, Cranfield MK43 0AL, UK
[3] Faculty of Agriculture, Suez Canal University, Ismailia 41522, Egypt
* Correspondence: said.nawar@ugent.be; Tel.: +32-9-264-6202

Received: 10 August 2017; Accepted: 20 October 2017; Published: 24 October 2017

Abstract: Accurate and detailed spatial soil information about within-field variability is essential for variable-rate applications of farm resources. Soil total nitrogen (TN) and total carbon (TC) are important fertility parameters that can be measured with on-line (mobile) visible and near infrared (vis-NIR) spectroscopy. This study compares the performance of local farm scale calibrations with those based on the spiking of selected local samples from both fields into an European dataset for TN and TC estimation using three modelling techniques, namely gradient boosted machines (GBM), artificial neural networks (ANNs) and random forests (RF). The on-line measurements were carried out using a mobile, fiber type, vis-NIR spectrophotometer (305–2200 nm) (AgroSpec from tec5, Germany), during which soil spectra were recorded in diffuse reflectance mode from two fields in the UK. After spectra pre-processing, the entire datasets were then divided into calibration (75%) and prediction (25%) sets, and calibration models for TN and TC were developed using GBM, ANN and RF with leave-one-out cross-validation. Results of cross-validation showed that the effect of spiking of local samples collected from a field into an European dataset when combined with RF has resulted in the highest coefficients of determination (R^2) values of 0.97 and 0.98, the lowest root mean square error (RMSE) of 0.01% and 0.10%, and the highest residual prediction deviations (RPD) of 5.58 and 7.54, for TN and TC, respectively. Results for laboratory and on-line predictions generally followed the same trend as for cross-validation in one field, where the spiked European dataset-based RF calibration models outperformed the corresponding GBM and ANN models. In the second field ANN has replaced RF in being the best performing. However, the local field calibrations provided lower R^2 and RPD in most cases. Therefore, from a cost-effective point of view, it is recommended to adopt the spiked European dataset-based RF/ANN calibration models for successful prediction of TN and TC under on-line measurement conditions.

Keywords: on-line vis-NIR measurement; total nitrogen; total carbon; spiking; gradient boosted machines; artificial neural networks; random forests

---

1. Introduction

Estimation of carbon and nitrogen status in the soil is crucial from both agricultural and environmental points of view. It is well known that soil total carbon (TC) and total nitrogen (TN)

are vital factors for soil fertility and crop production [1,2]. Traditional laboratory analysis methods for TN and TC are laborious, time-consuming, costly and destructive [3,4]. Therefore, proximal soil sensing (PSS) techniques, in particular visible and near infrared (vis-NIR) reflectance spectroscopy can be considered as a cost-effective and alternative technique for estimating TN and TC [5,6].

On-line (tractor-driven) vis-NIR spectroscopy offers the possibility of collecting high spatial resolution data, compared with conventional laboratory analyses. However, on-line spectroscopic measurements are affected by ambient and experimental conditions that need to be overcome for accurate prediction to be achieved. One way to reduce these negative influences is by adopting advanced multivariate calibrations techniques, particularly those approaches that account for nonlinearity between NIR spectral response and soil properties [5]. Furthermore, overlapping of absorption bands of those properties and scatter effects result in complex absorption patterns, which cannot be derived using simple correlation or linear techniques [7].

Non-linear regression has been introduced in the literature as the best option to model spectroscopic data [8,9]. Among those models, support vector machines (SVM) [5,9], artificial neural networks (ANNs) [10,11], boosted regression trees [12], multivariate adaptive regression splines (MARS) [9,13] and random forests (RF) [14,15] were proven to provide improved prediction performances as compared to the linear partial least squares regression (PLSR) for modelling nonlinear phenomena like soil properties [8,11]. Neural networks, specifically multilayer perceptrons (MLPS), are mathematical models that use learning algorithms inspired by the brain to store information [16]. They have been examined in the field of spectroscopy using simulated data [17]. They have been used successfully to model a complex spectral library including over 1100 soil samples for large-scale study [14], and were used to predict OC based on on-line vis-NIR measurements, outperforming PLSR with ratio of prediction deviation (RPD) of and 2.28 [11]. However, overfitting is a major problem for ANN analysis, which has required special data pretreatment [18].

Recently, RF has received growing attention in vis-NIR spectral analyses in different domains. It is an ensemble learning technique, introduced by Breiman [19], as a combination of tree predictors that is robust and rarely overfits; it hence yields highly accurate predictions [19–21]. Accordingly, RF can handle nonlinear and hierarchical behaviors when introducing variability to the general spectral library for predicting local samples. Boosting trees (BT) characterized by the stochastic that enhances predictive performance, decreases the variance of the final model, by utilizing only a arbitrary subset of data to match each new tree [22]. Viscarra Rossel and Behrens [14] have applied BT to predict soil OC, pH and clay content using non-mobile (laboratory-based) spectroscopy measurement. Gradient boosted machines (GBM) is a hybrid method that incorporates both boosting and bagging approaches [22,23]. It performs boosting through choosing, at each step, the arbitrary sample of the data ultimately causing a progressive enhancement of the model performance [23]. GBM has been used successfully in digital mapping of OC [23–25]. Despite the importance of RF and GBM, no study on the use of both modelling methods for on-line spectroscopy measurement of soil properties can be found in the literature. The hypothesis of this study is that both GBM and RF outperform ANN for the on-line prediction of soil TN and TC.

The main goal of this paper is to compare the performance of GBM, ANNs and RF for the on-line prediction of TN and TC based on local (single field) dataset from two target fields and spiking of local samples of these two target fields into an European dataset.

2. Materials and Methods

2.1. Experimental Sites

Two experimental fields were used in this study, namely, Hessleskew and Hagg with total area of about 12 ha and 21 ha, respectively, both located in Yorkshire (Hessleskew, longitudes $-0.590°$ and $-0.586°$ W, and latitudes $53.844°$ and $53.844°$ N; Hagg, longitudes of $-1.172°$ and $-1.166°$ W, and latitudes of $53.936°$ and $53.941°$ N), The United Kingdom. Hessleskew field is cultivated with cereal crops in rotation, where Hagg field is cultivated with vegetable crops (e.g., carrots, cabbage, onions and leeks). The soil texture for the Hessleskew and Hagg fields is clay and sandy loam, respectively, according to United States Department of Agriculture (USDA) textural soil classification system [26].

2.2. On-Line Soil Measurement and Collection of Soil Samples

Both fields were scanned using the on-line system designed and developed by Mouazen [27]. This is a multi-sensor platform consists of a subsoiler, which penetrates the soil to any depth (5–50 cm), creating a trench, whoever bottom part is smoothened with the downwards forces acting on the subsoiler. The subsoiler has been retrofitted with the optical probe and attached to a frame. It was installed into the three point hitch of the tractor. The optical measurement was performed using an AgroSpec mobile, fibre type, vis-NIR spectrophotometer (Tec5 Technology for Spectroscopy, Geramany) with spectral range of 305–2200 nm. A differential global positioning system (DGPS) (EZ-Guide 250, Trimble, Sunnyvale, CA, USA) was utilized to record the positioning associated with on-line measured spectra along with sub-meter precision (Figure 1). The on-line measurement had been completed after previous crop harvest in summer of 2015 and 2016 for Hesslelekew and Hagg fields, respectively. The subsoiler was dragged at parallel transects of 12 m apart, setting the subsoiler tip at about 15 cm deep. A total of 122 and 149 soil samples were collected during the on-line measurement from the former and latter fields, respectively. These samples were used for calibration and validation of the vis-NIR sensor.

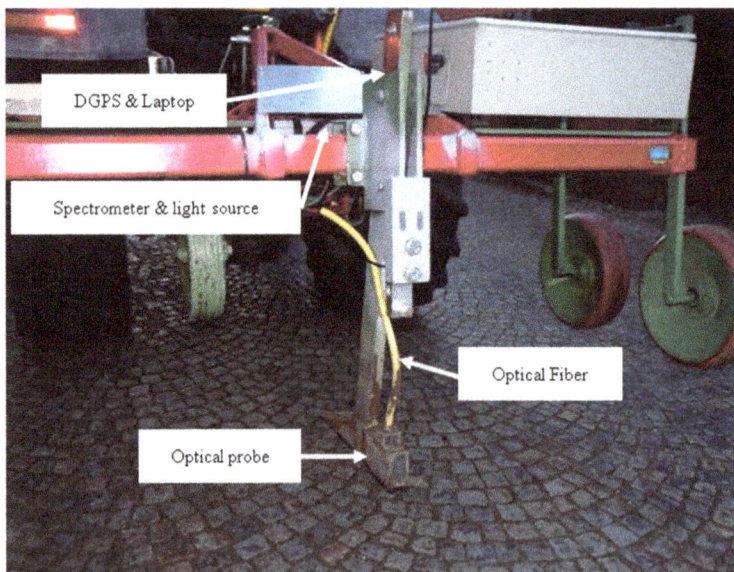

Figure 1. The on-line visible and near infrared (vis-NIR) spectroscopy sensor developed by Mouazen [27].

2.3. Laboratory Chemical and Optical Measurements

Fresh soil samples were used in the laboratory spectral and chemical analyses. Each soil sample was placed in a glass container and mixed well then divided into two parts. The first part was used to fill three Petri dishes of 2 cm in diameter 2 cm deep, representing three replicated measurements. Each soil sample were packed into plastic Petri dishes for soil scanning using the same spectrometer used in the on-line measurements. To obtain optimal diffuse reflection, and hence a good signal-to-noise ratio, all plant and pebble particles were manually removed and the surface was pressed gently with a spatula to be smooth before scanning. A total of ten scans were collected from each replicate, and these were averaged into one spectrum. The second part of each sample was air dried before it was analyzed for total carbon (TC) using the combustion method. This was done by oxidizing the carbon to carbon dioxide (CO_2) by heating the soil to at least 900 °C on a flow of oxygen-containing gas that is free from carbon dioxide. The amount of carbon dioxide released is then measured by a thermal conductivity detector (TCD). When the soil is heated to a temperature of at least 900 °C, any carbonates present are completely decomposed [28]. The total nitrogen was determined using the Dumas method by heating the soil to a temperature of at least 900 °C in the presence of oxygen gas. During oxidized combustion, mineral and organic nitrogen compounds produce the oxidation products NOx, in addition to molecular nitrogen (N_2). Copper in the reduction tube quantitatively reduces these nitrogen oxides to N_2 and binds excess oxygen. The amount of nitrogen is then measured by a TCD [29].

2.4. Spectra Pretreatment

The raw average spectra of the on-line and laboratory scanning were subjected to pre-processing, including successively, noise cut, maximum normalization, first derivative and smoothing using the prospectr-R package [30]. First, the spectral range outside 370–1979 nm was cut to remove the noise at both edges. Then, a moving average with five successive wavelengths was used to reduce noise. Maximum normalization followed, which is typically used to get all data to approximately the same scale, with maximum values of 1. The maximum normalisation led to better results for the measurement of TC and TN as compared to the other pre-treatment options tested, including mean and peak normalization. Spectra were then subjected to first derivation using gap–segment derivative (gapDer) algorithms [31] with a second-order polynomial approximation. This method enables the first or higher-order derivatives, including a smoothing factor, to be computed, which determines how many adjacent variables will be used to estimate the polynomial approximation used for derivatives. This gapDer resulted in a better performance than second derivative that increased the noise and reduced the quality of models' prediction performance. Finally, smoothening with the Savitzky–Golay technique was carried out to remove noise from the spectra.

2.5. Dataset Set Selection and Modelling Techniques

The following two data sets were considered in this study:

1. Local dataset: where samples collected from two fields (Hessleskew, $n = 122$; Hagg, $n = 149$),
2. European dataset ($n = 528$), where a total of 528 samples collected from five European countries, namely, Germany (150 samples from two fields), Denmark (147 samples from five fields), the Netherlands (43 samples from one field), Czech Republic (99 samples from four fields), and the UK (89 samples from four fields) were collected.

The Kennard–Stone algorithm [32] was used to select the calibration set (75%), and the rest of the samples (25%) were assigned for the prediction set. The Kennard–Stone algorithm allows to select points (samples) with a uniform distribution over the predictor space. It begins through selecting the pair of samples that are the farthest apart. They are assigned to the calibration set and removed from the dataset. Then, the procedure assigns remaining samples to the calibration set by computing the distance between each unassigned samples i_0 and selected samples i and finding the sample i_0 for which:

$$d_{selected} = \max_{i_0}(\min_i(d_{i,\,i_0}))$$

(1)

This essentially selects sample i_0, which is the farthest apart from its closest neighbors i in the calibration set based on the the Mahalanobis distance (H), which can be defined as the dissimilarity measure matrix (H) between samples in a given matrix X and can be computed as follows:

$$H\left(x_i, x_j\right) = \sqrt{\left(x_i, x_j\right)^{M-1}\left(x_i, x_j\right)^T} \tag{2}$$

where M is the variance–covariance matrix and vector T. The algorithm uses the H that can be achieved by performing a PCA analysis on the input data and computing H as follows:

$$H_{ij}^2 = \sum_{a=1}^{A} \left(\hat{t}_{ia} - \hat{t}_{ja}\right)^2 / \hat{\lambda}_{ia} \tag{3}$$

where $\hat{t}_{ia}$ is the ath principal component score of sample i, $\hat{t}_{ja}$ is the corresponding value for sample j, $\hat{\lambda}_{ia}$ is the eigenvalue of principal component a, and A is the number of principal components included in the computation [30].

Spiking was used to introduce the local variability of the two experimental fields into the European dataset. A total of 85 and 110 samples that have been selected from the Hessleskew and Hagg fields, respectively, using the Kennard–Stone algorithm were spiked into the European dataset.

Before running the analysis, the entire dataset of each target field (Hesselskew or Hagg) was divided into 75% for calibration, and 25% for prediction as described above. This was done for both the laboratory and on-line collected soil spectra. The 75% soil samples were used for developing the calibration models for the local dataset (single field), and the spiked European dataset model. To evaluate these models, cross-validation technique with leave-one-out cross-validation (LOOV) was performed on the training data (75%). For independent validation, the laboratory reference measurement values of the prediction set (25%), e.g., 37 and 39 samples from Hessleskew and Hagg, respectively, were compared with the laboratory and on-line predicted concentration values at the same positions.

2.5.1. Random Forests Regression

Random forests (RF) is an ensemble learning method developed by Breiman [19], which can be described as follows:

Suppose we have a calibration set $C = \{C_1, \ldots, C_n\}$ with $C_i \equiv (x_i, y_i)$ and an independent test case C_0 with predictor x_0, the following steps can be carried out:

(1) Sample the calibration set C with replacement to generate bootstrap resamples $B_1, \ldots, B_M$
(2) For each resample B_m, $m = 1, \ldots, M$, grow a regression tree T_m.
(3) For predicting the test case C_0 with covariate x_0, the predicted value by the whole RF is obtained by combining the results given by individual trees. Let $\hat{f}_m^*(x_0)$ denote the prediction of C_0 by mth tree, the RF prediction for regression problems can then be written [33] as:

$$\frac{1}{M} \sum_{m=1}^{M} \hat{f}_m^*(x_0) \tag{1a}$$

RF is generally used for data classification and regression. The algorithm works by growing an ensemble of regression trees based on binary recursive partitioning, where the algorithm first begins with a number of bootstrap samples (ntree) from the predictor space (original data) [34]. Each bootstrap sample will then grow regression tree with a modifying operation, in which subsequently a number of the predictors (mtry) are randomly sampled, and the algorithm chooses the best split from among those sampled variables rather than considering all variables. The default mtry value is the square root of the total number of variables [35]. Therefore, the number of trees (ntree) needs to be set sufficiently high. Consequently, RF hardly overfits when more trees are added [19], but produce a limited generalisation error [20,36]. The final prediction can be obtained as the mean value of the individual predictions made by each decision tree. RF does not need complicated data pretreatment and runs very fast when compared to other machines learning algorithms such as ANNs and GBM [37], which is a very important factor in the field of on-line and in situ measurements. Figure 2 shows the main processes of the RF algorithm. In this work, an ntree of 100 and an mtry of 2 were used to develop the TN and TC models. These parameters were determined by the tune RF function implemented in the R software package, named random forest version 4.6–12 [38]. The same split of datasets described above (75% calibration, 25% prediction) were utilised for RF analysis.

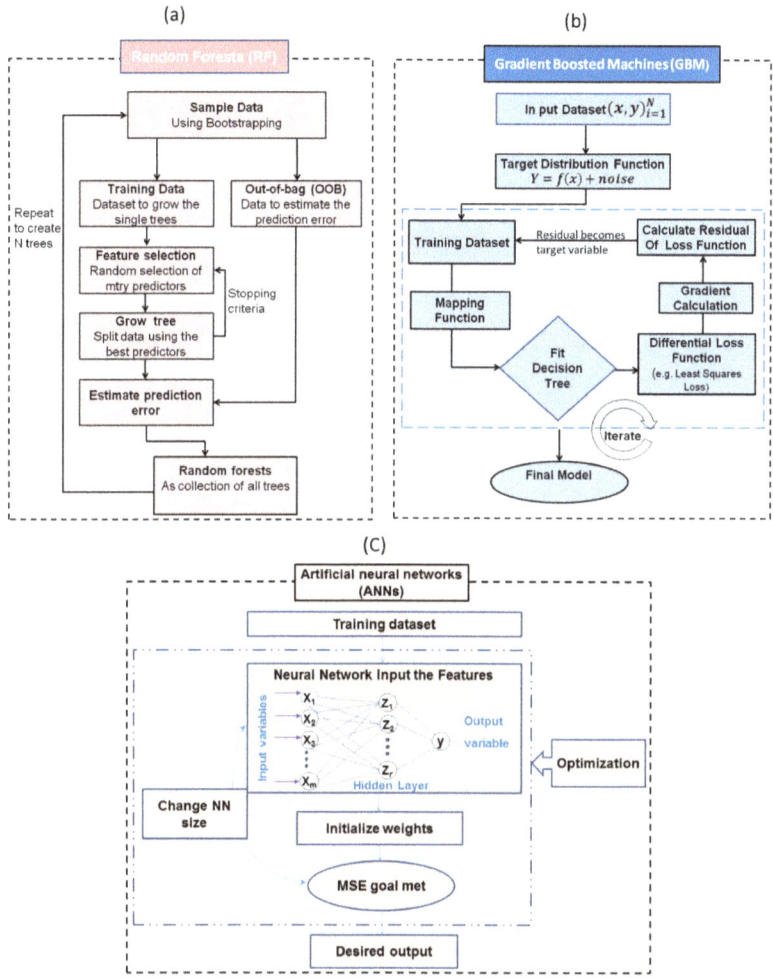

Figure 2. Overview flowchart listing the main steps of the multivariate analysis methods. (**a**) Random forests regression; (**b**) gradient boosted machines (GBM); and (**c**) artificial neural networks (ANNs).

2.5.2. Gradient Boosted Machines (GBM)

Boosting is a method based on the idea of combining a set of weak learners and delivers superior predictive performance whose always highly accurate [39]. In GBM, the learning procedure sequentially fits new models to the training data, utilizing suitable techniques (loss function, weak learner and additive model) progressively to increase emphasis on observations modelled poorly through the existing collection of trees. This particular enhancement can be achieved through constructing the new base-learners to become maximally related using the negative gradient of the loss function, linked to the entire ensemble (Figure 2). Boosting draws bootstrap samples of the predictor data, fits a tree, and subtracts the prediction from the original data. The trees tend to be iteratively suited to the residuals and the predictions summed up [40]. Steps to avoid overfitting are essential because the sequential nature of boosting allows trees to be added before the model is entirely overfitted [41]. According to Hastie [38] the GBM algorithm can be described as follows:

$$1 - \text{ Initialize } f_0(x) = \arg \min_\gamma \sum_{i=1}^{N} L(\mathcal{Y}_i, \gamma). \tag{2a}$$

This initializes the optimal constant model, which is just a single terminal node tree.
For $m = 1$ to M:

(a) For $i = 1, 2, \ldots, N$ compute

$$r_{im} = -\left[\frac{\partial L(\mathcal{Y}_i, f(x_i))}{\partial f(x_i)} \right]_{f=f_{m-1}} \tag{2b}$$

The components of the negative gradient are referred to the generalized residuals r_{im} of the current model on the ith observation evaluated at $f = f_{m-1}$.

(b) Fit a regression tree to the targets r_{im} giving terminal regions

$$R_{jm}, \ j = 1, 2, \ldots \ldots . J_m.$$

(c) For $j = 1, 2, \ldots, J_m$ compute

$$\gamma_{jm} = \arg \min_\gamma \sum_{x_i \in R_{jm}} L(\mathcal{Y}_i, f_{m-1}(x_i) + \gamma) \tag{2c}$$

γ parameterizes the split variables and split points at the internal nodes, and the predictions at the terminal nodes. In the gbm package ϵ is shrinkage with default 0.001 that to allow at least for 1000 trees. The best fits the current residuals is added to the expansion at each step as in step (d). This produces $fm(x)$, and the process is repeated. At each iteration m, one solves for the optimal lose function and add to the current expansion $f_{m-1}(x)$.

The boosting models were fitted by using the code published by Elith et al. [42], which is based on the package gbm in R software. There is a range of tuning parameters for GBM model; shrinkage reduces the participation of each tree to the final model. Shrinkage setup was recommend to be small enough (0.01–0.001) to allow at least for 1000 trees [42]. Hence, for all procedures, shrinkage was set to the lower end of the recommendations (0.001). The subsampling rate "bag fraction" that specifies the ratio of the data to be used at each iteration was set up to the default (0.50). The number of trees (ntree) is more relevant than for random forest, as gradient boosting overfits if ntree is excessive. Hence, ntree was determined for each individual modeling case. The same split of datasets described above (75% calibration, 25% prediction) were utilised for GBM analysis.

2.5.3. Artificial Neural Networks (ANNs)

ANNs are a machine learning framework that attempts to mimic the learning pattern of natural biological neural networks and are based on their ability to "learn" throughout a training procedure exactly where they're given inputs and a set of anticipated results. ANNs are a machine learning framework that attempts to mimic the learning pattern of natural biological neural networks and are based on their ability to "learn" throughout a training procedure exactly where they're given inputs and a set of anticipated results. The neural network used in this study was a multilayer perceptron (MLP) neural network. It typically consists of an input layer (i.e., spectral data or principal components), one or more hidden layers, where the real processing is performed via a system of weighted 'connections', and an output layer (prediction), where the answer is output (Figure 2).

They function by linking the input neurons to output neurons, through the connections (weights). The ANNs algorithm with single layer can be described [43] as follows:

First, r different linear combinations of the x-variables are built

$$y_j = a_{0j} + a_{1j}x_1 + \ldots + a_{mj}x_m \quad \text{for } j = 1, \ldots, r \tag{3a}$$

and then a nonlinear function s—often the sigmoid function—is applied:

$$z_j = \sigma(y_j) = \frac{1}{1 + \exp(-y_j)} \quad \text{for } j = 1, \ldots, r \tag{3b}$$

Equations (3a) and (3b) constitute a neuron with several inputs x and one output z. The new variables z_j can be used in different ways to produce the final output y: (a) as inputs of a neuron with output y, (b) in a linear regression model,

$$y = b_0 + b_1 z_1 + b_2 z_2 + \ldots + b_r z_r + e \tag{3c}$$

and (c) in a nonlinear regression model

$$y = b_0 + b_1 f_1(z_1) + b_2 f_2(z_2) + \ldots + b_r f_r(z_r) + e \tag{3d}$$

The most straightforward approach was used to build the ANNs model. This is performed using the training and test sets. Samples in the training data sets were the same as those in the calibration sets used in the RF and GBM analyses, whereas the test sets were the same as the prediction set. Leave-one-out cross-validation was used to avoid over-fitting and to monitor the training error. The input layer has the same number of input nodes to the number of soil samples used in each calibration set. The output layer has one node of TN and TC. The number of nodes in the hidden layer was adjusted during the training from 6 to 20 to get the optimised network structure, which resulted in the lowest training error. The training algorithm was selected as stochastic gradient descent, and the training time was set to 1000 times. Exponential and logistic functions were selected for the hidden and the output layers, respectively. The performance of the resultant models were chosen according to the following evaluation parameters: high R^2 (both in calibration and prediction), and low root mean square error of prediction (RMSEP). The *caret* package [44] has been used to perform the ANN models in R software [45].

2.6. Evaluation of Model Accuracy

Model performance for the prediction of TN and TC were evaluated by means of R^2, RMSEP and RPD, which can be defined as follows:

$$RMSE = \sqrt{\frac{\sum_{i=1}^{n}(\hat{y}_i - y_i)^2}{n - 1}} \tag{4a}$$

$$R^2 = 1 - \frac{SS_{error}}{SS_{total}} \tag{4b}$$

where the SS_{total} and SS_{error} are the variance of measured values and the sum of squared residuals, respectively:

$$SS_{total} = \sum_{i=1}^{n}(y_i - \bar{y})^2 \tag{4c}$$

$$SS_{error} = \sum_{i=1}^{n}(y_i - \hat{y}_i)^2 \tag{4d}$$

$$RPD = SD/RMSE \tag{4e}$$

Viscarra Rossel et al. [46] classified the *RPD* values referring to accuracy of modelling into six classes: excellent (*RPD* > 2.5), very good (*RPD* = 2.5–2.0), good (*RPD* = 2.0–1.8), fair (*RPD* = 1.8–1.4), poor (*RPD* = 1.4–1.0), and very poor model (*RPD* < 1.0). In this study, we adopted this model classification criterion to compare between different calibration models in cross-validation and in laboratory and on-line predictions.

3. Results

3.1. Laboratory Measured Soil Properties

The descriptive statistics for measured TN and TC in both fields are shown in Table 1 and Figure 3. It can be observed that TN concentration is low, with mean and maximum values of 0.25% and 0.34%, respectively, whereas the mean and maximum values of TC are 2.12% and 3.67%, respectively, in the Hessleskew field. Both TN and TC in the Hagg field are even smaller than in the Hessleskew field, with mean values of 0.21% and 1.92%, respectively (Table 1). The small range of variability in TN and TC implies these fields are certainly not the optimal case study, as the smaller the variability is, the less successful results can be expected for the prediction capability of the vis-NIR spectroscopy calibration models established [6].

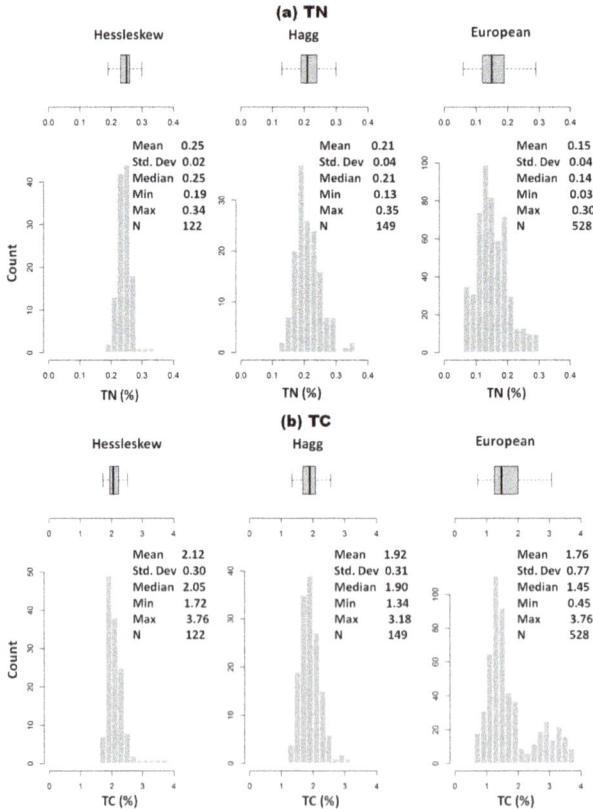

Figure 3. Histograms, box-plots and descriptive statistics of (**a**) soil total nitrogen (TN) and (**b**) total carbon (TC) for Hessleskew and Hagg fields, and European dataset.

The mean and median of TN are equal in Hessleskew and Hagg, indicating that TN fallows normal distribution, meanwhile European dataset shows left skewed with the mean being greater than the median (0.15 and 0.14%, respectively). The sample distribution of TC is similar to TN in both Hessleskew and Hagg with unimodal as the mean and median values are comparable, whereas the distribution of TC in European dataset shows non-modality and left skewed with the mean being larger than the median (1.67% and 1.45%, respectively).

Table 1. Descriptive statistics for soil total carbon (TC) and total nitrogen (TN) for Hessleskew, Hagg fields, and European datasets.

		Min	1st Qu.	Median	Mean	3rd Qu.	Max	St.dev
Hessleskew	(*n* = 122)							
TN (%)		0.19	0.23	0.25	0.25	0.26	0.34	0.02
TC (%)		1.72	1.94	2.05	2.12	2.22	3.67	0.30
Hagg	(*n* = 149)							
TN (%)		0.13	0.19	0.21	0.21	0.24	0.35	0.04
TC (%)		1.34	1.68	1.90	1.92	2.08	3.18	0.31
European	(*n* = 528)							
TN (%)		0.03	0.11	0.14	0.15	0.17	0.30	0.04
TC (%)		0.45	1.22	1.45	1.67	1.70	3.76	0.77

3.2. Performance of the Calibration Models for Predicting TN

Table 2 and Figures 4 and 5 show the cross-validation, laboratory and on-line prediction results for TN calibration models developed with local and European datasets. In cross-validation, RF outperformed both GBM and ANN, successively, for modelling TN. The best results achieved with RF are based on the spiked European dataset with $R^2 = 0.97$, RMSECV = 0.01%, and RPD = 5.58 for the Hagg field, and $R^2 = 0.96$, RMSECV = 0.01%, and RPD = 4.83 for the Hessleskew field (Table 2). The lowest results are obtained with ANN based on local dataset with $R^2 = 0.35$, RMSE = 0.03%, and RPD = 1.25 for the Hagg field, and $R^2 = 0.62$, RMSE = 0.01%, and RPD = 1.62 for the Hessleskew field.

The performance of the laboratory prediction shows a different trend to that of the cross-validation, where the GBM based on the spiked European dataset models generally provide the best results with $R^2 = 0.87$, RMSE = 0.02%, and RPD = 2.79 in the Hesselskew field (Table 2; Figure 4), followed by RF model based on the spiked European dataset with $R^2 = 0.84$, RMSE = 0.02%, and RPD = 2.51 in the Hagg field (Table 2; Figure 5). However, GBM-local dataset-based models in particular for the Hessleskew field has resulted in the least significant results ($R^2 = 0.60$, RMSE = 0.01%, and RPD = 1.60), and $R^2 = 0.62$, RMSE = 0.02%, and RPD = 1.65 in the Hagg field, shown in Table 2). ANN outperforms GBM in modeling based on local dataset with R^2 of 0.69 and 0.66, RMSE of 0.01% and 0.02%, and RPD of 1.81 and 1.74 for the Hesselskew and Hagg field, respectively. However, RF local model shows better performances for laboratory prediction of TN, compared to both ANN and GMB based on the corresponding datasets, particularly in the Hesselskew field.

Like for the cross-validation, the best results of on-line prediction are obtained with RF, followed successively by GBM and ANN (Table 2). This is true in the Hagg field, where the highest results of RF model obtained with the spiked European dataset with $R^2 = 0.83$, RMSE = 0.02%, and RPD = 2.40, followed by RF model based on local dataset with $R^2 = 0.79$, RMSE = 0.02%, and RPD = 2.20 in the Hagg field (Table 2; Figure 5). The local-ANN based model has the lowest results in the Hagg field ($R^2 = 0.11$, RMSE = 0.04%, and RPD = 1.07) and the Hessleskew field ($R^2 = 0.26$, RMSE = 0.02%, and RPD = 1.18), followed by GBM local model for the the Hessleskew field ($R^2 = 0.53$, RMSE = 0.02%, and RPD = 1.48, Table 2). However, the model prediction performance varies between the two studied fields. Although the best performing local dataset was with RF, ANN provides a better prediction with the spiked dataset in the Hesselskew field, although the differences are small compared to RF and GBM.

Table 2. Hessleskew and Hagg fields results in cross-validation, laboratory and on-line predictions using local, and spiked European dataset based on gradient boosted machines (GBM), artificial neural networks (ANNs) and random forests (RF) models.

			Hessleskew Local			Hessleskew European					Hagg Local			Hagg European		
	n.trees	TN/TC	RMSE	R^2	RPD	RMSE	R^2	RPD	TN/TC	n.trees	RMSE	R^2	RPD	RMSE	R^2	RPD
GBM																
Cross-	100	TN	0.01	0.63	1.64	0.01	0.96	4.81	TN	100	0.02	0.63	1.66	0.01	0.96	5.01
	100	TC	0.16	0.67	1.75	0.06	0.98	6.48	TC	100	0.19	0.65	1.70	0.12	0.98	6.49
Lab Prediction	100	TN	0.01	0.60	1.60	0.02	0.87	2.79	TN	100	0.02	0.62	1.65	0.02	0.79	2.21
	100	TC	0.23	0.60	1.59	0.20	0.82	2.40	TC	100	0.21	0.61	1.61	0.19	0.83	3.03
On-line Prediction	100	TN	0.02	0.53	1.48	0.02	0.66	1.80	TN	100	0.02	0.59	1.58	0.02	0.77	2.11
	100	TC	0.26	0.54	1.49	0.24	0.66	1.78	TC	100	0.22	0.52	1.46	0.20	0.79	2.95
ANN	size									size						
Cross- validation	2	TN	0.01	0.62	1.62	0.01	0.77	2.08	TN	2	0.03	0.35	1.25	0.03	0.73	1.92
	2	TC	0.21	0.44	1.34	0.15	0.86	2.69	TC	2	0.18	0.70	1.82	0.18	0.86	2.79
Lab Prediction	2	TN	0.01	0.69	1.81	0.01	0.71	2.02	TN	2	0.02	0.66	1.74	0.02	0.68	1.87
	2	TC	0.25	0.51	1.45	0.20	0.83	2.44	TC	2	0.25	0.47	1.40	0.21	0.84	2.59
On-line Prediction	2	TN	0.02	0.26	1.18	0.01	0.68	1.78	TN	2	0.04	0.11	1.07	0.03	0.59	1.59
	2	TC	0.34	0.19	1.13	0.25	0.78	2.14	TC	2	0.27	0.75	1.95	0.20	0.85	2.63
RF	ntree									ntree						
Cross- validation	100	TN	0.01	0.83	2.45	0.01	0.96	4.83	TN	100	0.01	0.84	2.50	0.01	0.97	5.58
	100	TC	0.12	0.82	2.38	0.06	0.98	6.48	TC	100	0.13	0.84	2.52	0.10	0.98	7.54
Lab Prediction	100	TN	0.01	0.82	2.40	0.02	0.81	2.33	TN	100	0.02	0.78	2.18	0.02	0.84	2.51
	100	TC	0.25	0.75	2.02	0.23	0.78	2.16	TC	100	0.18	0.81	2.35	0.14	0.88	3.49
On-line Prediction	100	TN	0.01	0.72	1.93	0.04	0.55	1.52	TN	100	0.02	0.79	2.20	0.02	0.83	2.40
	100	TC	0.21	0.69	1.82	0.20	0.75	2.13	TC	100	0.15	0.77	2.13	0.14	0.86	3.24

n.trees = total number of trees to fit, size = number of units in the hidden layer, ntree = number of trees.

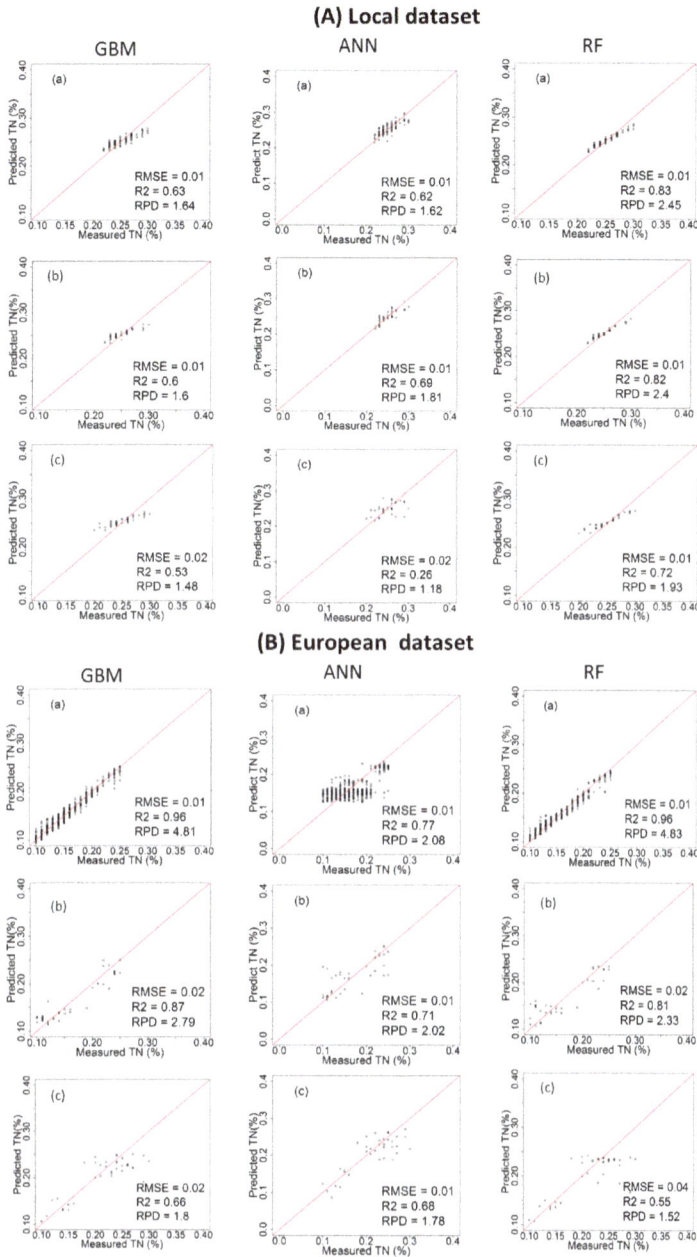

Figure 4. Scatter plots of visible and near infrared (vis-NIR)-predicted versus laboratory-analysed total nitrogen (TN) in Hessleskew field in cross validation (**a**); lab prediction (**b**) and on-line prediction (**c**); using local dataset (**A**) and spiked European dataset (**B**); comparing between gradient boosted machines (GBM), artificial neural network (ANNs) and random forests (RF) models.

(A) Local dataset

(B) European dataset

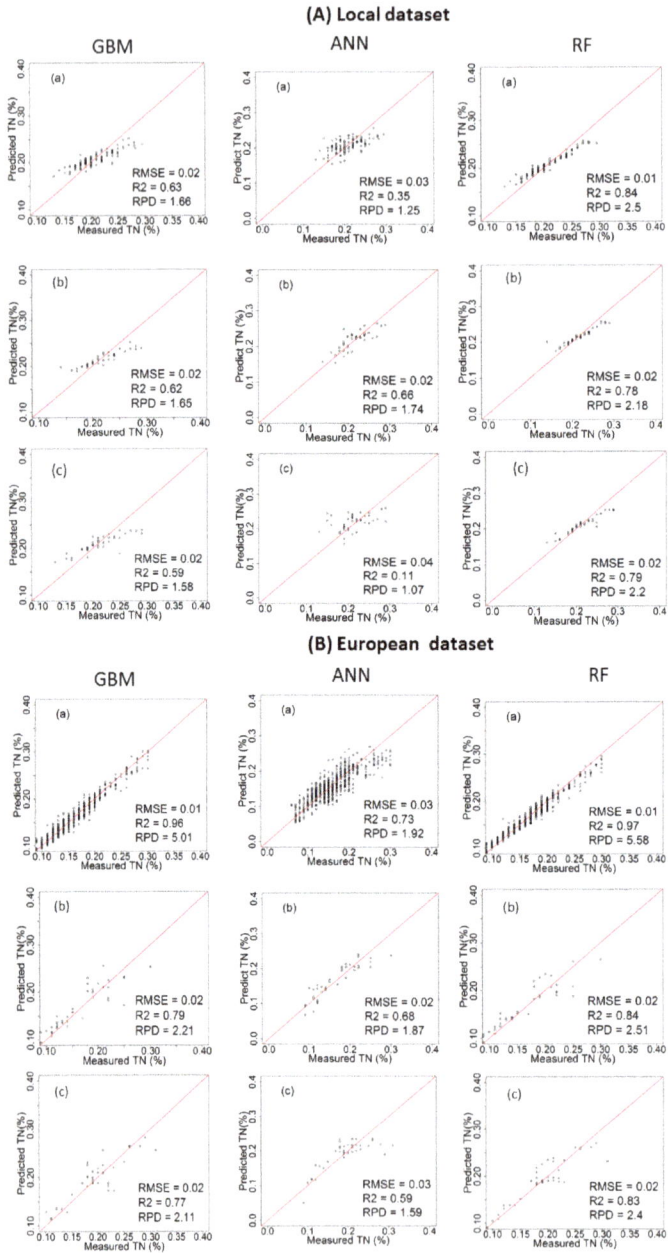

Figure 5. Scatter plots of visible and near infrared (vis-NIR)-predicted versus laboratory-analysed total nitrogen (TN) in Hagg field in cross validation (**a**); lab prediction (**b**) and on-line prediction (**c**), using local dataset (**A**) and spiked European dataset (**B**); comparing between gradient boosted machines (GBM); artificial neural network (ANNs) and random forests (RF) models.

3.3. Performance of the Calibration Models for Predicting TC

Table 2 and Figures 6 and 7 show the results of cross-validation, laboratory and on-line predictions. For cross-validation, the RF outperformed both GBM and ANN models. The best results are achieved with RF using the spiked European dataset in both fields with R^2 = 0.98, RMSECV = 0.10% and RPD = 7.54 for the Hagg field, and R^2 = 0.98, RMSECV = 0.06% and RPD = 6.48 for the Hesselskew field. However, the performance of the RF model in the Hessleskew field is identical to that of GBM. While the results of local-ANN based model is the poorest in the Hessleskew field with R^2 = 0.44, RMSE = 0.21% and RPD = 1. 34, followed by local-GBM based model in the Hagg field with R^2 = 0.65, RMSECV = 0.19% and RPD = 1.70. Overall, the cross-validation results for TC is identical to that for TN reported above.

The laboratory prediction of TC behaves similarly to the cross-validation stage, where RF over-performs both GBM and ANN (Table 2), with the best results obtained for the RF model based on the spiked European dataset (R^2 = 0.88, RMSE = 0. 14% and RPD = 3.49 in the Hagg field, followed by GBM model based on the spiked European dataset (R^2 = 0.83, RMSE = 0. 15% and RPD = 3.16 in the Hagg field). However, ANN model outperforms both RF and GBM in the Hessleskew field with the spiked European dataset (R^2 = 0.83, RMSE = 0.20% and RPD = 2.44). RF models outperform both GBM and ANN models based on the local dataset in both Hesselsekew and Hagg fields (Table 2 and Figure 8).

Similarly to the laboratory prediction, the best results for the on-line prediction are achieved using RF based on the spiked European dataset in the Hagg field (R^2 = 0.86, RMSE = 0.14% and RPD = 3.24), followed by GBM based on the spiked European dataset in the Hagg field also (R^2 = 0.85, RMSE = 0. 20% and RPD = 2.95). Again, ANN based on the spiked European dataset outperform both RF and GBM in the Hessleskew field (R^2 = 0.78, RMSE = 0.25% and RPD = 2.14). For the local dataset based models, RF outperforms both GBM and ANN models in both the Hagg and Hessleskew fields (Table 2; Figure 8).

4. Discussion

4.1. Comparison of Model Performance

In this work we compared the accuracy of the GBM, ANN and RF methods for the prediction of TN and TC based on local and European datasets. The variations of R^2, RPD as well as RMSE values obtained from cross-validation, laboratory and on-line prediction are shown in Table 2 and Figures 8 and 9.

Although RF models have resulted in the highest prediction performance followed by GBM in cross-validation, this was the case for the laboratory and on-line predictions in the Hagg field only, whereas ANN models based on the spiked European dataset has provided improved results for the laboratory (for TC only) and on-line (for both TC and TN) predictions in the Hessleskew field only. This means that the laboratory prediction followed the same trend as for cross-validation in the Hagg field only, where RF outperform both GBM and ANN. Sorenson et al. [15] found RF to outperform ANN for the prediction of OC and TN for non-mobile measurement, reporting RMSE = 0.62 and 1.56%, and RPD = 2.1 and 0.90 for RF and ANN, respectively, for OC and RMSE of 0.60 and 0.12%, and RPD of 2.1 and 1.0 for RF and ANN, respectively for TN. Viscarra Rossel and Behrens [14] reported better prediction results for RF compared to BT, but was less performing than ANN using the discrete wavelet transform as the predictors (RMSEP of 0.99%, 0.93% and 0.75%, and R^2 of 0.83, 0.84, and 0.89 for DWT-BT, DWT-RF, and DWT-ANN, respectively). This points out that, depending on the geographic region, one method may outperform several others [47].

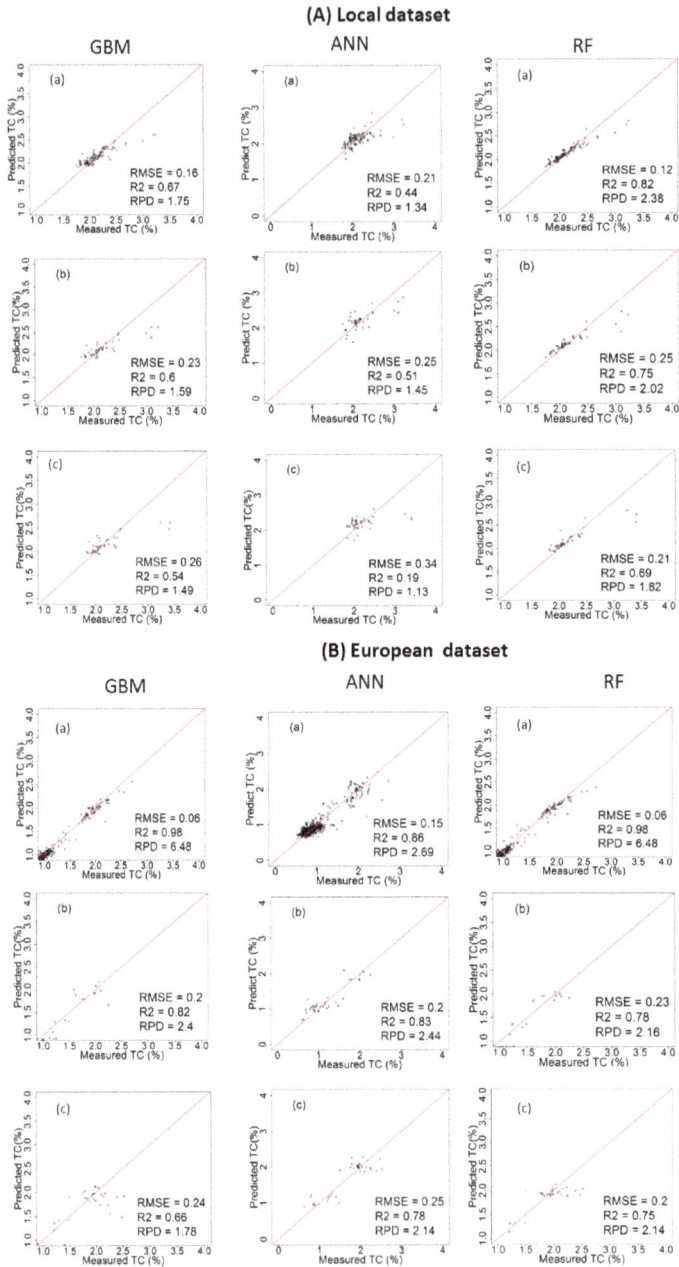

Figure 6. Scatter plots of visible and near infrared (vis-NIR)-predicted versus laboratory-analysed total carbon (TC) in the Hesselskew field in cross validation (**a**); lab prediction (**b**) and on-line prediction (**c**); using local dataset (**A**) and spiked European dataset (**B**); comparing between gradient boosted machines (GBM), artificial neural networks (ANNs) and random forests (RF) models.

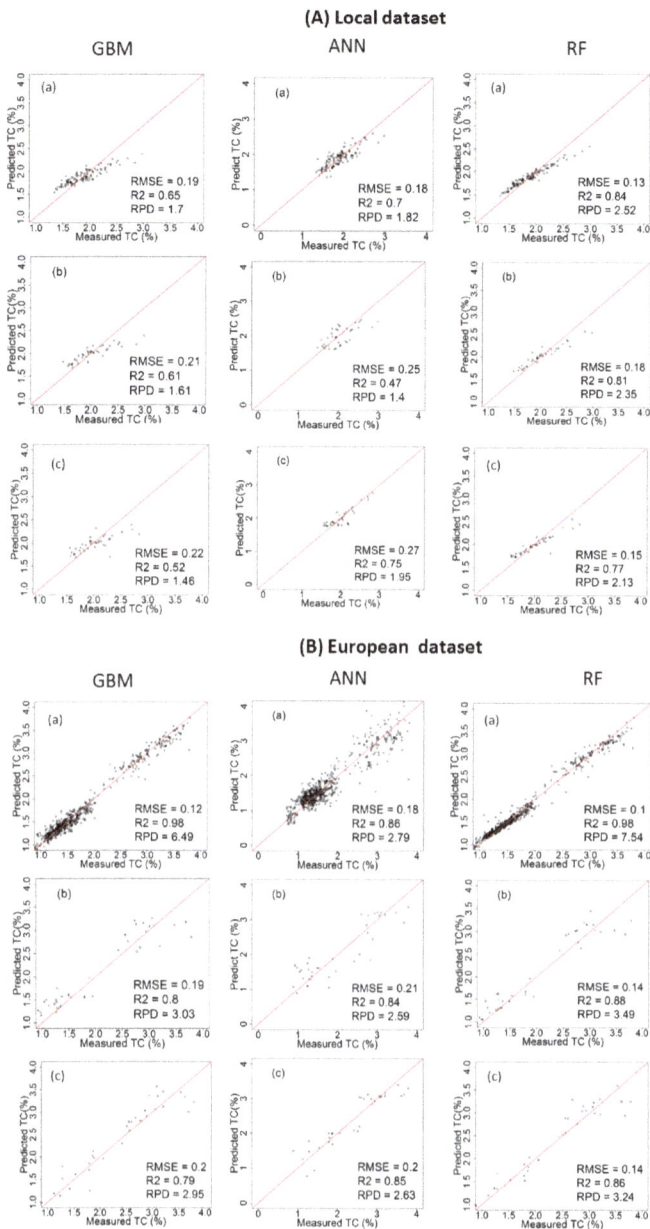

Figure 7. Scatter plots of visible and near infrared (vis-NIR)-predicted versus laboratory-analysed total carbon (TC) in the Hagg field in cross validation (**a**); lab prediction (**b**) and on-line prediction (**c**); using local dataset (**A**) and spiked European dataset (**B**); comparing between gradient boosted machines (GBM), artificial neural networks (ANNs) and random forests (RF) models.

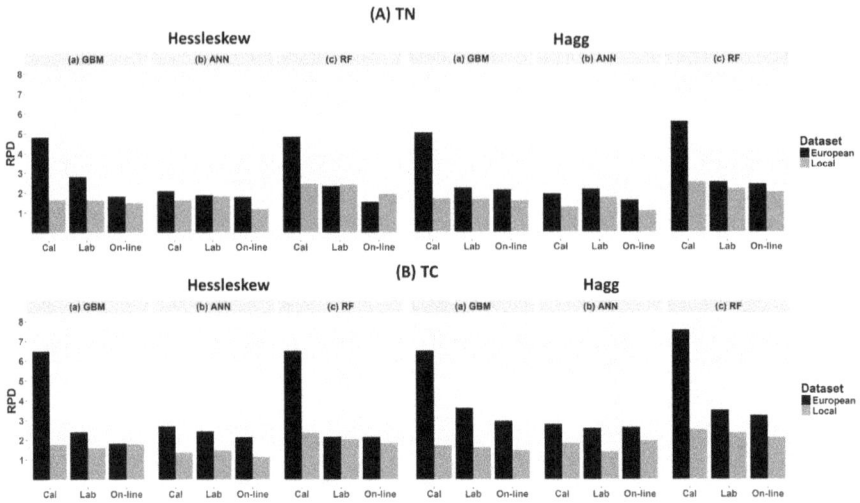

Figure 8. Comparison of residual prediction deviation (RPD) values for (**A**) total nitrogen (TN) and (**B**) total carbon (TC) predictions obtained with (**a**) gradient boosted machines (GBM); (**b**) artificial neural networks (ANNs) and (**c**) random forests (RF) analyses in cross-validation (Cal), laboratory prediction (Lab) and on-line prediction (Online). Results were generated with local field datasets of 122 and 149 samples for the Hessleskew and Hagg fields, respectively, and a spiked European dataset (528 samples).

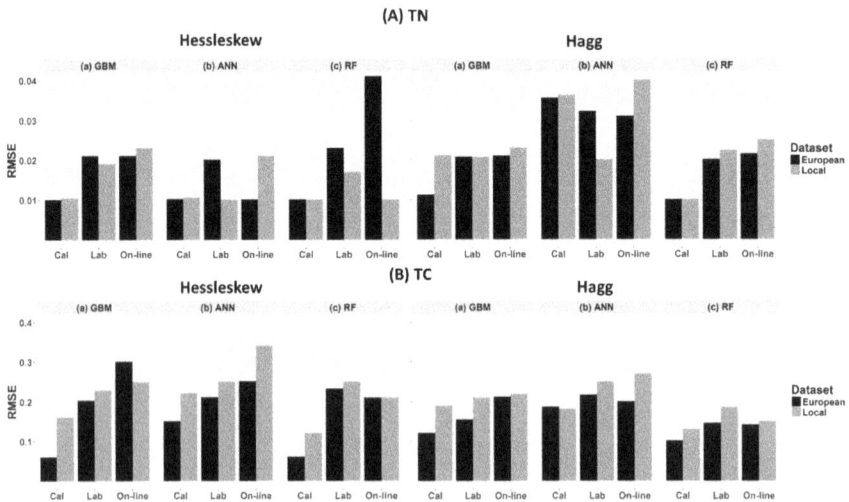

Figure 9. Comparison of root mean square error (RMSE) values for (**A**) total nitrogen (TN) and (**B**) total carbon (TC) predictions obtained with (**a**) gradient boosted machines (GBM); (**b**) artificial neural networks (ANNs) and (**c**) random forests (RF) analyses in cross-validation (Cal), laboratory prediction (Lab) and on-line prediction (Online). Results were generated with local field datasets of 122 and 149 samples for the Hessleskew and Hagg fields, respectively, and a spiked European dataset (528 samples).

Similar to the on-line predictions, the results of RF for the Hagg field are better than those in the Hessleskew field, which is in line with the results reported by Nawar and Mouazen [5] for on-line measurement based on the spiked European dataset (RMSEP of 0.03–0.19% and RPD of 5.21–5.94 for TN and TC, respectively). However, results in this research are better than those reported by Kuang and Mouazen [48] using PLSR, with RPD of 2.52 and 2.33 for TN and OC, respectively and better than ANN models [11] for on-line prediction of OC (RPD = 2.28, compared to RPD of 2.52 of the current research). RF outperforms both GBM and ANN for on-line predictions in the Hagg field only, whereas ANN replaces RF in being the best performing for on-line prediction in the Hessleskew field. From the quality of on-line prediction of the studied two soil properties it can be concluded that ANN might perform equally as RF, which rejects the hypothesis of the current work that both RF and GBM outperform ANN for on-line prediction of TN and TC.

4.2. Influence of Dataset on Models' Performance

The influence of dataset size and concentration range showed great influences on the performance in calibration and prediction. The results associated with spiking local samples into the European dataset more often enhances the overall model performance, especially for cross-validation, in comparison with those obtained using the local dataset (Table 2 and Figure 8), which is in agreement with the results presented by Brown [49] and Sankey et al. [50] for non-mobile measurements, and Kuang and Mouazen [48] for on-line measurements. The improvement was mainly expressed as improved R^2 and RPD, and RMSEP in laboratory and on-line predictions (Table 2 and Figure 9). This finding is in agreement with Kuang and Mouazen [48], who reported improvement in R^2 and RPD for predictions of TN and TC by adding local samples into a general library. Furthermore, Nawar and Mouazen [5] reported that the spiking of local soil samples into European datasets turned out to be a competent method to enhance the prediction associated with target field samples. Compared to published results, using the spiking of target field samples into European samples obtained with PLSR analyses [48] for TN (RPD = 1.96–2.52) and OC (RPD = 1.88–2.38), the results of on-line prediction based on RF in the current research is better for TN and TC, and better than those results reported for on-line measurement of OC by Kuang et al. [11] using ANN analysis with RPD and RMSE values of 2.28 and 1.25%, respectively. Taking into account the small variation range of TN and TC in the two scanned fields (Table 1), spiking of the European dataset with local samples seems to provide the best scenario to improve on-line prediction performance. This was also proved to be true for laboratory-scanned (non-mobile) soil spectra spiked into global or European datasets [51,52].

A possible explanation for the high performance of both the laboratory and on-line predictions with the spiked European data set is the wider concentration ranges (larger variability) within the datasets for both properties compared to the narrow range of the local datasets. This wide range or variability is indeed a fundamental factor in the calibration of the vis-NIR spectroscopy which is essential for successful modelling of data, particularly in fields with narrow concentration ranges. In fact, if the concentration range in a field is too narrow, no calibration models can be established at all, and it will be essential at this point to spike selected samples from a target field into existing spectral library with wide concentration range. This implies that the overall model performance may depend to a large extent of variability exist in the dataset [53]. This is the reason why researchers have concluded that calibration models should be established based on libraries that capture wide concentration range and soil types [54]. Kuang and Mouazen [6] reported that fields with small variations in concentrations of a given soil property will properly lead to inferior model performance (small R^2 and RPD). Furthermore, Nawar and Mouazen [5] found that spiked local sample with small variation (small range) into an European dataset with wide concentration range improved the on-line prediction (in terms of improved R^2 and RPD and decreased RMSEP) of TN and TC compared to local datasets.

5. Conclusions

In this study the performance of generalized boosted machines (GBM), artificial neural networks (ANNs), and random forests (RF) methods was compared for the visible and near infrared spectroscopy prediction of soil total nitrogen (TN) and total carbon (TC) under laboratory (non-mobile) and on-line (mobile) scanning conditions in two selected fields in the UK (Hessleskew and Hagg fields). We have tested the performance of these modelling methods using local and European datasets, spiked with samples from the two target fields. Generally, the performance of the GBM, ANN and RF models varied according to the dataset used. Results showed the majority of the RF models to outperform the corresponding GBM and ANN models in cross-validation, laboratory and on-line predictions. Results in cross-validation showed improved performance with the spiked European dataset that were collected from 16 fields in five European countries. Nevertheless, the performance of laboratory and on-line predictions does not necessarily behave similarly to cross-validation. The ANN model based on the spiked European dataset showed better performance than RF and GBM in laboratory (for TC only) and on-line prediction (for TC and TN) in the Hessleskew field only. The highest on-line prediction results were observed with RF models in the Hagg field based on the spiked European dataset.

From the results obtained in this work, it is observed that calibrations obtained with the spiked European dataset is the most successful option for on-line predictions of the TN and TC, compared to field local calibration. The spiked European calibrations based on 528 samples provided a larger coefficient of determination (R^2) and residual prediction deviation (RPD) compared to the local calibration models for TN and TC in both fields. Future work needs to focus on optimizing the selection of an optimal dataset to be spiked into the European dataset. This needs to test distance matrices and sample selection algorithms for potential improvement in the prediction quality of resulted models compared to random sample selection based modelling.

Acknowledgments: Authors acknowledge the financial support received through Tru-Nject project (Nr. 36428-267209), which was jointly sponsored by Innovate UK and Biotechnology and Biological Sciences Research Council (BBSRC). Authors also acknowledge the FWO funded Odysseus SiTeMan Project (Nr. G0F9216N).

Author Contributions: Said Nawar and Abdul Mouazen conceived and designed the experiments; Said Nawar performed the experiments; Said Nawar analyzed the data; Said Nawar and Abdul Mouazen wrote the paper. Both authors have read and approved the manuscript.

Conflicts of Interest: The authors declare no conflict of interest.

References

1. Kucharik, C.J.; Brye, K.R.; Norman, J.M.; Foley, J.A.; Gower, S.T.; Bundy, L.G. Measurements and Modeling of Carbon and Nitrogen Cycling in Agroecosystems of Southern Wisconsin: Potential for SOC Sequestration during the Next 50 Years. *Ecosystems* **2001**, *4*, 237–258. [CrossRef]
2. Muñoz, J.D.; Kravchenko, A. Soil carbon mapping using on-the-go near infrared spectroscopy, topography and aerial photographs. *Geoderma* **2011**, *166*, 102–110. [CrossRef]
3. McDowell, M.L.; Bruland, G.L.; Deenik, J.L.; Grunwald, S.; Knox, N.M. Soil total carbon analysis in Hawaiian soils with visible, near-infrared and mid-infrared diffuse reflectance spectroscopy. *Geoderma* **2012**, *189*, 312–320. [CrossRef]
4. Wang, D.; Chakraborty, S.; Weindorf, D.C.; Li, B.; Sharma, A.; Paul, S.; Ali, M.N. Synthesized use of VisNIR DRS and PXRF for soil characterization: Total carbon and total nitrogen. *Geoderma* **2015**, *243–244*, 157–167. [CrossRef]
5. Nawar, S.; Mouazen, A.M. Predictive performance of mobile vis-near infrared spectroscopy for key soil properties at different geographical scales by using spiking and data mining techniques. *Catena* **2017**, *151*, 118–129. [CrossRef]
6. Kuang, B.; Mouazen, A.M. Calibration of visible and near infrared spectroscopy for soil analysis at the field scale on three European farms. *Eur. J. Soil Sci.* **2011**, *62*, 629–636. [CrossRef]
7. Martens, H.; Naes, T. *Multivariate Calibration*; Wiley: Hoboken, NJ, USA, 1991; ISBN 0471930474.

8. Morellos, A.; Pantazi, X.-E.; Moshou, D.; Alexandridis, T.; Whetton, R.; Tziotzios, G.; Wiebensohn, J.; Bill, R.; Mouazen, A.M. Machine learning based prediction of soil total nitrogen, organic carbon and moisture content by using VIS-NIR spectroscopy. *Biosyst. Eng.* **2016**, *152*, 1–13. [CrossRef]
9. Nawar, S.; Buddenbaum, H.; Hill, J.; Kozak, J.; Mouazen, A.M. Estimating the soil clay content and organic matter by means of different calibration methods of vis-NIR diffuse reflectance spectroscopy. *Soil Tillage Res.* **2016**, *155*, 510–522. [CrossRef]
10. Mouazen, A.M.; Kuang, B.; De Baerdemaeker, J.; Ramon, H. Comparison among principal component, partial least squares and back propagation neural network analyses for accuracy of measurement of selected soil properties with visible and near infrared spectroscopy. *Geoderma* **2010**, *158*, 23–31. [CrossRef]
11. Kuang, B.; Tekin, Y.; Mouazen, A.M. Comparison between artificial neural network and partial least squares for on-line visible and near infrared spectroscopy measurement of soil organic carbon, pH and clay content. *Soil Tillage Res.* **2015**, *146*, 243–252. [CrossRef]
12. Brown, D.J.; Shepherd, K.D.; Walsh, M.G.; Dewayne Mays, M.; Reinsch, T.G. Global soil characterization with VNIR diffuse reflectance spectroscopy. *Geoderma* **2006**, *132*, 273–290. [CrossRef]
13. Shepherd, K.D.; Walsh, M.G. Development of Reflectance Spectral Libraries for Characterization of Soil Properties. *Soil Sci. Soc. Am. J.* **2002**, *66*, 988–998. [CrossRef]
14. Rossel Viscarra, R.A.; Behrens, T. Using data mining to model and interpret soil diffuse reflectance spectra. *Geoderma* **2010**, *158*, 46–54. [CrossRef]
15. Sorenson, P.T.; Small, C.; Tappert, M.C.; Quideau, S.A.; Drozdowski, B.; Underwood, A.; Janz, A. Monitoring organic carbon, total nitrogen, and pH for reclaimed soils using field reflectance spectroscopy. *Can. J. Soil Sci.* **2017**, *97*, 241–248. [CrossRef]
16. Marini, F. Neural Networks. In *Comprehensive Chemometrics*; Elsevier: Amsterdam, The Netherlands, 2010; Volume 3, pp. 477–505.
17. Long, J.R.; Gregoriou, V.G.; Gemperline, P.J. Spectroscopic calibration and quantitation using artificial neural networks. *Anal. Chem.* **1990**, *62*, 1791–1797. [CrossRef]
18. Diamantaras, K.; Duch, W.; Iliadis, L.S. Artificial Neural Networks. In Proceedings of the ICANN 2010: 20th International Conference, Thessaloniki, Greece, 15–18 September 2010; pp. 31–32.
19. Breiman, L. Random forests. *Mach. Learn.* **2001**, *45*, 5–32. [CrossRef]
20. Prasad, A.M.; Iverson, L.R.; Liaw, A. Newer classification and regression tree techniques: Bagging and random forests for ecological prediction. *Ecosystems* **2006**, *9*, 181–199. [CrossRef]
21. Ishwaran, H. Variable importance in binary regression trees and forests. *Electron. J. Stat.* **2007**, *1*, 519–537. [CrossRef]
22. Friedman, J.H. Stochastic gradient boosting. *Comput. Stat. Data Anal.* **2002**, *38*, 367–378. [CrossRef]
23. Forkuor, G.; Hounkpatin, O.K.L.; Welp, G.; Thiel, M. High Resolution Mapping of Soil Properties Using Remote Sensing Variables in South-Western Burkina Faso: A Comparison of Machine Learning and Multiple Linear Regression Models. *PLoS ONE* **2017**, *12*, e0170478. [CrossRef] [PubMed]
24. Martin, M.P.; Wattenbach, M.; Smith, P.; Meersmans, J.; Jolivet, C.; Boulonne, L.; Arrouays, D. Spatial distribution of soil organic carbon stocks in France. *Biogeosciences* **2011**, *8*, 1053–1065. [CrossRef]
25. Martin, M.P.; Orton, T.G.; Lacarce, E.; Meersmans, J.; Saby, N.P.A.; Paroissien, J.B.; Jolivet, C.; Boulonne, L.; Arrouays, D. Evaluation of modelling approaches for predicting the spatial distribution of soil organic carbon stocks at the national scale. *Geoderma* **2014**, *223–225*, 97–107. [CrossRef]
26. Natural Resources Conservation Service, USDA. *Soil Taxonomy: A Basic System of Soil Classification for Making and Interpreting Soil Surveys*; Agricultural Handbook 436; Natural Resources Conservation Service, USDA: Washington, DC, USA, 1999.
27. Mouazen, A.M. *Soil Sensing Device. International Publication, Published under the Patent Cooperation Treaty (PCT)*; World Intellectual Property Organization, International Bureau: Brussels, Belgium, 2006.
28. British Standards Institution. *BS 7755-3.8:1995 ISO 10694:1995 Part 3: Chemical Methods—Section 3.8 Determination of Organic and Total Carbon after Dry Combustion (Elementary Analysis)*; British Standards Institution: London, UK, 1995.
29. British Standards Institute. *BS EN 13654-2:2001: Soil Improvers and Growing Media. Determination of Nitrogen. Dumas Method*; British Standards Institution: London, UK, 2001.
30. Stevens, A.; Ramirez Lopez, L. An Introduction to the Prospectr Package. 2014. Available online: https://cran.r-project.org/web/packages/prospectr/vignettes/prospectr-intro.pdf (accessed on 22 April 2016).

31. Norris, K. Applying Norris Derivatives. Understanding and correcting the factors which affect diffuse transmittance spectra. *NIR News* **2001**, *12*, 6–9. [CrossRef]

32. Kennard, R.W.; Stone, L.A. Computer Aided Design of Experiments. *Technometrics* **1969**, *11*, 137–148. [CrossRef]

33. Segal, M.; Xiao, Y. Multivariate random forests. *WIREs Data Mining Knowl Discov.* **2011**, *1*, 80–87. [CrossRef]

34. Cutler, A.; Cutler, D.R.; Stevens, J.R. Random Forests. In *Ensemble Machine Learning*; Springer: Boston, MA, USA, 2012; pp. 157–175.

35. Abdel Rahman, A.M.; Pawling, J.; Ryczko, M.; Caudy, A.A.; Dennis, J.W. Targeted metabolomics in cultured cells and tissues by mass spectrometry: Method development and validation. *Anal. Chim. Acta* **2014**, *845*, 53–61. [CrossRef] [PubMed]

36. Peters, J.; De Baets, B.; Verhoest, N.E.C.; Samson, R.; Degroeve, S.; De Becker, P.; Huybrechts, W. Random forests as a tool for ecohydrological distribution modelling. *Ecol. Modell.* **2007**, *207*, 304–318. [CrossRef]

37. Caruana, R.; Niculescu-Mizil, A. An Empirical Comparison of Supervised Learning Algorithms | Machine Learning | Support Vector Machine. Available online: https://www.scribd.com/document/113006633/2006-An-Empirical-Comparison-of-Supervised-Learning-Algorithms# (accessed on 17 September 2017).

38. Liaw, A.; Wiener, M. Breiman and Cutler's Random Forests for Classification and Regression. 2015. Available online: https://cran.r-project.org/web/packages/randomForest/randomForest.pdf (accessed on 28 April 2016).

39. Schapire, R.E. The Boosting Approach to Machine Learning: An Overview. In *Nonlinear Estimation and Classification*; Springer: New York, NY, USA, 2003; pp. 149–171.

40. Hastie, T.; Tibshirani, R.; Friedman, J. *The Elements of Statistical Learning*; Springer Series in Statistics; Springer New York: New York, NY, USA, 2009; Volume 20, ISBN 978-0-387-84857-0.

41. Dormann, C.F.; McPherson, J.M.; Araújo, M.B.; Bivand, R.; Bolliger, J.; Carl, G.; Davies, R.G.; Hirzel, A.; Jetz, W.; Kissling, W.D.; et al. Methods to account for spatial autocorrelation in the analysis of species distributional data: A review. *Ecography* **2007**, *30*, 609–628. [CrossRef]

42. Elith, J.; Leathwick, J.R.; Hastie, T. A working guide to boosted regression trees. *J. Anim. Ecol.* **2008**, *77*, 802–813. [CrossRef] [PubMed]

43. Varmuza, K.; Filzmoser, P. *Introduction to Multivariate Statistical Analysis in Chemometrics*; CRC Press: Boca Raton, FL, USA, 2009; ISBN 9781420059472.

44. Kuhn, M.; Wing, J.; Weston, S.; Williams, A.; Keefer, C.; Engelhardt, A.; Cooper, T.; Mayer, Z.; Benesty, M.; Lescarbeau, R.; et al. Package "Caret" Title Classification and Regression Training Description Misc Functions for Training and Plotting Classification and Regression Models. Available online: https://cran.r-project.org/web/packages/caret/caret.pdf (accessed on 9 August 2017).

45. R Core Team. R: A Language and Environment for Statistical Computing. R Foundation for Statistical Computing: Vienna, Austria, 2016. Available online: https://www.r-project.org/ (accessed on 9 August 2017).

46. Viscarra Rossel, R.A.; Walvoort, D.J.J.; McBratney, A.B.; Janik, L.J.; Skjemstad, J.O. Visible, near infrared, mid infrared or combined diffuse reflectance spectroscopy for simultaneous assessment of various soil properties. *Geoderma* **2006**, *131*, 59–75. [CrossRef]

47. Yu, C.; Grunwald, S.; Xiong, X. *Transferability and Scaling of VNIR Prediction Models for Soil Total Carbon in Florida*; Springer: Singapore, 2016; pp. 259–273.

48. Kuang, B.; Mouazen, A.M. Effect of spiking strategy and ratio on calibration of on-line visible and near infrared soil sensor for measurement in European farms. *Soil Tillage Res.* **2013**, *128*, 125–136. [CrossRef]

49. Brown, D.J. Using a global VNIR soil-spectral library for local soil characterization and landscape modeling in a 2nd-order Uganda watershed. *Geoderma* **2007**, *140*, 444–453. [CrossRef]

50. Sankey, J.B.; Brown, D.J.; Bernard, M.L.; Lawrence, R.L. Comparing local vs. global visible and near-infrared (VisNIR) diffuse reflectance spectroscopy (DRS) calibrations for the prediction of soil clay, organic C and inorganic C. *Geoderma* **2008**, *148*, 149–158. [CrossRef]

51. Wetterlind, J.; Stenberg, B. Near-infrared spectroscopy for within-field soil characterization: Small local calibrations compared with national libraries spiked with local samples. *Eur. J. Soil Sci.* **2010**, *61*, 823–843. [CrossRef]

52. Guerrero, C.; Zornoza, R.; Gómez, I.; Mataix-Beneyto, J. Spiking of NIR regional models using samples from target sites: Effect of model size on prediction accuracy. *Geoderma* **2010**, *158*, 66–77. [CrossRef]

Sensors **2017**, *17*, 2428

53. Stenberg, B.; Viscarra Rossel, R.A.; Mouazen, A.M.; Wetterlind, J. *Visible and Near Infrared Spectroscopy in Soil Science*; Sparks, D.L., Ed.; Academic Press: Burlington, VT, USA, 2010; Volume 107, pp. 163–215.
54. Bonett, J.P.; Camacho-Tamayo, J.H.; Ramírez-López, L. Mid-infrared spectroscopy for the estimation of some soil properties. *Agron. Colomb.* **2015**, *33*, 99–106. [CrossRef]

sensors

MDPI

Article

Fast Detection of Striped Stem-Borer (*Chilo suppressalis* Walker) Infested Rice Seedling Based on Visible/Near-Infrared Hyperspectral Imaging System

Yangyang Fan [1,2], Tao Wang [1,2], Zhengjun Qiu [1,2,*], Jiyu Peng [1,2], Chu Zhang [1,2,*] and Yong He [1,2]

1   College of Biosystems Engineering and Food Science, Zhejiang University, Hangzhou 310058, China; fanyangy@zju.edu.cn (Y.F.); wt0330@zju.edu.cn (T.W.); jypeng@zju.edu.cn (J.P.); yhe@zju.edu.cn (Y.H.)
2   Key Laboratory of Spectroscopy Sensing, Ministry of Agriculture, Hangzhou 310058, China
*   Correspondence: zjqiu@zju.edu.cn (Z.Q.); chuzh@zju.edu.cn (C.Z.); Tel.: +86-571-8898-2728 (Z.Q.)

Received: 20 September 2017; Accepted: 24 October 2017; Published: 27 October 2017

Abstract: Striped stem-borer (SSB) infestation is one of the most serious sources of damage to rice growth. A rapid and non-destructive method of early SSB detection is essential for rice-growth protection. In this study, hyperspectral imaging combined with chemometrics was used to detect early SSB infestation in rice and identify the degree of infestation (DI). Visible/near-infrared hyperspectral images (in the spectral range of 380 nm to 1030 nm) were taken of the healthy rice plants and infested rice plants by SSB for 2, 4, 6, 8 and 10 days. A total of 17 characteristic wavelengths were selected from the spectral data extracted from the hyperspectral images by the successive projection algorithm (SPA). Principal component analysis (PCA) was applied to the hyperspectral images, and 16 textural features based on the gray-level co-occurrence matrix (GLCM) were extracted from the first two principal component (PC) images. A back-propagation neural network (BPNN) was used to establish infestation degree evaluation models based on full spectra, characteristic wavelengths, textural features and features fusion, respectively. BPNN models based on a fusion of characteristic wavelengths and textural features achieved the best performance, with classification accuracy of calibration and prediction sets over 95%. The accuracy of each infestation degree was satisfactory, and the accuracy of rice samples infested for 2 days was slightly low. In all, this study indicated the feasibility of hyperspectral imaging techniques to detect early SSB infestation and identify degrees of infestation.

Keywords: rice; striped stem-borer; hyperspectral imaging; texture feature; data fusion

---

1. Introduction

Rice is one of the most important foods for more than half of the global population. Pest infestation is one of the severe threats to rice growth, and it usually leads to serious loss of yield and quality [1]. Striped stem-borer (SSB) is one of the destructive rice pests in many rice-growing countries [2]. The traditional detection method for SSB is manual inspection according to conspicuous symptoms, such as a dead heart at tillering age and a white head at booting age [3]. As striped stem-borer is a boring insect and feeds on plant tissue in the stem wall [4], the stalk characteristics will change earlier than the canopy characteristics. Accurate SSB statistics need to dissect rice in the laboratory, which demands expert knowledge of the pest. This procedure is time-consuming and labor-intensive, and will decrease the detection efficiency and delay the appropriate controlling time. Hence, an efficient and effective detection method is necessary for early detection of SSB infestation in rice.

The optical properties of the plant refer to the absorption, reflectance and transmittance of light when the plant surface interacts with radiant energy. Reflectance can be influenced by the plant's physiological properties and, thus, has been utilized by spectral technology to detect plant disease [5], fruit quality [6], agricultural product characteristics [7] and so on. Pest infestation can cause external and internal damage to the plant, such as the destruction of cell structure by the nibbling of tissue [8] and the loss of photosynthetic pigments by the piercing and sucking of sap [9], etc. The reflectance of plants in the visible waveband is associated with pigments, while reflectance in the near-infrared waveband provides information about plant water content and physical structure [10]. Thus, the spectral technique has the potential to be used to detect pest infestation in plants by measuring changes in spectral reflectance [11–13].

Besides changes in spectral characteristics, the external features of a plant would also change along with the changes of its physiological properties induced by pest feeding, excretion and other activities. These changes, such as yellowing, rot, defect, etc., could be captured by machine vision, and the image features would vary with the level of aggravation of the infestation. The extracted image features could be used to establish detection models with the reference data. Hence, imaging techniques have been applied in order to detect pest infestation [14–16].

Hyperspectral imaging is a technique integrating spectroscopy and imaging techniques, which can acquire both spectral and spatial information at high resolution. Hyperspectral imaging has been investigated as a potential technique in crop protection [17]. Sytar et al. [18] have reviewed the studies of hyperspectral imaging techniques to detect plant changes caused by salt stress and have shown the potential of this technique to detect salinity in soil. Thomas et al. [19] have reviewed research about plant disease detection based on the hyperspectral imaging technique and have shown the potential to detect and identify plant diseases before visible symptoms appear.

Zhao et al. [20] have distinguished Chinese cabbage infested by aphids from healthy cabbage based on hyperspectral imaging technology, and obtained the highest accuracy rate of 90%. Wu et al. [21] have employed the hyperspectral imaging technique to detect *Pieris rapae* larvae on cabbage leaves, and acquired classification accuracy above 96%. Thus, these studies indicated that the hyperspectral imaging technique has great potential for detecting pest infestation.

The main purpose of this study was to detect early striped stem-borer infestation in rice and identify degrees of infestation based on a visible/near-infrared hyperspectral imaging system. The specific objectives were to: (1) establish back-propagation neural network (BPNN) models to identify healthy rice samples and samples infested to different degrees; (2) select characteristic wavelengths by successive projection analysis (SPA); (3) extract textural features based on the gray-level co-occurrence matrix (GLCM); (4) improve detection performance by combining characteristic wavelengths and texture features.

2. Materials and Methods

2.1. Rice Samples Preparation

A total of 114 rice plants (Y Liangyou689, non-glutinous rice) were grown in an outdoor environment under insect-proof screen in Zhejiang University, Hangzhou, China. Eggs of the striped stem-borer were bought from the Shennong Biotechnology Company, Hangzhou, China, and were hatched on a moistened filter paper in a petri dish at a temperature of 28–30 °C and under illumination of 3000 lux for 10 h. When the rice was at the tillering stage, one first-instar striped stem-borer larva was placed on the rice after 2 h of starvation [22]; 69 rice plants were inoculated as the experimental group; and 45 rice plants were kept as the control group without inoculation.

The hyperspectral images of control and infested rice plants were acquired every two days from 22 July 2016 (two days after infestation) to 30 July. The acquisition terminated on the eleventh day after infestation because the symptoms of top yellowing and stem-rotting lesions were already serious enough for there to be no need for detection by hyperspectral technology. The degrees of infestation (DI) were divided into DI1, DI2, DI3, DI4, DI5 according to the infested days, as shown in Figure 1; and the control group with healthy rice plants was referred as DI0. DI1 referred to the samples infested for two days, DI2 referred to the samples infested for 4 days, and so on. The number of infested samples on the last day decreased to 44 because of the loss caused by aggravated infestation. Thus, a total 365 samples (the number of DI0, DI1, DI 2, DI3, DI4, DI5 were 45, 69, 69, 69, 69, and 44 respectively) were acquired in this study.

Figure 1. Samples of six degrees of infestation: (**a**) DI0; (**b**) DI1; (**c**) DI2; (**d**) DI3; (**e**) DI4; (**f**) DI5.

2.2. Hyperspectral Imaging System and Image Acquisition

2.2.1. Hyperspectral Imaging System

The visible/near-infrared hyperspectral imaging system, with 512 bands in the spectral ranges of 380–1030 nm, includes an imaging spectrograph (ImSpectorV10E; Spectral Imaging Ltd., Oulu, Finland); a 672 × 512 CCD camera (C8484-05, Hamamatsu Photonics, Hamamatsu, Japan); a camera lens (OLES23; Specim, Spectral Imaging Ltd., Oulu, Finland); two 150 W tungsten halogen lamps (Fiber-Lite DC950 Illuminator; Dolan Jenner Industries Inc., Boxborough, MA, USA) placed on both sides of the camera at a 45° angle; and a conveyer belt driven by a stepping motor (IRCP0076, Isuzu OpticsCrop, Zhubei, Taiwan). The system is controlled by a computer with Spectral Image-V10E software (Isuzu Optics Corp, Zhubei, Taiwan).

2.2.2. Image Acquisition and Calibration

Before image acquisition, the parameters of the hyperspectral imaging system should be adjusted to acquire a clear and non-distorted image. The height between the samples and the lens was 35 cm; the conveyer belt's moving speed was 3.00 mm/s; and the exposure time of the camera was 0.08 s. The sample was placed flat on the conveyer when collecting the image. To reduce noise and avoid the influence of dark current, the raw hyperspectral image should be calibrated according to the following formula:

$$I_c = \frac{I_{raw} - I_{dark}}{I_{white} - I_{dark}} \tag{1}$$

where the I_c is the calibrated hyperspectral image; I_{raw} is the raw hyperspectral image; I_{dark} is the dark reference image with 0% reflectance; and I_{white} is the white reference image with 99.9% reflectance.

2.3. Spectral Information Extraction

The region of interest (ROI) was predefined as the stalk region of rice in the image. As each pixel in the hyperspectral image corresponds to a spectral curve in full bands, the spectrum of all pixels in the ROI was averaged as representative of the sample. The samples were divided into a calibration set and a validation set by the Kennard–Stone (KS) algorithm [23] in a ratio of 2:1. There were 243 samples in the calibration set and 122 samples in the prediction set.

2.4. Texture Feature Extraction

Textural features have been utilized to reflect a plant's physical characteristics such as firmness, color and roughness, which are related to the spatial arrangement of pixel intensity in an image. SSB infestation would influence not only the spectral characteristic but also the textural features of rice stalk. The gray-level co-occurrence matrix (GLCM) [24], as one of the most commonly used textural features in hyperspectral imaging, is defined as the relative frequency of occurrence of pixel pairs in a certain distance (D) and direction (θ) [25]. Eight descriptors of GLCM were chosen in this study, including mean, variance, homogeneity, contrast, dissimilarity, entropy, second moment, and correlation. Mean is the average grey level in the chosen image. Variance reflects the grey-level standard deviation. Homogeneity measures the closeness of the distribution of elements in the GLCM to the GLCM diagonal. Contrast is a measure of the degree of spread of the grey levels or the average grey-level difference between neighboring pixels. Dissimilarity is similar to contrast, but increases linearly as the difference between two pixels increases. Entropy measures the degree of disorder in an image. Second moment measures the textural uniformity or pixel-pair repetitions. Correlation is a measure of grey-level linear dependencies in the image [26,27].

Each band in a hyperspectral image corresponds to a gray-scale image, and one hyperspectral image contains 512 images. The textural features set will be huge and difficult to calculate if GLCM features are extracted based on full bands. Thus, principal component analysis (PCA) was employed to transform hundreds of images into principal component (PC) images, and the textural features were extracted based on the first few PC images that contained enough valid information. The extraction of spectral and textural features from hyperspectral images was performed on ENVI 4.6 (ITT, Visual Information Solutions, Boulder, CO, USA).

2.5. Data Analysis

2.5.1. Characteristic Wavelength Selection

The hyperspectral data of each sample contains 512 variables in a full band. The redundancy and collinearity of the huge dataset would inevitably disturb the detection accuracy. A selection of characteristic wavelengths which have the most influence on the degree of infestation is essential in order to reduce the data dimensions and improve the detecting efficiency.

The successive projection algorithm (SPA) was employed in this study to choose the characteristic wavelengths. SPA is a common and effective method for reducing the variables of hyperspectral data, which can minimize the collinearity effects of raw input [28,29]. SPA is a forward variable selection method [30] by optimizing the multiple linear-regression (MLR) model, which includes two phases.

Phase 1 is to project the input X ($N \times K$) matrix, and generate K chains with M variables, $M = \min(N - 1, K)$. This process includes six steps: Step 1 is to initialize z^i with x_k, where i is the iteration counter, and initialize x_j^i with x_j, where $j = 1, \dots, k$. Step 2 is to calculate the matrix P of projection onto the orthogonal subspace to z^i. Step 3 is to calculate the projected vector x_j^{i+1}. Step 4 is to determine the index j_{max} of the largest projected vector and store the index. Step 5 is to initialize z^{i+1} with x_{jmax}^{i+1}. Step 6 is to return to Step 2 and start another iteration if $i < M$.

Phase 2 is to choose the best variable subset from the candidate subsets extracted from the K chains according to the minimum root mean square error (RMSE) obtained by applying the MLR model to the validation set [31].

2.5.2. Chemometrics Algorithm

To identify degrees of infestation, the chemometrics algorithm needs to be employed as a classifier in order to accept the extracted information as input. The BPNN is a multi-layer feed-forward neural network with great capacity for non-linear mapping, and has been applied in hyperspectral imaging analysis in many studies [32–34]. The basic BPNN consists of an input layer, a hidden layer and an output layer. The connection between layers depends on the nodes of each layer. There are three main steps to train the BPNN model, including feed-forward computation, errors back-propagation, and weights updates [35]. Feed-forward computation aims to calculate and transmit the value of nodes in the order from the input layer to the output layer. Errors back-propagation aims to calculate the errors between the output and the reference and transmit the errors back successively. The weights are then updated until the error meets the target error or the training times reach the requirements. After comparing multiple network structures with different parameters, the optimal parameters of the number of nodes, the learning rate, the target error and the training times were set as 5, 0.6, 1×10^{-5}, and 1000, respectively. Identification accuracy and run time were employed to evaluate the BPNN performance with different datasets. The SPA and BPNN algorithms were executed on Matlab R2011b (The Math Works, Natick, MA, USA).

3. Results

3.1. Spectra Features

The head and end ranges of wavebands contain a large proportion of noise. Therefore, the first 82 bands and the last 22 bands were removed to improve the signal-noise ratio. Spectra in the range of 480–1000 nm were pre-processed by Savitzky–Golay smoothing before analysis. The average spectra of each degree are shown in Figure 2. It was found that the general trends of six curves were similar. Significant differences of reflectance were observed in the range of 530–700 nm and 750–940 nm. The differences between the first three degrees were smaller than the differences between the last three degrees, which could be explained by the fact that damage symptoms of the samples in the first four days were mild and the stem structures were not destroyed too seriously. In the visible range of 570–700 nm, the reflectance was higher with the increase of infestation severity. This was because of the destruction of chlorophyll located in the chloroplast of the rice stem's cortex cell [36]. In the near-infrared range of 750–1000 nm, the reflectance of DI0 was higher than that of the infested samples. The more severe the sample was infested, the lower the reflectance. The reduction of reflectance with the increase of severity was mainly due to the destruction of stem structure, which led to photon scattering [37].

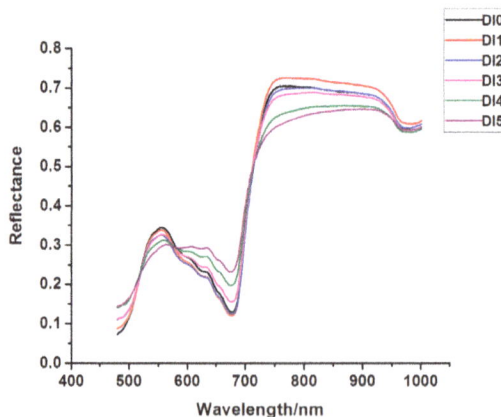

Figure 2. Average spectral curves of samples in six degrees of infestation.

As illustrated in Figure 2, the reflectance of DI1 and DI2 was lower than that of DI0 in the range of 570–700 nm, and the reflectance of DI1 and DI2 was higher than that of DI0 in the range of 750–1000 nm. This phenomenon may be explained by the compensation effect while pest infestation was in the incipient stage. In the early stage of striped stem-borer infestation, rice can generate a series of compensating responses to the injury, such as increasing the photosynthesis rate of healthy leaves, translocating photoassimilates from the damaged tillers to healthy tillers, and increasing productive tiller numbers [38,39]. But if the SSB cannot be controlled in a timely way, the compensating responses will not counteract the damage that SSB causes to rice, leading to serious yield loss.

3.2. Qualitative Analysis by PCA

PCA was employed in this study to investigate qualitatively the clustering trend of the samples based on full spectra. PCA can orthogonally transform the original possibly correlated variables into more uncorrelated variables that display the internal structure of the data [40]. The first principal component explained the largest variance, and the explained variables of the following PCs decreased successively. Thus, the first few PCs usually explain the most variances. In this study, PC1, PC2, and PC3 totally explained 98.3% of the variables, which were chosen to investigate the distribution pattern.

The three-dimensional scores' scatter plot is displayed in Figure 3. In general, there was an obvious separation trend between the first four degrees and the last two degrees. The samples of the first four degrees were closely distributed while the samples of the last two degrees had a scattered distribution. The overlaps were serious between different degrees, but the samples of the first three degrees were seldom confused with the samples of the last two degrees. These phenomena indicated that the spectral characteristics would evidently change after DI3, and the chemometric method is necessary to identify accurately the degree of infestation.

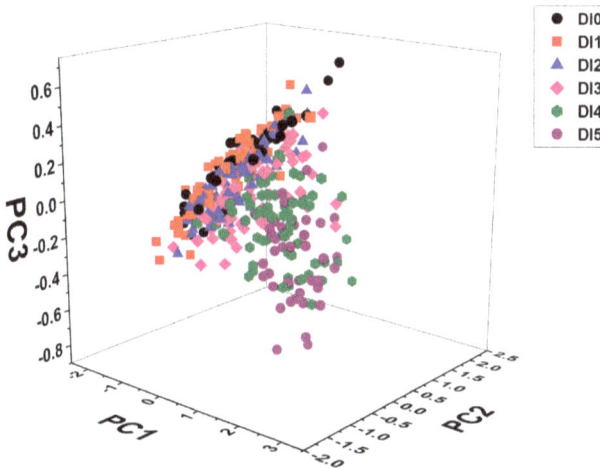

Figure 3. Scores' scatter plots of samples in six degrees of infestation.

3.3. Identification Results Based on Full Spectra

Full spectra were used as input to the BPNN model in order to identify different degrees of infestation. The results are given in Table 1. The overall accuracy was satisfactory, with classification accuracy over 90% for both the calibration and prediction sets. The classification accuracy of each degree was over 90% for both the calibration and prediction sets, except for DI0. The detection accuracy of DI1 was higher than DI0 but lower than DI2, DI3, DI4 and DI5. The main errors were attributed to the misclassification of the adjacent degree. A total of 9 samples and 6 samples were confused between DI0 and DI1 in the calibration and prediction sets, respectively.

Table 1. Detection accuracy of six infestation degrees by the BPNN model based on full spectra.

Model	Actual Value	Calibration Set							Prediction Set						
		DI [1]0	DI1	DI2	DI3	DI4	DI5	Accuracy	DI0	DI1	DI2	DI3	DI4	DI5	Accuracy
	DI0	23	6	0	0	1	0	76.67%	9	5	0	0	1	0	60%
	DI1	3	43	0	0	0	0	93.48%	1	22	0	0	0	0	95.65%
	DI2	0	0	46	0	0	0	100%	0	0	23	0	0	0	100%
BPNN	DI3	1	0	0	44	0	1	95.65%	0	0	0	22	0	1	95.65%
	DI4	0	0	0	0	46	0	100%	0	0	0	0	23	0	100%
	DI5	0	0	0	0	0	29	100%	0	0	0	0	0	15	100%
	Total							95.06%							93.44%

[1] Degree of infestation.

It can be concluded from the results that the spectra combined with the BPNN algorithm was effective in identifying the degree of SSB infestation. Furthermore, early infestation was comparatively difficult to identify, which might because few changes of spectral characteristics occurred in this degree.

3.4. Characteristic Wavelengths Selection

Hyperspectral images with hundreds of variables will result in information collinearity and redundancy, and slow the calculation efficiency. Thus, selecting the most informative variables will reduce the variables, obviously, and hence simplify the analysis. SPA was implemented in this study and selected a total of 17 variables (481, 497, 505, 532, 539, 564, 588, 638, 655, 681, 696, 762, 830, 958, 979, 998, and 1000 nm) from the entire 490 variables. The selected wavelengths are shown in Figure 4. The number of variables has decreased by more than 96%, which can simplify the detection models and improve the calculation efficiency.

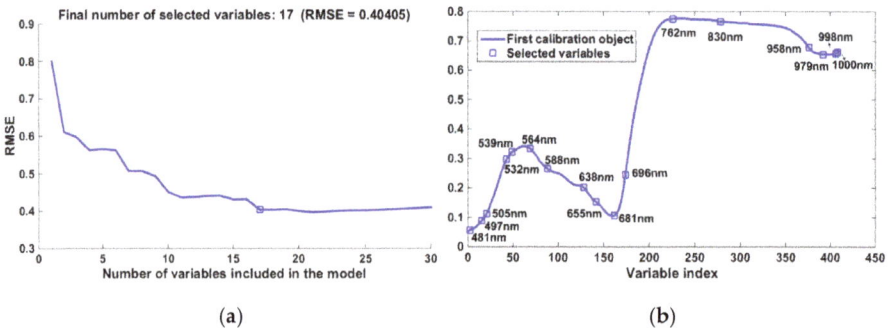

(a)

(b)

Figure 4. Selection of characteristic wavelengths by the successive projection algorithm (SPA): (**a**) numbers of characteristic wavelengths with the minimum root mean square error (RMSE); (**b**) distribution of characteristic wavelengths in the full band.

As shown in Figure 4b, there were 11 wavelengths in the visible range and 6 wavelengths in the near-infrared range. The characteristic wavelengths in the visible range were mainly due to the alteration of photosynthetic pigments; for example, 532 nm was related to xanthophyll and 696 nm was largely sensitive to variation in chlorophyll content [41,42]. Furthermore, the characteristic wavelengths in the near-infrared range had a close relationship with the water content and internal structure of the rice [36]. These were correlated with destruction caused by the SSB, further indicating that characteristic wavelengths selected by SPA were meaningful in this study.

3.5. Identification Results Based on Characteristic Wavelengths

The results of the BPNN model based on characteristic wavelengths are shown in Table 2. On the whole, the overall accuracy of both calibration and prediction sets slightly decreased compared with the model based on full spectra. A better performance was achieved for detecting DI0. The results of each degree were promising, as accuracy was all beyond 90% except for DI1. The accuracy of DI0 and DI1 was lower than the rest of the degrees. Samples in DI2 and DI3, as well as in DI4 and DI5, were more easily confused with each other. However, misclassification between DI3 and DI4 occurred rarely, which was consistent with the clustering trend in the PCA analysis. These results indicated that the selection of characteristic wavelengths by SPA was effective at both maintaining performance and reducing variables. The spectral characteristics of the samples in adjacent degrees was similar, and this would increase the difficulty in identifying the degree of infestation accurately.

Table 2. Detection accuracy of infestation degrees by the BPNN model based on characteristic wavelengths.

Model	Actual Value	Calibration Set							Prediction Set						
		DI [1] 0	DI1	DI2	DI3	DI4	DI5	Accuracy	DI0	DI1	DI2	DI3	DI4	DI5	Accuracy
	DI0	24	2	1	0	3	0	80%	13	0	1	0	1	0	86.67%
	DI1	5	40	0	0	1	0	86.96%	4	18	0	0	1	0	78.26%
	DI2	0	0	45	1	0	0	97.83%	0	0	22	1	0	0	95.65%
BPNN	DI3	0	0	1	44	0	1	95.65%	0	0	0	22	0	1	95.65%
	DI4	0	1	0	0	44	1	95.65%	0	0	0	0	22	1	95.65%
	DI5	0	0	0	0	1	28	96.55%	0	0	0	0	1	14	93.33%
	Total							92.59%							90.98%

[1] Degree of infestation.

3.6. Identification Results Based on Textural Features

The hyperspectral image contains 512 gray-scale images according to the dimension of wavelengths, and it would be better to compress these into fewer images. Therefore, PCA was performed on the ROI hyperspectral image to extract the most informative PC images. The PC1 and PC2 images comprised a total 93.42% of the eigenvalues, and the rest of the PC images contained more noise than information, which would disturb the detection. Thus, only the PC1 and PC2 images were retained to extract the GLCM features. There were a total of 8 features for each image including mean, variance, homogeneity, contrast, dissimilarity, entropy, second moment, and correlation; thus, a new features set was formed for 16 features of each sample, as shown in Figure 5. Multiple linear-regression analysis was executed to inspect the relationship between textural features and infestation degrees. The multiple correlation coefficient R was 0.811 and the coefficient of determination R^2 was 0.658, which meant the prediction was satisfactory. These demonstrated that the GLCM features contained useful information, which could be helpful for identification of SSB infestation.

The new dataset was used as input for the BPNN, and the results are shown in Table 3. The identification accuracy of the DI0 and DI4 and DI5 was all over 80% in the prediction set, while the accuracy of the DI1-DI3 was relatively low. Furthermore, the rice samples in medium infestation degrees were similar with respect to the textural features, and hence had a greater possibility of being confused with each other. However, the overall results proved that the GLCM features were worthy of further exploration combined with spectral data.

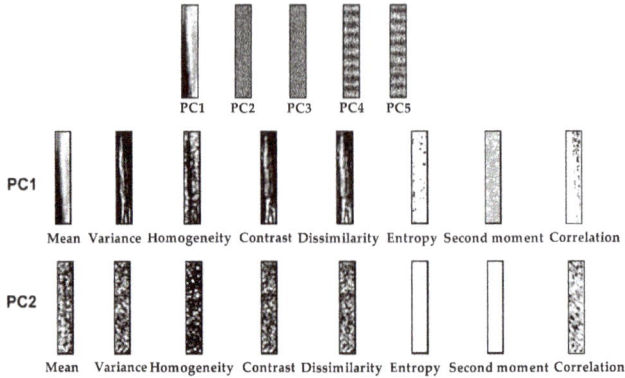

Figure 5. Texture feature images and PC images of a rice sample.

Table 3. Detection accuracy of infestation degrees by the BPNN model based on the GLCM features.

Model	Actual Value	Calibration Set							Prediction Set						
		DI [1]0	DI1	DI2	DI3	DI4	DI5	Accuracy	DI0	DI1	DI2	DI3	DI4	DI5	Accuracy
BPNN	DI0	30	0	0	0	0	0	100%	15	0	0	0	0	0	100%
	DI1	9	25	3	5	3	1	54.35%	3	12	2	3	3	0	52.17%
	DI2	0	1	32	10	1	2	69.57%	0	0	16	4	1	2	69.57%
	DI3	0	5	10	26	2	3	56.52%	0	2	6	14	0	1	60.87%
	DI4	0	0	2	1	38	5	82.61%	0	0	1	0	19	3	82.61%
	DI5	0	0	0	4	2	23	79.31%	0	0	0	2	0	13	86.67%
	Total							71.60%							72.95%

[1] Degree of infestation.

3.7. Identification Results Based on Data Fusion

The fusion of spectral data and textural features has been explored by many studies into food quality and plant-disease detection based on hyperspectral imaging technology [43–45]. Data fusion can be performed at three levels: pixel-level fusion, feature-level fusion, and decision-level fusion [46]. Data fusion at the feature level means extracting a feature from different data sets and fuses statistical approaches such as arithmetic combinations and filters [47]. This study adopted feature-level fusion to fuse the characteristic wavelengths with textural features. Spectral and textural features were also normalized in the same dimension before fusion.

The results of the BPNN model using data fusion are shown in Table 4. The identification results of both calibration and prediction sets were excellent, with classification accuracy over 95%. Meanwhile, the performance in detecting each infestation degree was satisfactory, with classification accuracy over 95%, except for DI1. The cause of this error was misclassification with DI0, which was consistent with the phenomenon discussed in the above sections.

Table 4. Detection accuracy of infestation degrees by the BPNN model based on data fusion.

Model	Actual Value	Calibration Set							Prediction Set						
		DI [1]0	DI1	DI2	DI3	DI4	DI5	Accuracy	DI0	DI1	DI2	DI3	DI4	DI5	Accuracy
BPNN	DI0	29	1	0	0	0	0	100%	15	0	0	0	0	0	100%
	DI1	4	42	0	0	0	0	82.61%	4	19	0	0	0	0	82.61%
	DI2	0	1	45	0	0	0	95.65%	0	1	22	0	0	0	95.65%
	DI3	0	0	1	44	1	0	100%	0	0	0	23	0	0	100%
	DI4	0	1	0	2	43	0	95.65%	0	0	0	1	22	0	95.65%
	DI5	0	0	0	0	1	28	100%	0	0	0	0	0	15	100%
	Total							95.06%							95.10%

[1] Degree of infestation.

3.8. Comparision of BPNN Models Based on Different Datesets

In this study there were four data sets—including full spectra, characteristic wavelengths, textural features and data fusion—as input for the BPNN model to detect different degrees of SSB infestation. The total accuracy of the BPNN models based on full spectra and characteristic wavelengths was higher than models based on textural features. This indicated that the spectral features contributed more to the identification of infestation than the textural features. The accuracy of the BPNN models based on characteristic wavelengths decreased in comparison with the BPNN models based on full spectra, which might be connected with the loss of certain useful information after extracting 17 wavebands from the full 408 wavebands. The BPNN models based on data fusion acquired the highest total accuracy among the four data sets, which could not only remedy the deficiency of the set of individual characteristic wavelengths or textural features, but also improve the calculation efficiency, as the run time decreased by about 90% compared with models based on the full spectra set in Table 5. The accuracy of individual degrees was all elevated by the fusion of characteristic wavelengths and textural feature; even the accuracy of DI1 exceeded 80%. These results indicate that the spectral and textural features were complementary in internal and external aspects, which could reflect integrated changes of plant characteristics. The fusion of spectral and textural features could take full advantage of hyperspectral imaging technology, and was effective at detecting SSB infestation and identifying different degrees of it.

Table 5. Run time of the BPNN models based on different data sets.

Data Set	Run Time
Full spectra	16.91s
Characteristic wavelength	3.86 s
Texture features	1.64 s
Data fusion	1.80 s

4. Conclusions

This study explored the feasibility of using a visible/near-infrared hyperspectral imaging system to detect early SSB infestation and identify degrees of infestation in rice. We selected 17 characteristic wavelengths by SPA from full spectral data, and extracted 8 GLCM features from the PC images transformed from hyperspectral images. BPNN models were established using different data sets, including full spectra, characteristic wavelengths, textural features and features fusion. The BPNN model based on feature fusion acquired the best results, with overall accuracy over 95%. The identification accuracy of each infestation degree was over 95%, except for DI1, which was also improved compared to models using characteristic wavelengths and textural features alone. The run time decreased significantly as a result of variables selection. DI1 was easily confused with DI0, which increased the difficulty of the early detection of SSB infestation. In total, these results proved that the fusion of spectral and textural features from hyperspectral images combined with BPNN was feasible for identifying degrees of SSB infestation in rice. In future studies, we will develop more stable and universal models with more rice cultivars and more species of pests for laboratory-based detection. Field-based research will also be conducted on the basis of this laboratory-based research in order to expand the application of hyperspectral imaging combined with chemometrics in the field. Small and portable pest-detection equipment will be developed based on future work.

Acknowledgments: The authors are thankful for the support of the China national key research and development program (2016YFD0700304) and the China Postdoctoral Science Foundation (2017M610370).

Author Contributions: Zhengjun Qiu, Chu Zhang and Yong He conceived the experiments and established guidance for the writing of the manuscript. Yangyang Fan and Tao Wang performed the experiments. Yangyang Fan analyzed the data and wrote the whole paper. Jiyu Peng prepared the experimental materials and contributed to data analysis. All authors reviewed the manuscript.

Conflicts of Interest: The authors declare no conflict of interest.

References

1. Chen, H.; Lin, Y.; Zhang, Q. Review and prospect of transgenic rice research. *Chin. Sci. Bull.* **2009**, *54*, 4049–4068. [CrossRef]
2. Yin, C.L.; Liu, Y.; Liu, J.; Xiao, H.; Huang, S.; Lin, Y.; Han, Z.; Li, F. Chilodb: A genomic and transcriptome database for an important rice insect pest chilo suppressalis. *Database* **2014**, *2014*, 92–108. [CrossRef] [PubMed]
3. Dale, D. *Insect Pests of the Rice Plant—Their Biology and Ecology*; Wiley: Hoboken, NJ, USA, 1994.
4. Fademi, O.A. Chemical control of the striped stem borer, chilo-suppressalis (walker) in rice. *Trop. Pest Manag.* **1985**, *31*, 292–293. [CrossRef]
5. Zhu, H.; Chu, B.; Zhang, C.; Liu, F.; Jiang, L.; He, Y. Hyperspectral imaging for presymptomatic detection of tobacco disease with successive projections algorithm and machine-learning classifiers. *Sci. Rep.* **2017**, *7*, 4125. [CrossRef] [PubMed]
6. Sun, Y.; Wang, Y.; Xiao, H.; Gu, X.; Pan, L.; Tu, K. Hyperspectral imaging detection of decayed honey peaches based on their chlorophyll content. *Food Chem.* **2017**, *235*, 194–202. [CrossRef] [PubMed]
7. Skjelvareid, M.H.; Heia, K.; Olsen, S.H.; Stormo, S.K. Detection of blood in fish muscle by constrained spectral unmixing of hyperspectral images. *J. Food Eng.* **2017**, *212*, 252–261. [CrossRef]
8. War, A.R.; Paulraj, M.G.; Ahmad, T.; Buhroo, A.A.; Hussain, B.; Ignacimuthu, S.; Sharma, H.C. Mechanisms of plant defense against insect herbivores. *Plant Signal. Behav.* **2012**, *7*, 1306–1320. [CrossRef] [PubMed]
9. Ni, X.Z.; Quisenberry, S.S.; Heng-Moss, T.; Markwell, J.; Higley, L.; Baxendale, F.; Sarath, G.; Klucas, R. Dynamic change in photosynthetic pigments and chlorophyll degradation elicited by cereal aphid feeding. *Entomol. Exp. Appl.* **2002**, *105*, 43–53. [CrossRef]
10. Huang, W.; Lamb, D.W.; Niu, Z.; Zhang, Y.; Liu, L.; Wang, J. Identification of yellow rust in wheat using in-situ spectral reflectance measurements and airborne hyperspectral imaging. *Precis. Agric.* **2007**, *8*, 187–197. [CrossRef]
11. Huang, J.R.; Sun, J.Y.; Liao, H.J.; Liu, X.D. Detection of brown planthopper infestation based on spad and spectral data from rice under different rates of nitrogen fertilizer. *Precis. Agric.* **2015**, *16*, 148–163. [CrossRef]
12. Yang, Z.; Rao, M.N.; Elliott, N.C.; Kindler, S.D.; Popham, T.W. Differentiating stress induced by greenbugs and russian wheat aphids in wheat using remote sensing. *Comput. Electron. Agric.* **2009**, *67*, 64–70. [CrossRef]
13. Nansen, C.; Sidumo, A.J.; Martini, X.; Stefanova, K.; Roberts, J.D. Reflectance-based assessment of spider mite "bio-leaves" to maize leaves and plant potassium content in different irrigation regimes. *Comput. Electron. Agric.* **2013**, *97*, 21–26. [CrossRef]
14. Rupanagudi, S.R.; Ranjani, B.S.; Nagaraj, P.; Bhat, V.G.; Thippeswamy, G. A Novel Cloud Computing Based Smart Farming System for Early Detection of Borer Insects in Tomatoes. In Proceedings of the International Conference on Communication, Information & Computing Technology, Mumbai, India, 15–17 January 2015; pp. 1–6.
15. Bhadane, G.; Sharma, S.; Nerkar, V.B. Early pest identification in agricultural crops using image processing techniques. *Int. J. Electr. Electron. Comput. Eng.* **2013**, *2*, 77–82.
16. Huddar, S.R.; Gowri, S.; Keerthana, K.; Vasanthi, S.; Rupanagudi, S.R. Novel Algorithm for Segmentation and Automatic Identification of Pests on Plants Using Image Processing. In Proceedings of the Third International Conference on Computing Communication & Networking Technologies (ICCCNT), Coimbatore, India, 26–28 July 2012; pp. 1–5.
17. Lowe, A.; Harrison, N.; French, A.P. Hyperspectral image analysis techniques for the detection and classification of the early onset of plant disease and stress. *Plant Methods* **2017**, *13*, 80. [CrossRef] [PubMed]
18. Sytar, O.; Brestic, M.; Zivcak, M.; Olsovska, K.; Kovar, M.; Shao, H.; He, X. Applying hyperspectral imaging to explore natural plant diversity towards improving salt stress tolerance. *Sci. Total Environ.* **2016**, *578*, 90. [CrossRef] [PubMed]
19. Thomas, S.; Kuska, M.T.; Bohnenkamp, D.; Brugger, A.; Alisaac, E.; Wahabzada, M.; Behmann, J.; Mahlein, A.-K. Benefits of hyperspectral imaging for plant disease detection and plant protection: A technical perspective. *J. Plant Dis. Prot.* **2017**, 1–16. [CrossRef]

20. Zhao, Y.; Yu, K.; Feng, C.; Cen, H.; He, Y. Early detection of aphid (myzus persicae) infestation on chinese cabbage by hyperspectral imaging and feature extraction. *Trans. ASABE* **2017**, *60*, 1045–1051. [CrossRef]
21. Wu, X.; Zhang, W.; Qiu, Z.; Cen, H.; He, Y. A novel method for detection of pieris rapae larvae on cabbage leaves using nir hyperspectral imaging. *Appl. Eng. Agric.* **2016**, *32*, 311–316.
22. Hu, L.; Ye, M.; Li, R.; Zhang, T.; Zhou, G.; Wang, Q.; Lu, J.; Lou, Y. The rice transcription factor wrky53 suppresses herbivore-induced defenses by acting as a negative feedback modulator of mitogen-activated protein kinase activity. *Plant Physiol.* **2015**, *169*, 2907–2921. [PubMed]
23. Macho, S.; Rius, A.; Callao, M.P.; Larrechi, M.S. Monitoring ethylene content in heterophasic copolymers by near-infrared spectroscopy-standardisation of the calibration model. *Anal. Chim. Acta* **2001**, *445*, 213–220. [CrossRef]
24. Haralick, R.M.; Shanmugam, K.; Dinstein, I.H. Textural features for image classification. *IEEE Trans. Syst. Man Cybern.* **1973**, *smc-3*, 610–621. [CrossRef]
25. Hui, H.; Li, L.; Ngadi, M.O. Assessment of intramuscular fat content of pork using nir hyperspectral images of rib end. *J. Food Eng.* **2017**, *193*, 29–41.
26. Tetuko, J. Analysis of co-occurrence and discrete wavelet transform textures for differentiation of forest and non-forest vegetation in very high resolution optical-sensor imagery. *Int. J. Remote Sens.* **2008**, *29*, 3417–3456.
27. Ozdemir, I.; Norton, D.A.; Ozkan, U.Y.; Mert, A.; Senturk, O. Estimation of tree size diversity using object oriented texture analysis and aster imagery. *Sensors* **2008**, *8*, 4709–4724. [CrossRef] [PubMed]
28. Galvao, R.K.H.; Araujo, M.C.U.; Silva, E.C.; Jose, G.E.; Soares, S.F.C.; Paiva, H.M. Cross-validation for the selection of spectral variables using the successive projections algorithm. *J. Braz. Chem. Soc.* **2007**, *18*, 1580–1584. [CrossRef]
29. Sun, Y.; Gu, X.; Sun, K.; Hu, H.; Xu, M.; Wang, Z.; Tu, K.; Pan, L. Hyperspectral reflectance imaging combined with chemometrics and successive projections algorithm for chilling injury classification in peaches. *LWT Food Sci. Technol.* **2017**, *75*, 557–564. [CrossRef]
30. Araujo, M.C.U.; Saldanha, T.C.B.; Galvao, R.K.H.; Yoneyama, T.; Chame, H.C.; Visani, V. The successive projections algorithm for variable selection in spectroscopic multicomponent analysis. *Chemom. Intell. Lab. Syst.* **2001**, *57*, 65–73. [CrossRef]
31. Galvão, R.K.H.; Araújo, M.C.U.; Fragoso, W.D.; Silva, E.C.; José, G.E.; Soares, S.F.C.; Paiva, H.M. A variable elimination method to improve the parsimony of mlr models using the successive projections algorithm. *Chemom. Intell. Lab. Syst.* **2008**, *92*, 83–91. [CrossRef]
32. Zhang, C.; Ye, H.; Liu, F.; He, Y.; Kong, W.; Sheng, K. Determination and visualization of pH values in anaerobic digestion of water hyacinth and rice straw mixtures using hyperspectral imaging with wavelet transform denoising and variable selection. *Sensors* **2016**, *16*, 244. [CrossRef] [PubMed]
33. Tao, F.; Peng, Y.; Gomes, C.L.; Chao, K.; Qin, J. A comparative study for improving prediction of total viable count in beef based on hyperspectral scattering characteristics. *J. Food Eng.* **2015**, *162*, 38–47. [CrossRef]
34. Guo, D.; Xie, R.; Qian, C.; Yang, F.; Zhou, Y.; Deng, L. Diagnosis of ctv-infected leaves using hyperspectral imaging. *Intell. Autom. Soft Comput.* **2015**, *21*, 269–283. [CrossRef]
35. Al-Allaf, O.N.A. Improving the performance of backpropagation neural network algorithm for image compression/decompression system. *J. Comput. Sci.* **2010**, *6*, 834–838.
36. Zhao, Y.-R.; Li, X.; Yu, K.-Q.; Cheng, F.; He, Y. Hyperspectral imaging for determining pigment contents in cucumber leaves in response to angular leaf spot disease. *Sci. Rep.* **2016**, *6*, 27790. [CrossRef] [PubMed]
37. Prabhakar, M.; Prasad, Y.G.; Thirupathi, M.; Sreedevi, G.; Dharajothi, B.; Venkateswarlu, B. Use of ground based hyperspectral remote sensing for detection of stress in cotton caused by leafhopper (hemiptera: Cicadellidae). *Comput. Electron. Agric.* **2011**, *79*, 189–198. [CrossRef]
38. Jiang, M.X.; Cheng, J.A. Interactions between the striped stem borer chilo suppressalis (walk.) (lep., pyralidae) larvae and rice plants in response to nitrogen fertilization. *J. Pest Sci.* **2003**, *76*, 124–128. [CrossRef]
39. Rubia, E.G.; Heong, K.L.; Zalucki, M.; Gonzales, B.; Norton, G.A. Mechanisms of compensation of rice plants to yellow stem borer scirpophaga incertulas (walker) injury. *Crop Prot.* **1996**, *15*, 335–340. [CrossRef]
40. Jolliffe, I.T.; Cadima, J. Principal component analysis: A review and recent developments. *Philos. Trans.* **2016**, *374*, 20150202. [CrossRef] [PubMed]
41. Penuelas, J.; Filella, I. Visible and near-infrared reflectance techniques for diagnosing plant physiological status. *Trends Plant Sci.* **1998**, *3*, 151–156. [CrossRef]

42. Sims, D.A.; Gamon, J.A. Relationships between leaf pigment content and spectral reflectance across a wide range of species, leaf structures and developmental stages. *Remote Sens. Environ.* **2002**, *81*, 337–354. [CrossRef]
43. Zhang, C.; Guo, C.; Liu, F.; Kong, W.; He, Y.; Lou, B. Hyperspectral imaging analysis for ripeness evaluation of strawberry with support vector machine. *J. Food Eng.* **2016**, *179*, 11–18. [CrossRef]
44. Xie, C.Q.; He, Y. Spectrum and image texture features analysis for early blight disease detection on eggplant leaves. *Sensors* **2016**, *16*, 676. [CrossRef] [PubMed]
45. Fan, S.; Zhang, B.; Li, J.; Liu, C.; Huang, W.; Tian, X. Prediction of soluble solids content of apple using the combination of spectra and textural features of hyperspectral reflectance imaging data. *Postharvest Biol. Technol.* **2016**, *121*, 51–61. [CrossRef]
46. Pohl, C.; Genderen, J.L.V. Review article multisensor image fusion in remote sensing: Concepts, methods and applications. *Int. J. Remote Sens.* **1998**, *19*, 823–854. [CrossRef]
47. Liu, D.; Pu, H.; Sun, D.-W.; Wang, L.; Zeng, X.-A. Combination of spectra and texture data of hyperspectral imaging for prediction of ph in salted meat. *Food Chem.* **2014**, *160*, 330–337. [CrossRef] [PubMed]

sensors

MDPI

Article

A Real-Time Smooth Weighted Data Fusion Algorithm for Greenhouse Sensing Based on Wireless Sensor Networks

Tengyue Zou *, Yuanxia Wang, Mengyi Wang and Shouying Lin

College of Mechanical and Electronic Engineering, Fujian Agriculture and Forestry University, Fuzhou 350002, China; fafuwxm@163.com (Y.W.); yzzdyxlll@sina.com (M.W.); linshouying@fafu.edu.cn (S.L.)
* Correspondence: zouty@fafu.edu.cn; Tel.: +86-591-8378-9374; Fax: +86-591-8378-9374

Received: 19 September 2017; Accepted: 3 November 2017; Published: 6 November 2017

Abstract: Wireless sensor networks are widely used to acquire environmental parameters to support agricultural production. However, data variation and noise caused by actuators often produce complex measurement conditions. These factors can lead to nonconformity in reporting samples from different nodes and cause errors when making a final decision. Data fusion is well suited to reduce the influence of actuator-based noise and improve automation accuracy. A key step is to identify the sensor nodes disturbed by actuator noise and reduce their degree of participation in the data fusion results. A smoothing value is introduced and a searching method based on Prim's algorithm is designed to help obtain stable sensing data. A voting mechanism with dynamic weights is then proposed to obtain the data fusion result. The dynamic weighting process can sharply reduce the influence of actuator noise in data fusion and gradually condition the data to normal levels over time. To shorten the data fusion time in large networks, an acceleration method with prediction is also presented to reduce the data collection time. A real-time system is implemented on STMicroelectronics STM32F103 and NORDIC nRF24L01 platforms and the experimental results verify the improvement provided by these new algorithms.

Keywords: greenhouse; wireless sensor network; data fusion; dynamic weight

1. Introduction

In recent years, wireless sensor networks (WSNs) have been widely used to monitor the environment [1–3], such as the temperature, humidity, gas concentration, gas composition, dust and so on, particularly for agricultural production purposes [4–6]. A greenhouse is an agricultural facility designed to extend the production season and improve the quality of agricultural products [7]. Because greenhouses generally contain climate-regulating equipment, the internal sensing data of greenhouses must be comprehensive and accurate. WSNs are composed of hundreds or thousands of sensor nodes that are used to acquire parameters under a range of conditions and transmit these parameters to a base station or sink node [8–10], enabling an information-based decision to be made in an automated manner. They are often used as the system for fire alarm in forests or used to monitor the environment in the field. In a greenhouse, the number of sensor nodes may be less. But if the greenhouse is large, it may also need several hundreds of sensor nodes. Furthermore, in complex and inhomogeneous environments, noise often corrupts the sensing data. Thus, a data fusion algorithm is required for selecting correct reports from mass data to identify accurate values from the measurements [11–13].

Figure 1 shows an illustration of a sensor network deployed in a greenhouse containing a heating unit and windows. When the heating unit is working, the temperature of the area nearest to the unit may become hotter than other regions far from the heater because the heating effect is related to the distance from the actuator. This effect can be termed the 'actuator effect' and tends to disrupt sensor reports by introducing inhomogeneous information into the dataset. The actuator influence is common in agricultural and industrial environments; for example, when sunlight shines through windows in the roof of a greenhouse, the light intensity value detected in small regions increases but most parts of the greenhouse remain dark. If the automatic control system follows the high light intensity value, subsequent incorrect operation is challenging to avoid. Thus, a data fusion algorithm is expected to reduce the influence of the actuator effect or other disturbing factors during the sensing procedure; the subsequent analysis of correct data is meaningful for the automation of the agriculture industry.

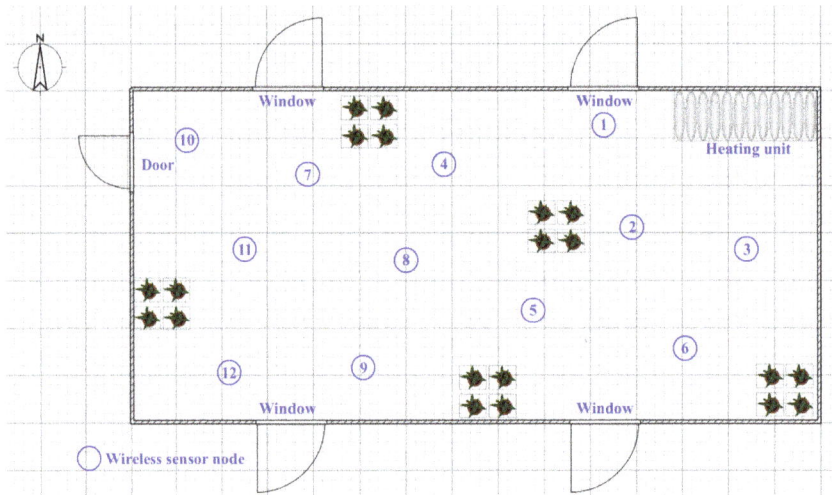

Figure 1. Illustration of a sensor network deployed in a greenhouse.

To improve the accuracy of sensing and reduce the burden of data transmission, researchers are now focusing on data fusion mechanisms [14–16]. A data fusion scheme based on a grey model and an extreme learning machine has been proposed to reduce redundant transmissions and extend the lifetime of the network [17]. The performance of this technique has been demonstrated through simulation. The Peeling algorithm [18] was developed to improve the performance of serial data fusion and the simulation results highlight the effectiveness of this algorithm in reducing energy consumption and time responsiveness. A distributed data fusion algorithm, which aims to minimize energy cost, was also deployed in the active network paradigm using WSNs [19]. The optimal linear estimation method was also derived to achieve data fusion in multi-rate sensor networks [20]. The fusion of quantized and un-quantized sensor data has been investigated to improve estimates [21]. Task-oriented distributed fusion functions have been introduced to adapt the dynamics of tasks and the topology of self-organized networks [22]. Rough-set and back-propagation neural networks have also been adopted to raise the accuracy of prediction [23]. Maximum a priori probability [24] and vector-space-based methods [25] can be used to reduce the influence of noisy or conflicting data in sample values. Neural networks have also been used to assist the data fusion for sensing [26,27]. The neural networks can help to select important features from datasets and improve the fusion work. They can also be used to optimize the localization for WSNs [28]. Feature selection and integration are also important techniques for data fusion. The adaptive weight and feedback control can help to find out the key points from the datasets efficiently [29]. However, few researchers have addressed the inhomogeneous effect of environmental elements influencing the sampled data of WSNs. Furthermore, most of the studies typically prove new algorithms through simulation only and not through real-time hardware and experimentation. Thus, this study focuses on the solution of noisy data produced by the inhomogeneous effect in certain application circumstances and the mechanism to speed up calculations. An embedded hardware platform is built to evaluate the effectiveness of these new algorithms and simulations are run on a large-scale dataset.

The contributions of this study are as follows: (1) An adjacent graph is introduced to describe the regional relationship among sensor nodes. The value assigned to the arc of the graph is used to indicate the distance between two neighboring nodes; (2) A stable area-searching method based on Prim's algorithm is designed to locate regions that are not influenced by the actuator effect; (3) A voting mechanism based on dynamic weight is introduced to complete the data fusion and the self-healing function is designed to distinguish the actuator effect from normal sensing; (4) An acceleration mechanism with adaptive threshold is proposed to speed up the process of decision making in real-time applications.

The remainder of this paper is organized as follows: Section 2 presents a detection method for stable areas and the data fusion procedure based on a voting mechanism. In Section 3, an acceleration method is designed to speed up the computational process in real-time applications. Experiments are performed on embedded hardware and the results are presented in Section 4. Finally, the conclusions are provided in Section 5.

2. Data Fusion and Self-Learning

To improve the accuracy of the final decision, a data fusion mechanism is required to reduce the influence of an actuator. The actuator effect is related to the position under a certain circumstance. Once the location of a sensor node is set, the accuracy of the node's sensing data is approximately confirmed. The improved method must detect the nodes influenced by the actuator and reduce their contribution via a weighting method in the data fusion algorithm.

2.1. Adjacent Graph and Stable Area

To find the stable area in the map, the spatial relationships of the sensor nodes should first be determined. An adjacent graph is used in this work to describe the regional relationship between neighboring nodes. The adjacent graph is a type of undirected graph that uses arc connections to express the relationship between neighboring nodes. As shown in Figure 2, a node in the adjacent map represents a sensor node whose location is recorded by the engineer manually or using its locating device. A weight value w_{i-j} $(i < j)$ is assigned to each arc according to the distance between node i and j, which is normalized as shown in Equation (1).

$$w_{i-j} = \frac{w_{i-j\text{-}org} - w_{min}}{w_{max} - w_{min}} \tag{1}$$

where $w_{i-j\text{-}org}$ represents the original straight-line distance between sensor node i and j, w_{min} denotes the minimum distance between two sensor nodes in the sensor network and w_{max} is the maximum value.

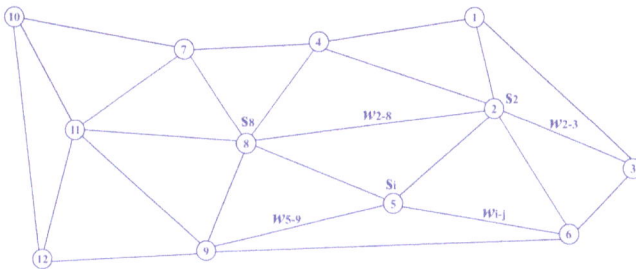

Figure 2. Illustration of an adjacent graph for a sensor network.

Furthermore, a smoothing value S_k is calculated for each node in the graph following Equations (2) and (3) to express the uniformity of sensing data in the node's neighborhood. The actuator effect is gradually reduced as the distance from actuator device increases. Thus, along with the spreading paths of the effect, the sensing data in that region is not uniform. This key point can help determine the main scope of the actuator's effect in the map.

$$s_{k\text{-}org} = \sum_{i=1}^{t} (D_i - D_k)/t \tag{2}$$

$$s_k = \frac{s_{k\text{-}org} - s_{min\text{-}org}}{s_{max\text{-}org} - s_{min\text{-}org}} \tag{3}$$

where t is the number of neighboring nodes for the center node k according to the adjacent graph; D_k represents the data acquired from node k and D_i is the data from its neighboring node i; $s_{k\text{-}org}$ denotes the original smoothing value for node k; and $s_{min\text{-}org}$ and $s_{max\text{-}org}$ are the minimum and maximum smoothing values, respectively.

After the construction of the adjacent graph, the weight of each arc $w_{i\text{-}j}$ and the smoothing value S_k for every node are generated. As Algorithm 1 shows, a method based on Prim's algorithm is introduced in this work to search a stable area for data fusion. Prim's algorithm is a classical approach to compute a minimum spanning tree from a graph and a smoothing value threshold λ is set to avoid using nodes in unstable areas. Thus, the stable area can be marked by the node set U in the result; the time complexity of the algorithm is $O(|V|^2)$. Using only the sensing data from stable areas, the final fusion decision can be made and the result is recorded as history that can influence future decision making.

Algorithm 1. Procedure to find a stable area via a search algorithm

---

Input: The adjacent graph ($\{V, S, E, W\}$), in which $V = \{u_0, u_1, u_2, \dots \}$ is the set of nodes with corresponding
   $S = \{s_0, s_1, s_2, \dots \}$ as their smoothing values and E is the set of arcs with W as their weights.
Output: a set of nodes U in the stable area
Start:
Sort the set V from small to large by their smoothing value S to make V'
Initialize $U = \{u_0\}$, $O = \{\}$ and $TE = \{\}$, in which u_0 is the first vertex in V', O is a set for storing obsolete vertexes
during operation and TE is a set for storing valid arcs during operation.
while $U \cup O \neq V'$
   Find the arc (u_i, v_j) with minimum weight that satisfies $u_i \in U$, $v_j \in V'\text{-}U\text{-}O$
   if $s_j < \lambda$ (s_j is the smoothing value for vertex v_j)
      $U = U \cup \{v_j\}$; $TE = TE \cup \{(u_i, v_j)\}$
   else
      $O = O \cup \{v_j\}$
   end if
end while
end

2.2. Data Fusion Based on Dynamic Weight

After acquiring the stable area, the data fusion result can be generated according to that area. Because the reports are not the same in all applications, a principle should be designed to determine the extent to which each node participates in the final decision. The voting mechanism based on weight is a satisfactory choice for this principle. Equation (4) shows the procedure of voting, where $\hat{Y}$ denotes the data fusion result, Y_i represents the report value from sensor node i, ω_{ti} is the corresponding time-related weight for node i at time t and s_i is the corresponding smoothing value.

$$\hat{Y} = \sum_{i=1}^{N} \omega_{ti} s_i Y_i / \sum_{i=1}^{N} \omega_{ti} s_i \tag{4}$$

When an actuator works, low weights should be assigned for the sensor nodes near it, whereas normal weights should be used when the actuator is closed. Because of the challenges associated with continually describing the working time of an actuator, a dynamic weight with self-healing ability is proposed. Equation (5) shows the rule that determines the dynamic weight, where n indicates the continuous count of being outside the stable area, α is an adjustment parameter for the variation control and t denotes the time ticks after the setting process. When the stable area is confirmed at a new time tick, the weights of the sensor nodes outside the area are refreshed by Equation (5). For example, if a sensor node is outside the stable area at a time tick, this node's weight follows the curve of $1-2/(1 + e^{\alpha t})$. Furthermore, if the node is outside the stable area again before its weight recovers to 1, the curve should be refreshed by the function of $1-2/(1 + e^{0.5 \times \alpha t})$; otherwise, the curve follows the previous curve until the node's weight is restored to value 1. Thus, based on this mechanism, the sensor nodes adjacent to the actuator can gradually recover participation in data fusion over time to recover from the actuator effect. Continuous regions that remain outside the stable area enhance the effect of the decreased weight and the discontinuous regions do not influence the data significantly. Figure 3 shows the weight variation of a sensor node adjacent to an actuator in the application. The node enters the unstable area at time ticks 0, 3 and 10 and its weight is gradually reduced to avoid the actuator's influence.

$$\omega_{ti} = 1 - \frac{2}{1 + e^{\frac{1}{2^{n-1}} \cdot \alpha \cdot t}} \tag{5}$$

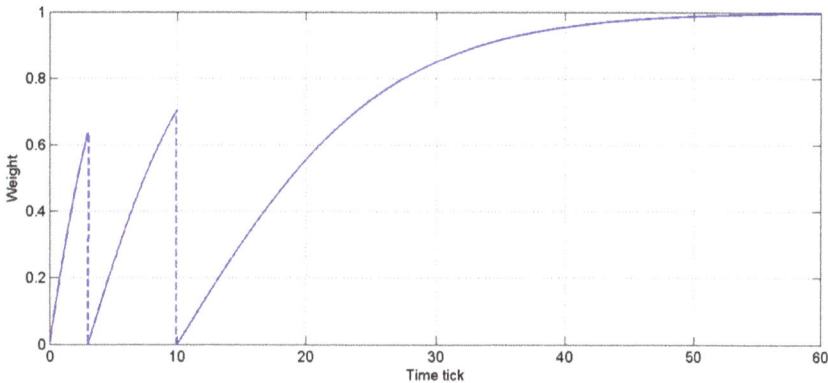

Figure 3. Example of the weight of a sensor node influenced by the actuator.

3. Acceleration Mechanism with an Adaptive Threshold

Since the data fusion mechanism is now influenced by the voting procedure, the precision of the result can be improved. The conventional voting procedure nevertheless requires that data are collected from all participants. This approach can lead to an increase in time latency for automatic decision and can become an obstacle to real-time application. To minimize the time consumption for decision making, an acceleration mechanism is proposed in this section to achieve a balance between accuracy and speed.

3.1. Procedure of the Acceleration Mechanism

According to the voting algorithm, the workstation must wait until all data, from all sensor nodes, are collected to make a final decision. However, communicational obstacles may exist for data transmission in large sensor networks or networks with faulty nodes that exhaust their energy or encounter problems with the corresponding equipment. In these extreme cases, the data fusion program must wait for the data from each node until timeout, which slows down data collection and increases the response time of the system in real-time applications. Thus, an acceleration mechanism with adaptive thresholds is designed to address this problem. As shown in Figure 4, the acceleration is based primarily on a forward prediction and a backward adaptive threshold adjustment. The whole procedure can be divided into two steps, the initialization and real-time sensing. At the initialization step, the value of the data fusion result at time 0 and the weighting for each sensor node are assigned according to the voting algorithm. Furthermore, at the real-time sensing step, the workstation starts to receive reports from sensor nodes and refreshes the fusion result. In circumstances with uniform parameters, only some of the reports from the sensor nodes (rather than all of them) are required due to the similarity of the reports.

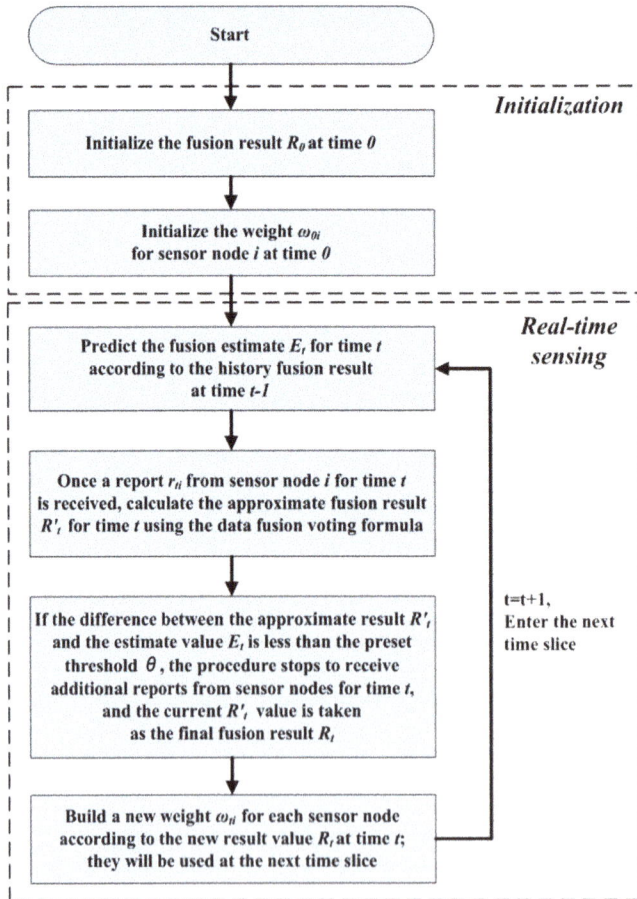

Figure 4. Procedure of the acceleration mechanism.

Sensors **2017**, *17*, 2555

Thus, the procedure generally uses the history value at time $t - 1$ to estimate a value for time t via a prediction algorithm before the formal data fusion occurs. Then, once the workstation receives a report r_{ti} from sensor node i for time t, an approximate fusion result R'_t is calculated for time t using the data fusion voting formula. Then, if the difference between the approximate result R'_t and the estimate value E_t is less than the pre-set threshold θ, the procedure stops to receive additional reports from sensor nodes for time t and the current R'_t value is taken as the final fusion result R_t. After that, the weight of each sensor node is refreshed according to the new result value R_t at time t and is used for the next time $t + 1$. Following this design, under the circumstance of uniform parameters that change linearly, the acceleration algorithm soon converges to a final decision with little error. However, under complex environments, the algorithm requires additional time to obtain a relatively accurate result. The balance between accuracy and speed is controlled by the prediction method and the pre-set threshold θ, whose large value allows for more error but uses less time to acquire a result. Equations (6) and (7) show the adaptive threshold recommended in this work. As the variation of measuring object increases, θ decreases to enhance restriction for data fusion, which may avoid larger errors in the result. Furthermore, intermittent use of acceleration mechanisms can also help maintain the balance between accuracy and speed. For example, after the acceleration procedure is used three times, we omit using it the fourth time, using instead the result calculated from all the reports to help eliminate the accumulated error.

$$\Delta x_k = |x_k - x_{k-1}| \tag{6}$$

$$\theta = E + \varepsilon \cdot (1/\sqrt{2\pi}\sigma)e^{-\frac{(\Delta x_k - \mu)^2}{2\sigma^2}} \tag{7}$$

where Δx_k represents the variation of measuring object from time $k - 1$ to time k; E is a fixed value assigned by the engineer and ε is an adjustment value selected from a Gaussian distribution.

3.2. Prediction Algorithm

The prediction algorithm is another important element for the acceleration mechanism. The physical parameters measured by sensor networks are regulated linearly or non-linearly and the prediction algorithm should be designed for various parameters with this factor in mind. This research focuses on the most general environmental parameters for agriculture, such as temperature, humidity, light and CO_2 concentration. Because these parameters do not typically change sharply, a Kalman filter is a suitable choice for prediction.

The purpose of the Kalman filter is to minimize the squared error of the estimated non-stationary signal in the noise. Each state update in the Kalman filter is recursively calculated from its previous estimate and the latest input data so that it is necessary to store only the previous estimate without all the past observations. Thus, the Kalman filter is widely used to reduce the influence of sensing and processing noise in applications that predict subsequent values. We consider the linear system described by the state space and measuring space:

$$\begin{cases} x(k) = A(k-1)x(k-1) + B(k)n(k) \\ y(k) = C^T(k)x(k-1) + D(k)n_1(k) \end{cases} \tag{8}$$

where $x(k)$ represents the $(N + 1) \times 1$-dimensional state variable vector; $y(k)$ is an observation signal vector; and $n(k)$ and $n_1(k)$ denote the process noise and observation noise, respectively. If M is the number of system inputs and L is the number of system outputs, then $A(k - 1)$, $B(k)$, $C(k)$ and $D(k)$ are coefficient matrixes with dimensions $(N + 1) \times (N + 1)$, $(N + 1) \times M$, $(N + 1) \times L$ and $L \times L$, respectively.

Let $\hat{x}(k|k - 1)$ represent the estimated value of $x(k)$ using the observation value until time $k - 1$ and let $\hat{x}(k|k)$ denote the estimated value of $x(k)$ using the observation value until time k. Thus, the corresponding estimated error can be defined as Equations (9) and

(10). Furthermore, the corresponding covariance matrix of these errors can be calculated using Equations (11) and (12).

$$e(k|k) = x(k) - \hat{x}(k|k) \tag{9}$$

$$e(k|k-1) = x(k) - \hat{x}(k|k-1) \tag{10}$$

$$R_e(k|k) = E\left[e(k|k)e^T(k|k)\right] \tag{11}$$

$$R_e(k|k-1) = E\left[e(k|k-1)e^T(k|k-1)\right] \tag{12}$$

$$R_n(k) = E\left[n(k)n^T(k)\right] \tag{13}$$

$$R_{n_1}(k) = E\left[n_1(k)n_1{}^T(k)\right] \tag{14}$$

Algorithm 2 shows the procedure for using the Kalman filter, where $K(k)$ is an $(N + 1) \times L$ matrix called the Kalman gain and $R_n(k)$ and $R_{n_1}(k)$ are defined in Equations (13) and (14), respectively. Using the $\hat{x}(k|k-1)$ as the prediction value for time k, the acceleration mechanism can be performed according to the adaptive threshold θ.

Algorithm 2. Kalman filter procedure

Initialization:
  $\hat{x}(0|0) = x(0)$ $R_e(0|0) = x(0)x^T(0)$
Calculation: when $k \geq 0$,
  $\hat{x}(k|k-1) = A(k-1)\hat{x}(k-1|k-1)$
  $R_e(k|k-1) = A(k-1)R_e(k-1|k-1)A^T(k-1) + B(k)R_n(k)B^T(k)$
  $K(k) = R_e(k|k-1)C(k)\left[C^T(k)R_e(k|k-1)C(k) + R_{n_1}(k)\right]^{-1}$
  $\hat{x}(k|k) = \hat{x}(k|k-1) + K(k)(y(k) - C^T(k)\hat{x}(k|k-1))$
  $R_e(k|k) = \left[I - K(k)C^T(k)\right]R_e(k|k-1)$

---

4. Simulation and Experiments

4.1. Simulation Results and Discussion

To verify the effect of the smooth weighted data fusion (SWDF) process introduced in this work, a simulation implemented in MATLAB is performed on a PC with a 3.4 GHz Intel Core CPU and 4 GB memory. In the simulation, the sensor nodes are uniformly deployed in a 400 m × 400 m rectangular field and 20 actuators are randomly settled in the field with an influencing range consisting of a 5 m radius circle. Four different algorithms are involved in the simulation: the GM-OP-ELM (GOE) method [17], double cluster head model (DCHM) [14], a cosine theorem-based method for identifying and fusing conflicting data (CTB) [25] and the SWDF introduced in this paper. GOE uses a grey model and an extreme learning machine to predict the data of the next period to accelerate the calculation process. Moreover, the DCHM selects two cluster heads in each group and adopts Bayesian data fusion to improve robustness. In CTB, a fusion algorithm based on the degree of mutual support is proposed to accommodate conflicting data. The parameters were set to $\varepsilon = 0.17$ in GOE and $\alpha = 1$, $\lambda = 0.3$, $E = 0.3$, $\mu = 0$, $\sigma = 1$ and $\varepsilon = 1$ in SWDF. The ratio of compromised sensor nodes is set to 30% in the DCHM. Figure 5 shows the simulation results for 100, 200, 400 and 800 sensor nodes when sensing the environmental temperature. These sensor nodes are uniformly deployed in the simulation field. Three hundred simulation tests are run for each group and the average results are calculated for presentation.

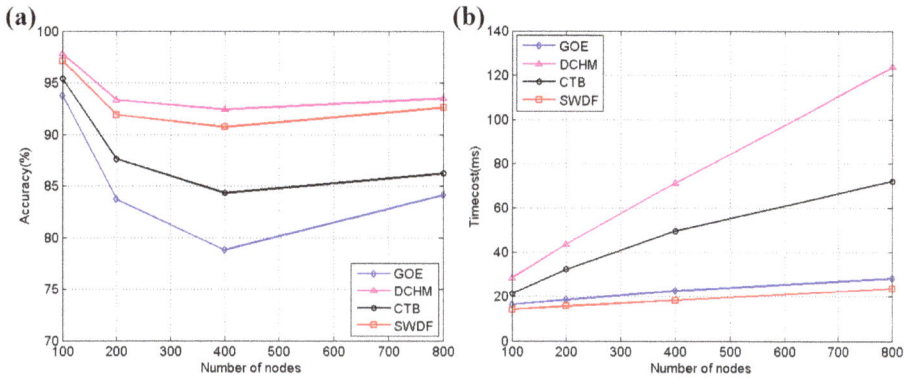

Figure 5. (a) The average accuracy for methods in the simulation; **(b)** The average time/cost for the methods in the simulation.

As shown in Figure 5a, the DCHM obtained the highest average accuracy in these four algorithms. The accuracy represents the degree of sensing data following the real one, as calculated by Equation (15). The accuracy decreases as the density of the sensor nodes increases, indicating that the actuator's effect is gradually increasing. When the number of nodes reaches 800, the accuracy exhibits a small rebound due to the influence of multiple nodes outside the influencing range of actuators. We found that the DCHM, CTB and SWDF all featured capabilities to reduce the influence of the actuator's effect on the simulation, while GOE did not obtain satisfactory performance because it lacked a robust way to accommodate distorted data. The DCHM, consisting of a weighted DBSCAN (Density-Based spatial clustering of applications with noise) algorithm [14], Bayesian data fusion and double cluster heads mechanism, is found to be a powerful approach to address actuator disturbance.

$$\left(1 - \frac{|d_s - d_r|}{d_r}\right) \times 100\% \tag{15}$$

where d_s represent the sensing value acquired by the WSN system and d_r denotes the real value.

However, the large calculation burden of the DCHM, as shown in Figure 5b, makes it challenging to deploy in real-time applications, particularly when the devices of the applications are equipped with a low-frequency MCU. Figure 5b shows the time/cost of each round of sensing calculations for these methods. GOE and SWDF are designed with a prediction-based accelerating method, which does not require the reports from all nodes to make a final decision. GOE and SWDF enable a low time/cost when the number of sensor nodes increases, while other algorithms do not efficiently control the time consumption. The DCHM requires the most calculation time in each group, due primarily to the cluster head re-selection procedure. In WSNs, the energy is an important factor of interest to engineers because sensor nodes often contain limited battery capacity that is not easily recharged. Figure 6a shows the average energy consumption in an hour for each testing group. SWDF is well suited for low-energy-consumption applications in real-time circumstances relative to other traditional methods, particularly in agricultural applications using low-cost hardware.

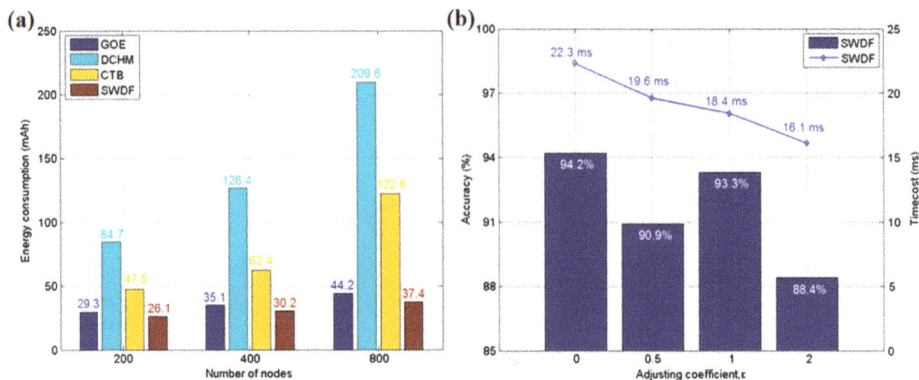

Figure 6. (**a**) The average energy consumption of each sensor node over an hour; (**b**) The average accuracy and time/cost for various adjustment coefficients ε.

The process of parameter setting is another area of concern in SWDF application. With fewer parameters, the SWDF is simple and sufficiently efficient to be used over a large-range sensing area. The adjustment coefficient ε in Equation (7) determines the value of the adaptive threshold θ, which influences the balance between accuracy and speed. When $\alpha = 1$, $\lambda = 0.3$, $E = 0.3$, $\mu = 0$ and $\sigma = 1$ and ε is set to 0, 0.5, 1, or 2, Figure 6b shows the effect of these various ε values in SWDF working with 200 sensor nodes. A different ε may lead to a different degree of involvement of the adaptive part of the tolerant threshold for prediction error. When $\varepsilon = 0$, the threshold is fixed and it costs 22.3 ms for each sensing round. As ε increases, the adaptive part has more effect in Equation (7) and leads to a different sensing time and accuracy. As shown in Figure 6b, the influence of ε is not monotonic. When $\varepsilon = 1$, the accuracy is larger than when $\varepsilon = 0.5$. Thus, this value should be assigned by the user according to the application environment. Brief on-site testing is recommended to determine a suitable ε value after the algorithm is deployed. A satisfactory choice of this value will lead to quick response of the system and (in some cases) high accuracy.

4.2. Implementation and Experiments

As shown in Figure 7a, to further verify the effect of the SWDF in real-time applications, the hardware of a WSN node was implemented on STMicroelectronics STM32F103 and NORDIC nRF24L01 platforms with a SHT20 temperature and humidity sensor from Sensirion Corporation. The STM32F103 is an ARM 32-bit Cortex-M3 microcontroller that is used to drive the sensor and execute calculations. The nRF24L01 is a single-chip 2.4 GHz transceiver with an embedded baseband protocol engine, which is used to establish the WSN in the sensing system. The SHT20 is a humidity and temperature sensor with fully calibrated digital output. Each chip should be calibrated by the producer before it comes to the market. Thus, it is not necessary to calibrate the sensor nodes before our experiments. The hardware is established on FR-4 PCB board with lithium battery for providing power. As Figure 8 shows, the experiments were performed in a 30 m × 40 m rectangular greenhouse containing 12 fans, shutters, outside shades and inside shades as actuators. The 48 sensor nodes were arranged in a line at 5 m intervals as shown in Figure 7b. The SWDF, DCHM and the general average algorithm (GAA), without any data fusion mechanisms, were evaluated in the experiments. The parameters for SWDF in experiments were set to $\alpha = 1$, $\lambda = 0.3$, $E = 0.3$, $\mu = 0$, $\sigma = 1$ and $\varepsilon = 1$ and the ratio of compromised sensor nodes in DCHM was set to 30%.

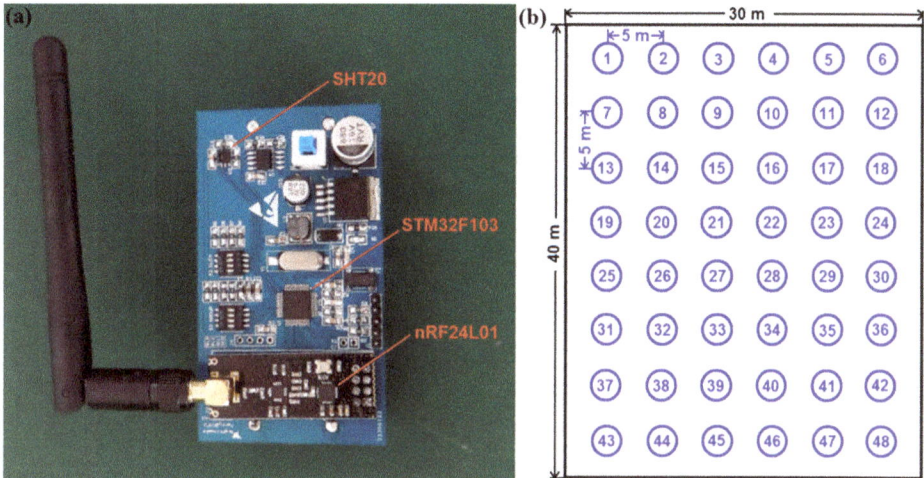

Figure 7. (**a**) Illustration of the wireless sensor node for experiments; (**b**) Illustration of the deployment of sensor nodes for experiments.

The real temperature values for comparison in experiments were calculated using specialized sensor nodes in the correct area of the greenhouse, as chosen manually. A correct area means a robust area for data sensing. The sensor nodes inside the correct area will never be influenced by actuators and therefore they will not report data imprecisely. The correct area can be found automatically by previous experiments. If the error between the sensing data is less than the set variance, the sensor node can be considered to be located within the correct area. Furthermore, the detection boundary of the outermost correct nodes can be considered as the boundary of the correct area. The specialized sensor nodes are generally chosen manually from correct area which can provide stable and precise parameter values for experiments. One way for choosing these specialized nodes is to select the nodes far away from the actuators on the map and ensure that they will not be affected by any actuator; another way is to observe the nodes over a period of time in previous experiments and find out the nodes with stable sensing data. Only several specialized sensor nodes are sufficient and the average value of their sensing data will be regarded as the estimated real temperature value in the environment. The experiments were run for ten rounds and at each round, a refreshment of 10.8 thousand data bytes were acquired for each algorithm. The temperature refreshing time was set to once per second and each round lasted 3 h. The experimental results for these three algorithms are shown in Figure 9.

Figure 8. (**a**) Illustration of the appearance of the experimental greenhouse; (**b**) Illustration of the inside structure for the experimental greenhouse.

(a)

(b)

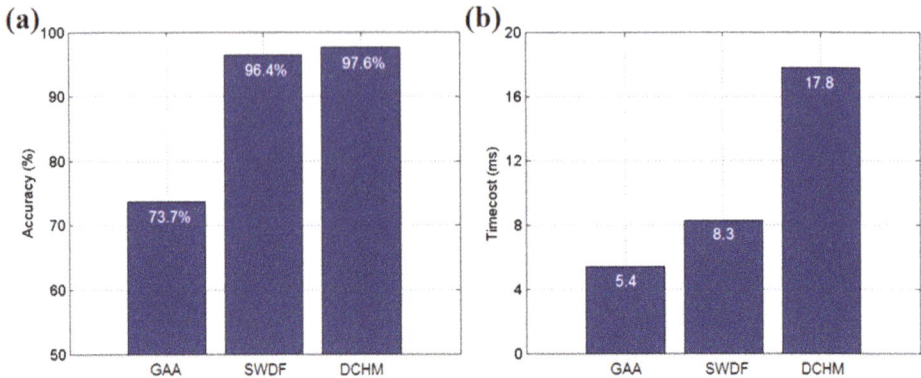

Figure 9. (**a**) Average accuracy of each algorithm in the experiments; (**b**) Average time cost of each algorithm in the experiments.

As shown in Figure 9, the accuracy of SWDF was similar to that of the DCHM in experiments. However, the GAA obtained the lowest accuracy because none of the data fusion mechanisms were deployed; thus, the average value of all of the reports from the sensor nodes was taken as the result. Moreover, the DCHM cost more than twice the time taken by SWDF and GAA used the least time/cost of the three algorithms. In summary, we found experimentally that SWDF is an efficient method that can use limited time to obtain a stable data fusion result. The SWDF algorithm, deployed in real-time applications, will work particularly well for the agricultural WSN with a low cost of hardware and multiple sensor nodes. Although the SWDF is not necessary for a greenhouse control, it can improve the robustness of the sensing system. Because the operation of the environmental control system in the greenhouse depends largely on the correctness of the sensing data, the SWDF can help to avoid the malfunction and overshoot of the actuator. That is very meaningful for greenhouse production. This algorithm can also be used in the fields outdoor but the effect is not as significant as in a closed greenhouse.

5. Conclusions

Accurate sensor results are important for automatic control in agricultural and industrial applications. However, sensor nodes near actuators may encounter noisy data. To increase the system robustness, we propose the SWDF method, in which voting data fusion (based on adjacent graph theory) is designed to reduce the influence of data corruption and dynamic weights are proposed to maintain a balance between accuracy and speed. To shorten the response time, an acceleration mechanism with a prediction algorithm is developed to reduce the number of redundancy reports from sensor nodes. A real-time hardware system is implemented on the STM32F103 MCU and nRF24L01 wireless platforms. Simulations and experiments in a greenhouse demonstrate the improvements achieved using the new set of algorithms.

Acknowledgments: This study was supported by the National Natural Science Foundation (grant No. 81701087) and the Fujian Provincial Natural Science Foundation (grant No. 2016J01096).

Author Contributions: Tengyue Zou and Shouying Lin are the main contributors of this paper. They developed the idea, wrote the paper and carried out the experiments and data analysis. Yuanxia Wang and Mengyi Wang assisted in the experiments for algorithms and provided some useful suggestions.

Conflicts of Interest: The authors declare no conflict of interest.

Sensors **2017**, *17*, 2555

References

1. Kampianakis, E.; Kimionis, J.; Tountas, K.; Konstantopoulos, C.; Koutroulis, E.; Bletsas, A. Wireless environmental sensor networking with analog scatter radio and timer principles. *IEEE Sens. J.* **2014**, *14*, 3365–3376. [CrossRef]
2. Iqbal, M.; Naeem, M.; Anpalagan, A.; Ahmed, A.; Azam, M. Wireless sensor network optimization: Multi-objective paradigm. *Sensors* **2015**, *15*, 17572–17620. [CrossRef] [PubMed]
3. Zou, T.; Lin, S.; Feng, Q.; Chen, Y. Energy-Efficient Control with Harvesting Predictions for Solar-Powered Wireless Sensor Networks. *Sensors* **2016**, *16*, 53. [CrossRef] [PubMed]
4. Polo, J.; Hornero, G.; Duijneveld, C.; García, A.; Casas, O. Design of a low-cost Wireless Sensor Network with UAV mobile node for agricultural applications. *Comp. Electron. Agric.* **2015**, *119*, 19–32. [CrossRef]
5. Jiang, J.A.; Wang, C.H.; Liao, M.S.; Zheng, X.Y.; Liu, J.H.; Chuang, C.L.; Hung, C.L.; Chen, C.P. A wireless sensor network-based monitoring system with dynamic convergecast tree algorithm for precision cultivation management in orchid greenhouses. *Precis. Agric.* **2016**, *17*, 766–785. [CrossRef]
6. Xu, L. Design of a RSSI location system for greenhouse environment. *Int. J. Distrib. Sens. Netw.* **2015**, *11*. [CrossRef]
7. Li, T.; Zhang, M.; Ji, Y.; Sha, S.; Jiang, Y.; Li, M. Management of CO_2 in a tomato greenhouse using WSN and BPNN techniques. *Int. J. Agric. Biol. Eng.* **2015**, *8*, 43.
8. Lai, Y.; Xie, J.; Lin, Z.; Wang, T.; Liao, M. Adaptive data gathering in mobile sensor networks using speedy mobile elements. *Sensors* **2015**, *15*, 23218–23248. [CrossRef] [PubMed]
9. Cayirpunar, O.; Kadioglu-Urtis, E.; Tavli, B. Optimal base station mobility patterns for wireless sensor network lifetime maximization. *IEEE Sens. J.* **2015**, *15*, 6592–6603. [CrossRef]
10. Yin, X.; Fang, D.; Wang, W.; Chen, X. EETC: To transmit or not to transmit in mobile wireless sensor networks. *Wirel. Netw.* **2016**, *22*, 635–646. [CrossRef]
11. Moshou, D.; Pantazi, X.E.; Kateris, D.; Gravalos, I. Water stress detection based on optical multisensor fusion with a least squares support vector machine classifier. *Biosyst. Eng.* **2014**, *117*, 15–22. [CrossRef]
12. Moshou, D.; Gravalos, I.; Bravo, D.K.C.; Oberti, R.; West, J.S.; Ramon, H. Multisensor fusion of remote sensing data for crop disease detection. In *Geospatial Techniques for Managing Environmental Resources*; Springer: Dordrecht, The Netherlands, 2011; pp. 201–219.
13. Felisberto, F.; Fdez-Riverola, F.; Pereira, A. A ubiquitous and low-cost solution for movement monitoring and accident detection based on sensor fusion. *Sensors* **2014**, *14*, 8961–8983. [CrossRef] [PubMed]
14. Fu, J.S.; Liu, Y. Double cluster heads model for secure and accurate data fusion in wireless sensor networks. *Sensors* **2015**, *15*, 2021–2040. [CrossRef] [PubMed]
15. Palafox-Albarran, J.; Jedermann, R.; Hong, B.; Lang, W. Cokriging for cross-attribute fusion in sensor networks. *Inf. Fusion* **2015**, *24*, 137–146. [CrossRef]
16. Soganli, A.; Ercetin, O.; Cetin, M. On the Quality and Timeliness of Fusion in a Random Access Sensor Network. *IEEE Signal Process. Lett.* **2015**, *22*, 1259–1263. [CrossRef]
17. Luo, X.; Chang, X. A novel data fusion scheme using grey model and extreme learning machine in wireless sensor networks. *Int. J. Control Autom. Syst.* **2015**, *13*, 539–546. [CrossRef]
18. Mostefaoui, A.; Boukerche, A.; Merzoug, M.A.; Melkemi, M. A scalable approach for serial data fusion in Wireless Sensor Networks. *Comp. Netw.* **2015**, *79*, 103–119. [CrossRef]
19. Lu, Z.; Tan, S.L.; Biswas, J. Fusion function placement for Active Networks paradigm in wireless sensor networks. *Wirel. Netw.* **2013**, *19*, 1525–1536. [CrossRef]
20. Yan, L.; Jiang, L.; Xia, Y.; Fu, M. State estimation and data fusion for multirate sensor networks. *Int. J. Adapt. Control Signal Process.* **2016**, *30*, 3–15. [CrossRef]
21. Saska, D.; Blum, R.S.; Kaplan, L. Fusion of Quantized and Unquantized Sensor Data for Estimation. *IEEE Signal Process. Lett.* **2015**, *22*, 1927–1930. [CrossRef]
22. He, H.; Zhu, Z.; Mäkinen, E. Task-oriented distributed data fusion in autonomous wireless sensor networks. *Soft Comp.* **2015**, *19*, 2305–2319. [CrossRef]
23. Rawat, S.; Rawat, S. Multi-sensor data fusion by a hybrid methodology-A comparative study. *Comp. Ind.* **2016**, *75*, 27–34. [CrossRef]
24. Ferrari, G.; Martalò, M.; Abrardo, A. Information fusion in wireless sensor networks with source correlation. *Inf. Fusion* **2014**, *15*, 80–89. [CrossRef]

25. Zhang, Z.; Liu, T.; Chen, D.; Zhang, W. Novel algorithm for identifying and fusing conflicting data in wireless sensor networks. *Sensors* **2014**, *14*, 9562–9581. [CrossRef] [PubMed]

26. Jing, L.; Wang, T.; Zhao, M.; Wang, P. An Adaptive Multi-Sensor Data Fusion Method Based on Deep Convolutional Neural Networks for Fault Diagnosis of Planetary Gearbox. *Sensors* **2017**, *17*, 414. [CrossRef] [PubMed]

27. Si, L.; Wang, Z.; Liu, X.; Tan, C.; Xu, J.; Zheng, K. Multi-sensor data fusion identification for shearer cutting conditions based on parallel quasi-newton neural networks and the Dempster-Shafer theory. *Sensors* **2015**, *15*, 28772–28795. [CrossRef] [PubMed]

28. Chuang, P.J.; Jiang, Y.J. Effective neural network-based node localization scheme for wireless sensor networks. *IET Wirel. Sens. Syst.* **2014**, *4*, 97–103. [CrossRef]

29. Tian, G.Y.; Gledhill, D. Visualisation based feedback control for multiple sensor fusion. In Proceedings of the Tenth International Conference on Information Visualisation (IV'06), London, UK, 5–7 July 2006.

sensors

MDPI

Article

FieldSAFE: Dataset for Obstacle Detection in Agriculture

Mikkel Fly Kragh [1,*,†], **Peter Christiansen** [1,†], **Morten Stigaard Laursen** [1], **Morten Larsen** [2], **Kim Arild Steen** [3], **Ole Green** [3], **Henrik Karstoft** [1] and **Rasmus Nyholm Jørgensen** [1]

[1]   Department of Engineering, Aarhus University, Aarhus N 8200, Denmark;
       repetepc@gmail.com (P.C.); msl@eng.au.dk (M.S.L.); hka@eng.au.dk (H.K.); rnj@eng.au.dk (R.N.J.)
[2]   Conpleks Innovation ApS, Struer 7600, Denmark; morten.larsen@conpleks.com
[3]   AgroIntelli, Aarhus N 8200, Denmark; kas@agrointelli.com (K.A.S.); olg@agrointelli.com (O.G.)
*   Correspondence: mkha@eng.au.dk; Tel.: +45-5176-1455
†   These authors contributed equally to this work.

Received: 28 September 2017; Accepted: 7 November 2017; Published: 9 November 2017

Abstract: In this paper, we present a multi-modal dataset for obstacle detection in agriculture. The dataset comprises approximately 2 h of raw sensor data from a tractor-mounted sensor system in a grass mowing scenario in Denmark, October 2016. Sensing modalities include stereo camera, thermal camera, web camera, 360° camera, LiDAR and radar, while precise localization is available from fused IMU and GNSS. Both static and moving obstacles are present, including humans, mannequin dolls, rocks, barrels, buildings, vehicles and vegetation. All obstacles have ground truth object labels and geographic coordinates.

Keywords: dataset; agriculture; obstacle detection; computer vision; cameras; stereo imaging; thermal imaging; LiDAR; radar; object tracking

1. Introduction

For the past few decades, precision agriculture has revolutionized agricultural production systems. Part of the development has focused on robotic automation, to optimize workflow and minimize manual labor. Today, technology is available to automatically steer farming vehicles such as tractors and harvesters along predefined paths using accurate global navigation satellite systems (GNSS) [1]. However, a human operator is still needed to monitor the surroundings and intervene when potential obstacles appear in front of the vehicle to ensure safety.

In order to completely eliminate the need for a human operator, autonomous farming vehicles need to operate both efficiently and safely without any human intervention. A safety system must perform robust obstacle detection and avoidance in real time with high reliability. Additionally, multiple sensing modalities must complement each other in order to handle a wide range of changes in illumination and weather conditions.

A technological advancement like this requires extensive research and experiments to investigate combinations of sensors, detection algorithms and fusion strategies. Currently, a few publicly known commercial R&D projects exist within companies that seek to investigate the concept [2–4]. In scientific research, projects investigating autonomous agricultural vehicles and sensor suites have existed since 1997, where a simple vision-based anomaly detector was proposed [5]. Since then, a number of research projects has experimented with obstacle detection and sensor fusion [6–14]. However, to our knowledge, no public platforms or datasets are available that address the important issues of multi-modal obstacle detection in an agricultural environment.

Within urban autonomous driving, a number of datasets has recently been made publicly available. Udacity's Self-Driving Car Engineer Nanodegree program has given rise to multiple challenge datasets including stereo camera, LiDAR and localization data [15–17]. A few research institutions such as the University of Surrey [18], Linköping University [19], Oxford [20], and Virginia Tech [21] have published similar datasets. Most of the above cases, however, only address behavioral cloning, such that ground truth data are only available for control actions of the vehicles. No information is thus available for potential obstacles and their location in front of the vehicles.

The KITTI dataset [22], however, addresses these issues with object annotations in both 2D and 3D. Today, it is the de facto standard for benchmarking both single- and multi-modality object detection and recognition systems for autonomous driving. The dataset includes high-resolution grayscale and color stereo cameras, a LiDAR and fused GNSS/IMU sensor data.

Focusing specifically on image data, an even larger selection of datasets is available with annotations of typical object categories such as cars, pedestrians and bicycles. Annotations of cars are often represented by bounding boxes [23,24]. However, pixel-level annotation or semantic segmentation has the advantage of being able to capture all objects, regardless of their shape and orientation. Some of these are synthetically-generated images using computer graphic engines that are automatically annotated [25,26], whereas others are natural images that are manually labeled [27,28].

In agriculture, only a few similar datasets are publicly available. The Marulan Datasets [29] provide multi-sensor data from various rural environments and include a large variety of challenging environmental conditions such as dust, smoke and rain. However, the datasets focus on static environments and only contain a few humans occasionally walking around with no ground truth data available. Recently, the National Robotics Engineering Center (NREC) Agricultural Person-Detection Dataset [30] was made publicly available. It contains labeled image sequences of humans in orange and apple orchards acquired with moving sensing platforms. The dataset is ideal for pushing research on pedestrian detection in agricultural environments, but only includes a single modality (stereo vision). Therefore, a need still exists for an object detection dataset that allows for investigation of sensor combinations, multi-modal detection algorithms and fusion strategies.

While some similarities between autonomous urban driving and autonomous farming are present, essential differences exist. An agricultural environment is often unstructured or semi-structured, whereas urban driving involves planar surfaces, often accompanied by lane lines and traffic signs. Further, distinction between traversable, non-traversable and processable terrain is often necessary in an agricultural context such as grass mowing, weed spraying or harvesting. Here, tall grass or high crops protruding from the ground may actually be traversable and processable, whereas ordinary object categories such as humans, animals and vehicles are not. In urban driving, however, a simplified traversable/non-traversable representation is common, as all protruding objects are typically regarded as obstacles. Therefore, sensing modalities and detection algorithms that work well in urban driving do not necessarily work well in an agricultural setting. Ground plane assumptions common for 3D sensors may break down when applied on rough terrain or high grass. Additionally, vision-based detection algorithms may fail when faced with visual ambiguous information from, e.g., animals that are camouflaged to resemble the appearance of vegetation in a natural environment.

In this paper, we present a flexible, multi-modal sensing platform and a dataset called FieldSAFE for obstacle detection in agriculture. The platform is mounted on a tractor and includes stereo camera, thermal camera, web camera, 360° camera, LiDAR and radar. Precise localization is further available from fused IMU and GNSS. The dataset includes approximately 2 h of recordings from a grass mowing scenario in Denmark, October 2016. Both static and moving obstacles are present including humans, mannequin dolls, rocks, barrels, buildings, vehicles and vegetation. Ground truth positions of all obstacles were recorded with a drone during operation and have subsequently been manually labeled and synchronized with all sensor data. Figure 1 illustrates an overview of the dataset including recording platform, available sensors, and ground truth data obtained from drone recordings. Table 1 compares our proposed dataset to existing datasets in robotics and agriculture. The dataset supports research into object detection and classification, object tracking, sensor fusion, localization and mapping. It can be downloaded from https://vision.eng.au.dk/fieldsafe/.

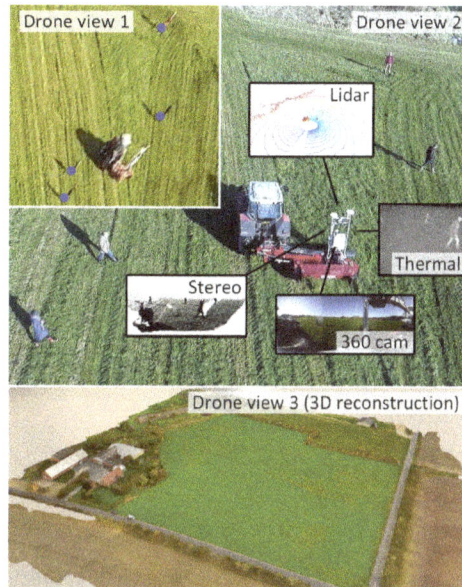

Figure 1. Recording platform surrounded by static and moving obstacles. Multiple drone views record the exact position of obstacles, while the recording platform records local sensor data.

Table 1. Comparison to existing datasets in robotics and agriculture.

Dataset	Environment	Length	Localization	Sensors	Obstacles	Annotations
KITTI [22]	urban	6 h	✓	stereo camera, LiDAR	cars, trucks, trams, pedestrians, cyclists	2D + 3D bounding boxes
Oxford [20]	urban	1000 km	✓	stereo camera, LiDARs, color cameras	cars, trucks, pedestrians, cyclists	none
Marulan [29]	rural	2 h	✓	lasers, radar, color camera, infra-red camera	humans, box, poles, bricks, vegetation	none
NREC [30]	orchards	8 h	✓	stereo camera	humans, vegetation	bounding boxes (only humans)
FieldSAFE (ours)	grass field	2 h	✓	stereo camera, web camera, thermal camera, 360° camera, LiDAR, radar	humans, mannequins, rocks, barrels, buildings, vehicles, vegetation	GPS position and labels

2. Sensor Setup

Figure 2 shows the recording platform mounted on a tractor during grass mowing. The platform was mounted on an A-frame (standard in agriculture) with dampers for absorbing internal engine vibrations from the vehicle. The platform consists of the exteroceptive sensors listed in Table 2, the proprioceptive sensors listed in Table 3 and a Conpleks Robotech Controller 701 used for data collection with the Robot Operating System (ROS) [31]. The stereo camera provides a timestamped left (color) and right (grayscale) raw and rectified image pair along with an on-device calculated depth image. Post-processing methods are further available for generating colored 3D point clouds. The web camera and 360° camera provide timestamped compressed color images. The thermal camera provides a raw grayscale image that allows for conversion to absolute temperatures. The LiDAR provides raw distance measurements and calibrated reflectivities for each of the 32 laser beams. Post-processing methods are available for generating 3D point clouds. The radar provides raw CAN messages with up to 16 processed radar detections per frame from mid- and long-range modes simultaneously. The radar detections consist of range measurements, azimuth angles and amplitudes. ROS topics and data formats for each sensor are available on the FieldSAFE website. Code examples for data visualization are further available on the corresponding git repository.

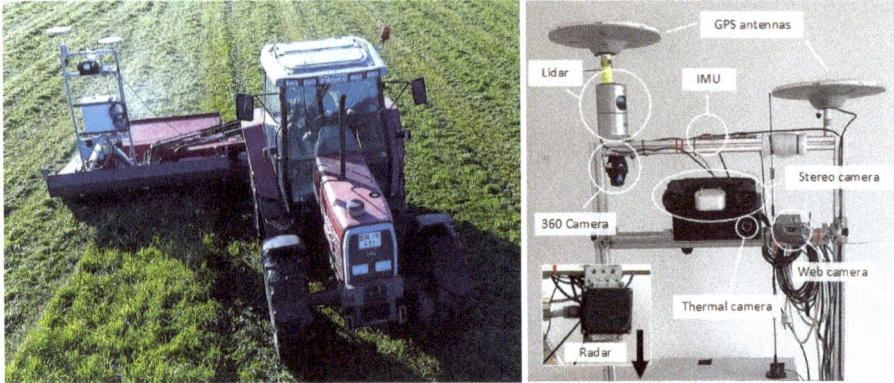

Figure 2. Recording platform.

Table 2. Exteroceptive sensors.

Sensor	Model	Resolution	FOV	Range	Acquisition Rate
Stereo camera	Multisense S21 CMV2000	1024 × 544	85° × 50°	1.5–50 m	10 fps
Web camera	Logitech HD Pro C920	1920 × 1080	70° × 43°	-	20 fps
360° camera	Giroptic 360cam	2048 × 833	360° × 292°	-	30 fps
Thermal camera	Flir A65, 13 mm lens	640 × 512	45° × 37°	-	30 fps
LiDAR	Velodyne HDL-32E	2172 × 32	360° × 40°	1–100 m	10 fps
Radar	Delphi ESR	16 targets/frame	90° × 4.2°	0–60 m	20 fps
		16 targets/frame	20° × 4.2°	0–174 m	20 fps

Table 3. Proprioceptive sensors.

Sensor	Model	Description	Acquisition Rate
GPS	Trimble BD982 GNSS	Dual antenna RTK GNSS system. Measures position and horizontal heading of the platform.	20 Hz
IMU	Vectornav VN-100	Measures acceleration, angular velocity, magnetic field and barometric pressure.	50 Hz

The proprioceptive sensors include GPS and IMU. An extended Kalman filter has been setup to provide global localization by fusing GPS and IMU with the robot_localization package [32] available in ROS. The localization code and resulting pose information are available along with the raw localization data.

Figure 3 illustrates a synchronized pair of frames from stereo camera, 360° camera, web camera, thermal camera, LiDAR and radar.

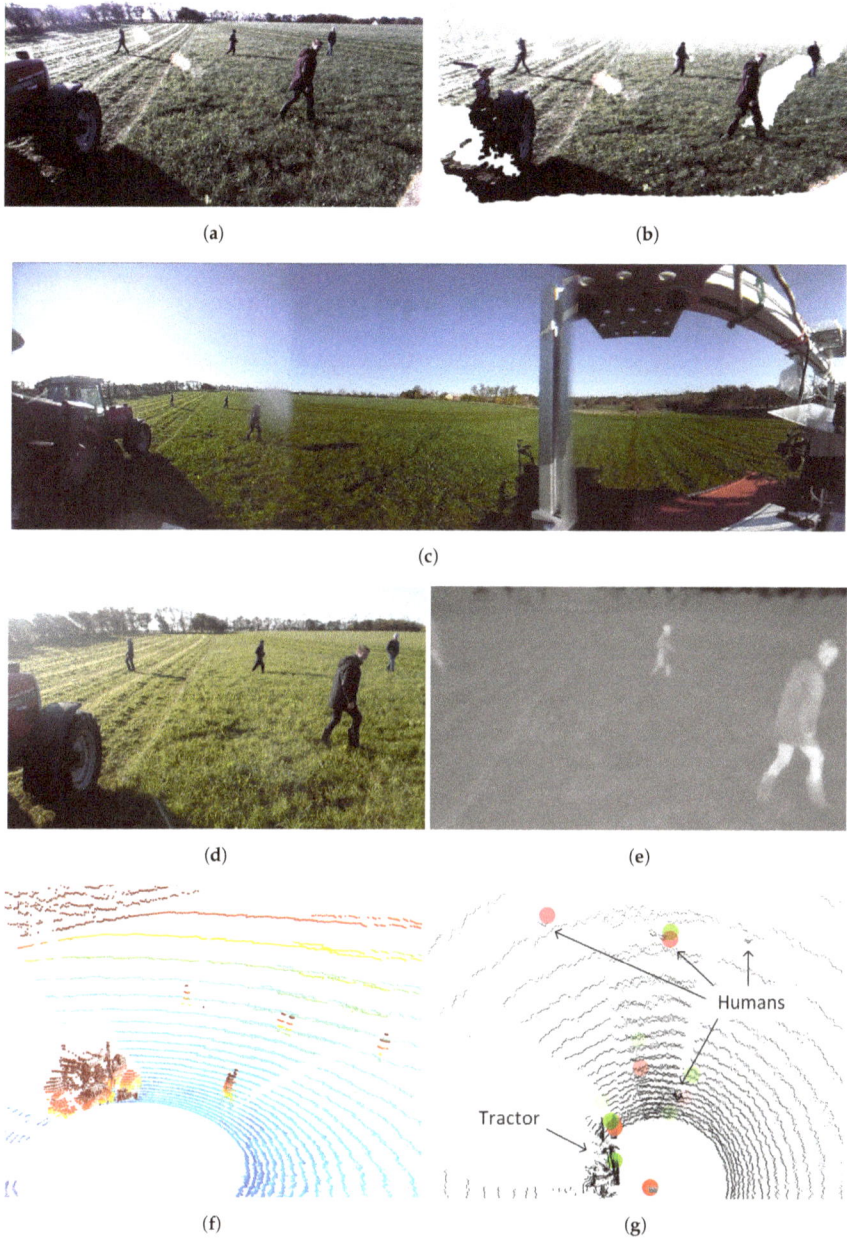

Figure 3. Example frames from the FieldSAFE dataset. (**a**) Left stereo image; (**b**) stereo pointcloud; (**c**) 360° camera image (cropped); (**d**) web camera image; (**e**) thermal camera image (cropped); (**f**) LiDAR point cloud (cropped and colored by height); (**g**) radar detections overlaid on LiDAR point cloud (black). Green and red circles denote detections from mid- and long-range modes, respectively.

Synchronization: Trigger signals for the stereo and thermal cameras were synchronized and generated from a pulse-per-second signal from an internal GNSS in the LiDAR, which allowed exact timestamps for all three sensors. The remaining sensors were synchronized in software using a best-effort approach in ROS, where the ROS system time was used to timestamp each message once it got delivered. However, best-effort message delivery does not provide any guarantees for delivery times, and the specific time delays for the different sensors therefore depend on the internal processing in the sensor, the transmission to the computer, network traffic load, the kernel scheduler and software drivers in ROS [33]. Time delays can therefore vary significantly and are not necessarily constant.

IMU and GNSS both use serial communication and therefore have very small transmission latencies. The same applies for radar that sends its data on the CAN bus. The web camera, however, uses a USB 2.0 interface and thus experiences a short delay in the transmission. A typical delay for the web camera has been measured as 100 ms. The 360° camera uses the TCP protocol and experiences a large amount of packet retransmissions. The delay has therefore been measured up to 4.5 s. The time delays are both specified in relation to the stereo camera, which is synchronized to the LiDAR and thermal camera.

Registration: All sensors were registered by estimating extrinsic parameters (translation and rotation). A common reference frame, base link, was defined at the mount point of the recording frame on the tractor. From here, extrinsic parameters were estimated either by hand measurements or using automated calibration procedures. Figure 4 illustrates the chain of registrations and how they were carried out. The LiDAR and the stereo camera were registered by optimizing the alignment of 3D point clouds from both sensors. For this procedure, the iterative closest point (ICP) was used on multiple static scenes. An average over all scenes was used as the final estimate. The stereo and thermal cameras were registered and calibrated using the camera calibration method available in the Computer Vision System Toolbox in MATLAB. Since the thermal camera did not perceive light in the visual spectrum, a custom-made visual-thermal checkerboard was used. For a more detailed description of this procedure, we refer the reader to [34]. The remaining sensors were registered by hand, by estimating extrinsic parameters of their positions. All extrinsic parameters are contained in the dataset. Instructions for how to extract these are available at the FieldSAFE website. Here, the estimated intrinsic camera parameters are further available for download.

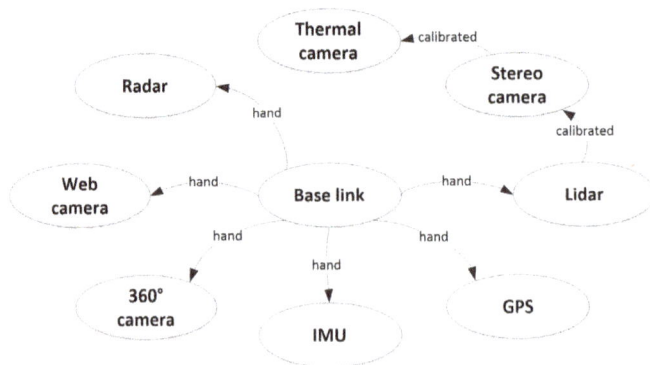

Figure 4. Sensor registration. "Hand" denotes a manual measurement by hand, whereas "calibrated" indicates that an automated calibration procedure was used to estimate the extrinsic parameters.

3. Dataset

The dataset consists of approximately 2 h of recordings during grass mowing in Denmark, 25 October 2016. The exact position of the field was 56.066742, 8.386255 (latitude, longitude). Figure 5a shows a map of the field with tractor paths overlaid. The field is 3.3 ha and surrounded by roads, shelterbelts and a private property.

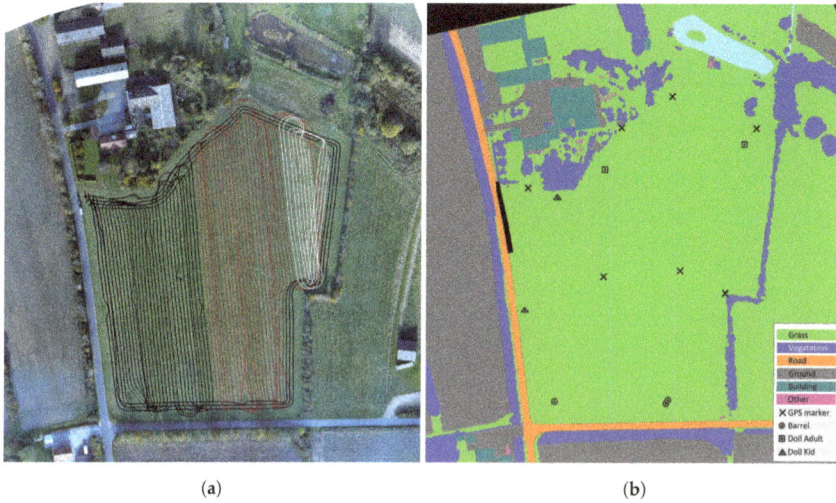

(a)   (b)

Figure 5. Colored and labeled orthophotos. (**a**) Orthophoto with tractor tracks overlaid. Black tracks include only static obstacles, whereas red and white tracks also have moving obstacles. Currently, red tracks have no ground truth for moving obstacles annotated. (**b**) Labeled orthophoto.

A number of static obstacles exemplified in Figure 6 were placed on the field prior to recording. They included mannequin dolls (adults and children), rocks, barrels, buildings, vehicles and vegetation. Figure 5b shows the placement of static obstacles on the field overlaid on a ground truth map colored by object classes.

Figure 6. Examples of static obstacles.

Additionally, a session with moving obstacles was recorded where four humans were told to walk in random patterns. Figure 7 shows the four subjects and their respective paths on a subset of the field. The subset corresponds to the white tractor tracks in Figure 5a. The humans crossed the path of the tractor a number of times, thus emulating dangerous situations that must be detected by a safety system. Along the way, various poses such as standing, sitting and lying were represented.

(**a**) Human 1   (**b**) Human 2   (**c**) Human 3   (**d**) Human 4

Figure 7. Examples of moving obstacles (from the stereo camera) and their paths (black) overlaid on the tractor path (grey).

During the entire traversal and mowing of the field, data from all sensors were recorded. Along with video from a hovering drone, a static orthophoto from another drone and corresponding manually-annotated class labels, these are all available from the FieldSAFE website.

4. Ground Truth

Ground truth information on object location and class labels for both static and moving obstacles is available as timestamped global (geographic) coordinates. By transforming local sensor data from the tractor into global coordinates, a simple look-up of the class label in the annotated ground truth map is possible.

Prior to traversing and mowing the field, a number of custom-made markers were distributed on the ground and measured with exact global coordinates using a handheld Topcon GRS-1 RTK GNSS. A DJI Phantom 4 drone was used to take overlapping bird's-eye view images of an area covering the field and its surroundings. Pix4D [35] was used to stitch the images and generate a high-resolution orthophoto (Figure 5a) with a ground sampling distance (GSD) of 2 cm. The orthophoto was manually labeled pixel-wise as either grass, ground, road, vegetation, building, GPS marker, barrel, human or other (Figure 5b). Using the GPS coordinates of the markers and their corresponding positions in the orthophoto, a mapping between GPS coordinates and pixel coordinates was estimated.

For annotating the location of moving obstacles, a DJI Matrice 100 was used to hover approximately 75 m above the ground while the tractor traversed the field. The drone recorded video at 25 fps with a resolution of 1920 × 1080. Due to limited battery capacity, the recording was split into two sessions of each 20 min. The videos were manually synchronized with sensor data from the tractor by introducing physical synchronization events in front of the tractor in the beginning and end of each session. Using the seven GPS markers that were visible within the field of view of the drone, the videos were stabilized and warped to a bird's-eye view of a subset of the field. As described above for the static orthophoto, GPS coordinates of the markers and their corresponding positions in the videos were then used to generate a mapping between GPS coordinates and pixel coordinates. Finally, the moving obstacles were manually annotated in each frame of one of the videos using the vatic video annotation tool [36]. Figure 7 shows the path of each object overlaid on a subset of the orthophoto. The second video is yet to be annotated.

5. Summary and Future Work

In this paper, we have presented a calibrated and synchronized multi-modal dataset for obstacle detection in agriculture. The dataset supports research into object detection and classification, object tracking, sensor fusion, localization and mapping. We envision the dataset to facilitate a wide range of future research within autonomous agriculture and obstacle detection for farming vehicles.

In future work, we plan on annotating the remaining session with moving obstacles. Additionally, we would like to extend the dataset with more scenarios from various agricultural environments while widening the range of encountered illumination and weather conditions.

Currently, all annotations reside in a global coordinate system. Projecting these annotations to local sensor frames inevitably causes localization errors. Therefore, we would like to extend annotations with, e.g., object bounding boxes for each sensor.

Acknowledgments: This research is sponsored by the Innovation Fund Denmark as part of the project "SAFE—Safer Autonomous Farming Equipment" (project No. 16-2014-0). The authors thank Anders Krogh Mortensen for his valuable help in processing all drone recordings and generating stitched, georeferenced orthophotos. Further, we thank the participating companies in the project, AgroIntelli, Conpleks Innovation ApS, CLAAS E-Systems, KeyResearch and RoboCluster, for their help in organizing the field experiment, providing sensor and processing equipment and promoting the project in general.

Author Contributions: M.F.K.and P.C. designed the sensor platform including interfacing, calibration, registration and synchronization. M.F.K. and P.C. conceived of and designed the experiments and provided manual ground truth annotations. M.S.L. and M.L. contributed with sensor interfacing, calibration and synchronization. K.A.S., O.G. and R.N.J. contributed with agricultural domain knowledge, provided test facilities and performed the experiments. H.K. contributed with insight into the experimental design and computer vision. M.F.K. wrote the paper.

Conflicts of Interest: The authors declare no conflict of interest.

References

1. Abidine, A.Z.; Heidman, B.C.; Upadhyaya, S.K.; Hills, D.J. Autoguidance system operated at high speed causes almost no tomato damage. *Calif. Agric.* **2004**, *58*, 44–47.
2. Case IH. Case IH Autonomous Concept Vehicle, 2016. Available online: http://www.caseih.com/apac/en-in/news/pages/2016-case-ih-premieres-concept-vehicle-at-farm-progress-show.aspx (accessed on 9 August 2017).
3. ASI. Autonomous Solutions, 2016. Available online: https://www.asirobots.com/farming/ (accessed on 9 August 2017).
4. Kubota, 2017. Available online: http://www.kubota-global.net/news/2017/20170125.html (accessed on 16 August 2017).
5. Ollis, M.; Stentz, A. Vision-based perception for an automated harvester. In Proceedings of the 1997 IEEE/RSJ International Conference on Intelligent Robot and Systems, Innovative Robotics for Real-World Applications (IROS '97), Grenoble, France, 11 September 1997; Volume 3, pp. 1838–1844.
6. Stentz, A.; Dima, C.; Wellington, C.; Herman, H.; Stager, D. A system for semi-autonomous tractor operations. *Auton. Robots* **2002**, *13*, 87–104.

7. Wellington, C.; Courville, A.; Stentz, A.T. Interacting markov random fields for simultaneous terrain modeling and obstacle detection. In Proceedings of the Robotics: Science and Systems, Cambridge, MA, USA, 8–11 June 2005; Volume 17, pp. 251–260.
8. Griepentrog, H.W.; Andersen, N.A.; Andersen, J.C.; Blanke, M.; Heinemann, O.; Madsen, T.E.; Nielsen, J.; Pedersen, S.M.; Ravn, O.; Wulfsohn, D. Safe and reliable: Further development of a field robot. *Precis. Agric.* **2009**, *9*, 857–866.
9. Moorehead, S.S.J.; Wellington, C.K.C.; Gilmore, B.J.; Vallespi, C. Automating orchards: A system of autonomous tractors for orchard maintenance. In Proceedings of the IEEE International Conference on Intelligent Robots and Systems, Workshop on Agricultural Robotics, Vilamoura, Portugal, 7–12 October 2012; p. 632.
10. Reina, G.; Milella, A. Towards autonomous agriculture: Automatic ground detection using trinocular stereovision. *Sensors* **2012**, *12*, 12405–12423.
11. Emmi, L.; Gonzalez-De-Soto, M.; Pajares, G.; Gonzalez-De-Santos, P. New trends in robotics for agriculture: Integration and assessment of a real fleet of robots. *Sci. World J.* **2014**, *2014*, doi:10.1155/2014/404059.
12. Ross, P.; English, A.; Ball, D.; Upcroft, B.; Corke, P. Online novelty-based visual obstacle detection for field robotics. In Proceedings of the IEEE International Conference on Robotics and Automation, Seattle, WA, USA, 26–30 May 2015; pp. 3935–3940.
13. Ball, D.; Upcroft, B.; Wyeth, G.; Corke, P.; English, A.; Ross, P.; Patten, T.; Fitch, R.; Sukkarieh, S.; Bate, A. Vision-based obstacle detection and navigation for an agricultural robot. *J. Field Robot.* **2016**, *33*, 1107–1130.
14. Reina, G.; Milella, A.; Rouveure, R.; Nielsen, M.; Worst, R.; Blas, M.R. Ambient awareness for agricultural robotic vehicles. *Biosyst. Eng.* **2016**, *146*, 114–132.
15. Didi. Didi Data Release #2—Round 1 Test Sequence and Training. Available online: http://academictorrents.com/details/18d7f6be647eb6d581f5ff61819a11b9c21769c7 (accessed on 8 November 2017).
16. Udacity. Udacity Didi Challenge—Round 2 Dataset. Available online: http://academictorrents.com/details/67528e562da46e93cbabb8a255c9a8989be3448e (accessed on 8 November 2017).
17. Udacity, Didi. Udacity Didi $100k Challenge Dataset 1. Available online: http://academictorrents.com/details/76352487923a31d47a6029ddebf40d9265e770b5 (accessed on 8 November 2017).
18. DIPLECS. DIPLECS Autonomous Driving Datasets, 2015. Available online: http://ercoftac.mech.surrey.ac.uk/data/diplecs/ (accessed on 31 August 2017).
19. Koschorrek, P.; Piccini, T.; Öberg, P.; Felsberg, M.; Nielsen, L.; Mester, R. A multi-sensor traffic scene dataset with omnidirectional video. Ground Truth—What is a good dataset? In Proceedings of the 2013 IEEE Conference on Computer Vision and Pattern Recognition Workshops (CVPRW), Portland, OR, USA, 23–28 June 2013.
20. Maddern, W.; Pascoe, G.; Linegar, C.; Newman, P. 1 Year, 1000 km: The Oxford RobotCar Dataset. *Int. J. Robot. Res.* **2017**, *36*, 3–15.
21. InSight. InSight SHRP2, 2017. Available online: https://insight.shrp2nds.us/ (accessed on 31 August 2017).
22. Geiger, A.; Lenz, P.; Stiller, C.; Urtasun, R. Vision meets Robotics: The KITTI Dataset. *Int. J. Robot. Res. (IJRR)* **2013**, *32*, 1231–1237.
23. Matzen, K.; Snavely, N. NYC3DCars: A dataset of 3D vehicles in geographic context. In Proceedings of the International Conference on Computer Vision, Sydney, Australia, 1–8 December 2013.
24. Caraffi, C.; Vojir, T.; Trefny, J.; Sochman, J.; Matas, J. A system for real-time detection and tracking of vehicles from a single Car-Mounted camera. In Proceedings of the 2012 15th International IEEE Conference on Intelligent Transportation Systems (ITSC), Anchorage, AK, USA, 16–19 September 2012; pp. 975–982.
25. Ros, G.; Sellart, L.; Materzynska, J.; Vazquez, D.; Lopez, A. The SYNTHIA Dataset: A Large collection of synthetic images for semantic segmentation of urban scenes. In Proceedings of the 2016 IEEE Conference on Computer Vision and Pattern Recognition (CVPR), Las Vegas, NV, USA, 27–30 June 2016.
26. Gaidon, A.; Wang, Q.; Cabon, Y.; Vig, E. Virtual worlds as proxy for multi-object tracking analysis. In Proceedings of the 2016 IEEE Conference on Computer Vision and Pattern Recognition (CVPR), Las Vegas, NV, USA, 27–30 June 2016.
27. Cordts, M.; Omran, M.; Ramos, S.; Rehfeld, T.; Enzweiler, M.; Benenson, R.; Franke, U.; Roth, S.; Schiele, B. The cityscapes dataset for semantic urban scene understanding. In Proceedings of the IEEE Conference on Computer Vision and Pattern Recognition (CVPR), Las Vegas, NV, USA, 27–30 June 2016.

28. Neuhold, G.; Ollmann, T.; Rota Bulò, S.; Kontschieder, P. The mapillary vistas dataset for semantic understanding of street scenes. In Proceedings of the International Conference on Computer Vision (ICCV), Venice, Italy, 22–29 October 2017.

29. Peynot, T.; Scheding, S.; Terho, S. The Marulan Data Sets: Multi-sensor perception in natural environment with challenging conditions. *Int. J. Robot. Res.* **2010**, *29*, 1602–1607.

30. Pezzementi, Z.; Tabor, T.; Hu, P.; Chang, J.K.; Ramanan, D.; Wellington, C.; Babu, B.P.W.; Herman, H. Comparing apples and oranges: Off-road pedestrian detection on the NREC agricultural person-detection dataset. *arXiv* **2017**, arXiv:1707.07169.

31. Quigley, M.; Conley, K.; Gerkey, B.P.; Faust, J.; Foote, T.; Leibs, J.; Wheeler, R.; Ng, A.Y. ROS: An Open-Source Robot Operating System. In Proceedings of the ICRA Workshop on Open Source Software, Kobe, Japan, 17 May 2009.

32. Moore, T.; Stouch, D. A Generalized extended kalman filter implementation for the robot operating system. In *Advances in Intelligent Systems and Computing*; Springer: Cham, Switzerland, 2016.

33. Lütkebohle, I. Determinism in Robotics Software. Conference Presentation, ROSCon, 2017. Available online: https://roscon.ros.org/2017/presentations/ROSCon%202017%20Determinism%20in%20ROS.pdf (accessed on 31 October 2017).

34. Christiansen, P.; Kragh, M.; Steen, K.A.; Karstoft, H.; Jørgensen, R.N. Platform for evaluating sensors and human detection in autonomous mowing operations. *Precis. Agric.* **2017**, *18*, 350–365.

35. Pix4D. 2014. Available online: http://pix4d.com/ (accessed on 5 September 2017).

36. Vondrick, C.; Patterson, D.; Ramanan, D. Efficiently scaling up crowdsourced video annotation. *Int. J. Comput. Vis.* **2013**, *101*, 184–204.

MDPI

St. Alban-Anlage 66

4052 Basel

Switzerland

Tel. +41 61 683 77 34

Fax +41 61 302 89 18

www.mdpi.com

Sensors Editorial Office

E-mail: sensors@mdpi.com

www.mdpi.com/journal/sensors

www.ingramcontent.com/pod-product-compliance
Lightning Source LLC
Chambersburg PA
CBHW051712210326
41597CB00032B/5449